ELECTRONIC DEVICES AND INTEGRATED CIRCUITS

ELECTRONIC DEVICES AND INTEGRATED CIRCUITS

ELECTRONIC DEVICES AND INTEGRATED CIRCUITS

SECOND EDITION

AJAY KUMAR SINGH
Senior Lecturer
Faculty of Engineering and Technology (FET)
Multimedia University (MMU)
Malaysia

PHI Learning Private Limited
Delhi-110092
2013

₹ 525.00

ELECTRONIC DEVICES AND INTEGRATED CIRCUITS, Second Edition
Ajay Kumar Singh

ISBN-978-81-203-4471-6

The export rights of this book are vested solely with the publisher.

Fourth Printing (Second Edition)   ⋯   ⋯   **October, 2013**

Published by Asoke K. Ghosh, PHI Learning Private Limited, Rimjhim House, 111, Patparganj Industrial Estate, Delhi-110092 and Printed by Baba Barkha Nath Printers, Bahadurgarh, Haryana-124507.

To
My Supervisor
Professor P. Khastgir

Chapter 3 TRANSPORT PHENOMENON IN SEMICONDUCTOR 75–134

Chapter 4 *p-n* JUNCTION 135–208

Chapter 7 METAL OXIDE SEMICONDUCTOR FIELD EFFECT TRANSISTOR (MOSFET) 264–363

Chapter 10 OPTOELECTRONIC DEVICES AND POWER DEVICES ... 504–556

Preface

A course in device physics is the first exposure for the students who have to pursue their career in the electronics engineering especially in integrated circuit design. This is also a place where students can enhance their skills related to design aspect.

The purpose of the Second Edition of this book is to provide a basis for understanding the characteristics, working principle, operation and limitations of semiconductor devices. The goal of this book is to bring together the fundamentals of the semiconductor materials and the semiconductor device physics. Throughout the book the applications and operating requirement imposed on semiconductor hardware by different applications are used to describe the important figure of merit of each device. In this way, the students can clearly see what properties a particular device must have to meet the system requirements for which it is designed.

The purpose of this book is also to discuss the basic concepts that govern the behaviour of CMOS and bipolar transistors. The emphasis is given on those parameters and performance factors that are important for very large scale integration (VLSI) devices of deep-submicron dimensions.

There are several excellent textbooks available in the market on Semiconductors, one might wonder why another book is needed. Though these books are unquestionably excellent, but they have not included many topics related to semiconductor materials and devices. In some of the textbooks, equations and relationships are simply stated with no or very little derivation. Therefore, one has to consult some more advanced books. However, I feel in the present book there is enough mathematics included to provide a good foundation for the basic understanding of the semiconductor devices. The primary aim of this book is to provide all the relevant topics on the semiconductor materials and semiconductor devices in a single volume. In this book many solved examples are given to explain the concepts clearly. Other than the solved examples, each chapter contains many unsolved numericals and reasoning based problems. The important concepts are highlighted in each chapter. Further, the present book provides a look at not only the state-of-the-art devices but also future approaches that go beyond the current technology.

The present book is organized in 11 chapters. The book starts with a presentation of atomic structure and theory of solids. This chapter is included to give an essential fundamental of solids and bond formation.

Chapters 2 and 3 discuss the basic properties of the semiconductor materials and their conduction processes. The main emphasis is given on the two most important semiconductors like silicon (Si) and compound semiconductors (e.g. GaAs). The concepts given in these two chapters will be used throughout this book.

Chapter 4 contains the basic physics of the *p-n* junction. The main goal of this chapter is to provide an intuitive understanding of the operation, band diagram and conduction processes in the *p-n* junction. This chapter also discusses the varactor diode and heterojunction.

In Chapter 5, a detailed discussion on the energy band diagram and current voltage characteristics of both Schottky as well as ohmic contact are provided. This material is necessary and sufficient for the understanding of the MOS transistors. Field effect transistors are discussed in Chapter 6 which includes the various types of FETs and their operations.

Chapter 7 presents a discussion on metal insulator semiconductor systems, particularly MOS devices, long and short channel MOSFETs. The chapter discusses in detail about the ideal MOS capacitor, electrostatic potential and charge distribution in silicon and non-ideal oxide capacitance behaviour. The subthreshold characteristics, substrate bias effect, MOSFET channel mobility and small signal model of MOSFET are also provided in this chapter. The present chapter also discusses the various short channel effects in the MOSFETs. CMOS scaling and challenges to further improvement of CMOS devices are discussed in Chapter 8. This chapter also discusses the need of the low power VLSI design and the future novel devices.

The bipolar transistor is reviewed in Chapter 9. This chapter mainly covers construction and operation of BJT and its current–voltage characteristics in terms of both real as well as ideal transistors, Ebers–Moll and Gummel–Poon models, and non-ideal and switching behaviour of BJT. The basic ac model of BJT and design consideration are also discussed. This chapter covers additional devices such as heterojunction bipolar transistors, thyristors, unijunction transistors and the Schottky-clamped transistor. The future BiCMOS technology has been covered in this chapter.

In Chapter 10, a discussion of optoelectronic devices and power devices used in light wave communication systems such as LEDs, LASERs, solar cells and photo detectors is presented. The book concludes with a discussion of integrated circuits and fabrication in Chapter 11.

Apart from chapters, the values of the various constants are given in Appendix A, and values of the material parameters for important semiconductors, Si and GaAs are provided in Appendix B. With the growing emphasis on custom integrated-circuit design, the use of computer simulation of integrated circuits has become an integral part of circuit design. The de facto simulator for integrated-circuit design is one of the versions of SPICE. Device models both from the device physics and empirical models provide a description of how a device will behave in a circuit. SPICE is discussed in Appendix C. PSPICE may be run on personal computers as well as workstations.

This book is designed as a textbook for the undergraduate students of engineering and science.

Although an immense amount of attention was taken to prepare the manuscript, this book may still have some errors and flaws. The author would welcome any suggestions and corrections from readers for the improvement of technical content of this book.

Ajay Kumar Singh

Acknowledgements

I would like to thank first and foremost, to my wife, Nisha and son, Anirban Kumar for their support, patience and understanding during this seemingly endless task. I would like to give my sincere thanks to my parents for their moral support throughout the project. I am grateful to my students (especially when I was in BITS-Pilani, India) for their valuable discussion and feedback on various topics related to semiconductors.

Ajay Kumar Singh

CHAPTER 1

Atomic Structure and Theory of Solids

1.1 INTRODUCTION

A question always arises in the mind that why one needs to study semiconductor material and devices. One probable reason is that the semiconductor devices are the foundation of the electronic industry. The ability to control the movement of the electrons in the solids is the basis of the semiconductor device engineering. In order to understand the electronic properties of these devices, it is necessary to understand the basic properties of these materials and how these properties are affected by adding impurities, temperature, illumination and applied voltages.

Semiconductor devices are small, but versatile units that can perform a variety of control functions in the electronic equipment. They have the ability to control the movement of the charges. They are basically used in rectifiers, detectors, amplifiers, electronic switches, modulators and so on. Semiconductor devices are small in size and light in weight. They consume less power and they are extremely rugged. These devices can be easily operated in many severe environmental conditions.

Without the proper knowledge of electron's behaviour in a crystal and its interaction with lattice, the understanding of the electron transport through a semiconductor device is difficult. In seeking to understand the physics of semiconductors at an atomic level, one must understand the quantum mechanical effect. Therefore, in the present chapter we have discussed quantum mechanical effects like quantum tunnelling, de Broglie wave equation and Schrödinger wave equation. We have also discussed Bohr's model and main bonding in the atoms.

1.2 THE BOHR MODEL

Niels Bohr made several assumptions to develop a model for hydrogen atom. He assumed that electrons exist in a certain specified circular stable orbits, called **energy levels**. The electrons cannot radiate energy as long as they orbit in the same energy level. Electrons can radiate energy only when they make a transition from one energy level to other, i.e., they absorb energy when they excite from lower energy level to higher energy level and emit energy when jump from higher energy level to lower energy level. As seen in Fig. 1.1, the transition energy in each case is equal to $h\nu$.

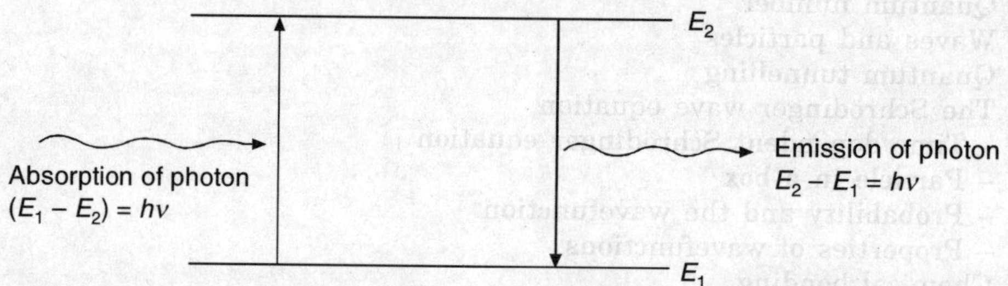

FIGURE 1.1 Schematic representation of emission and absorption of photon.

where, $\nu = c/\lambda$, and λ is the wavelength of the particular light.

Bohr also assumed that the angular momentum of the electron in a given orbit is an integral multiple of Planck's constant, i.e.,

$$P = n\hbar = n\frac{h}{2\pi} = mvr \qquad (1.1)$$

where

h = Planck's constant

n = 1, 2, 3, ...

Let us consider an electron orbiting around the nucleus in a circular path of radius r, as shown in Fig. 1.2.

Since mass of the nucleus is 1.67×10^{-27} kg, which is 1830 times heavier than the mass of electron, therefore, the nucleus is treated as the stationary body. This electron-nucleus system can be visualized as the neutral hydrogen atom.

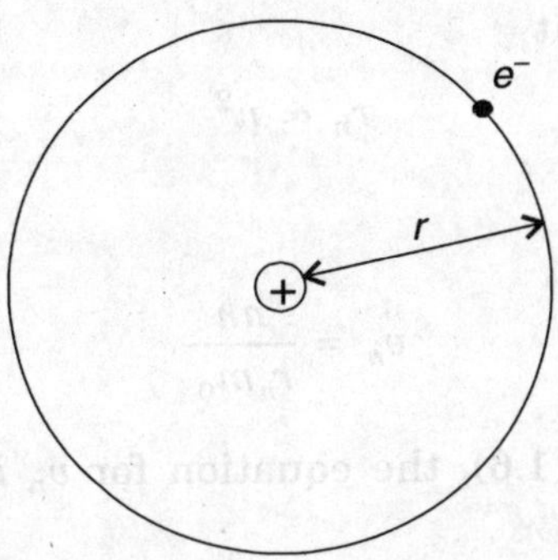

FIGURE 1.2 Hydrogen atom.

Treating electron and nucleus as the point charges, we can equate the electrostatic force acting between them to the centripetal force, i.e.

$$-\frac{q^2}{4\pi\varepsilon_0 r^2} = -\frac{m_0 v^2}{r} \tag{1.2}$$

where

m_0 = Free mass of electron = 9.31×10^{-31} kg
q = Electronic charge = 1.6×10^{-19} C
v = Velocity of electron

From Eq. (1.1), the momentum of the nth orbit electron is:

$$m_0 v_n r_n = n\hbar \tag{1.3}$$

where r_n is the radius of the nth orbit.
Squaring both sides of Eq. (1.3), we have

$$m_0^2 v_n^2 r_n^2 = n^2 \hbar^2$$

or

$$m_0^2 v_n^2 = \frac{n^2 \hbar^2}{r_n^2}$$

or

$$m_0 v_n^2 = \frac{n^2 \hbar^2}{r_n^2 m_0} \tag{1.4}$$

Substituting Eq. (1.4) into Eq. (1.2) and after simplification, we have

$$r_n = \frac{n^2 \hbar^2 4\pi\varepsilon_0}{m_0 q^2}$$

or

$$r_n = k\left(\frac{n^2 \hbar^2}{m_0 q^2}\right) \tag{1.5}$$

where $k = (4\pi\varepsilon_0)$. Equation (1.5) gives the radius of the nth orbit.

From Eq. (1.5), it is clear that

$$r_n \propto n^2$$

where $n = 1, 2, 3, \ldots$
From Eq. (1.1),

$$v_n = \frac{n\hbar}{r_n m_0} \tag{1.6}$$

Substituting Eq. (1.5) in Eq. (1.6), the equation for v_n is given as:

$$v_n = \frac{q^2}{kn\hbar} \tag{1.7}$$

Therefore, the kinetic energy of the orbiting electron in the nth orbit is:

$$\text{K.E.} = \frac{1}{2} m_0 v_n^2 = \frac{m_0 q^4}{2k^2 n^2 \hbar^2} \tag{1.8}$$

The potential energy is:

$$Ep_n = -\frac{q^2}{kr_n} = -\frac{m_0 q^4}{k^2 n^2 \hbar^2} \tag{1.9}$$

The total energy of the orbiting electron is given by the sum of potential energy and kinetic energy. Therefore, using Eqs. (1.8) and (1.9),

$$\text{Total energy } E_n = -\frac{m_0 q^4}{2k^2 n^2 \hbar^2} \tag{1.10}$$

or

$$E_n \propto \frac{1}{n^2}$$

Equation (1.10) clearly shows that the total energy of the orbiting electron is quantized. The negative sign only suggests that this energy keep electron bound; and equal amount of energy is transferred to electron to remove the electron from its orbit. Equation (1.10) suggests that higher the quantum number, i.e., farther the electron from the nucleus, electron will be loosely bound to nucleus and lower energy is required to remove the electron.

If

$$n \to \infty$$

$$E_n \to 0 \qquad \text{(means this is vaccum level)}$$

The energy level corresponds to $n \to \infty$, called vacuum in which electron is completely free.

Putting values of $m_0, q, k, \hbar^2$ in Eq. (1.10) for electron in hydrogen atom, we have

$$E_n = -\frac{13.6}{n^2} \text{ eV} \tag{1.11}$$

EXAMPLE 1.1

Determine the first three allowed electron energies in the Hydrogen atom.

Solution: Using Eq. (1.11),

$$E_n = -\frac{13.6}{n^2} \text{ eV}$$

We have:

For the First energy level $n = 1$,

$$E_1 = -13.6 \text{ eV}$$

For the Second energy level $n = 2$,

$$E_2 = -\left(\frac{13.6}{4}\right) = -3.39 \text{ eV}$$

For the Third energy level $n = 3$,

$$E_3 = -\left(\frac{13.6}{9}\right) = -1.51 \text{ eV}$$

1.3   QUANTUM NUMBERS AND THE PAULI'S EXCLUSION PRINCIPLE

According to Eq. (1.11), the energy of an orbiting electron is quantized, i.e., electron can occupy only some allowed discrete energy levels, $n = 1, 2, 3, \ldots$. This quantum number n is called the **Principal Quantum Number**. There are three other quantum numbers, defined as:

- Azimuthal quantum number which describes the ellipticity of an orbit, and is represented by l.
- Magnetic quantum number which describes the tilt of the orbit, and is denoted by m.
- And last one is spin quantum number which describes the spin direction of electron and is represented by s.

The possible values of these four quantum numbers are:

- Principal quantum number: $n = 1, 2, 3, \ldots$
- Azimuthal quantum number: $l = 0, 1, 2, \ldots, (n-1)$
- Magnetic quantum number: $m = -l, \ldots, +l$
- Spin quantum number: $s = \pm 1/2$

Therefore, each allowed energy state of electron in the hydrogen atom is uniquely described by four quantum numbers n, l, m, s. By knowing these four quantum numbers one can identify the various energy levels of the electron. As it is clear from

Eq. (1.11) that the energy of orbiting electron is primarily determined by the principal quantum number and other quantum numbers affect the energy slightly.

According to Pauli's exclusion principle, no two electrons in a same energy level can have the same spin quantum number, i.e., if one electron has spin +1/2 (i.e., upward) then other electron must has spin –1/2 (i.e., downwards). For example, electron with $n = 1$, called **first Bohr orbit**, has $l = 0$ and $m = 0$ but there are two spin states allowed with $s = +1/2$ and $s = -1/2$. The number of maximum electrons in a given energy state is given by $2n^2$ where n is the principal quantum number.

1.4 WAVES AND PARTICLES

Particles are localized phenomena which transport both mass and energy as they move; while waves are delocalized phenomena (i.e., they are spread out in space) which carry energy but no mass as they move, e.g., cricket ball is an example of particle nature whereas ripples on a lake are waves. In quantum mechanics, this net distinction is blurred. The evidence for the description of light as waves was well established. The photoelectric effect of light also introduced firm evidence of its particle nature. Similarly, de Broglie hypothesis well establish the particle nature of electron whereas the experimental evidence of Davisson and Germer experiment suggested the wave nature of the same electron. Such ideas led de Broglie to conclude that all the entities had both wave as well as particle behaviour. This is known as **principle of wave-particle duality**. This wave-particle duality is the basic of the quantum theory of light.

de Broglie was able to relate the momentum of a particle to the wavelength of the corresponding wave. According to de Broglie, equation

$$P = \text{momentum} = \frac{h}{\lambda} \tag{1.12a}$$

where h is Planck's constant.

Therefore, by using Eq. (1.12a), it is possible to calculate the quantum wavelength of a particle through the knowledge of its momentum. The quantum mechanical effects is noticeable at the microscopic level, i.e., small objects like electrons have wavelength comparable to the microscopic atomic structures when they encounter in solids. Thus, a quantum mechanical description, which includes their wave-like aspects, is essential to understand the behaviour of electrons in solids.

Note: The following are among the most important things which quantum mechanics can describe well where classical mechanics fail.

- Discreteness of energy
- The wave-particle duality of light and matter
- Quantum tunnelling
- The Heisenberg uncertainty principle
- Spin of a particle

EXAMPLE 1.2

Calculate the energy of an X-rays with wavelength of $\lambda = 0.708 \times 10^{-8}$ cm. Also express the result in eV.

Solution: Using relation

$$E = h\nu = \frac{hc}{\lambda}$$

we have

$$E = \left[\frac{(6.625 \times 10^{-34})(3 \times 10^{10})}{0.708 \times 10^{-8}}\right] = 2.81 \times 10^{-15} \text{ J}$$

$$E(\text{eV}) = \frac{2.81 \times 10^{-15}}{1.6 \times 10^{-19}} = 1.75 \text{ eV}$$

1.5 QUANTUM TUNNELLING

Consider one-dimensional potential well of height V_0 as shown in Fig. 1.3.

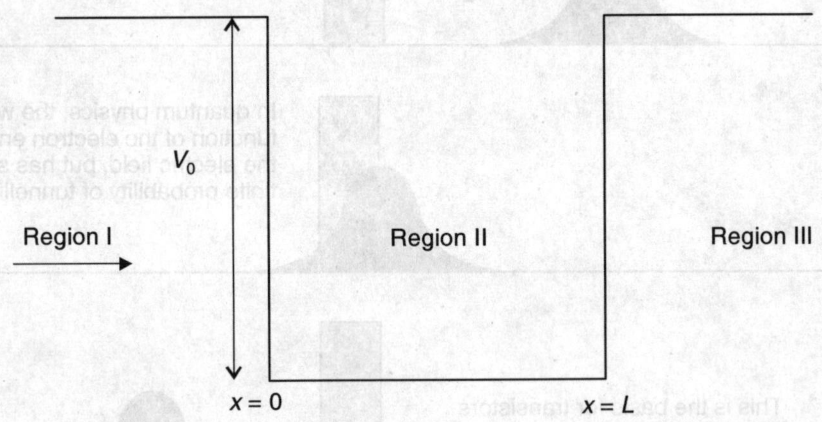

FIGURE 1.3 One-dimensional potential well of height V_0.

Consider a particle with energy E in the Region II of the potential well. According to classical mechanics, if $E < V_0$, where V_0 is the maximum height of the well, the particle will never come out of the well and always confined into the well. Particle will escape from the well only if $E > V_0$. But particle has finite probability to escape from the well even $E < V_0$, according to quantum mechanics. If the potential length is not very large (i.e., of order of nm), then the particle will tunnel through the potential barrier and emerge with the same energy E. The situation is shown in Fig. 1.4. This is known as **tunnelling**.

The phenomenon of tunnelling has important applications in the semiconductor devices like zener diode and MOSFET.

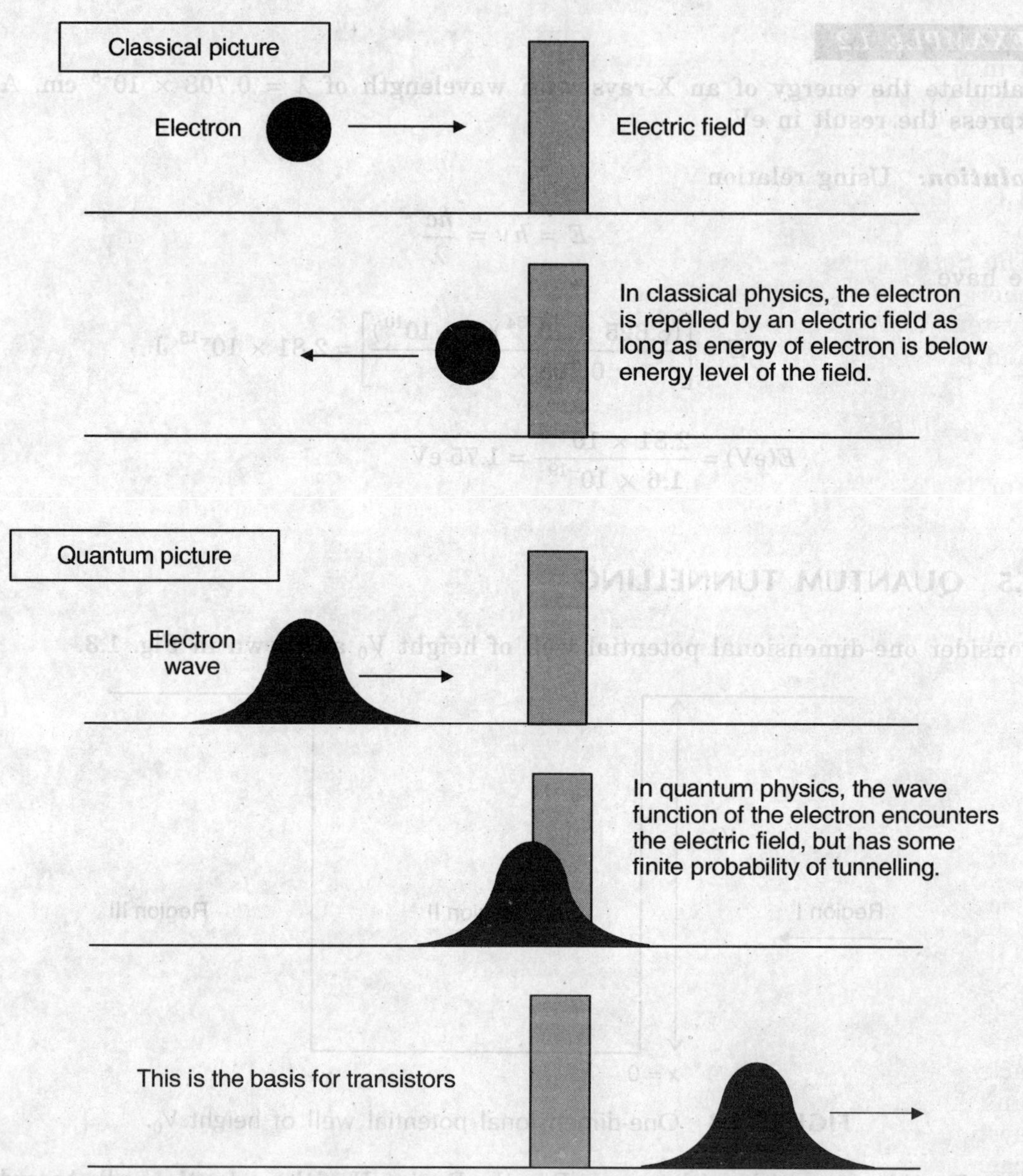

FIGURE 1.4 Classical and quantum pictures of tunnelling.

The transmission coefficient is defined as the ratio of flux transmitted to Region III to the incident flux in Region I. Due to tunnelling, there is finite probability that a particle implinging a potential barrier with energy $E(< V_0)$ penetrates the barrier and will appear in Region III. For $E < V_0$, the transmission probability decays exponentially with length of the barrier.

The quantum tunnelling phenomenon provides the basics of our understanding principle of Zener diode and how the electrons move from channel to gate or vice versa as one reduces the channel length as well as oxide thickness of the MOS devices.

1.6 THE SCHRÖDINGER WAVE EQUATION

The Schrödinger equation plays the role of Newton's laws and the conservation of energy in classical mechanics, i.e., it predicts the future behaviour of the dynamic system. It is a wave equation in terms of wavefunction which predicts analytically and precisely the probability of the occurring of events.

The total energy of a particle is expressed as:

Kinetic energy (K.E.) + Potential energy (P.E.) = Total energy

For Harmonic Oscillator example,

$$\frac{1}{2} mv^2 + \frac{1}{2} kx^2 = E \tag{1.12b}$$

Equation (1.12b) is derived on the basis of classical theory.

In quantum theory, conservation of energy is given by Schrödinger equation, i.e.,

$$H\psi = E\psi \tag{1.13}$$

where
  ψ = Wave function
  E = Energy eigenvalue for the system
  H = Hamiltonian operator and given as,

$$H \rightarrow -\frac{\hbar^2}{2m} \frac{\partial^2}{\partial x^2} + \frac{1}{2} kx^2 \tag{1.14}$$

Equation (1.14) gives the Hamiltonian operator for a quantum harmonic oscillator.

The kinetic and potential energies are transferred into the Hamiltonian which acts upon the wavefunction to generate the evaluation of the wavefunction in time and space. The Schrödinger wave equation gives the quantized energies of the system and also defines the wavefunction form so that the other properties can be calculated.

1.6.1 Time Dependent Schrödinger Equation

The one-dimensional time dependent Schrödinger equation is expressed as:

$$-\frac{\hbar^2}{2m} \frac{\partial^2 \psi(x,\,t)}{\partial x^2} + U(x)\,\psi(x,\,t) = i\hbar \frac{\partial \psi(x,\,t)}{\partial t} \tag{1.15}$$

where

$$\hbar = h/2\pi$$
$$h = \text{Planck's constant}$$
$$U(x) = \text{Potential energy of particle along } x\text{-direction}$$
$$m = \text{Mass of the particle}$$

For a free particle $U(x) = 0$, the time-dependent Schrödinger Eq. (1.15) takes the form of:

$$-\frac{\hbar^2}{2m}\frac{\partial^2 \psi(x, t)}{\partial x^2} = i\hbar \frac{\partial \psi(x, t)}{\partial t} \tag{1.16}$$

The solution of Eq. (1.16) can be given in the form of plane wave as:

$$\psi(x, t) = Ae^{i(kx - \omega t)} \tag{1.17}$$

which as a complex function can be expanded in the form:

$$\psi(x, t) = A \cos(kx - \omega t) + iA \sin(kx - \omega t) \tag{1.18}$$

Either the real or imaginary part of Eq. (1.18) could be appropriate for a given application.

The time independent Schrödinger equation for one dimension is of the form:

$$-\frac{\hbar^2}{2m}\frac{\partial^2 \psi(x)}{\partial x^2} + U(x)\,\psi(x) = E\psi(x) \tag{1.19}$$

where

$$U(x) = \text{Potential energy}$$
$$E = \text{Total energy of the system}$$

1.6.2 Particle in a Box

The idealized solution of a particle is an application of the Schrödinger equation. The wave-function must be zero at the walls, as shown in Fig. 1.5, and hence, the solution for the wavefunction yields just sine waves.

From Fig. 1.5, it is clear that the longest wavelength is:

$$\lambda = 2L \tag{1.20a}$$

The wavelength of the higher modes are given as:

$$\lambda_n = \frac{2L}{n} \tag{1.20b}$$

where $n = 1, 2, 3, \ldots$
and hence, the momentum of the particle is:

$$p = \frac{h}{\lambda} = \frac{nh}{2L} \tag{1.20c}$$

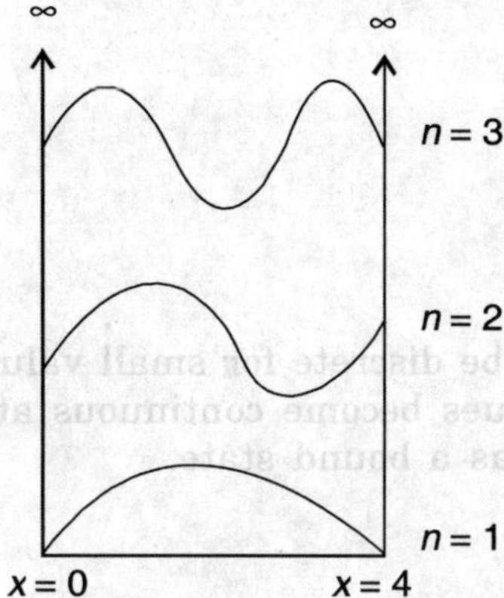

FIGURE 1.5 Schematic representation of particle in a box.

The kinetic energy associated with the particle, having momentum as given by Eq. (1.20c), is:

$$\frac{1}{2} mv^2 = \frac{p^2}{2m} = \frac{n^2 h^2}{L^2 m} = E_n \tag{1.21}$$

where Eq. (1.21) gives the energy of nth quantum state for the particle in a infinite box of length L.

Equation (1.21) clearly predicts

$$E_n \propto n^2 \qquad \text{where } n = 1, 2, 3, \ldots$$

means energy of a particle in a box is quantized and characterized by quantum number n.

The energy cannot be exactly zero. Smaller the confinement, the larger the energy required.

The energy of a particle confined in a three-dimensional box is:

$$E_n = \frac{(n_x^2 + n_y^2 + n_z^2)\, h^2}{8mL^2} \tag{1.22}$$

Equation (1.22) gives the more realistic expression for the available energies for contained particles. This expression is used in determining the density of possible energy states for electrons in solids. The energy values given by Eqs. (1.21) and (1.22) are called **Eigenvalues** of energy. For example, the energy eigenvalues of the quantum harmonic oscillator are given by

$$E_n = \left(n + \frac{1}{2}\right) \hbar\omega \tag{1.23}$$

where $n = 0, 1, 2, 3, \ldots$

$$\omega = \frac{2n}{v}$$

$$\hbar = \frac{h}{2\pi}$$

The energy eigenvalues may be discrete for small values of the energy, as shown in Fig. 1.6, but energy eigenvalues become continuous at higher energies because the system can no longer exists as a bound state.

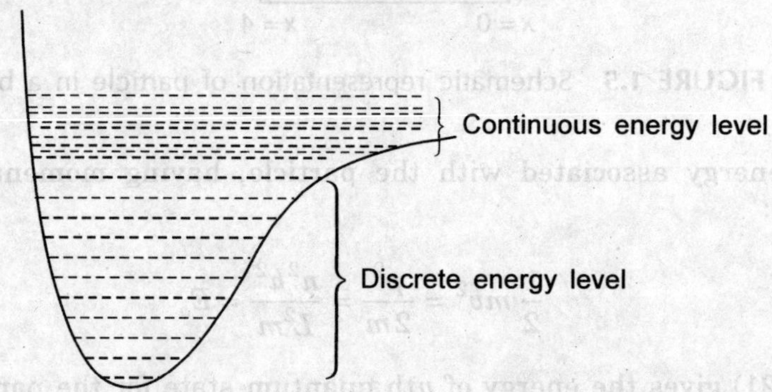

FIGURE 1.6   Energy eigenvalues of the quantum harmonic oscillator.

1.6.3   Probability and the Wavefunction

Each particle is represented by a wavefunction, which is given by $\psi(x, t)$, where ψ is a complex quantity. The product $\psi^*\psi$, where ψ^* is a complex conjugate of ψ, gives the probability of finding the particle at that position at that time, i.e., $P = \psi^*\psi = |\psi|^2$.

1.6.4   Properties of Wavefunction

1. Wavefunction contains all the measurable information about the particle.

2. $\int_0^1 \psi^*\psi \, dx = 1$, i.e., probability of finding the particle is one if particle exists.

3. Wavefunction is a continuous function.
4. Allows energy calculation via the Schrödinger equation.
5. Use to calculate the expectation value of a given variable.

EXAMPLE 1.3

The solution to Schrödinger's wave equation for a particular situation is given by

$$\psi(x) = \sqrt{\frac{2}{a_0}}\, e^{-(x/a_0)}$$

Determine the probability of finding the particle between $0 \leq x \leq (a_0/4)$.

Solution: Using probability $P = \int_{x_1}^{x_2} \psi(x)\, \psi^*(x)\, dx$

where $\Psi^*(x)$ is a complex conjugate.

Therefore,
$$P = \int_{0}^{a_0/4} \left(\sqrt{\frac{2}{a_0}}\, e^{-x/a_0}\right)\left(\sqrt{\frac{2}{a_0}}\, e^{-x/a_0}\right) dx = 0.393$$

EXAMPLE 1.4

What is the basic difference between the classical physics and quantum physics?

Solution: One major difference between the classical and quantum physics is that in classical mechanics, the position and energy of a particle or body can be determined precisely whereas in quantum mechanics, the position and energy of a particle are found in terms of probability.

1.7 CHEMICAL BONDING

In 1916, American Chemist Gilbert Lewis proposed that chemical bonds are formed between atoms because electron from atoms interact with each other. This interaction was established on the fact that elements are most stable when they contained eight electrons in their outermost orbit (called valence level). Any atom with fewer than eight valence electrons interact chemically with other atom to complete its valence shell. In this interaction either two atoms will share the electrons or one atom lose electron and other will accept the electron. In other words, we can say chemical bonds form to lower the energy of the system, and components of the system become more stable through the formation of the bonds.

There are several types of chemical bonds, namely

- Ionic bond
- Covalent bond
- Intermolecular bonding
  - Hydrogen bonding
  - Lewis forces
- Metallic bonding

In this chapter, we will discuss only ionic bond and covalent bond as shown in Fig. 1.7.

1.7.1   Ionic Bond

In ionic bonding electrons are completely transferred from one atom to other atom. In this process the reacting atoms either lose electrons or gain electrons and form ions. For example, an atom which loses electron will become positive ion and an atom who accepts electron become negatively charged ions. These two oppositely charged ions are attracted to each other by the electrostatic forces which are the basis of the ionic bond. For example, during the reaction of sodium with chlorine, sodium loses its one valence electron to chlorine to form ionic bond. In this process, sodium will become positive ion and chlorine negative ion. Positive ion tends to be smaller than their parent atoms while negative ions tend to be larger than their parent atoms.

Ionic bonding: One atom acquires and holds the electron(s) of an adjacent atom. Bonding is coulombic and strong.

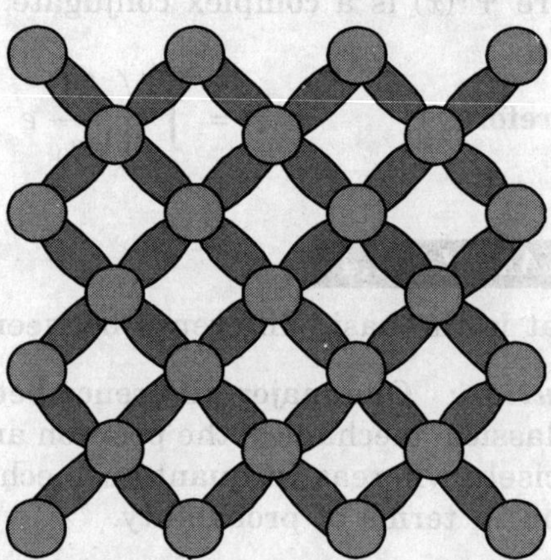

Covalent bonding: Atoms share electrons with the surrounding atoms. Bonding is moderately weak.

FIGURE 1.7 Ionic and covalent bonding in a crystal.

Important properties of ionic bonds are:

- Ionic bonds form between metals and non-metals.
- Ionic compounds dissolve easily in water and polar solvents.
- Ionic compounds are generally solids at room temperature and have high melting and boiling points.
- They are hard but brittle solids and poor conductors of electricity in the solid state.
- Ionic bonds are strong and omnidirectional.
- In solution, ionic compounds easily conduct electricity.

1.7.2   Covalent Bond

A covalent bonding occurs when two atoms (elements) share electrons. It occurs because the atoms in the compound have a similar tendency for electrons. This type of covalent bond occurs when two non-metals bond together. Both the non-metals want to gain electrons to complete their octet. For that these two non-metals share their valence electrons. Example of covalent bond is the bond between two hydrogen atoms. Covalent molecules are not strongly attached to one another. As a result covalent molecules move about freely and tend to exist as liquids or gases at room temperature. Covalent bond is further classified as polar and non-polar covalent bond. A non-polar covalent bond is formed between the two atoms having equal electron affinity. The bonding electrons are equally shared by two atoms, i.e. whenever two atoms of same element bond together, a non-polar bond is formed, e.g. hydrogen molecule. A polar bond is formed when electrons are unequally shared between two atoms. Polar covalent bonding occurs because one atom has stronger affinity for electrons than the other. In a polar covalent bond, the bonding electrons will spend a greater amount of time around the atom that has the stronger affinity for electrons. Most of the covalent bonds do not conduct electricity because of non-availability of free charge carriers. Depending on the number of shared electrons, bond length and bond energy can vary. When the number of shared electron is more, bond length decreases but bond energy increases. Covalent material is the hardest substance with high melting point.

EXAMPLE 1.5

Define electron affinity and ionization potential.

Solution: Electron affinity χ is defined as the energy difference between the vacuum level and the vacant state of the lowest energy.

Mathematically, $$\chi = E_{\text{vac}} - E_{\text{lowest}}$$

In the semiconductor material

$$E_{\text{lowest}} = E_c \text{ (i.e., conduction band edge)}$$

The electron affinity is a fundamental property of a semiconductor. It can not be measured accurately because to measure the electron affinity, excited electrons must pass through the semiconductor surface to a collector in a vacuum. The band properties at the surface can differ from those in the bulk because of the surface contamination. As might be expected, there is some scatter in published values for electron affinity.

Ionization energy γ is defined as the minimum energy required to excite an electron from the top of the valence band in the crystal to the vacuum level.

Mathematically, $$\gamma = E_{\text{vac}} - E_v$$

1.8 SUMMARY

The Bohr model of an atom consists of a nucleus that contains positively charged proton and neutral charge particle neutron. Due to positive charge on the proton, nucleus is positively charged. The nucleus is surrounded by electrons (negatively charged particles) which are distributed into distinct shells and travel in elliptical path rather than circular path. An atom is generally electrically neutral because in an atom there is equal number of protons and electrons. An atom becomes positive charged when it loses electrons and negative charged when gains electrons. The electron in the outer shell is known as valence electron and outershell is known as valence shell. The number of valence electrons will decide the position of atom in periodic table which in turn defines the chemical, electrical and optical properties of an atom. Total energy of the electron in a particular orbit is always negative. This shows that electron is bounded to nucleus by a force called **binding force** or **nuclear force**. If one wants to remove electron from the orbit then one should supply energy greater than its binding energy. The energy of an orbiting electron is quantized means certain energy levels are allowed for electrons to occupy. Electron shows the behaviour of wave nature as well as particle nature. In classical physics, electron is repelled by an electric field as long as energy of an electron is below potential barrier height energy. In quantum physics, the wavefunctions of electron encounter the potential barrier, and there is finite probability that electron can tunnel through the potential barrier if width of well is of order of few nm even the electron's energy is less than the potential barrier's height. The Schrödinger wave equation predicts the future behaviour of electron. It is a wave equation in terms of wavefunction which predicts analytically and precisely the probability of occurring the events. It was observed that when two atoms interact with each other then in this interaction either two atoms will share the electrons (covalent bond) or one atom will loose electrons and other will accept these electrons (ionic bond). This is due to the fact that elements are most stable when they contained eight electrons in their outermost orbit. In the bond formation only valence electrons take part.

REVIEW QUESTIONS

1. Why it is easier to remove the electron from the higher energy levels than the ground state energy level?

2. What do you mean by the quantization of energy levels?

3. What is the significance of $E_1 = -13.6$ eV?

4. Discuss the significance of different quantum numbers.

5. What is meant by a probability density function?

6. Discuss the consequence of the violation of Pauli's exclusion principle.

7. What is the importance of the equation $p = h/\lambda$?

8. What is the physical meaning of Schrödinger's wave equation?

9. Let us consider a case of one-dimensional potential well. Discuss classically and quantum mechanically, what will happen if a particle with energy $E < V_0$, (where V_0 is the maximum height of the well) is confined into the Region II of Fig. 1.3.

10. What is the difference between the energy levels and energy band?

11. What is the main reason of bond formation in the crystal? Why covalent bond is formed in the elementary semiconductors?

12. What is the relationship between a free electron's velocity and its wavelength?

NUMERICAL PROBLEMS

1. What is the de Broglie wavelength for a beam of electrons at room temperature whose kinetic energy is $(3/2)KT$?

2. Derive a relationship for electron wavelength in terms of the kinetic energy E and free electron mass m_0.

3. From the Bohr model, what emission wavelength would you expect for a transition in hydrogen from E_2 to E_1? Transition ending at $n = 1$ is called Lyman series. Similarly transition ending at $n = 2$ is called Balmer series. For $n = 3$ series is Paschen series.

4. What wavelength of light should you shine on hydrogen to cause electrons to go from E_1 to E_2 by optical absorption?

5. Show that the most probable value of the radius r for 1s electron in hydrogen atom is equal to the Bohr radius a_0.

6. The solution to Schrödinger's wave equation for a particular situation is given by

$$\psi(x) = \sqrt{\frac{2}{a_0}}\, e^{-x/a_0}$$

Determine the probability of finding the particle between

(a) $0 \leq x \leq (a_0/2)$
(b) $0 \leq x \leq (a_0)$
(c) $0 \leq x \leq (3a_0)/4$.

7. An electron is bound in a one-dimensional infinite potential well with a width of 50 A°. Determine the electron energy for the first three energy levels. Also find the momentum associated with the electron.

8. A one-dimensional infinite potential well with a width of 12 A° contains an electron. If an electron drops from the second energy level to the first, what is the wavelength of a photon that might be emitted?

9. Find the values of all other quantum numbers if $n = 2$ and $n = 3$.

10. An electron is moving with a velocity of 10^6 m/s. Determine its energy (in joule as well as in eV), momentum and de Broglie wavelength (in A°). If the de Broglie wavelength of an electron is 150 A° then find its energy in eV, momentum and velocity. Compare these two results.

11. Evaluate the transmission coefficient for an electron of energy 2.5 eV impinging on a potential barrier of height 8.0 eV and thickness of 10^{-9} m.

    (*Hint:* Use the relation $T \approx 16\,(E/V_0)\,(1 - (E/V_0))\exp(-2K_2 a)$, where a is the

    width of the barrier and $K_2 = \sqrt{\dfrac{2m(4\pi^2)(V_0 - E)}{h^2}}$)

CHAPTER
2

Semiconductor Material
and Its Band Diagram

LEARNING OBJECTIVES

- Classification of solid-state materials
  - Conductor
  - Insulator
  - Semiconductor (sc)
- Types of semiconductor
  - Elemental
  - Compound
- Crystal structure
- Band formation in semiconductor material
- Energy momentum diagram
  - Effective mass concept
- Intrinsic and extrinsic semiconductors
- Fermi-Dirac statistics
  - Fermi level
- Electron-hole concentration in
  - Intrinsic semiconductor
  - Extrinsic semiconductor
- Fermi-level position in
  - Intrinsic semiconductor
  - Extrinsic semiconductor

2.1 INTRODUCTION

Technology has completely revolutionized the society but technology itself is
revolutionizing by the solid-state electronics.

18

Although the crystalline, mechanical, thermal and optical properties of the solids are of greater interest but it is electrical properties of the materials that mainly differentiate the various solid-state materials. The single-most important electrical property of the material is its electrical conductivity which is defined as current flow per unit area of the material and given in siemens/cm.

In this chapter, we will classify the solid-state materials on the basis of their electrical conductivity as discussed in Section 2.2. The emphasis is on the detailed discussion of semiconductor materials and their properties. The mathematical expressions for carrier concentration in semiconductor materials are also discussed.

2.2 CLASSIFICATION OF SOLID-STATE MATERIALS

Solid-state materials can be classified into three major groups on the basis of their electrical conductivity like conductors (metals), insulators, semiconductors.

Conductors (metals) are highly conductive materials which offer very low resistance to the electrical current. The resistance of the metals increases with increase in temperature and hence, metal has positive temperature coefficient of resistance. Example of metals are silver, copper and aluminium. These materials have large number of free electrons available for conduction. In other words, we can say metals are sea of free electrons.

Insulators are those solid materials whose conductivity is zero at any temperature and offer virtually infinite resistance to the current flow. Example of insulators are glass, rubber, mica and plastics.

The third group of solids posses electrical conductivity that is neither very high nor very low, and are called **semiconductors**. The resistance of semiconductor decreases with increase in temperature and hence it has negative temperature coefficient of resistance. The semiconductor material behaves as insulator at absolute zero temperature (T = 0 K) and above absolute zero temperature, the material shows some conductivity behaviour. Examples of semiconductor materials are silicon, germanium, GaAs etc.

2.3 SEMICONDUCTORS

A relatively small group of elements and compounds which have important electrical behaviour is called **semiconductor**. These materials are neither good conductor nor good insulator but their ability to conduct electricity is intermediate in between the two, i.e., higher than insulator but lower than metal. Insulators have very low conductivity, of the order of 10^{-18} to 10^{-8} s/cm, and conductors have high conductivities, typically from 10^4 to 10^6 s/cm. Semiconductors have conductivities between those of insulators and conductors as shown in Fig. 2.1.

The conductivity of semiconductor material is generally sensitive to temperature, illumination, impurities etc. The sensitivity in conductivity makes the semiconductor one of the attractive materials for the electronic devices.

A semiconductor behaves as an insulator at absolute zero temperature (T = 0 K) because no free electrons are available for conduction and an appreciable conduction

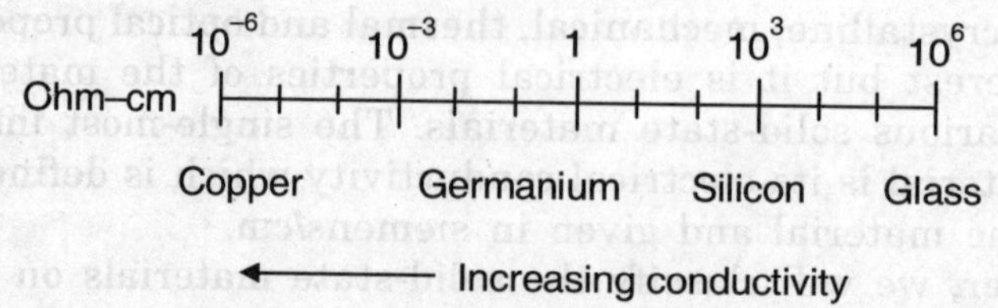

FIGURE 2.1 Classification of materials on the basis of conductivity.

occurs at temperature above 0 K. This shows that the resistance of semiconductor material decreases with increase in temperature. All electronic devices are made up of semiconductor material because their electrical conductivity can be easily controlled.

The study of semiconductor materials have started in early 19th century and over the years many semiconductor materials have been investigated as listed in Table 2.1.

Table 2.1 List of Elemental and Compound Semiconductor Materials

Group IV	: Elemental semiconductors	: Diamond, silicon, germanium
Group IV	: Compound semiconductors	: Silicon carbide (SiC),
		Silicon germanide (SiGe)
Group III–V	: Semiconductors	
	Aluminium antimonide (AlSb)	: Aluminium arsenide (AlAs)
	Aluminium nitride (AlN)	: Aluminium phosphide (AlP)
	Boron nitride (BN)	: Gallium antimonide (GaSb)
	Gallium arsenide (GaAs)	: Gallium nitride (GaN)
	Gallium phosphide (GaP)	: Indium antimonide (InSb)
	Indium arsenide (InAs)	: Indium nitride (InN)
	Indium phosphide (InP)	
Group III–V	: Ternary semiconductor alloys	
	Aluminium gallium arsenide ($Al_xGa_{1-x}As$)	
Group II–VI	: Semiconductors	
	Cadmium selenide (CdSe)	: Cadmium sulfide (CdS)
	Cadmium telluride (CdTe)	: Zinc oxide (ZnO),
		Zinc selenide (ZnSe)
	Zinc sulfide (ZnS)	: Zinetelluride (ZnTe)
Group I–VII	: Semiconductors	: Cuprous chloride (CuCl)
Group IV–VI	: Semiconductors	: Lead sulfide (PbS)
		Lead telluride (PbTe)
Group II–V	: Semiconductors	: Cadmium phosphide (Cd_3P_2) etc.
	Organic semiconductors	
	Magnetic semiconductors	

Semiconductor materials are classified as elementary semiconductor and compound semiconductor on the basis of their composition.

2.3.1 Elemental Semiconductor

The elemental semiconductors consist of crystal composed of single atomic element from group IVth of the periodic table. For example silicon (Si), germanium (Ge), carbon (C) and tin (Sn).

The two most common elemental semiconductor materials are silicon (Si) and germanium (Ge). The silicon and germanium have a diamond lattice structure. This structure belongs to the cubic crystal lattice family.

In the early 1950s, germanium was the major semiconducting material but in 1960s silicon has become the most commonly used electronic semiconductor material and become a practical substitute of germanium because silicon has excellent processing properties and relatively easy to fabricate in pure form. A layer of native oxide (SiO_2) can be easily grown. Silicon devices exhibit better properties at room temperature. There is also economic consideration because silicon is readily available in nature in form of silica and silicates.

Germanium has higher electrical conductivity than silicon and is used in most low and medium power diodes and transistors.

In the modern technology, silicon is one of the most studied elements in the periodic table and silicon technology is by far the most advanced among all semiconductor technologies.

2.3.2 Compound Semiconductor

Compound semiconductor materials are most important basic materials for optical applications. The compound semiconductor material composed of two or more different elements as given in Table 2.1. Most of which are formed from special combination of the third group and fifth group elements of the periodic table. For example, gallium arsenide (GaAs) which is a combination of gallium (Ga) from IIIrd group and arsenic (As) from Vth group of periodic table. The compound semiconductors are widely used in high speed devices and photonic applications.

Compound semiconductors offer high performance, i.e., optical characteristics, higher frequency and higher power than the elemental semiconductors. They also offer greater device design flexibility due to mixing of materials.

In addition to binary compounds, ternary compounds and quaternary compounds are made for special applications.

Examples of various compound semiconductors are:

Binary: GaAs, SiC, etc.
Ternary: $Al_xGa_{1-x}As$, $In_xGa_{1-x}N$ where $0 \leq x \leq 1$
Quaternary: $In_xGa_{1-x}As_yP_{1-y}$,   where $0 \leq x \leq 1$, $0 \leq y \leq 1$.

where x and y give the composition of Al and As respectively. For example, assume a compound semiconductor has 25% "atomic" concentrations of Ga, 25% *atomic* of In and 50% *atomic* of N. The chemical formula for compound material is:

$$Ga.25 \; In.25 \; N.5$$

and the correct reduced semiconductor formula is:

$$Ga_{0.05} \; In_{0.05} \; N$$

2.4 CRYSTAL STRUCTURE

Crystal structure refers to the internal arrangement of particles not the external appearance of the crystal. A crystal structure formed by repeating a three-dimensional pattern of atoms, molecules or ions. A crystal structure and symmetry play an important role in determining many properties of crystal like cleavage, electronic band structure and optical properties.

Crystalline solids are arranged in fixed geometric patterns or lattices. They have ordered arrangement of units maximizing the space they occupy and practically are incompressible. Ionic and atomic crystals are hard and brittle with high melting point. A crystalline solid is shown in Fig. 2.2(a).

It is not necessary that all the solids are crystalline. Some crystals have no periodicity at all as shown in Fig. 2.2(b). These solids are called amorphous solid. Glass and most of the plastics are the examples of amorphous solid.

A solid can be composed of many crystalline grains of varying size and different orientations as shown in Fig. 2.2(c). Such solids are known as Polycrystalline solids. Almost all the common metals and many ceramics are the examples of polycrystalline solids.

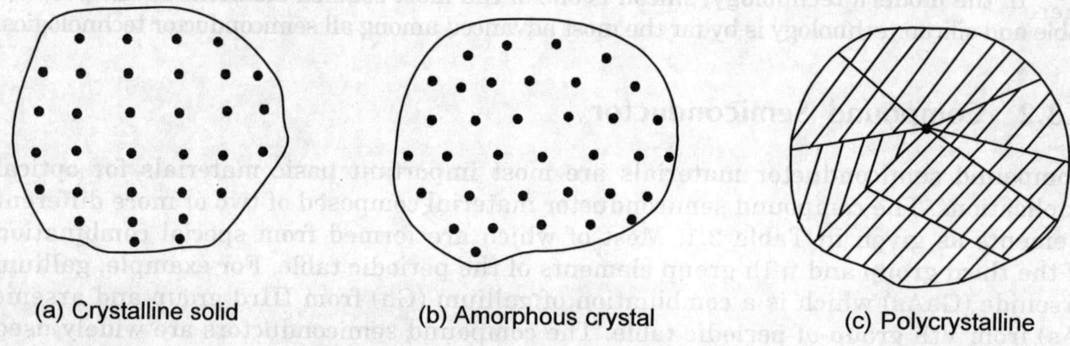

FIGURE 2.2 Different crystal structure.

Those crystals for which the electrical, mechanical and optical properties depend on the crystal axis and related to crystal orientation are called anisotropic crystals. The physical properties of the amorphous solids are identical in all directions along any axes. This is known as isotropic properties. At melting temperature (temperature at which solid converts into liquid), physical properties of the crystalline solids change sharply. Amorphous solids soften gradually when they are heated. A crystalline state is relatively more stable than the amorphous state. The transformation of amorphous substance into crystalline solid rarely takes place. The reverse transformation, i.e., crystalline solid to amorphous substance does not occur because it is energetically unfavoured.

2.4.1 Unit Cell

The Bravais lattice is the basic building block from which all crystals can be constructed. Mathematician Auguste Bravais discovered that there were 14 different collections of the groups of points, which are known as Bravais lattices. These lattices fall into seven different 'crystal systems' which are differentiated by the relationship between the angles (between sides of the 'unit cell') and the distance between points in the unit cell. Unit cell is the smallest collection of lattice points that can be repeated at regular intervals in three dimensions to create the crystalline solid. Unit cell can fill the entire space without creating void.

The 'lattice parameter' is the length between two points on the corners of a unit cell. Each of the various lattice parameters are designated by the letters a, b, and c. If two sides are equal, such as in a tetragonal lattice, then the lengths of the two lattice parameters are designated a and c, with b omitted. The angles are designated by the Greek letters α, β, and γ. For example, α is the angle between the b and c axes.

Primitive cell, as shown in Fig. 2.3, of a particular crystal structure is the smallest possible unit cell one can construct such that when tiled, it completely fills space. Every equivalent lattice point in the three-dimensional crystal can be found by using

$$\mathbf{R} = m\mathbf{a} + n\mathbf{b} + p\mathbf{c} \qquad (2.1)$$

where m, n and p are integers.

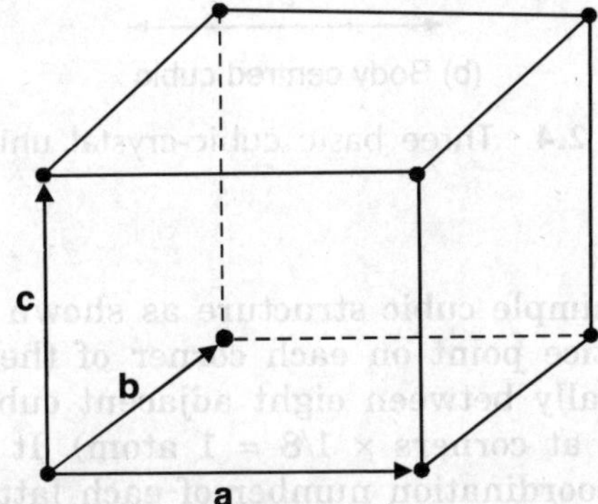

FIGURE 2.3 A generalized primitive unit cell.

The choice of primitive cell is not unique. All possible primitive unit cells have identical properties.

Note: The periodic arrangement of atoms in a crystal is known as lattice. One pattern is located at each lattice point and lattice is represented by set of translational vectors. A lattice is invariant after translation through any of these vectors or any sum of an integer number of these vectors.

2.4.2 Crystal Systems

There are seven unique crystal systems. These systems depend on the orientations (α, β, γ) and length of the basis vectors (**a**, **b**, **c**).

The simplest and most symmetric crystal is **cubic** (or **isomeric**) **system**. This has symmetry of cube, i.e., all the three axes are mutually perpendicular and of equal length ($\alpha = \beta = \gamma = 90°$ and $\mathbf{a} = \mathbf{b} = \mathbf{c}$). Common cubic crystals are diamond, gold, spinel, etc.

In the cubic crystal system, three types of arrangements are formed as shown in Fig. 2.4. These are:

- Simple cubic
- Body-centred cubic (bcc)
- Face-centred cubic (fcc)

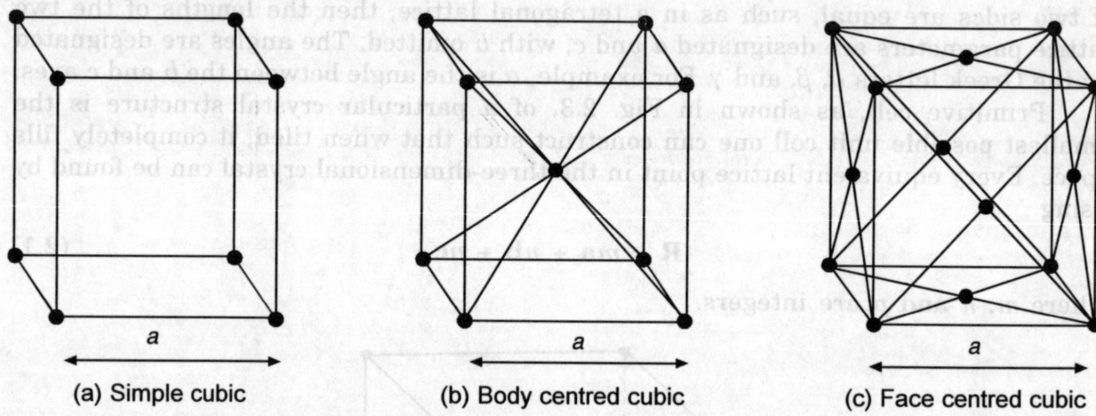

FIGURE 2.4   Three basic cubic-crystal unit cells.

Simple cubic

The simplest cubic cell is simple cubic structure as shown in Fig. 2.4(a). This cubic system consists of one lattice point on each corner of the cube. Each atom at the lattice point is shared equally between eight adjacent cubes and the unit cell and contains total one atom (8 at corners × 1/8 = 1 atom). It is very difficult to get a simple cubic system. The coordination number of each lattice point is equal to six.

Note: Coordination number is the number of nearest neighbours to any atom.

Body-centred cubic (bcc)

The body-centred cubic unit cell has atoms at each of the eight corners of a cube (like the cubic unit cell) plus one atom in the centre of the cube as shown in Fig. 2.4(b). Each of the corner atom is the corner of another cube so the corner atoms are shared among eight unit cells. **It is said to have a coordination number of 8**. The bcc unit cell has, 8 (at corners) × 1/8 + 1 (body centre) = 2 atoms. The volume of atoms iᵣ a cell per the total volume of a cell is called the **packing factor**. The bcc unit cell has a packing fact of 0.68. The bcc structure is often the high temperature form of metals that are closely-packed at lower temperatures.

Face-centred cubic (fcc)

The face-centred cubic structure has atoms located at each of the corners and the centres of all the cubic faces as shown in Fig. 2.4(c). Each of the corner atoms is the corner of another cube so the corner atoms are shared among eight unit cells. Additionally, each of its six face centred atoms is shared with an adjacent atom. Since 12 of its atoms are shared, it is said to have a coordination number of 12. The number of atoms per unit cell in fcc is calculated as: 8(centres) × (1/8) (for each corner) + (1/2) (for each face) × 6 = 1 + 3 = 4 atoms. In the fcc structure, the atoms can pack closer together than they can in the bcc structure. The atoms from one layer nest themselves in the empty space between the atoms of the adjacent layer. The packing factor (the volume of atoms in a cell per the total volume of a cell) is 0.74 for fcc crystals. Some of the metals that have the fcc structure include aluminium, copper, gold, iridium, lead, nickel, platinum and silver.

The other six systems, in order of decreasing symmetry are as follows:

Hexagonal

The hexagonal crystal system has four crystallographic axes. Three of the four axes are in one plane, intercept at 120° and are of same length. The fourth is either longer or shorter but must be at right angle toward the other corners. In brief,

$$\alpha = \beta = 90°, \ \gamma = 120° \text{ and } \mathbf{a} = \mathbf{b} \neq \mathbf{c}$$

Tetragonal

The tetragonal crystal will have square base and top, but taller height. All the three axes should be at right angle, i.e., $\alpha = \gamma = \beta = 90°$ and two basis vectors must be of equal length, $\mathbf{a} = \mathbf{b} \neq \mathbf{c}$.

Orthorhombic

All the three axes are perpendicular to each other with unequal lengths, i.e. $\alpha = \gamma = \beta = 90°$ and $\mathbf{a} \neq \mathbf{b} \neq \mathbf{c}$.

Rhombohedral

Also known as trigonal. This crystal is described by vectors of equal length and all the three axes are not mutually perpendicular, i.e.,

$$\mathbf{a} = \mathbf{b} = \mathbf{c} \text{ and } \alpha = \gamma = \beta \neq 90°$$

Monoclinic

Monoclinic crystal system is described by three vectors of unequal length. They form a rectangular prism with a parallelogram as its base. The two pairs of vectors are at right angle to each other and third pair makes an angle other than 90°, i.e.

$$\mathbf{a} \neq \mathbf{b} \neq \mathbf{c} \text{ and } \alpha = \gamma = 90° \neq \beta$$

Triclinic

The triclinic crystal system has three unequal crystallographic axes and all of which intersects at oblique angles, i.e.,

$$\mathbf{a} \neq \mathbf{b} \neq \mathbf{c} \text{ and } \alpha \neq \gamma \neq \beta$$

2.4.3 The Diamond and Zincblende Structures

The elemental semiconductors silicon and germanium have diamond lattice structure as shown in Fig. 2.5(a). This structure belongs to fcc crystal family and can be seen as two interpenetrating fcc sublattices with one sublattice displaced from other by one quarter of the distance along a diagonal of the cube (i.e., a displacement of $a/4$). In a diamond lattice all atoms are identical and each atom is surrounded by four equidistant nearest neighbours that lie at the corners of the tetrahedron.

The Zincblende structure is same as the diamond structure except one fcc sublattice has atom of one type and other fcc sublattice has atom of other type as shown in Fig. 2.5(b). For example, compound semiconductor GaAs or InP atoms are arranged in basic diamond structure except one fcc sublattice has Ga and other has As. Some of the II–IV group compounds have Wurtzite lattice.

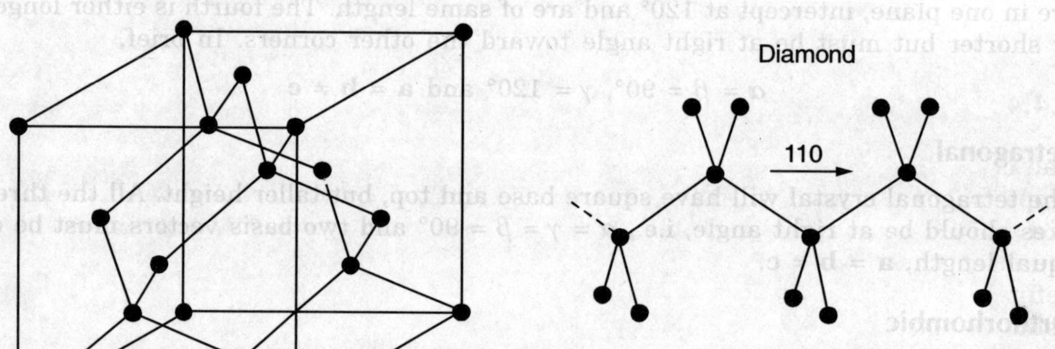

(a) Diamond structure

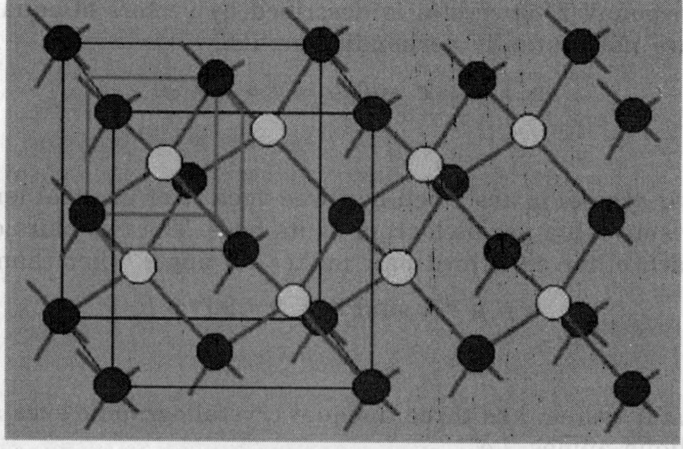

(b) Zincblende structure

FIGURE 2.5 Crystal structure of diamond and zincblende.

EXAMPLE 2.1

At T = 300 K, the number of silicon atoms per cubic centimetre is 5×10^{22} atoms/cm^3. Find the lattice constant and the density of silicon at room temperature T = 300 K.

Solution: There are eight atoms per unit cell. Therefore,

$$\frac{8}{a^3} = 5 \times 10^{22} \text{ atoms/cm}^3$$

or

$$a = 5.43 \text{ A}°$$

$$\text{Density} = \text{No. of atoms/cm}^3 \, \frac{\text{Atomic weight}}{\text{Avagadro's number}}$$

$$= 5 \times 10^{22} \, \frac{(28.09) \text{ gm/mol}}{6.02 \times 10^{23} \text{ (atoms/mol)}} = 2.33 \text{ gm/cm}^3$$

2.4.4 Miller Indices

The crystal properties along the different planes are different. Electrical and other properties of the materials are dependent on the crystal orientation. The orientation of a surface or a crystal plane may be defined by considering how any plane intersects the main crystallographic axes of the solid. By using Miller indices one can easily define the various planes as well as directions in the crystal. The Miller indices are defined as the smallest-possible integers which have the same ratio as the inverse of the intersections of the given plane with a set of axis (defined by the unit vectors of that crystal). These three indices are obtained in the following way:

1. Find the intercepts of the planes with the crystal axis and express these intercepts as the integral multiples of the basis vectors.
2. Take the reciprocals of these three numbers (as found in Step 1) and reduce them to smallest three integers having the same ratio.
3. For planes, the index is the reciprocal of the value of the intersection of the plane with a particular axes, as shown in Fig. 2.6.
4. For directions, the index is the axis coordinate of the end point of the vector, converted to the nearest whole number.

General principles to remember:
* Replace the negative integers with bar over the number, i.e., replacing $(-h)$ by $\bar{h}$.
* If the plane is parallel to an axis, means it cuts at ∞ and $1/\infty = 0$, i.e. Miller index is zero.
* If the plane passes through origin, one can translate the unit cell in a suitable direction.
* Directions are indicated by square bracket like [hkl] whereas family of directions are indicated by <hkl>. Planes are indicated by (hkl) and families of plane are represented by {hk}.

* When a Miller index is smaller, the plane is more parallel to axis.
* When a Miller index is bigger, the plane is more perpendicular to the axis.
* The angle θ between two directions $[h_1, k_1, l_1]$ and $[h_2, k_2, l_2]$ is given by

$$\text{Cos } \theta = \frac{h_1 h_2 + k_1 k_2 + l_1 l_2}{\sqrt{(h_1^2 + k_1^2 + l_1^2)(h_2^2 + k_2^2 + l_2^2)}}$$

EXAMPLE 2.2

The plane has intercept at $2a$, $4a$ and $6a$ along the three coordinates as shown in Fig. 2.6. Label the plane.

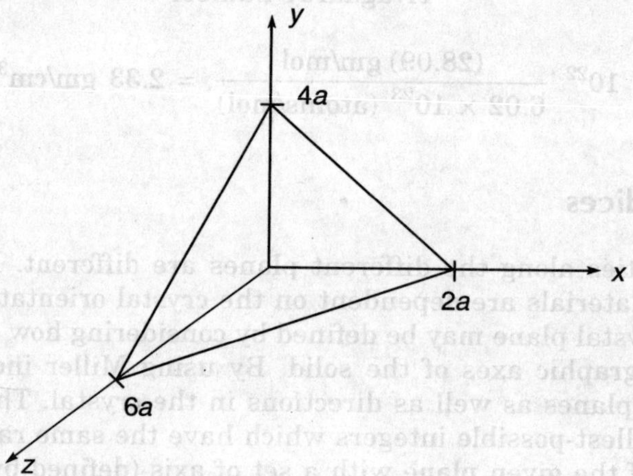

FIGURE 2.6   Figure for Example 2.2.

Solution:

Step 1: Intercepts are $(2a, 4a, 6a)$.

Step 2: Taking reciprocals of these intercepts, we have

$$\left(\frac{1}{2a}, \frac{1}{4a}, \frac{1}{6a} \right)$$

The three smallest integers having the same ratio is $(6, 3, 2)$. [This is obtained by multiplying each fraction by 12 because their LCM is 12].

Step 3: Thus, plane is referred as $(6, 3, 2)$-plane.

2.5 STRUCTURE OF ELEMENTAL SEMICONDUCTOR MATERIAL

As discussed in Section 2.4 elemental semiconductors (Si and Ge) have diamond like lattice structure. Each atom in a diamond lattice is surrounded by four nearest neighbours. Figure 2.7 shows a simplified two-dimensional bonding scheme for the

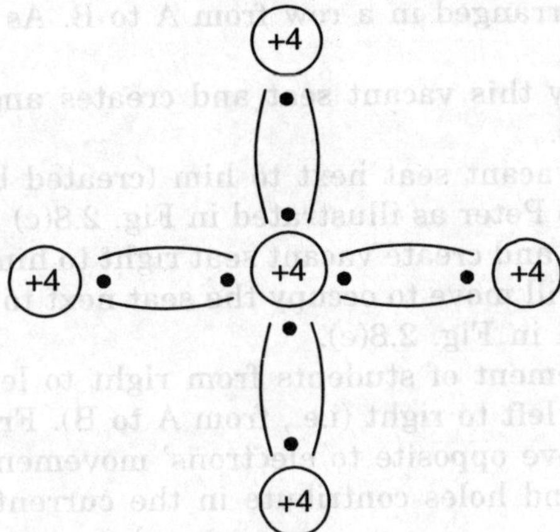

FIGURE 2.7 Two-dimensional bonding scheme for tetrahedron

tetrahedron. Since elemental semiconductor materials belong to 4th group of the periodic table, hence each atom has four electrons in the valence band. These valence electrons are shared in a paired manner as seen in Fig. 2.7 and this sharing of electrons is known as **covalent bond**. In the sharing process, both electrons are indistinguishable except that they must have opposite spin to satisfy Pauli's exclusion principle. Such a bond is very stable and holds the two atoms together very tightly, and requires very high energy to break this bond.

At $T = 0$ K, all the valence electrons are bound in their respective tetrahedron lattice and hence no free electrons are available for conduction. Therefore, all elemental semiconductors are insulators at low temperature.

Although the covalent bond results in stronger bond between the valence electron and their parent atom, it is still possible that valence electron can absorb sufficient energy either due to thermal excitation, illumination or electric field to break the covalent bond and assume a "free" state. At any temperature above $T = 0$ K, a finite number of electrons may acquire sufficient thermal energy to break the covalent bond and becomes free although the average thermal energy possessed by electron at room temperature ($T = 300$ K) is only 25.9 meV. These free electrons can move under the influence of applied electric field and results in current and elemental semiconductor behaves as metal (conductor). Each broken covalent bond gives one free electron. Deficiency of valence electron results another particle called *Hole*.

2.6 HOLE CONDUCTION IN SEMICONDUCTOR CRYSTAL

The movement of holes in a semiconductor material can be explained by considering the movement of students from occupied seats (analogous to electron) to unoccupied seats (analogous to hole). Imagine students Tom, Lee, Peter and Brian have occupied four seats and fifth seat of the right of Tom is unoccupied in a class room as shown in Fig. 2.8(a). Seats are arranged in a row from A to B. As clear, seat at position B is unoccupied.

Tom moves to occupy this vacant seat and creates another vacant seat left to him as seen in Fig. 2.8(b).

When Lee occupies vacant seat next to him (created by Tom's movement) the vacant seat moves right to Peter as illustrated in Fig. 2.8(c) and now Peter will move to occupy this vacant seat, and create vacant seat right to him as shown in Fig. 2.8(d).

Again at last Brian will move to occupy the seat next to him and create a vacant seat at position B as seen in Fig. 2.8(e).

This shows the movement of students from right to left (i.e. from B to A) has resulted vacant seat from left to right (i.e., from A to B). From this illustration, one can observe that holes move opposite to electrons' movement. In any semiconductor material, both electrons and holes contribute in the current.

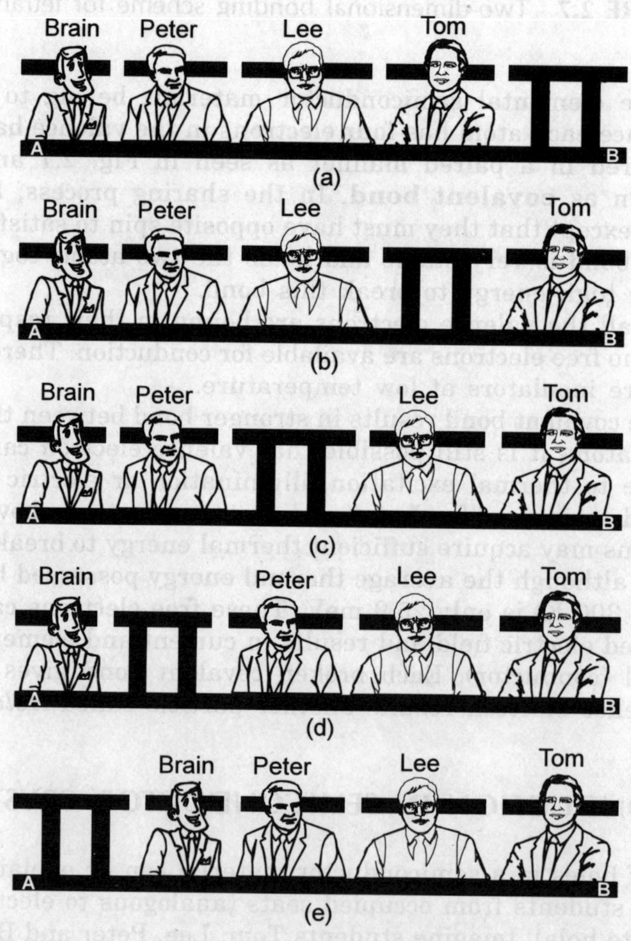

FIGURE 2.8 Hole conduction in semiconductor crystal.

2.7 BAND FORMATION IN A SEMICONDUCTOR MATERIAL

According to Bohr's quantum mechanics, for an isolated atom electrons can exist only in a specified energy level, i.e., electrons can take only discrete energy levels. The energy E_n of the nth orbit of the isolated atom (mainly hydrogen atom) is given by

$$E_n = -\frac{q^4 m_0}{8\varepsilon_0^2 h^2 n^2} \tag{2.2}$$

where

m_0 = Free-electron mass
q = Electronic charge on electron
ε_0 = Permittivity of free space
h = Planck's constant

The energy given by Eq. (2.2) is always measured in electron-volt and abbreviated as eV.

Substituting value of all the parameters, Eq. (2.2) can be further re-written as:

$$E_n = -\frac{13.6}{n^2} \text{ eV} \tag{2.3}$$

where n = 1, 2, 3, ..., are called **principal quantum number** and negative sign only indicates that electrons are bound in their respective energy levels.

It is also observed from detailed study that for higher principal quantum number ($n \geq 2$) energy levels are split according to their angular momentum quantum number (l = 0, 1, 2, ..., $n-1$).

Two identical atoms have identical electronic structure when they are isolated from each other so that there is no interaction of electron wave functions between them. When these two identical atoms are brought closer, the doubly degenerate energy level will split into two closely spaced energy levels, as shown in Fig. 2.9.

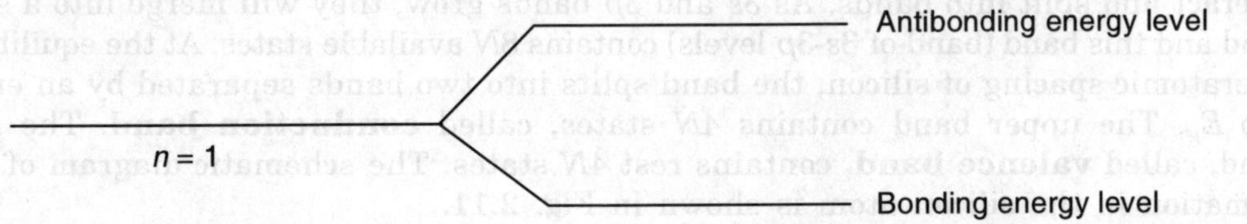

FIGURE 2.9 Splitting of single energy level into doubly degenerate energy level.

This splitting of energy level is due to interaction between their electron's wavefunctions.

As N isolated identical atoms are brought closer to form a solid, the orbits of the outer electrons of different atoms overlap and interact with each other. The interaction (including forces of attraction when electrons are far away from each other and repulsion, when electrons are closer) between them causes a split in their energy levels. Instead of two levels, N closely spaced energy levels are formed. If N is very large then the split energy levels form essentially continuous band of energies, which can extend over a few eV depending on the outer-atomic spacing for the crystal.

The actual splitting of energy levels in a semiconductor is much more complicated. Each isolated silicon atom has 14 electrons and their distribution in various energy level is shown in Fig. 2.10.

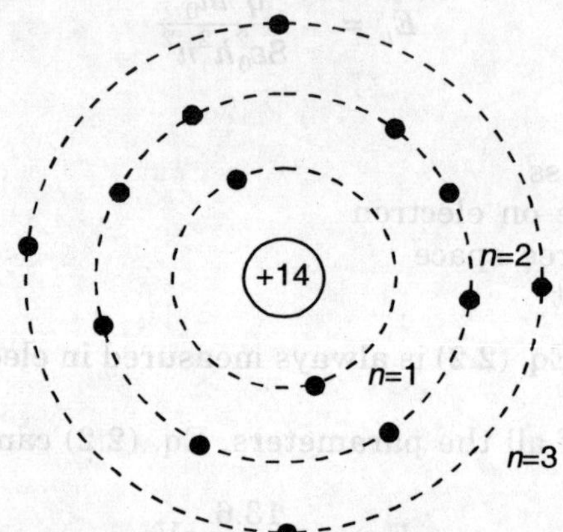

FIGURE 2.10 Electron distribution in various energy levels of silicon atom.

In the outer energy level ($n = 3$), there are four electrons which are loosely bound to nucleus and can take part in chemical reaction. These outer electrons are called **valence electrons**. The electronic distribution for an isolated silicon atom is like $1s^2\ 2s^2\ 2p^6\ 3s^2\ 3p^2$. This distribution shows that $3s$ subshell ($n = 3$, $l = 0$) has two allowed quantum states per atom and will contain two valence electrons at $T = 0$ K. The $3p$ subshell ($n = 3$, $l = 1$) has six allowed quantum states per atom and contains remaining two valence electrons. Let us consider N isolated silicon atoms. As the interatomic distance decreases, the $3s$ and $3p$ subshells of N isolated atoms will interact and split into bands. As $3s$ and $3p$ bands grow, they will merge into a single band and this band (band of $3s$-$3p$ levels) contains $8N$ available states. At the equilibrium interatomic spacing of silicon, the band splits into two bands separated by an energy gap E_g. The upper band contains $4N$ states, called **conduction band**. The lower band, called **valence band**, contains rest $4N$ states. The schematic diagram of band formation in the silicon atom is shown in Fig. 2.11.

At $T = 0$ K, electrons occupy the lowest energy states and hence all energy levels in the valence band are occupied whereas all the upper energy levels are empty.

The lowest energy level of conduction band is represented by E_c, called **conduction band edge**. Similarly, top most energy level of valence band is called **valence band edge** and is represented by E_v. The separation between the conduction band edge to valence band edge at absolute zero temperature is called **forbidden gap** or band gap and is represented by E_g.

i.e.,
$$E_g = E_c - E_v$$

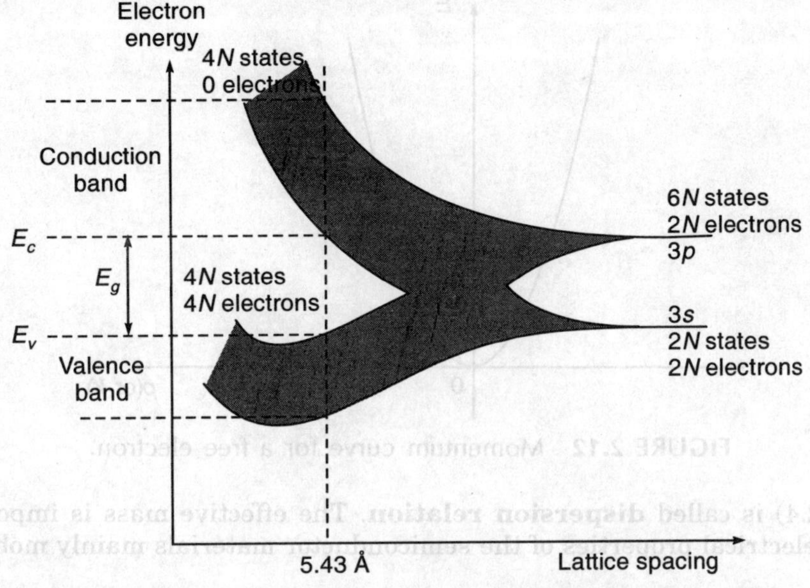

FIGURE 2.11   Bond formation in silicon crystal.

2.8   THE ENERGY-MOMENTUM DIAGRAM OR CONCEPT OF EFFECTIVE MASS

The relation between energy and momentum of a free electron is given by

$$E = \frac{p^2}{2m_0} \tag{2.4}$$

where

p = Momentum of free electron
m_0 = Mass of free-electron

Equation (2.4) shows that energy-momentum have parabolic relation as shown in Fig. 2.12.

Electrons with energy closer to band minima behaves as a free electron and accelerate in the presence of the field just like a free electron in a vacuum. But due to periodicity in the crystal, the electrons are not completely free but interact with the periodic potential of the lattice. As a result, their "wave particle" motion cannot be same as for electrons in free space and Eq. (2.4) cannot be valid.

Equation (2.4) can still be valid for electrons in a crystal, if free-electron mass m_0 is replaced by effective mass (m_n^* or m_p^*), i.e.,

$$E = \frac{p^2}{2m_{n \text{ or } p}^*} \tag{2.5}$$

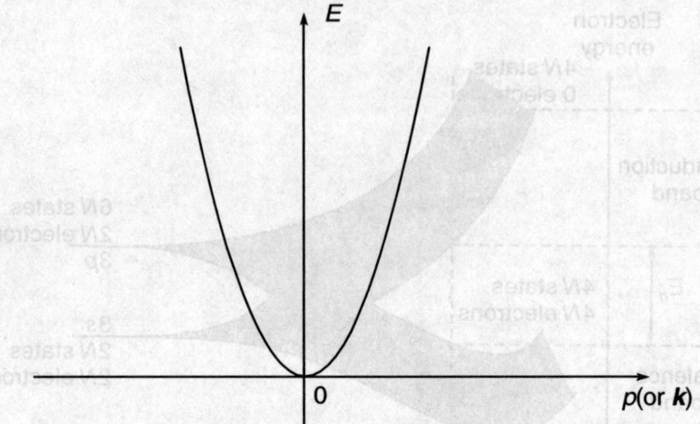

FIGURE 2.12   Momentum curve for a free electron.

Equation (2.4) is called **dispersion relation**. The effective mass is important as it affects the electrical properties of the semiconductor materials mainly mobility of the carriers.

Using de Broglie equation,

$$p = \hbar k = \left(\frac{h}{2\pi}\right) k \tag{2.6}$$

in Eq. (2.6), we have

$$E = \frac{\hbar^2 k^2}{2 m^*_{n\text{ or }p}} \tag{2.7}$$

where

$$\hbar = \frac{h}{2\pi} \tag{2.8a}$$

$$k = \frac{2\pi}{\lambda} \tag{2.8b}$$

Differentiating Eq. (2.7) w.r.t. k, one has

$$\frac{\partial E}{\partial k} = \frac{\hbar^2 k}{m^*_{n\text{ or }p}} \tag{2.9}$$

Again differentiating Eq. (2.9) w.r.t. k, we have

$$\frac{\partial^2 E}{\partial k^2} = \frac{\hbar^2}{m^*_{n\text{ or }p}}$$

or $\quad m^*_{n,p}$ = Effective mass of electrons (or holes) = $\dfrac{\hbar^2}{(\partial^2 E/\partial k^2)}$

$$= \dfrac{\hbar^2}{\text{Curvature of } E\text{-}k \text{ curve}} \tag{2.10}$$

Equation (2.10) clearly shows that the effective mass of the electron (or hole) is inversely proportional to the curvature of the E vs k curve. It can be concluded that narrower the parabola, larger the effective mass. A simplified E vs k diagram is shown in Fig. 2.13.

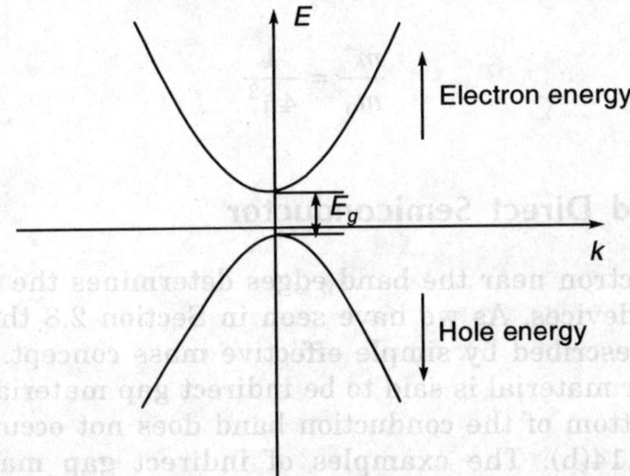

FIGURE 2.13 *E-k* curve for semiconductor material.

From Fig. 2.13, it is clear that the upper band (conduction band) is concave and so effective mass is a positive quantity whereas the lower band (valence band) is convex in nature and effective mass is negative. Thus, electrons near the top of the valence band have negative effective mass. The electron energy is measured upward and hole energy is measured downward as shown in Fig. 2.13. At the bottom of the conduction band, the total energy of electron is only potential energy and its total energy (kinetic energy plus potential energy) increases as one moves upwards from the conduction band edge. Similarly, at the valence band edge hole has only potential energy and its total energy (kinetic plus potential energy) increases as one moves downwards from the valence band edge.

EXAMPLE 2.3

For a crystal with $E = \dfrac{\hbar^2 k^2}{0.5\, m_0}$, find the effective mass.

Solution: Since $m^* = \dfrac{1}{(d^2 E/dk^2)}$

Differentiating $E = \dfrac{\hbar^2 k^2}{0.5\, m_0}$ w.r.t. k, we have

$$\frac{dE}{dk} = \frac{2\hbar^2 k}{0.5\, m_0}$$

Again, differentiating w.r.t. k, we get

$$\frac{d^2 E}{dk^2} = \frac{2\hbar^2}{0.5\, m_0} = \frac{4\hbar^2}{m_0}$$

Hence

$$\frac{m^*}{m_0} = \frac{1}{4\hbar^2}$$

2.8.1 Indirect and Direct Semiconductor

The behaviour of electron near the band edges determines the properties (electrical, optical) of the most devices. As we have seen in Section 2.8 that electrons near the band edges can be described by simple effective mass concept.

A semiconductor material is said to be indirect gap material when the top of the valence band and bottom of the conduction band does not occur at same value of k, as shown in Fig. 2.14(b). The examples of indirect gap material are elemental semiconductor Silicon and Germanium.

Semiconductor materials in which the band extrema (i.e. top of valence band and bottom of the conduction band) are aligned with k, called direct band gap material. The dispersion curve of direct band gap material is shown in Fig. 2.14(a). The example of direct band gap material is compound semiconductor.

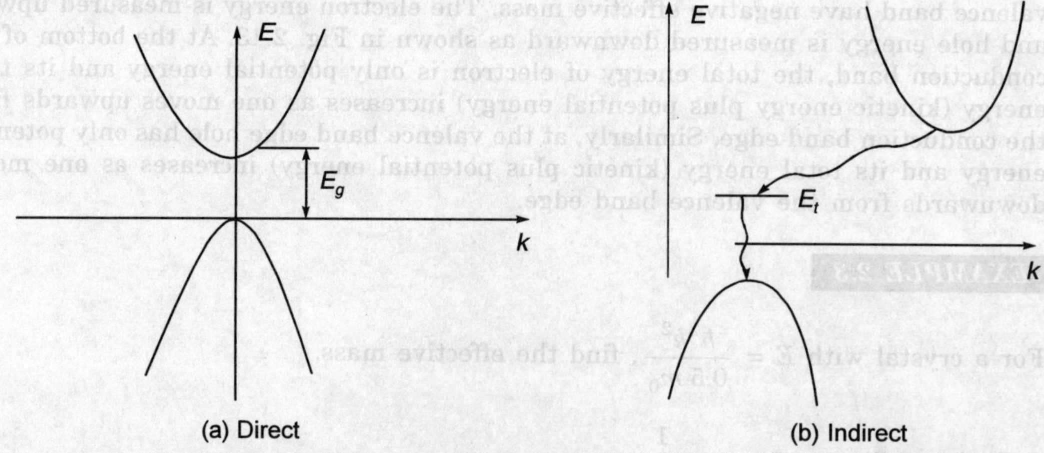

FIGURE 2.14 *E-k* curves for direct and indirect semiconductor materials.

Since, the electron in the indirect bandgap material cannot rejoin the valence band by radiative recombination due to violation of conservation of momentum, hence conduction electron typically last quite some time before recombining through any defect states or trap states in the material. Due to this reason silicon is generally not suitable material for light emitting diodes or laser diodes. The absorption of light in an indirect gap is much weaker than at a direct one. Since, in emission process, laws of conservation of momentum and energy must be satisfied, the only way to promote an electron from top of valence band to bottom of conduction band is to simultaneously emit or absorb phonon which should compensate the loss of momentum. The probability of such combined transition is very low. This shows silicon is not best choice for solar material compared to the direct bandgap material. The momentum-energy curve for direct and indirect band gap material is shown in Fig. 2.15. Direct bandgap materials are best choice for optoelectronic devices because they are much more efficient light emitters.

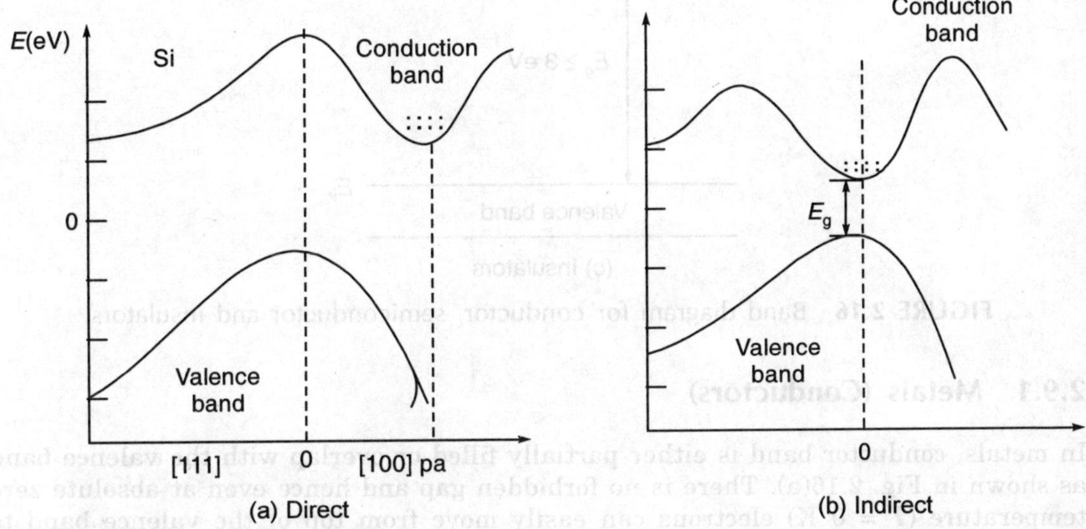

FIGURE 2.15 Momentum-energy curves for direct and indirect semiconductor materials.

In some indirect band gap materials, value of band gap is negative. This means top of the valence band is higher than the bottom of the conduction band in energy. Such materials are known as semimetals.

2.9 METALS SEMICONDUCTORS AND INSULATORS (on the basis of band diagram)

The different electrical behaviour of materials can be explained qualitatively in terms of their energy band diagrams because each solid has its own characteristics energy band structure as shown in Fig. 2.16.

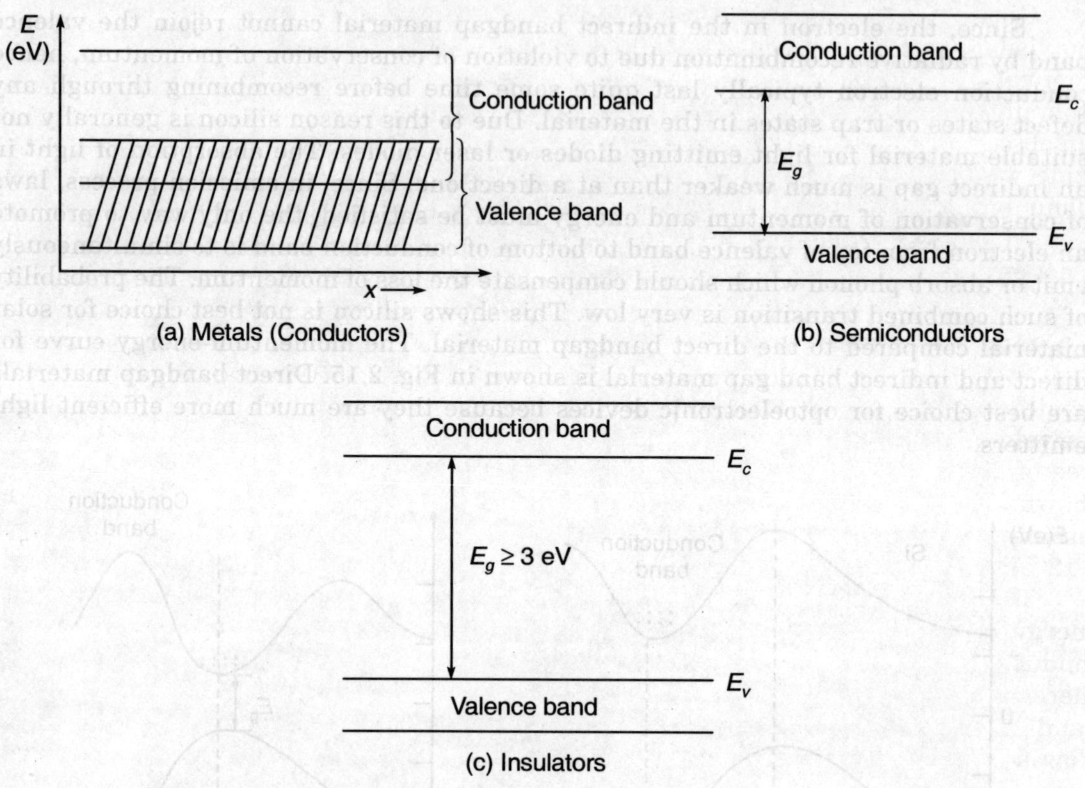

FIGURE 2.16   Band diagram for conductor, semiconductor and insulators.

2.9.1   Metals (Conductors)

In metals, conductor band is either partially filled or overlap with the valence band as shown in Fig. 2.16(a). There is no forbidden gap and hence even at absolute zero temperature ($T = 0$ K) electrons can easily move from top of the valence band to bottom of the partially filled conduction band. They can move freely in partially filled conduction band in the presence of the electric field. Therefore, current can easily flow in the metal at any temperature including zero absolute temperature.

2.9.2   Insulators

Insulators are non-conducting materials at any temperature. In terms of the band theory of solids, in insulator, there is a large forbidden gap ($E_g \geq 3$ eV) as shown in Fig. 2.16(c). There are higher energy levels available in an insulator but to move the electrons in these higher energy levels, one needs higher energy, i.e., ≥ 3 eV than the usual practical value. Thermal energy or energy acquired from applied electric field is not sufficient to move the electron from valence band edge to conduction band edge. Therefore, the conductivity of an insulator is very-very low or zero at any temperature. Silicon-dioxide (SiO_2) is an example of insulator.

Glass is an insulating material which may be transparent to visible light because the visible light photons do not have enough quantum energy to raise the electrons from valence band edge to conduction band edge. A very small percentage of impurity atom in the glass can give it colour by providing specific available energy levels which can absorb certain colours of visible light. For example, the ruby (corundum) is aluminium oxide with a small amount (about 0.05%) of chromium, and gives its pink or red colour by absorbing green and blue light.

While the doping of insulators can drastically change their optical properties, but it is not enough to overcome the large band gap to make them good conductor of electricity.

2.9.3 Semiconductors

Semiconductor behaves as an insulator at $T = 0$ K because all the energy levels in the valence band is completely occupied by electrons and conduction band is completely free of electrons. The band gap diagram of semiconductor is similar to insulator as shown in Fig. 2.16(b) except with the smaller energy band gap than insulator. The forbidden gap of three semiconductor materials at room temperature ($T = 300$ K) are 1.12 eV for silicon, 0.72 for pure germanium (Ge) and 1.43 eV for pure GaAs.

There is finite probability that some of valence electrons get sufficient thermal energy to break the covalent bond at temperature above $T = 0$ K and excited to conduction band. These excited electrons will create hole in the valence band. Since, there are many empty states in the conduction band, these free electrons in conduction band can easily move in the presence of small applied field and results in a current. Now semiconductor behaves as metal. As the temperature increases, probability of jumping electrons from valence band to conduction band increases. In semiconductors, both electrons as well as holes contribute in the conduction process. The band gap of semiconductor material decreases slighly as the temperature increases. The experimental variation of bandgap with temperature for two important semiconductor materials are given as:

$$E_g(T) = 1.170 - 4.73 \times 10^{-4}\ T^2\ (T + 636)\ \text{eV} \mid \text{for Si}$$

$$E_g(T) = 1.519 - 5.405 \times 10^{-4}\ T^2\ (T + 204)\ \text{eV} \mid \text{for GaAs}$$

The variation of energy gap for these two materials with temperature is shown in Fig. 2.17.

The conductivity of the semiconductor at a given temperature can also be enhanced by adding some impurities elements. The conductivity of a pure semiconductor can be increased by a factor greater than a billion by doping heavily. Hence, a semiconductor material can be classified as:

(a) Intrinsic semiconductor
(b) Extrinsic semicondutor

Intrinsic semiconductor

The word intrinsic which means *no foreign elements*.

A semiconductor in which no foreign elements are added, is called **intrinsic semiconductor**.

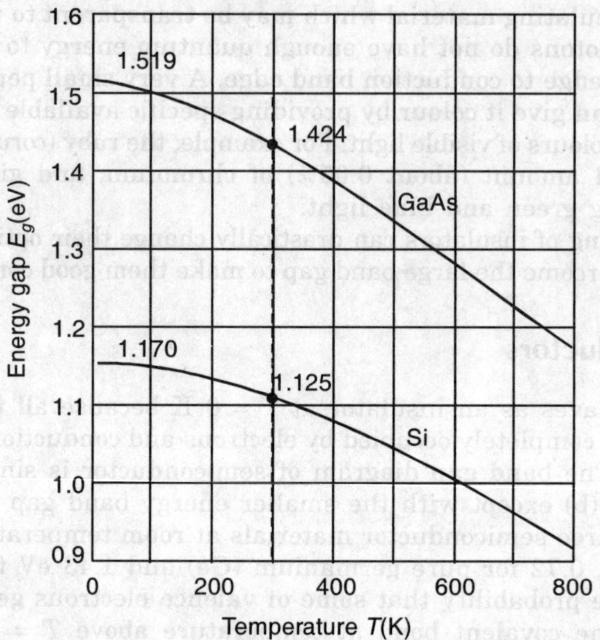

FIGURE 2.17 Energy gap variation with temperature for GaAs and Si.

In an intrinsic semiconductor, electrons are either thermally or optically excited from valence band to conduction band, and create same number of holes in the valence band. This process is called **electron-hole pair (EHP) generation process**. In other words,

$$n_0 = p_0 = n_i \qquad (2.11)$$

where

n_0 = Electron concentration in the conduction band at thermal equilibrium
p_0 = Hole concentration in the valence band at thermal equilibrium
n_i = Intrinsic carrier concentration

Therefore, at any temperature in intrinsic semiconductor material, number of electrons in conduction band and holes in valence band are equal. These carriers are called **intrinsic carrier concentration** and is represented by n_i. Intrinsic silicon has extremely low free carrier concentration at room temperature ($n_i \sim 1.5 \times 10^{10}$/cm^3). Therefore, the conductivity of a pure Si semiconductor material is very low at room temperature. In practice, it is very difficult to get a intrinsic semiconductor material at room temperature because it would require materials with an extremely high purity. The intrinsic carrier concentration strongly depends on temperature. The bond structure of pure semiconductor material is shown in Fig. 2.18.

At $T = 0$ K, intrinsic semiconductor behaves as an insulator due to absence of free electrons in the conduction band as explained earlier. Increasing temperature leads to an increase in number of carriers in either band because the thermal excitation will raise electrons from fully occupied valence band to empty conduction band, which

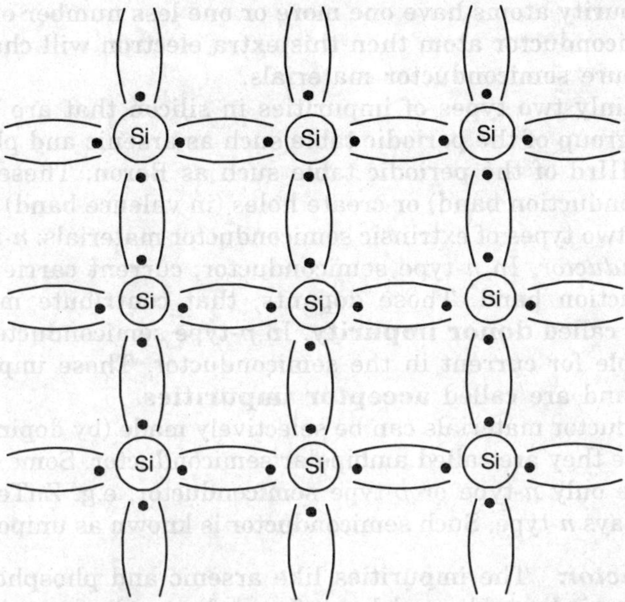

FIGURE 2.18 Two-dimensional band representation in intrinsic semiconductor.

leaves an equal number of holes in the valence band. Therefore, the conductivity of the intrinsic semiconductor material increases with temperature. This shows that the intrinsic carrier concentration as well as intrinsic conductivity are strongly dependent on the temperature. The probability of the excitation of carriers from bottom of valence band to top of conduction band is proportional to $e^{-E_g/kT}$, where, E_g is the Energy band gap, T is the Absolute temperature and k is Boltzmann's constant.

Extrinsic semiconductor

Semiconductors are useful in electronics because their electrical properties can be greatly altered in a controlled manner by adding small amount of foreign elements, called **impurities**. These impurities are also called **dopant**. The process of injection of impurities into pure semiconductor (intrinsic) is called **doping**. Doping is a process by which engineers enhance the conductivity of a pure semiconductor material at any temperature.

One way to alter the lattice structure of a pure semiconductor material is to add a small amount of impurities. These impurity atoms should have different atomic structure than Si and Ge. By adding these impurities, the basic electrical properties of a pure semiconductor material can be enhanced and controlled. A semiconductor material with added impurities are called **extrinsic semiconductor**.

When impurity atoms are added into the pure semiconductor material then impurity atoms take the place of semiconductor atoms in the lattice structure. If the impurity atoms have the same number of valence electrons as the atoms of pure semiconductor, they fit nearly into the lattice, forming the required number of electron pair bonds with semiconductor atoms, and thus the electrical properties of the semiconductor material are essentially unchanged.

When the impurity atoms have one more or one less number of valence electrons than the pure semiconductor atom then this extra electron will change the electrical behaviour of the pure semiconductor materials.

There are mainly two types of impurities in silicon that are electrically active. One from the Vth group of the periodic table such as arsenic and phosphorus and the other from group IIIrd of the periodic table such as Boron. These impurities either add electrons (in conduction band) or create holes (in valence band) in the pure silicon crystal and results two types of extrinsic semiconductor materials: *n-type semiconductor* and *p-type semiconductor*. In *n*-type semiconductor, current carriers are mainly free electrons in conduction band. Those dopants, that contribute mobile electrons in conduction band is called **donor impurity**. In *p*-type semiconductor holes in valence band are responsible for current in the semiconductor. Those impurities that cause holes in valence band are called **acceptor impurities**.

Most semiconductor materials can be selectively made (by doping) either *n*-type or *p*-type in which case they are called ambipolar semiconductor. Some semiconductor can be selectively made only *n*-type or *p*-type semiconductor, e.g. ZnTe is always *p*-type, whereas CdS is always *n*-type. Such semiconductor is known as unipolar semiconductor.

n-type semiconductor: The impurities like arsenic and phosphorus belong to the Vth group of the periodic table and have five valence electrons in their outermost orbit. When these impurities are added to the pure silicon semiconductor then four electrons of the impurity atoms will form covalent bond with their four neighbouring silicon atoms and fifth impurity electron is left unbonded as shown in Fig. 2.19. This fifth electron is considered to be similar to the hydrogen atom except that it is embedded in a dielectric material with a permittivity ε_{Si} and having effective mass.

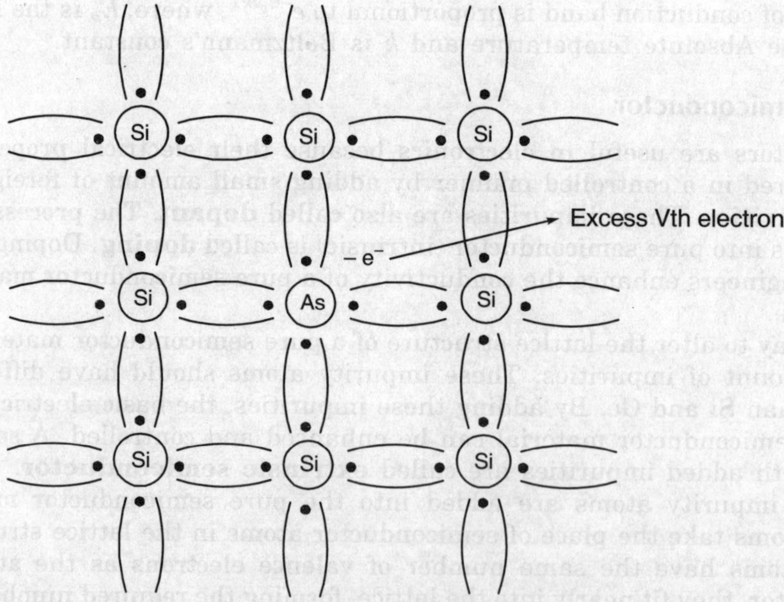

FIGURE 2.19 Two-dimensional representation of *n*-type semiconductor.

The fifth unbonded electron of the impurity atom is loosey bound to the nucleus and hence in most of the cases, the thermal energy at room temperature is sufficient to ionize the impurity atom and free one extra electron to the conduction band. In other words, we can say that the excess fifth impurity electron is donated to the conduction band. Such type of impurities are called **donor impurities** and atoms become positively charged after ionization. The required energy to free the fifth extra electron or ionize the impurity atom is given by

$$E_d = 13.6 \left(\frac{m_n^*/m_0}{(\varepsilon_s/\varepsilon_0)^2} \right) \text{eV} \tag{2.12}$$

where E_d is ionization energy. The ionization energy for donors in silicon is given as 0.032 eV and for GaAs, it is 0.005 eV.

When these donor impurities are ionized, they are represented by N_d^+ with the symbol ⊕.

Since, the excitation of the fifth electron into conduction band does not result hole in the valence band, therefore, the number of electrons in the conduction band exceeds the number of holes in the valence band of the semiconductor material. After doping of pure semiconductor material with Vth group dopant, the resulting semiconductor material has excess electron concentration in the conduction band than its thermal equilibrium concentration. Therefore, a pure silicon semiconductor material doped with Vth group elements or donor impurities is called **n-type semiconductor** and its electrical conductivity is dominated by majority carriers free electrons in conduction band. It is worth to note that the number of negative charge of the electrons is balanced by an equilvalent positive charge on the centre of impurity atom. Therefore, the net electrical charge of the semiconductor material is not disturbed.

The donor impurities introduce a new energy level just below the conduction band edge as shown in Fig. 2.20, designated as E_d, called **donor energy level**. At absolute zero temperature all the donor levels are occupied by donor impurities and these levels are neutral.

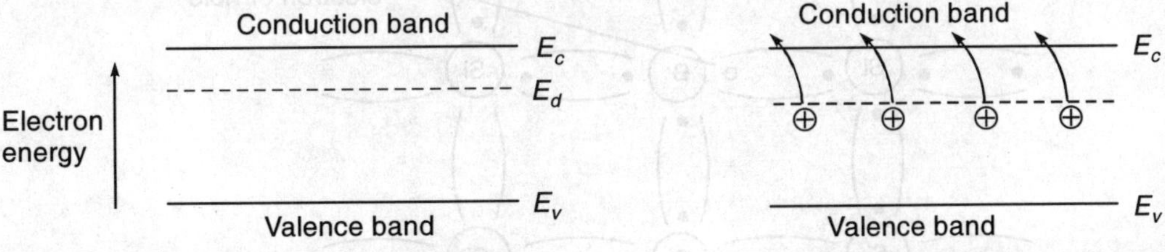

(a) Discrete donor energy states at T = 0 K

(b) The effect of some donor state being ionized creating free electrons in conduction band is at *T* > 0 K.

FIGURE 2.20   Energy band diagram for *n*-type semiconductor material at *T* = 0 K and *T* > 0 K.

As seen from Fig. 2.20(b), if a small amount of thermal energy (at $T > 0$ K) is added to fifth impurity electron (called donor electron) then it can be easily elevated into the conduction band, leaving behind positively charged immobile donor impurity ion. Therefore, the donor levels become positively charged. The electron in the conduction band is free to move through the crystal generating current, while positively charged ion is fixed in the crystal and does not take part in the conduction process. The ionization energy for donor impurities is denoted by $(E_c - E_d)$.

p-type semiconductor: When an impurity atom of IIIrd group of periodic table, (number of valence electrons are three), is substituted in the silicon crystal then all the valence electrons of the impurity atom form electron-pair bonds (covalent bond) with electrons of neighbouring silicon atom but one of the bonds in the lattice structure cannot be completed because of the deficiency of one electron in the impurity atom. Due to deficiency of one valence electron in the impurity atom, a hole exists in the lattice as shown in Fig. 2.21. An electron from the adjacent covalent bond may absorb enough energy to break its covalent bond and move through lattice to fill the hole. It is equivalent to saying that impurity atom is ionized by accepting one electron to complete its covalent bond. Such impurity is called **acceptor impurity**, and is represented by N_a^-. When a sufficiently large number of acceptor impurities are added into the pure silicon crystal then the number of holes greatly out number the thermally-excited electrons in the conduction band (n_0) because the acceptor atoms can generate holes in the valence band without generating electrons in the conduction band. Thus, holes become majority carriers, and electrons are minority carriers. A pure silicon materials doped with IIIrd group impurity atoms are called ***p-type semiconductor***.

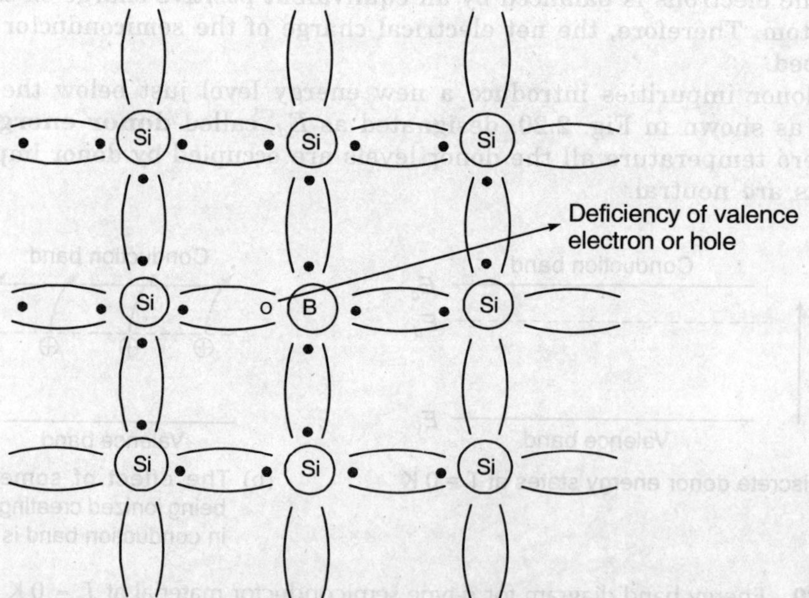

FIGURE 2.21 Two-dimensional representation of *p*-type semiconductor.

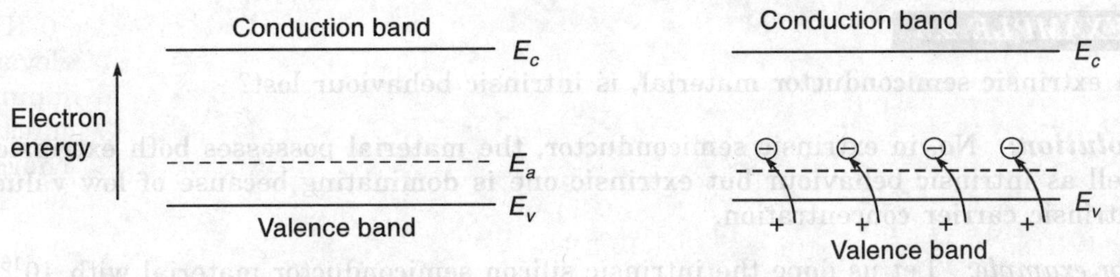

FIGURE 2.22   The energy band diagram of *p*-type semiconductor at $T = 0$ K and $T > 0$ K.

An added acceptor impurity in a pure silicon atom introduces a new energy level just above the valence band edge in forbidden gap. This new energy level is called **acceptor energy level** and is represented by E_a as shown in Fig. 2.22. The acceptor level is neutral when unoccupied at $T = 0$ K and becomes negative charge when occupied at $T > 0$ K. The ionization energy is measured as $(E_a - E_v)$. The ionization energy for acceptors on silicon is 0.059 eV and 0.037 eV for GaAs.

The ⊖ve sign in Fig. 2.22(b) represents the immobile acceptor ionized. Immobile acceptor impurities do not take part in conduction process and remain fixed in the lattice.

Note: Normally, the donor (E_d) and acceptor (E_a) energy levels are represented by broken lines in the energy band diagram to demonstrate that donor and acceptor states are localized in space within the semiconductor, since the concentration of donor and acceptor atoms is much smaller than density of semiconductor material atoms.

Arsenic and phosphorous are commonly used donors for *n*-type semiconductor whereas Boron is commonly used acceptor impurity for *p*-type semiconductor because of their highest solid solubility among all the impurities; which make them the most important dopant in VLSI technology.

EXAMPLE 2.4

In a pure semiconductor material at $T = 300$ K, there are 10^{10} electrons/cm^3 in conduction band and 10^{10} holes/cm^3 in valence band, from where these carriers are coming?

Solution: At $T = 300$ K, some of electrons from the valence band (because there are 4×10^{22} electrons/cm^3 in valence band of silicon crystal) acquire sufficient thermal energy and probability of crossing the forbidden gap after breaking covalent bond increases. At room temperature the average thermal energy possessed by electrons are 26 meV. The breaking of covalent bond creates one electron in conduction band and one hole in valence band. Out of 10^{22} covalent bonds in silicon atoms, only 10^{10} bonds are broken and the same number of free electrons are excited to conduction band thermally.

EXAMPLE 2.5

In extrinsic semiconductor material, is intrinsic behaviour lost?

Solution: No, in extrinsic semiconductor, the material possesses both extrinsic as well as intrinsic behaviour but extrinsic one is dominating because of low value of intrinsic carrier concentration.

For example: Let us dope the intrinsic silicon semiconductor material with 10^{16} As atoms/cm^3. In this case, out of 10^{22} silicon atom/cm^3, only 10^{16} silicon atom are going to be replaced by As atoms and hence result only 10^{16} free electrons in the conduction band whereas rest of silicon atoms still having covalent bond and exhibit the intrinsic behaviour.

2.10   FERMI-DIRAC STATISTICS

Electrons are indistinguishable and identical particles with half-integer spin and obey the Pauli's exclusion principle. The energy distribution of electrons in a solid is governed by the Fermi-Dirac statistics. The probability of occupying any electronic state E by an electron is given by Fermi-Dirac distribution function as:

$$f(E) = \frac{1}{1 + \exp\left(\dfrac{E - E_F}{kT}\right)} \tag{2.13}$$

where

   k = Boltzmann's constant
   T = Absolute temperature
   E_F = Reference energy, called Fermi level

   The Fermi-Dirac distribution function is generally called the **Fermi function**. Consider two cases at $T = 0$ K.

Case I: $E < E_F$, i.e., the desired energy level is below the Fermi level and hence Eq. (2.13) reduces to

$$f(E) = 1$$

This clearly shows that the probability of occupying any energy level below the Fermi level by carriers is 100% at $T = 0$ K, i.e., all the states below the Fermi level is completely occupied by electrons at absolute zero temperature.

Case II: $E > E_F$:   The Fermi function given by Eq. (2.13) reduces to

$$f(E > E_F, T = 0 \text{ K}) = 0$$

This suggests that all the energy levels above the Fermi level is completely unoccupied by the carrier at $T = 0$ K.

Therefore, Fermi level can be defined as the reference level above which all the levels are unoccupied by the carriers and below which all levels are completely occupied by carriers at $T = 0$ K. In other words, probability of occupying the states below Fermi level is very high and above Fermi level is very low at $T = 0$ K.

This shows that the distribution takes the simple rectangular form at $T = 0$ K as shown in Fig. 2.23.

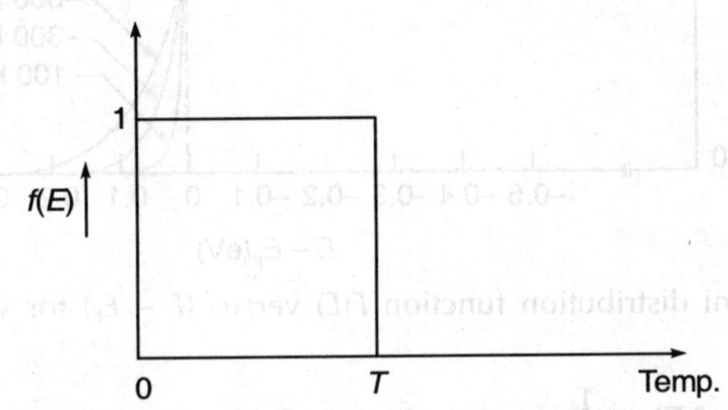

FIGURE 2.23 Fermi-Dirac distribution at $T = 0$ K.

Let us consider $E = E_F$ at any temperature then Fermi-Dirac distribution function f given by Eq. (2.13) becomes

$$f(E = E_F) = \frac{1}{2} = 50\%$$

Therefore, the Fermi energy is the energy at which the probability of occupation by an electron is exactly one-half at any temperature. In other words, the probability of occupying Fermi energy level at any temperature is always 50%. This mathematical expression $f(E = E_F) = 1/2$ defines the Fermi energy mathematically. According to this, Fermi energy is that energy at which Fermi-function is equal to 1/2. In other words Fermi energy is the highest energy that the electgron assumes (or occupies) at $T = 0$ K. The Fermi distribution at different temperatures is shown in Fig. 2.24.

Figure 2.24 also suggests that at finite temperature, some states above the Fermi level are filled and some states below it become empty. We can say that at any temperature above 0 K, there is finite probability to get carriers in the energy states above Fermi level. The Fermi function distribution is symmetrical about F_F for all temperatures.

The symmetry distribution of empty and filled states about E_F makes the Fermi level a natural choice of reference point to calculate the concentration of carriers in their respective bands.

In many cases, when the energy is at least several kT above or below the Fermi level, Eq. (2.13) can be approximated as:

$$f(E) \simeq e^{-\frac{(E - E_F)}{kT}} \qquad \text{for } (E - E_F) > 3kT \qquad (2.14a)$$

$$f(E) \simeq 1 - e^{-\frac{(E_F - E)}{kT}} \qquad \text{for } (E - E_F) < 3kT \qquad (2.14b)$$

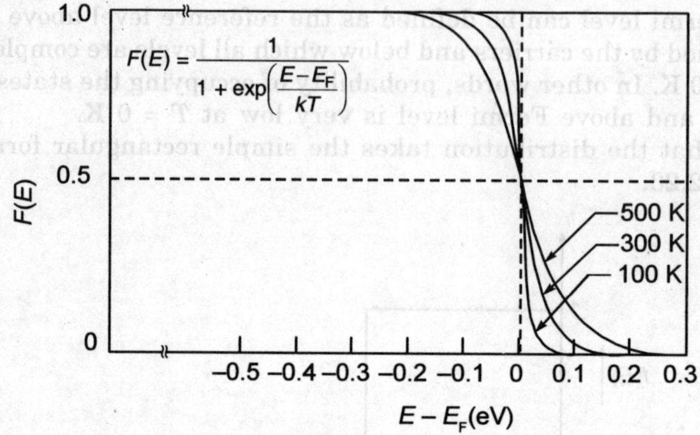

FIGURE 2.24 Fermi distribution function $F(E)$ versus $(E - E_F)$ for various temperatures.

$$f(E) = \frac{1}{2} \qquad \text{for } E = E_F \qquad (2.14c)$$

Equation (2.14b) can be interpreted as the probability that a hole occupies a state located at energy E.

Equation (2.14a) tells us that the population of electrons decreases exponentially at energies much larger than kT above the Fermi level, and similarly, population of holes decreases exponentially at energies more than kT below the Fermi level as seen from Eq. (2.14b).

2.10.1 Electron and Hole Concentration at Equilibrium

The Fermi-Dirac statistics can be used to calculate the concentration of electrons and holes in a semiconductor if the density of states in the conduction band and valence band are known. Therefore, the electron concentration in the conduction band may be written as:

$$n_0 = \int_{E_c}^{\infty} N(E) f(E) \, dE \qquad (2.15)$$

where E_c represents the conduction band edge.

$N(E)dE$ is the density of states (cm^{-3}) in the energy range dE, and is given as:

$$N(E) \, dE = \frac{1}{2\pi^2} \left(\frac{2 \, m_n^*}{\hbar^2} \right)^{3/2} (E - E_c)^{1/2} \qquad (2.16)$$

Hence, putting Eqs. (2.16) and (2.13) in Eq. (2.15), we have

$$n_0 = \frac{4\pi(2m_n^*)^{3/2}}{h^3} \int_{E_c}^{\infty} \sqrt{E - E_c} \; \frac{1}{1 + e^{(E - E_F)/kT}} \, dE \qquad (2.17)$$

To solve Eq. (2.17), let $\left(\dfrac{E - E_C}{kT}\right) = \xi$ and $\left(\dfrac{E_F - E_C}{kT}\right) = \eta$ and hence Eq. (2.17) can be rewritten as:

$$n_0 = \frac{4}{h^3}\,(2\,kT\,m_n^*)^{3/2}\int_0^\infty \frac{\sqrt{\xi}}{1 + \exp\,(\eta - \xi)}\,d\xi \qquad (2.18)$$

The integration can be easily evaluated by assuming

$$\frac{E - E_F}{kT} \gg 1 \quad \text{or} \quad (\xi - \eta) \gg 1$$

so that Fermi function $\dfrac{1}{1 + \exp\left[\dfrac{E - E_F}{kT}\right]}$ i.e., Fermi-Dirac statistics reduces to Boltzmann approximation and Eq. (2.18) becomes

$$n_0 = \frac{4\pi}{h^3}\,(2\,kT\,m_n^*)^{3/2}\,\exp\,(\eta)\int_0^\infty \sqrt{\xi}\,\exp\,(-\xi)\,d\xi \qquad (2.19)$$

Using the standard integral equation, called **Gamma function**,

$$\int_0^\infty \sqrt{\xi}\;e^{-\xi}\,d\xi = \frac{\sqrt{\pi}}{2}$$

We have

$$n_0 = 2\left(\frac{2\pi m_n^* kT}{h^2}\right)^{3/2} e^{-(E_c - E_F)/kT}$$

or $$n_0 = N_c\,e^{-(E_C - E_F)/kT} \qquad (2.20)$$

where $$N_c = 2\left(\frac{2\pi m_n^* kT}{h^2}\right)^{3/2} \qquad (2.21)$$

called **effective density** of states in conduction band. Equation (2.21) shows that density of states depends on temperature and proportional to $T^{3/2}$. N_c values for two semiconductor materials are:

$$N_c = 2.86 \times 10^{19}/\text{cm}^3 \quad \text{for Si at } T = 300 \text{ K}$$

$$N_c = 4.7 \times 10^{17}/\text{cm}^3 \quad \text{for GaAs at } T = 300 \text{ K}$$

Since, $f(E)$ gives the probability of occupying the states by an electron in conduction band, hence, the probability that state is not occupied by an electron, i.e., occupied by hole is given by

$$[1 - f(E)]$$

Therefore, the concentration of holes in the valence band is:

$$p_0 = \int_{-\infty}^{E_v} [1 - f(E)]\, N(E)\, dE \tag{2.22}$$

where E_v represents the valence band edge and

$$N(E)\, dE = \left(\frac{1}{2\pi^2}\right)\left(\frac{2m_p^*}{h^2}\right)^{3/2} (E_v - E)^{1/2} \tag{2.23a}$$

gives the density of states for holes in the valence band. Assuming $(E_F - E_v) >> kT$, Fermi-Dirac statistics reduces to Maxwell Boltzmann distribution, means

$$[1 - f(E)] \simeq e^{-(E_F - E_v)/kT}$$

and hence Eq. (2.22) can be written as:

$$p_0 = \int_{-\infty}^{E_v} e^{-(E_F - E)/kT} \sqrt{E_v - E}\, N(E)\, dE \tag{2.23b}$$

Again solving in the same way as done for n_o, we get

$$p_0 = N_v\, e^{-(E_F - E_v)/kT} \tag{2.24}$$

where

$$N_v = 2\left(\frac{2\pi\, m^*pkT}{h^2}\right)^{3/2} \tag{2.25}$$

is called **effective density** of states in the valence bands and proportional to $T^{3/2}$.

$$N_v = \begin{cases} 2.66 \times 10^{19}/\text{cm}^3 & \text{for Si} \\ 7 \times 10^{18}/\text{cm}^3 & \text{for GaAs} \end{cases} \quad \text{at } T = 300\text{ K}$$

EXAMPLE 2.6

(a) Calculate the probability that an energy state in the conduction band at $E = E_c + kT$ is occupied by an electron. (b) Also, calculate the thermal equilibrium electron concentration in silicon at $T = 300$ K. Use, $E_c - E_F = 0.20$ eV, $N_c = 2.8 \times 10^{19}/\text{cm}^3$ for Si at $T = 300$ K.

Solution: (a) $f(E_E + kT)$ = Probability of occupying energy state = $E_c + kT$ by an electron

$$= \frac{1}{1 + \exp\left(\dfrac{E_c + kT - E_F}{kT}\right)} \simeq \exp\left[-\frac{E_c + kT - E_F}{kT}\right]$$

$$\simeq \exp\left[-\left(\frac{0.2 \text{ eV} + 26 \text{ meV}}{26 \text{ meV}}\right)\right] \simeq 1.63 \times 10^{-4}$$

$$n_o = 2.8 \times 10^{19} \ \exp\left[-\left(\frac{0.2 \text{ eV}}{26 \text{ meV}}\right)\right] \simeq 1.24 \times 10^{16}/\text{cm}^3$$

This exercise only tells that the probability of occupying any energy level above the conduction band edge by an electron is very very small.

(b) Calculate the probability that an energy state in the valence band at $E = E_v - kT$ is empty of an electron and calculate the thermal equilibrium hole concentration in Si semiconductor at $T = 350$ K. Given: $E_F - E_v = .25$ eV, $N_v = 1.04 \times 1^{19}/\text{cm}^3$ for Si at $T = 300$ K.

Solution: $N_v(T = 350 \ K) = 1.04 \times 10^{19} \left(\dfrac{350}{300}\right)^{3/2} = 1.31 \times 10^{19}/\text{cm}^3$

and

$$kT = (0.0259) \left(\frac{350}{300}\right) = 0.0302 \text{ eV}$$

The probability that an energy state in the valence band at $E = E_v - kT$ is occupied by hole.

$$\Rightarrow 1 - f(E_v - kT) \simeq \exp\left[-\frac{(E_F - E_v + kT)}{kT}\right] \simeq 9.34 \times 10^{-5}$$

The thermal equilibrium hold conc. is

$$p_0 = N_v \ \exp\left[-\frac{(E_F - E_v)}{kT}\right] = 1.31 \times 10^{19} \ \exp\left(-\frac{0.25 \text{ eV}}{26 \text{ meV}}\right)$$

$$= 3.33 \times 10^{15}/\text{cm}^3$$

Probability of occupying the energy state below the valence band edge by hole is very very low.

2.10.2 Intrinsic Carrier Concentration at Equilibrium

In intrinsic semiconductor material,

$$n_0 = p_0 = n_i \tag{2.26a}$$

Equating Eqs. (2.20) and (2.24), one has

$$N_c\, e^{-(E_c - E_F)/kT} = N_v\, e^{-(E_F - E_v)/kT}$$

or

$$\frac{N_c}{N_v} = e^{-(2E_F - E_c - E_v)/kT}$$

or

$$\frac{N_v}{N_c} = e^{(2E_F - E_c - E_v)/kT}$$

or

$$\frac{2E_F - E_c - E_v}{kT} = \ln\left(\frac{N_v}{N_c}\right)$$

or

$$E_F = \frac{kT}{2}\,\ln\left(\frac{N_v}{N_c}\right) + \frac{1}{2}\,(E_c + E_v) \tag{2.26b}$$

This is the Fermi level position of intrinsic semiconductor material, and is represented by E_i.

$$E_i = \frac{kT}{2}\,\ln\left(\frac{N_v}{N_c}\right) + \frac{1}{2}\,(E_c + E_v)$$

Substituting values of N_v and N_c, we have

$$E_i = \frac{3kT}{4}\,\ln\left(\frac{m^*p}{m^*n}\right) + \frac{1}{2}\,(E_c - E_v + 2E_v)$$

$$= \frac{3kT}{4}\,\ln\left(\frac{m^*p}{m^*n}\right) + \frac{1}{2}\,(E_g) + E_v$$

or

$$E_i - E_v = \frac{3kT}{4}\,\ln\left(\frac{m^*p}{m^*n}\right) + \frac{E_g}{2} \tag{2.27a}$$

Similarly

$$E_c - E_i = (E_g/2) + \frac{3kT}{4}\,\ln\left(\frac{m^*p}{m^*n}\right) \tag{2.27b}$$

In general,

$$E_i = (E_g/2) + \frac{3}{4}\,kT\,\ln\left(\frac{m^*p}{m^*n}\right) \tag{2.28}$$

Since $m_n^* \neq m_p^*$, hence, the intrinsic Fermi level cannot lie at the middle of the forbidden gap of the material. Consider two cases:

Case I: $T = 0$ K, $m^*n \neq m^*p$, $E_i = (E_g/2)$ (2.29a)

Case II: $m^*n = m^*p$, $T \neq 0$ K, $E_i = (E_g/2)$ (2.29b)

Therefore, it is concluded that Fermi level in the intrinsic semiconductor material lies exactly at the middle of the forbidden gap either at $T = 0$ K or when both carriers (electrons/holes) have equal effective masses.

The band diagram for intrinsic semiconductor material at $T = 0$ K is shown in Fig. 2.25.

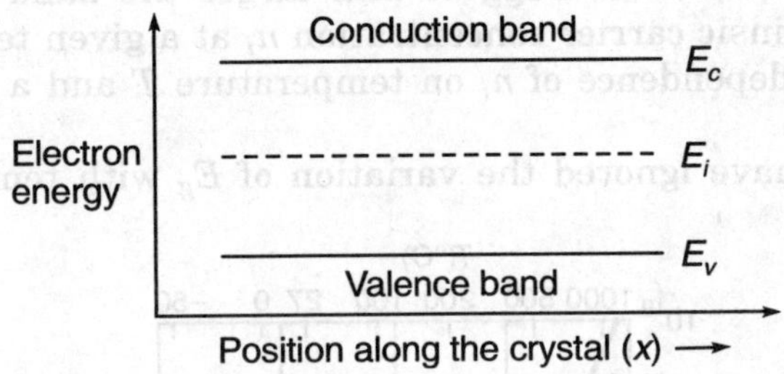

FIGURE 2.25 Band diagram for intrinsic semiconductor.

From Eq. [2.26a].

$$n_0 = n_i$$

$$p_0 = n_i$$

Hence,

$$n_0 p_0 = n_i^2 \tag{2.30}$$

Equation (2.30) is called **mass action law**, and is valid for both intrinsic semiconductor as well as extrinsic semiconductor at thermal equilibrium. If the concentration of one carrier is known then concentration of second carrier can be easily obtained by using Eq. (2.30) because at any particular temperature and for a given semiconductor material, the intrinsic carrier concentration (n_i) is constant.

Using Eqs. (2.20), (2.24) and (2.30), we have

$$n_i^2 = N_c N_v \exp\left[-\left(\frac{E_c - E_v}{kT}\right)\right]$$

or

$$n_i = \sqrt{N_c N_v} \exp\left(-\frac{E_g}{2kT}\right) \tag{2.31}$$

where $E_g = (E_c - E_v)$ is called **forbidden gap** or **energy gap** of the semiconductor material. The energy gap represents the minimum energy required to generate an electron-hole pair. The detail of energy band and band gap depends on the detailed quantum mechanical solutions for the semiconductor crystal structure.

Equation (2.31) is independent of Fermi energy and hence, is independent of the changes in carier concentration n_i is concentration of electrons at the conduction band edge in a pure semiconductor. Likewise, n_i is also the concentration of holes at the valence band edge in a pure semiconductor.

Equation (2.31) can be further written as:

$$n_i = 2\left(\frac{2\pi kT}{\hbar^2}\right)^{3/2} (m_n^* \, m_p^*)^{3/4} \, e^{-E_g/2kT} \tag{2.32}$$

Equations (2.31) and (2.32) suggest that larger the band gap of the material, lower the value of intrinsic carrier concentration n_i at a given temperature. Equation (2.32) shows a strong dependence of n_i on temperature T and a plot of $\ln n_i$ vs $10^3/T$ is shown in Fig. 2.26.

In this plot, we have ignored the variation of E_g with temperature.

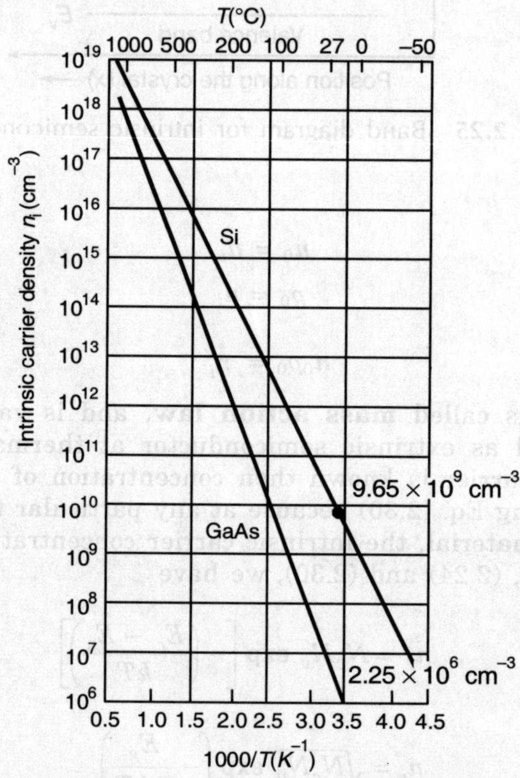

FIGURE 2.26 Plot of ln n_i with temperature.

EXAMPLE 2.7

(a) Find the intrinsic carrier concentration for silicon semiconductor material at room temperature (T = 300 K).

Given: Effective mass of electron = m^*n = $1.09m_0$
Effective mass of hole = m^*p = $1.15m_0$
N_c = 2.86 × 10^{19}/cm^3
N_c = 3.10 × 10^{19}/cm^3 at T = 300 K
Mass of free electron m_0 = 9.1 × 10^{-31} kg

Solution: Using Eq. (2.32),

$$n_i = 2\left(\frac{2\pi \times .026}{h^2}\right)^{3/2} (1.09 \times 1.15)^{3/4} \, (m_0^2)^{3/4} \, e^{-\frac{1.12}{2 \times .026}}$$

and on solving,

$$n_i = 1.08 \times 10^{10}/\text{cm}^3$$

(b) Also find electron and hole concentration at thermal equilibrium?

Solution: Since $n_i = 1.08 \times 10^{10}/\text{cm}^3$

Therefore,

$$n_0 = n_i = 1.08 \times 10^{10} \text{ electrons/cm}^3 \text{ in conduction band}$$
$$p_0 = n_i = 1.08 \times 10^{10} \text{ holes/cm}^3 \text{ in valence band}$$

2.10.3 Fermi Level Position in Extrinsic Semicondutor Material

Using Eqs. (2.20), (2.24) and (2.26a), the intrinsic carrier concentration for electrons and holes can be expressed as:

$$n_i = N_c \, e^{-(E_c - E_i)/kT} \tag{2.33a}$$

$$p_i = N_v \, e^{-(E_i - E_v)/kT} \tag{2.33b}$$

where i stands for intrinsic semiconductor material. Since, electrons and holes concentration at equilibrium are given by Eqs. (2.20) and (2.24). Therefore, using Eqs. (2.33a) and (2.20), electron concentration at equilibrium can be expressed in terms of intrinsic carrier concentration and E_i as:

$$n_0 = n_i \, e^{(E_F - E_i)/kT} \tag{2.34a}$$

Similarly, using Eq. (2.24) and (2.33b), hole concentration at equilibrium in terms of intrinsic concentration n_i and intrinsic Fermi level E_i can be expressed as:

$$p_0 = p_i \, e^{(E_i - E_F)/kT} \tag{2.34b}$$

Since, for intrinsic material, $n_i = p_i$ and, therefore, Eq. (2.34b) can be further rewritten as:

$$p_0 = n_i \, e^{(E_i - E_F)/kT} \tag{2.34c}$$

Equation (2.34a) clearly suggests that the electron concentration increases exponentially from the value of intrinsic carrier concentration n_i if Fermi level E_F moves away from the intrinsic Fermi level E_i. In other words, as Fermi level E_F moves closer to conduction band edge, the electron carrier concentration in conduction band increases.

Similarly, from Eq. (2.34c) it is clear that the hole concentration increases exponentially from its intrinsic value as the Fermi level E_F moves below the intrinsic Fermi level E_i. In other words, the hole concentration increases exponentially as the Fermi level E_F moves closer to valence band edge E_v.

In brief: If electron concentration is increasing from its intrinsic concentration then Fermi level moves closer to conduction band edge to accommodate this increase. Similarly, Fermi level E_F moves closer to valence band edge to accommodate any increase in hole concentration from its intrinsic value.

In contrast to intrinsic semiconductor material, Fermi level E_F in the extrinsic semiconductor material does not lie at the middle of the forbidden gap. The Fermi level E_F is close to conduction band edge for n-type semiconductor, and close to valence band edge for p-type semiconductor as shown in Fig. 2.27.

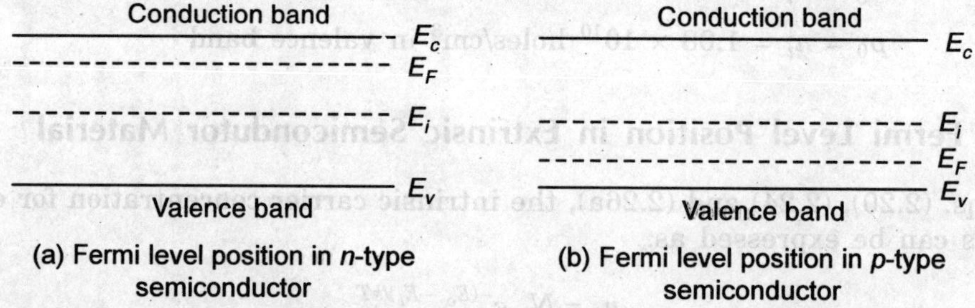

FIGURE 2.27 Fermi level position in extrinsic semiconductor material.

The exact position of Fermi level in extrinsic semiconductor depends on ionization energy as well as concentration of dopants. Let us consider a n-type semiconductor material with a donor impurity concentration of N_d then the charge neutrality condition in a semiconductor requires

$$n_0 - p_0 = N_d^+ \tag{2.35a}$$

where N_d^+ is called **ionized donors density**, and is given by

$$N_d^+ = N_d[1 - f(E_d)] = N_d \left[1 - \frac{1}{1 + 1/2\ e^{(E_d - E_F)/kT}} \right] \tag{2.35b}$$

The factor $(1/2)$ is introduced to take the account of spin degeneracy (up or down) of the available states associated with the ionized donor levels.

Substituting values of n_0, p_0 from Eqs. (2.20) and (2.24) in Eq. (2.35a), we have

$$N_c\ e^{-(E_c - E_F)/kT} - Nv\ e^{-(E_F - E_v)/kT} = N_d \left[1 - \frac{1}{1 + 1/2\ e^{(E_d - E_F)/kT}} \right] \tag{2.35c}$$

For shallow donor impurities with low to moderate doping at room temperature, thermal energy is sufficient enough to ionize all the donor impurities.

i.e.,
$$n_0 = N_d^+ = N_d \tag{2.36}$$

This condition is called **complete ionization**. Figure 2.28 shows the complete ionization where donor level E_d is measured w.r.t. conduction band edge. It is also seen from Figure that there are equal number of mobile electrons in conduction band and positive (+) immobile donor ions in the donor energy level.

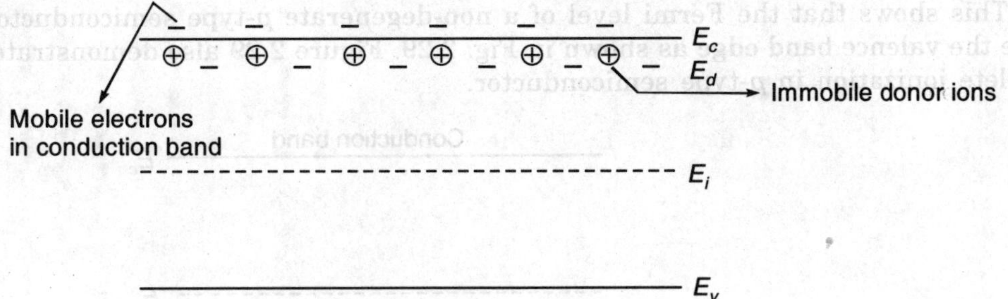

FIGURE 2.28 Representation of complete ionization.

Since in n-type semiconductor holes are minority carrier, i.e., $n_0 \gg p_0$ and hence neglecting IInd term of the L.H.S. of Eq. (2.35c), and assuming that

$$\left(\frac{N_d}{N_c}\right) \exp\left[(E_c - E_d)/kT\right] \ll 1$$

Eq. (2.35c) reduces to

$$N_C\, e^{-(E_c - E_F)/kT} = N_d \tag{2.37}$$

Equation (2.37) is true for shallow donor impurities with low to moderate doping density at $T = 300$ K. If the difference between the ionization level and the corresponding valence/conduction band is less than about $3\,kT$ ($\simeq 0.075$ eV at $T = 300$ K), the impurities are called shallow energy level dopant. These impurities are essentially completely ionized at $T = 300$ K.

This expression can be directly written by using Eq. (2.36) from Eq. (2.29),

$$E_C - E_F = kT \ln\left(\frac{N_C}{N_d}\right)$$

or

$$E_F = E_C - kT \ln\left(\frac{N_C}{N_d}\right) \tag{2.38}$$

Equation (2.38) confirms that in the case of n-type non-degenerate semiconductor material, Fermi level E_F should be below the conduction band edge.

Similar argument is true for shallow acceptor impurities. For complete ionization in case of acceptor impurities doping

$$p_0 = N_a = N_a^- \tag{2.39}$$

And therefore, Fermi level position for p-type semiconductor,

$$E_F = E_v + kT \ln \left(\frac{N_v}{N_a} \right) \tag{2.40}$$

This shows that the Fermi level of a non-degenerate p-type semiconductor lies above the valence band edge as shown in Fig. 2.29. Figure 2.29 also demonstrates the complete ionization in p-type semiconductor.

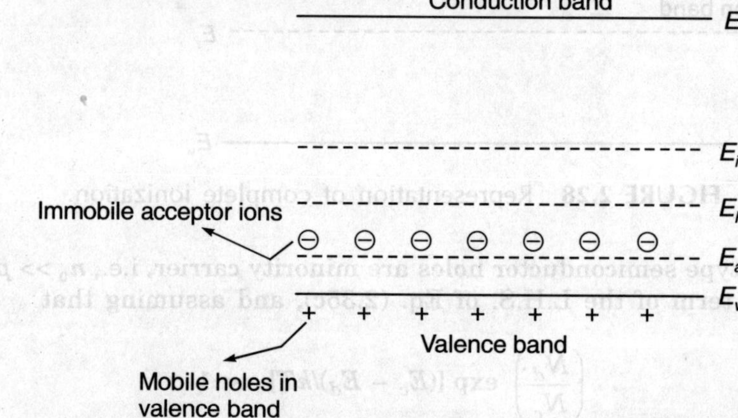

FIGURE 2.29   Band diagram of a non-degenerate p-type semiconductor materials.

In both cases, Fermi level E_F should be at least a few kT below the conduction band edge in n-type semiconductor, and a few kT above the valence band edge in p-type semiconductor.

Note: A semiconductor material is said to be non degenerate semiconductor if electron and hole concentration is assumed to be much lower than the effective density (i.e., $n_0 \ll N_v$).

Consider a n-type semiconductor material ($N_d = n_0 \gg n_i$), in which electrons are majority carriers and holes are minority carriers at thermal equilibrium means $n_0 \gg p_0$. Similar argument is true for p-type semiconductor ($N_a = p_0 \gg n_i$) where holes are majority carriers and electrons are minority carriers ($p_0 \gg n_0$). The product of $n_0 p_0$ at equilibrium is constant, independent of the type of dopant and position of Fermi level.

According to mass-action law, for n-type of semiconductor,

$$\text{Minority carrier hole concentration } p_0 = \frac{n_i^2}{n_0} \tag{2.41a}$$

for p-type of semiconductor

$$\text{Minority carrier electron concentration } n_0 = \frac{n_i^2}{p_0} \tag{2.41b}$$

The schematic band diagrams for density of states, Fermi-Dirac distribution and carrier concentration in case of intrinsic semiconductor as well as extrinsic semiconductor are shown in Fig. 2.30 at thermal equilibrium.

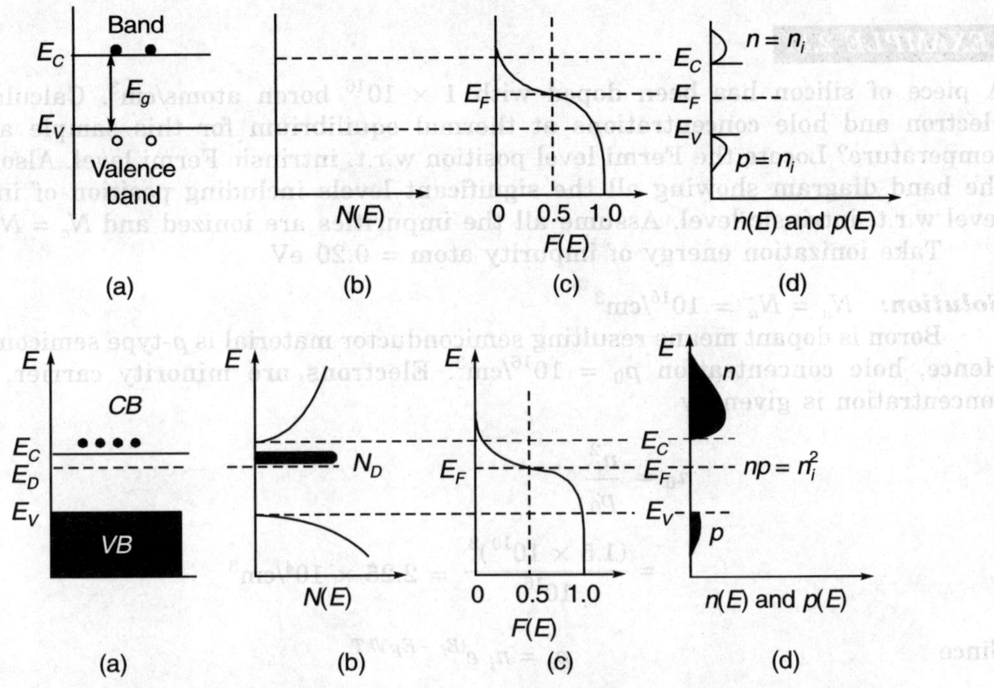

FIGURE 2.30 Schematic band diagram for density of states, Fermi-Dirac distribution and carrier concentration in intrinsic and extrinsic semiconductor materials.

From Eqs. (2.34a) and (2.34c), we have a more useful relations of E_F in terms of E_i and n_i as:

For n-type semiconductor,

$$E_F - E_i = kT \ln\left(\frac{N_d}{n_i}\right) \tag{2.42a}$$

and for p-type semiconductor,

$$E_i - E_F = kT \ln\left(\frac{N_a}{n_i}\right) \tag{2.42b}$$

Equations (2.42a) and (2.42b) predict that the distance between the Fermi level E_F and intrinsic Fermi level is a logarithmic function of doping concentration.

Note: Thermal equilibrium in semiconductor materials means that there is no transfer of carriers from one point to other point or the probability of finding the carriers at any point is same throughout the state.

Mathematically,
$$\frac{dE_F(x)}{dx} = 0$$

or
$$E_F(x) = \text{constant}$$

EXAMPLE 2.8

A piece of silicon has been doped with 1×10^{16} boron atoms/cm^3. Calculate the electron and hole concentrations at thermal equilibrium for this sample at room temperature? Locate the Fermi level position w.r.t. intrinsic Fermi level. Also sketch the band diagram showing all the significant levels including position of impurity level w.r.t. intrinsic level. Assume all the impurities are ionized and $N_c = N_v$.

Take ionization energy of impurity atom = 0.20 eV.

Solution: $N_a = N_a^- = 10^{16}$/cm^3

Boron is dopant means resulting semiconductor material is p-type semiconductor. Hence, hole concentration $p_0 = 10^{16}$/cm^3. Electrons are minority carrier, whose concentration is given by

$$n_0 = \frac{n_i^2}{p_0}$$

$$= \frac{(1.5 \times 10^{10})^2}{10^{16}} = 2.25 \times 10^4 \text{/cm}^3$$

Since
$$p_0 = n_i \, e^{(E_i - E_F)/kT}$$

On solving,
$$10^{16} = 1.5 \times 10^{10} \, e^{(E_i - E_F)/26 \text{ meV}}$$

$$E_i - E_F = 0.344 \text{ eV}$$

$$E_F - E_v = (E_F - E_i) + (E_i - E_v) = (E_g/2) + (E_F - E_i)$$

$$= \left(\frac{1.12}{2}\right) \text{eV} + (-\,0.344 \text{ eV})$$

$$= 0.56 \text{ eV} - 0.344 \text{ eV} = 0.2113 \text{ eV}$$

Since $N_c = N_v$, therefore, intrinsic Fermi level lies at the middle of band gap.

$$E_i - E_a = (E_i - E_v) + (E_v - E_a)$$

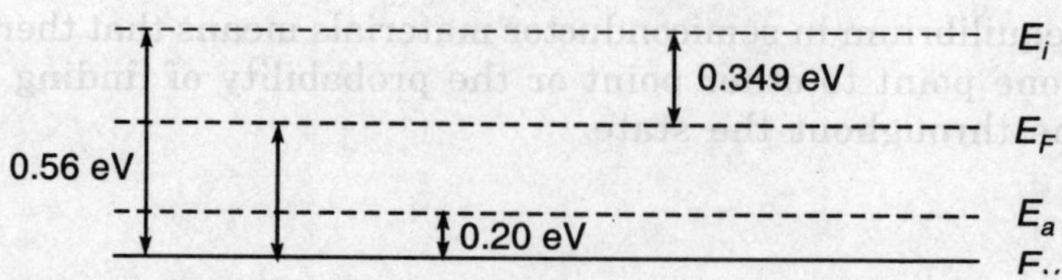

Since	$E_i - E_v = 0.56$ eV
and	$E_v - E_a = -0.20$ eV (Given)
Therefore,	$E_i - E_a = (0.56$ eV$) - (0.20$ eV$) = 0.36$ eV

EXAMPLE 2.9

Find the thermal equilibrium electron and hole concentration at $T = 300$ K and position of Fermi level w.r.t. E_c in silicon semiconductor for 1×10^{15} electrons/cm^3.

Solution: At $T = 300$ K,

$$E_g = 1.12 \text{ eV for silicon}$$
$$n_i = 1.5 \times 10^{10}/\text{cm}^3$$
$$n_0 = 10^{15}/\text{cm}^3, \quad kT = 26 \text{ meV}$$

Therefore,

$$p_0 = \frac{n_i^2}{n_0}$$

$$= \frac{2.25 \times 10^{20}}{10^{15}} = 2.25 \times 10^5/\text{cm}^3$$

This shows that $n_0 \gg p_0$. Using,

$$n_0 = n_i \, e^{(E_F - E_i)/kT}$$
$$10^{15} = 1.5 \times 10^{10} \, e^{(E_F - E_i)/kT}$$

or

$$(E_F - E_i) = 11.107 \, kT$$
$$= 0.29 \text{ eV}$$

Since $E_c - E_i = (E_g/2) = 0.56$ eV
Therefore,

$$E_c - E_F = (E_c - E_i) - (E_F - E_i)$$
$$= 0.56 \text{ eV} - 0.29 \text{ eV}$$
$$= 0.27 \text{ eV}$$

EXAMPLE 2.10

Find the electron and hole concentrations in GaAs doped with $N_a = 10^{16}/\text{cm}^3$, and locate the position of Fermi level.

Given: $E_g = 1.43$ eV at $T = 300$ K
$$N_c = 4.4 \times 10^{17}/\text{cm}^3$$
$$N_v = 8.3 \times 10^{18}/\text{cm}^3$$

Solution: Using

$$n_i = \sqrt{N_c N_v}\; e^{-E_g/2kT}$$

$$= \sqrt{(4.4 \times 10^{17})\,(8.3 \times 10^{19})}\; e^{-\frac{1.43}{2 \times .026}}$$

$$= 2.2 \times 10^6/\text{cm}^3$$

Since, $N_a \gg n_i$, therefore

$$p_0 = n_a = 10^{16}/\text{cm}^3$$

$$E_F - E_v = kT\, \ln\!\left(\frac{N_v}{p_0}\right), \qquad \text{(using Eq. 2.40)}$$

$$= 0.026 \text{ eV}\, \ln\!\left[\frac{8.3 \times 10^{18}}{10^{16}}\right] = 0.17 \text{ eV}$$

Electron concentration $n_0 = \dfrac{n_i^2}{p_0} = \dfrac{(2.2 \times 10^6)^2}{10^{16}}$

$$= 4.8 \times 10^{-4}/\text{cm}^3$$

EXAMPLE 2.11

Consider a silicon semiconductor material at $T = 300$ K, in which Fermi level is 0.25 eV above the valence-band edge. Find the thermal equilibrium concentration of electrons and holes.

   Given: E_g for silicon (at $T = 300$ K) = 1.12 eV

   Take Fermi energy 0.87 eV below the conduction band edge, $N_v = 1.04 \times 10^{19}/\text{cm}^3$, $N_c = 2.8 \times 10^{19}/\text{cm}^3$ at $T = 300$ K.

Solution: Since Fermi level is closer to valence band edge, and hence, material is p-type semiconductor. Concentration of majority carriers are:

$$p_0 = N_v\, e^{-(E_F - E_v)/kT}$$

Using Eq. (2.24),

$$p_0 = (1.04 \times 10^{19}) \exp\!\left[-\frac{0.25}{0.026}\right] = 6.68 \times 10^{14}/\text{cm}^3$$

Using Eq. (2.20),

$$n_0 = (2.8 \times 10^{19}) \exp\!\left[-\frac{0.87}{0.0260}\right] \simeq 7.23 \times 10^{14}/\text{cm}^3$$

EXAMPLE 2.12

Find the energy by which intrinsic Fermi level E_i is offset from midgap for Si at room temperature.

Solution: From Eq. (2.28),

$$E_i = (E_{\text{midgap}}) + \frac{3}{4}\, kT \ln\left(\frac{m_p^*}{m_n^*}\right)$$

$$E_g - E_{\text{midgap}} = \frac{3}{4}\, kT \ln\left(\frac{1.15}{1.09}\right) = 1.05 \text{ meV}$$

which is small compared to $E_g = 1.12$ eV.

2.10.4 Degenerate Semiconductor

A semiconductor material is said to be degenerate semiconductor if the impurity concentration N_d or N_a exceed the effective density of states N_c or N_v; so that $E_F > E_C$ for n-type and $E_F < E_v$ for p-type according to Eqs. (2.36) and (2.40). In other words, Fermi level moves onto the conduction band for heavily doped n-type semiconductor. A heavily doped n-type semiconductor ($n_0 \geq 10^{19}/\text{cm}^3$) is represented by n^+.

Similarly, the Fermi level moves into valence band for heavily doped p-type semiconductor ($p_0 > 10^{19}/\text{cm}^3$) and represented by p^+. When impurity concentration is higher than $10^{19}/\text{cm}^3$, the donor or acceptor levels broaden into respective bands. This results a decrease in the ionization energy. When the impurity band merges with the conduction band edge or valence band edge, the ionization energy becomes the zero and semiconductor behaves as a conductor (metal).

At high doping concentration the band gap energy of semiconductor decreases. The magnitude of reduction increases with increase in doping concentration. This effect is known as band gap narrowing or band gap shrinkage. The most important reason for narrowing is many body effects of free carriers which lowers the electron energy compared to a non-interacting carrier system. Many body effects describe the interaction of free carriers. This interaction becomes important as the separation between carriers reduces. In this situation electron can interact with each other either by their long range coulomb potential or via their spin. The band gap reduction ΔE_g for silicon at room temperature is given by

$$\Delta E_g = 22 \left(\frac{N}{10^{18}}\right) \text{ meV} \tag{2.43}$$

where doping is in per centimetre cube.

2.10.5 Compensated Semiconductor

If in a semiconductor material both donor and acceptor impurities are present simultaneously then such type of semiconductor is known as compensated semiconductor. The type of conductivity (either n-type or p-type) is determined by the impurity which is present in a larger or excess concentration and Fermi level adjust itself to preserve the charge neutrality in a homogeneous semiconductor.

Let us assume a semiconductor in which both impurities are present but $N_a > N_d$. Since, acceptor impurity concentration is larger than the donor impurity concentration therefore, the resulting material will behave as a *p*-type semiconductor and Fermi level will be closer to valence band edge as shown in Fig. 2.31.

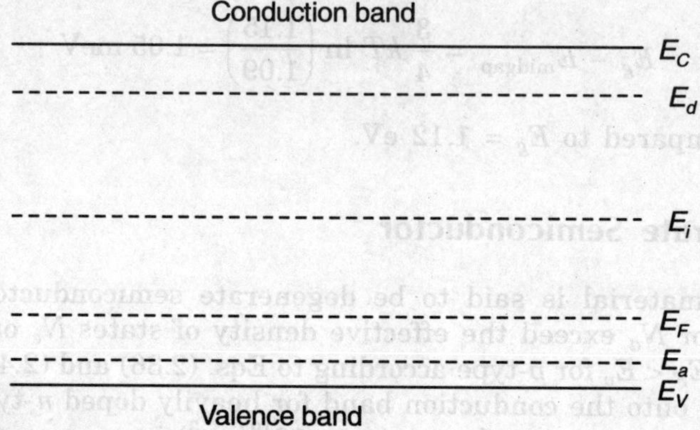

FIGURE 2.31 Band diagram for a compensated semiconductor.

According to charge neutrality condition,

Total negative charges (electrons + ionized acceptor ions) = Total positive charges (holes + ionized donor ions)

or
$$n_0 + N_a^- = p_0 + N_d^+ \tag{2.44}$$

Assuming complete ionization, i.e., $N_d^+ = N_d$

$$N_a^- = N_a$$

Equation (2.44) can be rewritten as:

$$n_0 + N_a = p_0 + N_d \tag{2.45}$$

The charge neutrality condition is applicable at thermal equilibrium as well as steady-state condition (non-equilibrium case). Arranging Eq. (2.45), we have,

$$n_0 - p_0 = N_d - N_a \tag{2.46}$$

Using mass-action law, $p_0 = \dfrac{n_i^2}{n_0}$ in Eq. (2.46), we have

$$n_0 - \frac{n_i^2}{n_0} = (N_d - N_a)$$

or
$$n_0^2 - n_0(N_d - N_a) - n_i^2 = 0 \tag{2.47}$$

On solving the quadratic Eq. (2.47) and taking only positive sign

$$n_0 = \frac{(N_d - N_a) + \sqrt{(N_d - N_a)^2 + 4n_i^2}}{2} \tag{2.48a}$$

Typical concentration $(N_d - N_a)$ is generally much larger than n_i, i.e.,

$$(N_d - N_a) >> n_i^2 \qquad (2.48b)$$

but this condition is not true at very high temperature and for low band gap material where n_i is very large.

Under the condition of (2.48b), Eq. (2.48a) reduces to

$$n_0 \simeq (N_d - N_a) \qquad (2.49a)$$

Electron concentration in a compensated semiconductor at thermal equilibrium can be given as:

$$n_{n0} \simeq (N_d - N_a) \qquad (2.49b)$$

where n_{n0} represents concentration of electron in n-type compensated material at thermal equilibrium.

Minority carrier hole concentration in n-type compensated semiconductor at thermal equilibrium is:

$$p_{n0} \cong \frac{n_i^2}{(N_d - N_a)} \qquad (2.49c)$$

Similarly, majority carrier hole concentration in a p-type compensated semiconductor at thermal equilibrium (under the condition given by [2.48(b)]) is given as:

$$p_{p0} \simeq (N_a - N_d) \qquad (2.49d)$$

and minority carrier electron concentration at thermal equilibrium in p-type compensated semiconductor is:

$$n_{p0} \simeq \frac{n_i^2}{(N_a - N_d)} \qquad (2.49e)$$

The compensated ratio is defined as:

$$\beta_r \simeq \frac{N_d}{N_a} \simeq \frac{N_a}{N_d} \qquad (2.50)$$

for either material and typical value of $\beta_r \le 1$.

For strongly n-type extrinsic semiconductor material,

$$(N_d - N_a)^2 >> 4n_i^2 \quad \text{and} \quad N_a = 0$$

Therefore, the majority carrier concentration at thermal equilibrium is given by;

$$n_{n0} \simeq N_d \qquad (2.51a)$$

Similarly, the majority hole carrier concentration for a strongly p-type extrinsic semiconductor is:

$$p_{p0} \simeq N_a \qquad (2.51b)$$

Note: It is important to note that mass action relation $n_0 p_0 = n_i^2$ holds only at thermal equilibrium condition.

Let us assume a semiconductor material is doped with both impurities (donor as well acceptor) and $N_d > N_a$. It is assumed that all the acceptor impurities are ionized, i.e., $N_a^- = N_a$. To compensate N_a number of these holes same number of electrons out of N_d, will jump from conduction band to valence band and recombine with holes.

Therefore,

Number of electrons left in the conduction band

= No. of electrons initially present − No. of electrons recombine with holes

i.e., $$n_{n0} = N_d - N_a \text{ instead of } N_d$$

This process is called **compensation** and material is said to be compensated semiconductor material.

If a semiconductor material is doped with both impurities in such a way that $N_d = N_a$, then all the electrons from the conduction band recombine with holes in the valence band. Then, there is no donated electrons remain left in the conduction band. In such compensated material, $n_0 = n_i = p_0$ and material becomes intrinsic material. Any further doping of acceptor impurities will create more holes in the valence band than the electrons in the conduction band. Now, material will behave as a p-type semiconductor and majority carrier hole concentration is given by Eq. (2.49d).

Figure 2.32 shows the variation of electron density at thermal equilibrium in silicon semiconductor material with temperature.

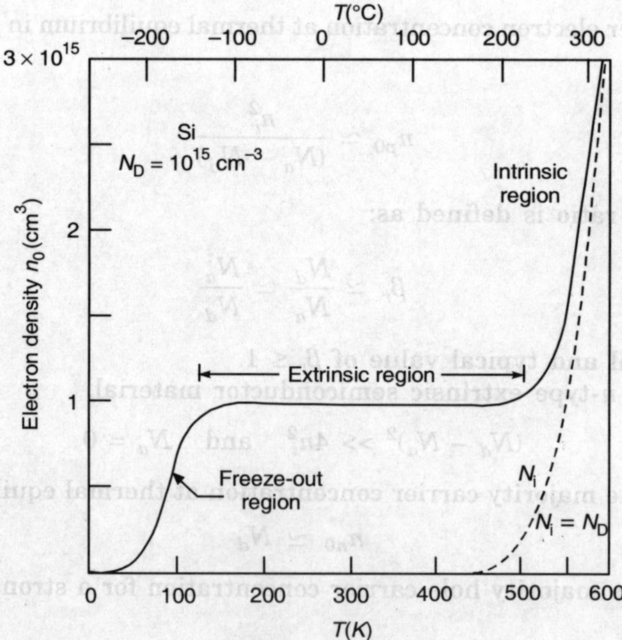

FIGURE 2.32 Variation of electron density with temperature for silicon semiconductor material.

At a very low temperature mostly electrons are bound to the nucleus of parent atom because thermal energy is not sufficient to ionize the donor impurity atoms and electron density is less than the donor concentration. This range of low temperature is called **Freeze-out region**. At higher temperature than the freeze-out region, the condition of complete ionization is reached, i.e., $n_0 = N_d = N^+_d$. Any further increase in temperature will not cause any change in electron concentration over a wide range of temperature. This is extrinsic region. And at a very high temperature, direct thermal excitation from the valence band to conduction band swaps the extrinsic density. Then, there is equal number of electrons in the conduction band and holes in the valence band. Now, material behaves as intrinsic semiconductor. The temperature at which material becomes intrinsic semiconductor depends on the impurity concentration and the band gap value.

EXAMPLE 2.13

Find the equilibrium electrons and hole concentration and location of Fermi level w.r.t. intrinsic level in silicon at $T = 300$ K, if the silicon contains 8×10^{16}. As atoms/cm^3 and 2×10^{16} B atoms/cm^3. Assume complete ionization.

Solution: $N_d = N^+_d = 8 \times 10^{16}/\text{cm}^3$
$$N_a = N^-_a = 2 \times 10^{16}/\text{cm}^3$$

Since N_d and N_a are of same order, so this is not the case of strong extrinsic material.

Also
$$(N_d - N_a)^2 >> 4n_i^2$$

and $N_d > N_a$, hence resulting material is n-type silicon.
Electron concentration in n-type silicon $= n_{n0} = (N_d - N_a)$
$$= 6 \times 10^{16}/\text{cm}^3$$

Minority hole concentration $p_{n0} = \dfrac{n_i^2}{n_{n0}} = 3750/\text{cm}^3$

$$n_{n0} = n_i \exp [(E_F - E_i)/kT] \Rightarrow 1.5 \times 10^{10} \exp [(E_F - E_i)/kT]$$

$$4 \times 10^6 = \exp [(E_F - E_i)/kT]$$

or $(E_F - E_i) = 0.395$ eV

EXAMPLE 2.14

A sample of silicon is doped everywhere with a background concentration of $N_A = 4 \times 10^{16}/\text{cm}^3$. Then $10^{17}/\text{cm}^3$ donors are added. Find the thermal equilibrium concentration of electrons and holes in the original and final materials and draw the energy band diagram for each.

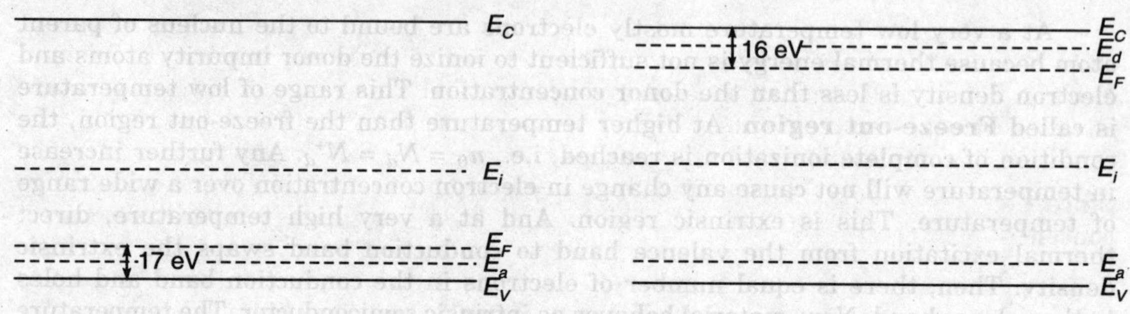

Solution: Original material is p-type semiconductor $p_0 = 4 \times 10^{16}/\text{cm}^3$

$$n_0 = \frac{n_i^2}{p_0} = \frac{2.25 \times 10^{20}}{4 \times 10^{16}} = 0.5625 \times 10^4/\text{cm}^3$$

$$= 5.625 \times 10^3/\text{cm}^3$$

$$E_f - E_v = kT \ln\left(\frac{N_v}{p_0}\right) = 0.17 \text{ eV}$$

This is **greater than 2.3** kT, so the material is **non-degenerate. After further doping:**

$$N_a = 4 \times 10^{16}/\text{cm}^3$$

$$N_d = 10^{17}/\text{cm}^3$$

$N_d > N_a$ hence, resulting material is n-type.

$$n_{n0} = (N_d - N_a) = 6 \times 10^{16}/\text{cm}^3$$

$$p_{n0} = \frac{n_i^2}{n_{n0}} = \frac{2.25 \times 10^{20}}{6 \times 10^{16}} = 0.375 \times 10^4 = 3.75 \times 10^3/\text{cm}^3$$

$$E_c - E_f = kT \ln\left(\frac{N_c}{N_d}\right) = 0.16 \text{ eV}$$

EXAMPLE 2.15

A silicon power device with n-type material is to be operated at $T = 475$ K. At this temperature, the intrinsic carrier concentration must contribute not more than 3% of the total electron concentration. Determine the minimum doping concentration require to meet this specification.

Solution: At $T = 475$ K,

$$n_i = \sqrt{N_c N_v}\, e^{-E_g/2kT}$$

$$= \sqrt{(2.8 \times 10^{19})(1.021 \times 10^{19})\left(\frac{475}{300}\right)^2 e^{-\frac{1.12}{0.0259} \times \frac{300}{475}}}$$

$$= 3.99 \times 10^{13}/cm^3$$

Since, according to problem, n_i must contribute not more than 3% of the total electron concentration, means

$$n_0 = 1.03 \times N_d$$

Using Eq. (2.40a) for $N_a = 0$,

$$n_0 = 1.03 \, N_d = \frac{N_d + \sqrt{N_d^2 + 4n_i^2}}{2}$$

On solving,

$$N_d = 2.27 \times 10^{14}/cm^3$$

2.11 SUMMARY

Electrical engineers classify the solid-state material either into conductor, insulator or semiconductor based on the electrical behaviour of the materials. Material scientist classify the materials based on the chemical bond. They classify the solids into four groups; molecular solids, covalent solids, ionic solids and metallic solids. Materials can be also classified as crystalline, polycrystalline, amorphous etc. based on the crystal structure. An imperfection in the solid is due to (a) lattice thermal vibration, (b) point defects, (c) line defects. There are mainly two types of impurities in a solid. It is either substitutional or interstitial. The electrical properties of the crystal can be altered by adding impurities. Impurities can be added into the crystal either by diffusion or ion implantation method. The semiconductor material is an 'attractive' material for the electronic device because their electrical conductivity can be easily controlled. A semiconductor material can be classified as elemental semiconductor (Si, Ge, etc.) and compound semiconductor (GaAs, InP, etc.) on the basis of their composition. Elemental semiconductors are indirect band gap material whereas compound semiconductor are direct band gap material. The properties of semiconductors are determined to a large extent by the crystal structure. Miller indices are mainly used to describe the crystal surfaces and crystal orientation. A change in temperature or by adding impurities the conductivity of semiconductor materials can be enhanced. A semiconductor is either intrinsic semiconductor (pure semiconductor) or extrinsic semiconductor (doped with foreign elements). Acceptor and donor impurities are most effectively used when the energy required to generate a carrier is small. In the case of small ionization energy, the energy levels associated with impurities lie within the band gap close to their respective bands, i.e. donor ionization energy is close to the conduction band and acceptor ionization energy is close to the valence band. A heavily doped semiconductor material is called **degenerate semiconductor**. A semiconductor is said to be *compensated* if both type of impurities are present.

REVIEW QUESTIONS

1. Define the Fermi level in a conductor.

2. Why hole is not available in the conduction band?

3. Why Fermi level should not vary spatially at thermal equilibrium?

4. Define thermal equilibrium.

5. Why the energy gap of pure Ge and Si decreases monotonically with temperature? Discuss qualitatively only.

6. Whether mass-action law holds for thermal equilibrium or steady state, explain briefly.

7. The temperature dependence of electron concentration at thermal equilibrium in a doped semiconductor is shown in Fig. 2.33.

   (a) Name the three regions with suitable argument.
   (b) Which region will you choose for device operation and why?

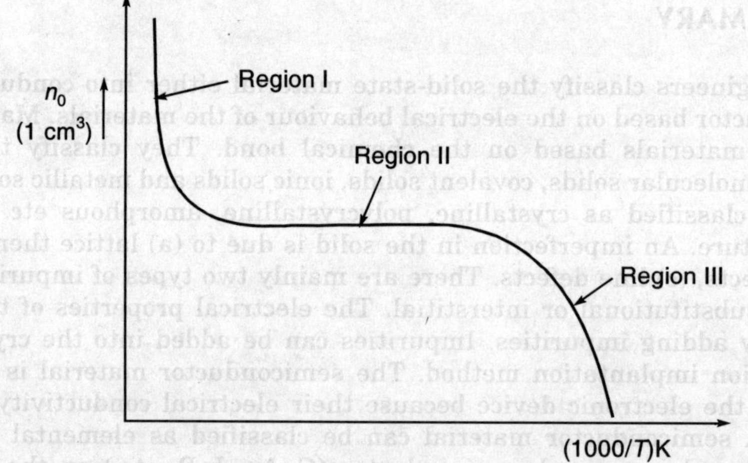

FIGURE 2.33 Temperature dependence of electron concentration at thermal equilibrium.

8. The E vs k diagram for two semiconductors with $n_0 = 10^{17}/cm^3$ at $T = 300$ K is shown in Fig. 2.34.

   (a) Explain whether both materials have same effective density of states N_c in conduction band or not. If not which material has lowest effective density of states? Explain your answer.
   (b) Explain, in which case will the Fermi level locate closest to the conduction band.

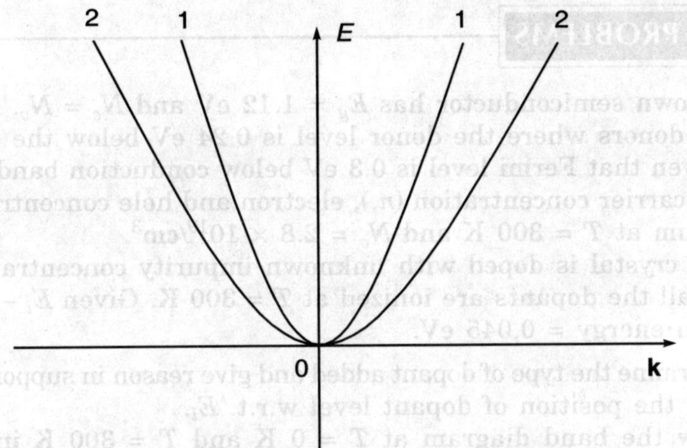

FIGURE 2.34 *E* vs *k* diagram.

9. Ge has lower energy band gap than Si but still Si is more preferable semiconductor material than Ge. Why?

10. What do you mean by compensation and compensated semiconductor material?

11. Why a heavily doped (degenerate) semiconductor material is not preferred for device?

12. Why Si and Ge are called indirect semiconductors?

13. Explain qualitatively the difference between effective mass of electron and free mass of electron.

14. Near the conduction band edge, the electron energy can be expressed as:

$$E_0 = E_C + Ak^2$$

where *A* is a constant.

(a) Find the effective mass of the electron in conduction band.

(b) If *A* is negative quantity, how the effective mass of the electron will get affected?

15. Explain qualitatively, when the probability of occupying the state by electron is maximum.

16. Why Fermi-Dirac statistics is applicable to get electron probability in any given state?

17. How heavily would you dope the silicon semiconductor material with donors to violate the assumption of non-degeneracy?

18. What is the total energy of electron at the conduction band edge and holes at the valence band edge? Explain why electron energy increases as one moves upwards from conduction band edge.

NUMERICAL PROBLEMS

1. An unknown semiconductor has E_g = 1.12 eV and $N_c = N_v$. It is doped with $10^{15}/cm^3$ donors where the donor level is 0.24 eV below the conduction band edge. Given that Fermi level is 0.3 eV below conduction band edge. Calculate intrinsic carrier concentration (n_i), electron and hole concentration at thermal equilibrium at T = 300 K and $N_c = 2.8 \times 10^{19}/cm^3$.

2. A silicon crystal is doped with unknown impurity concentration of $10^{16}/cm^3$. Assume all the dopants are ionized at T = 300 K. Given $E_i - E_f$ = 0.2 eV and ionization energy = 0.045 eV.

   (a) Determine the type of dopant added and give reason in support of your answer.
   (b) Find the position of dopant level w.r.t. E_i.
   (c) Draw the band diagram at T = 0 K and T = 300 K indicating all the energy levels clearly.
   (d) Find the majority and minority thermal equilibrium concentration.

3. A silicon sample at T = 300 K contains an acceptor impurity concentration of 10^{16} atoms/cm^3. Determine the concentration of substitutional impurity which must be added so that Fermi level is 0.20 eV below the conduction band edge. Also find the type of substitutional impurity.

4. Calculate the carrier density in Ge and Si at 300 K, 400 K, 500 K and 600 K.

5. Calculate the shift in intrinsic Fermi level from $E_g/2$ Si at room temperature T = 300 K if the effective mass values of electrons and holes are 1.10 m_0 and 0.56 m_0 respectively.

6. A silicon sample at thermal equilibrium contains $3 \times 10^{16}/cm^3$ ionized arsenic atoms, 2×10^{16} atoms/cm^3 ionized boron atoms and 4×10^{16} (atoms/cm^3) ionized phosphorus atoms, at T = 300 K.

   (a) Find majority and minority carrier concentrations.
   (b) Draw the energy band diagram at T = 300 K.

7. In a semiconductor sample, donor and acceptor levels are 0.3 eV apart from each other. If 80% of the acceptors are ionized at room temperature, find the fraction of ionized donors.

8. For a n-type silicon sample with $10^{16}/cm^3$ Phosphorus donor impurities and donor level at E_s = 0.45 eV, find the ratio of the neutral donor density to ionized donor density at 77 K where the Fermi level is 0.0459 eV below the conduction band edge.

9. Find the probability of occupancy of state at the bottom of the conduction band in the intrinsic silicon at room temperature.

10. Find the electron and hole concentrations, and locate Fermi-level position when GaAs is doped with $N_a = 10^{16}/cm^3$.

    Take T = 300 K, E_g = 1.43 eV
    $m_n^* = 0.067\ m_0,\ m_p^* = 0.48\ m_0$

    where m_0 is free mass of electron.

11. Find the donor impurity concentration in a compensated semiconductor material, assuming $N_s \gg N_a$, such that the electron concentration at thermal equilibrium is 10% greater than N_d.

12. Find the probability of occupying a state

    (a) which is 0.2 eV above the Fermi level.
    (b) which is 0.3 eV below the Fermi level.
    (c) which is at Fermi level.

13. Find the respective impurity concentration to elevate the Fermi level in silicon semiconductor material to 2.3 kT above E_i. Also identify the nature of doping concentration.

14. The probability of occupying an energy state in conduction band edge is of order of 10^{-4}.

    (a) Is provided information sufficient to know the type of semiconductor, i.e., intrinsic, n-type or p-type?
    (b) If yes, whether the material is intrinsic or extrinsic?
    (c) Find (N_d/N_a).

15. Let us consider two unknown semiconductor materials A and B. Both materials have same density of states effective mass. The energy gap of material B is twice of material A. Find the ratio of their intrinsic carrier concentrations. Which material will you choose for device operation at room temperature? Explain briefly.

16. If an instrinsic silicon semiconductor material is doped with both type of impurities such that $N_d = N_a = 10^{16}/\text{cm}^3$.

    (a) What will be the electron and hole concentration in case of resulting silicon material?
    (b) Locate the Fermi level.

17. An unknown semiconductor has $E_g = 1.1$ eV and $N_c = N_v$. It is doped with $10^{15}/\text{cm}^3$ donors where the donor level is 0.2 eV below E_c. Given that E_F is 0.25 eV below E_c. Calculate n_i and concentrations of electrons and holes in the semiconductor at $T = 300$ K.

18. Consider a silicon at $T = 300$ K doped with donor impurity of phosphorus to a concentration of $10^{19}/\text{cm}^3$.

    (a) Ignore the concentration dependence of the donor ionization energy. Calculate the concentration of ionized phosphorus and neutral phosphorus. What is free hole concentration?
    (b) At what phosphorous concentration, should the ionization energy go to zero?
    (c) If E_d goes to zero, what will happen to free electron concentration?

19. In a pure silicon semiconductor material at the room temperature, the electron energy near the conduction band edge is given by

$$E = E_c + 0.56 \, m_0 \, K^2$$

and the hole effective mass is 1.5 times the effective mass of electron, where m_0 is free electron mass. Find the intrinsic electron as well as hole concentration. At what temperature intrinsic carrier concentration will become double of the value calculated at room temperature?

20. Calculate the intrinsic carrier concentration for silicon semiconductor material at $T = 300$ K, 500 K, 600 K for $m_n^* = 0.5 \ m_0$ and $m_p^* = 1.5 \ m_0$, neglect the temperature variation of energy gap.

21. Is the effective mass of carriers affecting the position of Fermi level w.r.t. conduction band for n-channel material at a fixed temperature? Find the position of Fermi level w.r.t. conduction band edge for electron for two cases when it is doped with 10^{16} donor impurities atoms/cm^3.

Case I: $\qquad\qquad m_n^* = 0.56 \ m_0 \qquad m_p^* = 2 \ m_e^*$

where m_0 is free-electron mass, $T = 300$ K

Case II: $\qquad m_n^* = m_p^* = 0.5 \ m_0$ and $T = 400$ K.

CHAPTER 3

Transport Phenomenon in Semiconductor

- **Introduction of carrier drift**
  - Mobility concept
  - Scattering mechanisms
- **Resistivity and conductivity of semiconductor material**
  - Integrated circuit resistor
- **Hall effect**
- **Carrier diffusion process**
  - Induced electric field: Built-in electric field
  - Einstein relation
- **Carrier generation and recombination processes in semiconductor**
  - Concept of excess carrier generation
  - Direct and indirect recombination
  - Shockley-Hall-Read recombination theory
  - Auger recombination
- **Continuity equation**
- **Quasi Fermi levels**
- **Shockley-Hall experiment**

3.1 INTRODUCTION

In this chapter a detailed study of transport mechanism of carriers in a semiconductor material is carried out. The present chapter deals with the mobility concept, diffusion and drift phenomenon Quasi Fermi level and Shockley experiment etc. A detail discussion was made on resistivity and conductivity of the semiconductor materials. The mathematical expression were derived to clear the understanding of concepts. Hall effect, induced electric field in a non-homogeneous semiconductor material are also included in the chapter. At last the various generation and recombination processes in the semiconductor material were discussed.

3.2 CARRIER DRIFT (MOBILITY CONCEPT)

Consider a n-type semiconductor material at thermal equilibrium which is doped uniformly with donor impurities. Electrons in the conduction band are treated as a free electron because they are not associated with any particular lattice or donor site. The influence of the crystal lattice is incorporated by replacing free mass of conduction electron by the effective mass. For a non-degenerate semiconductor material, the average thermal energy of conduction electron at equilibrium is $1/2\ kT$ per particle per degree of freedom. Since, the electrons in the conduction band can move about in a three-dimensional space and have three degrees of freedom, therefore, the kinetic energy of the electron is given as:

$$\frac{1}{2}\ m_n^*\ v_{Th}^2 = \frac{3}{2}\ kT \tag{3.1}$$

where v_{Th} is thermal velocity of electron at room temperature and it is equal to 1.5×10^7 cm/sec for Si.

In the absence of any applied electric field (i.e., $E_x = 0$), electrons move randomly in all directions due to thermal energy and colliding with lattice, dopant impurities and other scattering centre. In this process, electrons constantly change their direction of motion and velocity. The random motion of electron in absence of electric field is shown in Fig. 3.1(a).

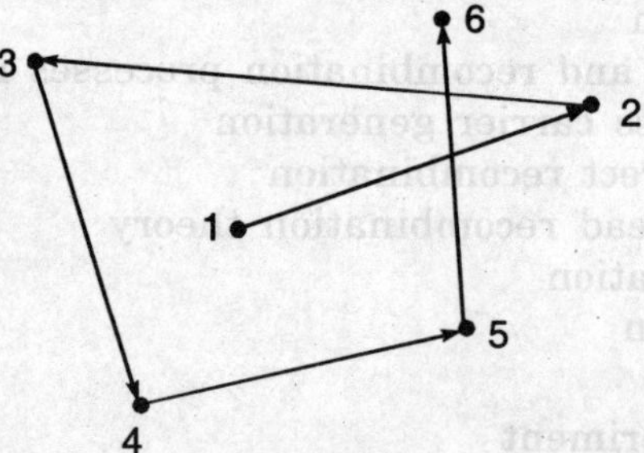

FIGURE 3.1(a)   Random motion of electron in absence of field.

If there are large number of carriers then the probability of carriers coming back to its original position is very small. Due to thermal motion (which causes random motion), each particle will collide either with other carriers or lattice or dopant impurities and hence carriers will either gain or loose energy in this collision. Therefore, one can say that the random motion of carriers lead to net zero displacement of carrier over a sufficiently long period of time and no current will flow in the semiconductor material in absence of applied field. Average distance covered by carrier between collisions is called **mean free path**, and average time between collision is called **mean free time (τ)**.

When an electric field is applied to the semiconductor, each electron will experience a net force $-qE$ from the field and accelerate, in opposite direction to the applied field, in between collisions. In this process, electrons acquire a drift velocity superimposed on their random thermal motion. The drift velocity of electrons is opposite to the field and drift velocity of holes is in the direction of applied field. The combined displacement of an electron in presence of thermal motion as well as drift velocity is shown in Fig. 3.1(b).

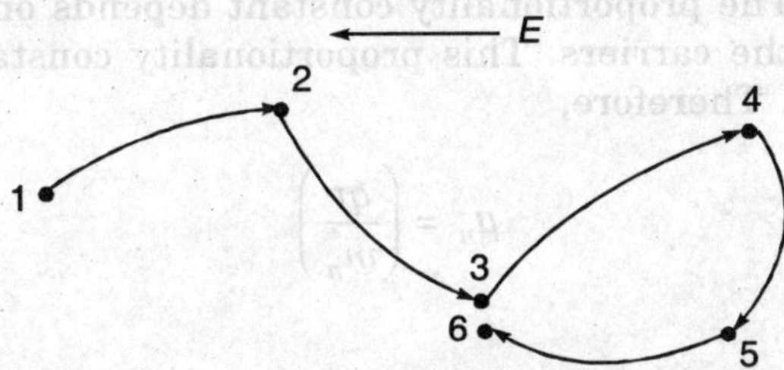

FIGURE 3.1(b) Displacement of electron due to thermal motion and drift velocity.

Therefore, the total force acting on the electron is:

$$F = qE + m\frac{dv}{dt} \tag{3.2}$$

where first term on the RHS of Eq. (3.2) represents the force acting on the electron in presence of small field and second term gives change in momentum due to collisions. Replacing m by effective mass m_n^* of electron in Eq. (3.2), we have

$$F = qE + m_n^*\frac{dv}{dt} \tag{3.3}$$

or

$$F = qE + m_n^*\frac{<v>}{\tau} \tag{3.4}$$

where $\dfrac{<v>}{\tau}\,m_n^*$ gives the average loss of momentum due to collisions. Equation (3.4) can be further written as:

$$m_n^*\frac{d}{dt}<v> = qE + m_n^*\frac{<v>}{\tau} \tag{3.5}$$

where $<v>$ represents the average drift velocity during collision. Assuming steady state condition,

i.e.,
$$\frac{d}{dt}<v> = 0$$

Equation (3.5) reduces to

$$qE = - \, m_n^* \frac{<v>}{\tau}$$

or
$$<v> = - \left(\frac{q\tau}{m_n^*} \right) E \tag{3.6}$$

$$= (\text{constant}) \; E$$

Equation (3.6) states that the electron's drift velocity is proportional to the applied field for a small field. The proportionality constant depends on the mean free time τ and effective mass of the carriers. This proportionality constant is called **mobility** and represented by μ_n. Therefore,

$$\mu_n = \left(\frac{q\tau}{m_n^*} \right) \tag{3.7a}$$

Similarly, for holes

$$\mu_p = \left(\frac{q\tau}{m_p^*} \right) \tag{3.7b}$$

The unit of mobility is cm^2/V-sec.

Mobility is a measure of the responsiveness of the carrier motion within a semiconductor crystal to an electric field that exerts a force $-qE$ on the electron or qE on the hole. Therefore, Eq. (3.6) can be written as:

$$v_d = - \, \mu_n E \quad \text{for electrons} \tag{3.8a}$$

and
$$v_d = \mu_p E \quad \text{for holes} \tag{3.8b}$$

Since, mobility is inversely proportional to the effective mass of the carriers and since effective mass of hole is larger than the effective mass of electron ($m_{\text{hole}}^* > m_n^*$), therefore, mobility of hole is lower than mobility of electron.

i.e.,
$$\mu_p < \mu_n$$

and it was found that the electron mobility is approximately 2.5 times the mobility of holes,

$$\mu_n \simeq 2.5 \; \mu_p \tag{3.9}$$

It is, therefore, concluded that hole moves slower than electron. Due to high mobility of electrons, for most of the devices n-type semiconductor is preferred. Equations [3.8(a) and (b)] are valid only for the weak electric field.

Equation (3.7) also tells us that mobility is directly proportional to the mean free time between the collisions, which is determined by various scattering mechanisms like

- Lattice scattering
- Ionized impurity scattering
- Optical scattering
- Surface roughness scattering
- Dipole or neutral surface state scattering
- Neutral impurity scattering which is weaker than ionized impurity scattering

The two most dominating scattering mechanisms in silicon are lattice scattering and impurity scattering.

3.2.1 Lattice Scattering or Phonon Scattering

Lattice scattering results from thermal vibrations of the lattice atom at any temperature above absolute zero. These vibrations disturb the lattice periodic potential and allow energy to be transferred between the carriers and the lattice due to collision between them. As the temperature increases, lattice vibration amplitude increases. This will result in increase in probability of scattering of carriers from lattice and reduces mobility. Because as temperature increases amplitude of lattice vibration increases which results the increase in collision probability between carriers and lattice (Phonon). Since, lattice vibration increases with temperature, therefore, the lattice scattering dominates at higher temperature and at low doping. Scattering by acoustic phonons (quantized lattice vibrating waves) has often limit the mobility of carriers in any semiconductor at room temperature.

$$\text{Lattice scattering rate } \left(\frac{1}{\tau_L}\right) \propto T^{3/2}$$

or

$$\mu_L \propto T^{-3/2} \tag{3.10}$$

This shows that mobility due to acoustical phonon scattering decreases with temperature as $T^{-3/2}$.

3.2.2 Ionized Impurity Scattering

A second important scattering mechanism is ionized impurity scattering. This scattering is a consequence of electrostatic (coulombic) field generated by ionized impurities within the crystal. The ionized impurities are introduced into the crystal due to added extrinsic dopant. The carriers experience a coulombic force as they travel past an ionized impurity atom and get scattered as shown in Fig. 3.2. The probability of scattering of carriers from ionized impurities depend on the concentration of extrinsic dopant in the crystal. As the doping concentration increases beyond 10^{15}–10^{16}/cm^3 at a given temperature, the probability of carrier's scattering (or collision) with ionized impurity atoms increases which results in reduction of carrier mobility. In general, carrier mobility decreases as doping concentration increases at the given temperature. Although donor and acceptor impurities are oppositely ionized within the lattice,

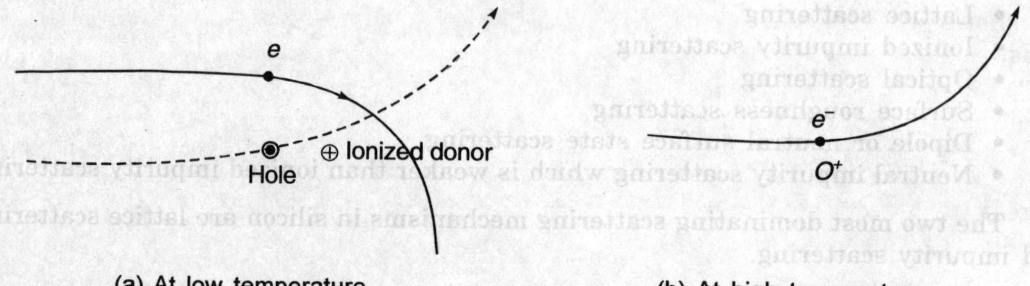

FIGURE 3.2   Ionized impurity scattering at low and high temperature.

overall effects of electrostatic scattering do not depend on the sign of the charge. Therefore, both donor and acceptor atoms can be expected to have similar influence on carrier mobility.

Since, thermal energy of carriers are very low at low temperature and hence there is larger probability that carriers remain near the ionized impurity for long time and experience a strong coulombic interaction force as seen in Fig. 3.2(a). In other words at low temperature due to low thermal velocity, carrier will take longer time to pass the ionized impurity and hence there is more probability that carrier motion will be influenced by coulombic force of ionized impurity.

At the higher temperature, the scattering probability through ionized impurity decreases because carriers move faster and remain near the ionized impurity centre for a shorter duration and hence less effectively scattered as shown in Fig. 3.2(b).

$$\text{Ionized scattering rate,} \quad \frac{1}{\tau_I(T)} \propto T^{-3/2}$$

Hence, $$\mu_I \propto T^{3/2} \tag{3.11}$$

The mobility near room temperature due to ionized impurity scattering varies with temperature as well as with impurity concentration N_I. Mathematically, it is expressed as:

$$\mu_I \propto \frac{T^{-3/2}}{N_I} \tag{3.12}$$

where $N_I = N_d^+ + N_a^-$ is total ionized impurity in semiconductor. In case of phonon scattering $N_I = N_d^+ - N_a^-$ or $N_I = N_a^- - N_d^+$. Equation (3.12) also tells us that if the ionized impurity centre increases then mobility of carrier reduces. Although acceptor and donor impurities compensate each other in terms of extrinsic doping, they still invariably contribute to reduction of carrier mobility and hence carrier mobility in compensated semiconductor will be lower than impure semiconductor.

If τ_L, τ_I, τ_o ... are the time between collisions due to lattice scattering, ionized impurity scattering, optical phonon scattering, etc., and assuming all scattering processes are independent of each other then the total probability of scattering event occurring in a short interval of time dt is given by the sum of individual events or in other words,

$$\frac{dt}{\tau} = \frac{dt}{\tau_I} + \frac{dt}{\tau_L} + \frac{dt}{\tau_o} + \dots$$

where τ is mean free time between any two collisions.

or

$$\frac{1}{\tau} = \frac{1}{\tau_I} + \frac{1}{\tau_L} + \frac{1}{\tau_o} + \dots$$

Putting values of τ_I, τ_L, τ_o, ... we have

$$\left(\frac{q}{m_n^*}\right)\left(\frac{1}{\mu}\right) = \left(\frac{q}{m_n^*}\right)\left(\frac{1}{\mu_I}\right) + \left(\frac{q}{m_n^*}\right)\left(\frac{1}{\mu_L}\right) + \left(\frac{q}{m_n^*}\right)\left(\frac{1}{\mu_o}\right) + \dots$$

or

$$\frac{1}{\mu} = \frac{1}{\mu_L} + \frac{1}{\mu_I} + \frac{1}{\mu_o} + \dots \tag{3.13}$$

where μ_I, μ_L and μ_o, ..., are components of mobilities due to lattice scattering, impurity scattering, optical-phonon scattering, etc. Since, lattice as well as ionized impurity scattering phenomenons are discussed in this chapter, and hence resultant mobility due to these two scattering processes are:

$$\frac{1}{\mu} = \frac{1}{\mu_I} + \frac{1}{\mu_L} \tag{3.14}$$

Equation (3.14) shows that the net inverse mobility is given by the sum of the inverse of the mobilities of the two scattering processes. This relation [Eq. (3.13)] is known as **Mathiessen's Rule**. This relation shows that smallest mobility dominates.

Figures 3.3(a) and (b) show the variation of electron/hole mobilities with temperature as well as with total ionized impurity concentration N_I.

Mobility of electron and hole for different semiconductor materials at room temperature $T = 300$ K is given in Table 3.1.

Table 3.1 Mobilities of Semiconductor Materials

Semiconductor	Electron mobility(μ_n) cm²/v-s	Hole mobility (μ_p)
Ge	3900	1900
Si	1500	450
GaSb	5000	850
GaAs	8500	400
GaP	110	75
InAs	33000	460
InP	46000	150

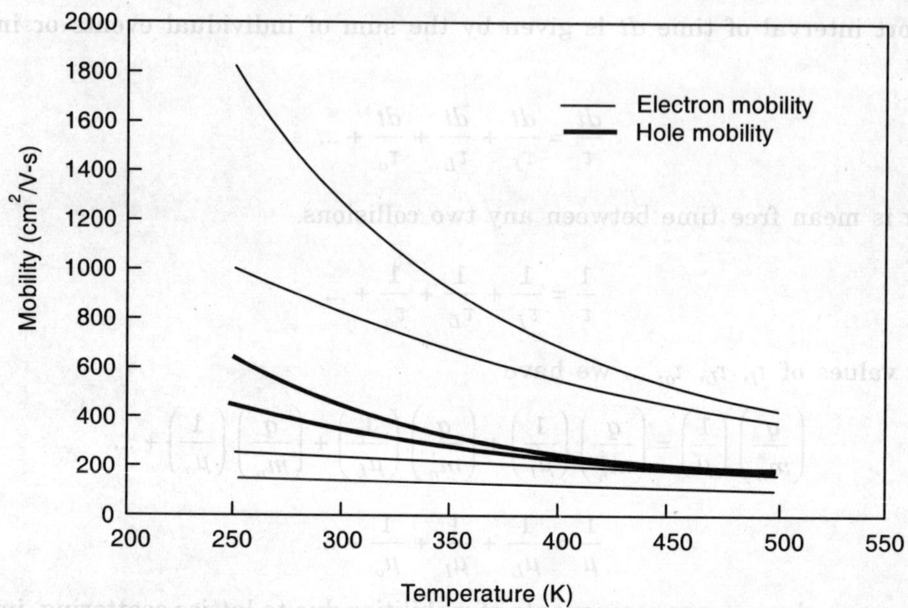

FIGURE 3.3(a) Variation of mobility with temperature for electron and hole.

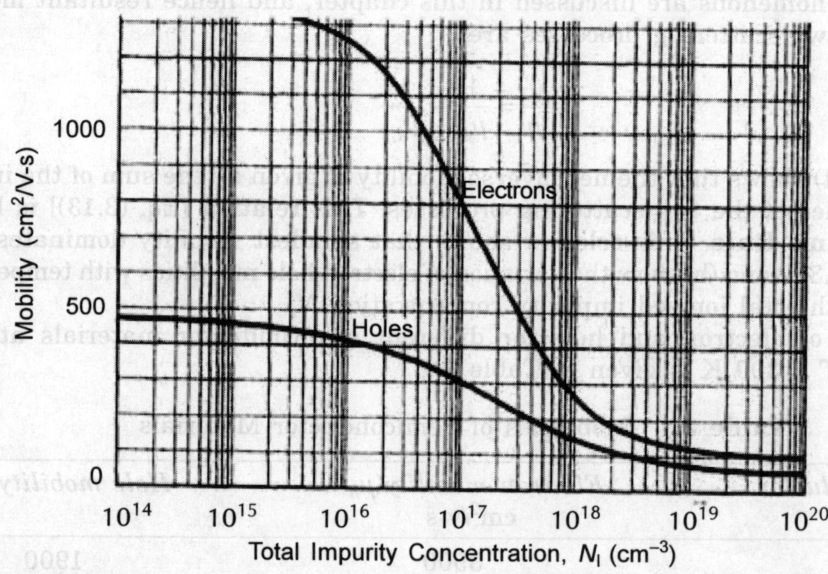

FIGURE 3.3(b) Variation of mobility with impurity concentration for electron and hole.

EXAMPLE 3.1

Find the mobilities of electron and hole in a silicon semiconductor material at room temperature. Assuming the mean free time between scattering τ_n and τ_p are 2×10^{-13} s and 1×10^{-13} s respectively. Take $m_n^* = 0.26\, m_0$ and $m_p^* = 0.36\, m_0$.

Solution: Using Eqs. (3.7a) and (3.7b), we have

$$\mu_n = \frac{q\tau_n}{m_n^*}$$

$$= \frac{1.6 \times 10^{-19} \times 2 \times 10^{-13}}{0.26 \times 9.11 \times 10^{-31}} = 1330 \text{ cm}^2/\text{v-s}$$

$$\mu_p = \frac{q\tau_p}{m_n^*}$$

$$= \frac{1.6 \times 10^{-19} \times 10^{-13}}{0.36 \times 9.11 \times 10^{-31}} \simeq 495 \text{ cm}^2/\text{v-s}.$$

EXAMPLE 3.2

Calculate the mean free time and mean free path of an electron having a mobility of 1000 cm^2/v-s at room temperature. Assuming $m_n^* = 0.26\ m_0$.

Solution: Using Eq. (3.7a),

$$\tau_n = \frac{\mu_n m_n^*}{q} = \frac{1000 \times 10^{-4} \times 0.26 \times 10^{-19} \times 9.1}{1.6 \times 10^{-19}} = 1.48 \times 10^{-13} \text{ s}$$

Mean free path $l = \tau_n v_{Th}$

$$= 1.48 \times 10^{-13} \times 10^7 = 14.8 \text{ nm}$$

Note: This equation $v_d = \mu E$ is generally applicable when $E < 1000$ V/cm and at higher electric field, carrier velocity is no longer proportional to the electric field but tends toward a constant value called "saturated drift velocity". The field at which carrier velocity saturates is known as "critical electric field (E_C)".

Explanation: At low electric field, thermal motion is dominating the carrier motion. At high electric field ($\geq 10^5$ V/cm), carrier's motion dominates and thermal random motion reduces to zero. As the field increases the average carrier motion as well as average carrier energy increases. As the energy increases beyond the optical phonon energy, the probability of emitting an optical phonon increases abruptly: means this excess energy (average energy – phonon energy) is given to lattice instead of acceleration of carrier and this causes the velocity to saturate at high field as shown in Fig. 3.4. The energy of carrier is much larger than zero field energy, i.e., thermal energy ($3/2kT$). These electrons are called hot electrons and they suffer increased scattering and velocity saturates.

3.3 RESISTIVITY

Consider a homogeneous semiconductor material of cross-sectional area A, length L as shown in Fig. 3.5(a) with a free electron density n.

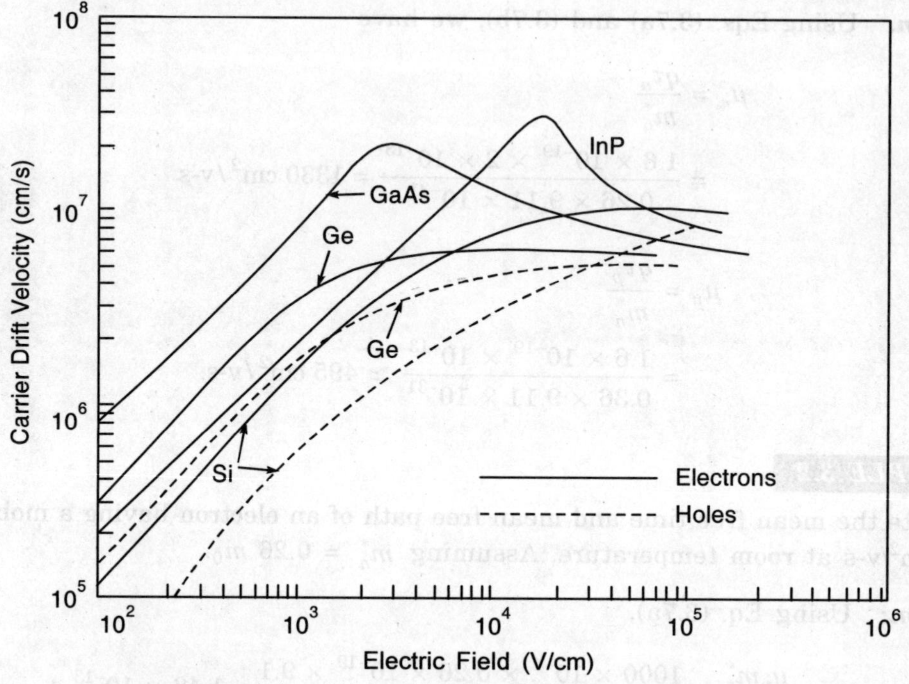

FIGURE 3.4 Variation of velocity with electric field.

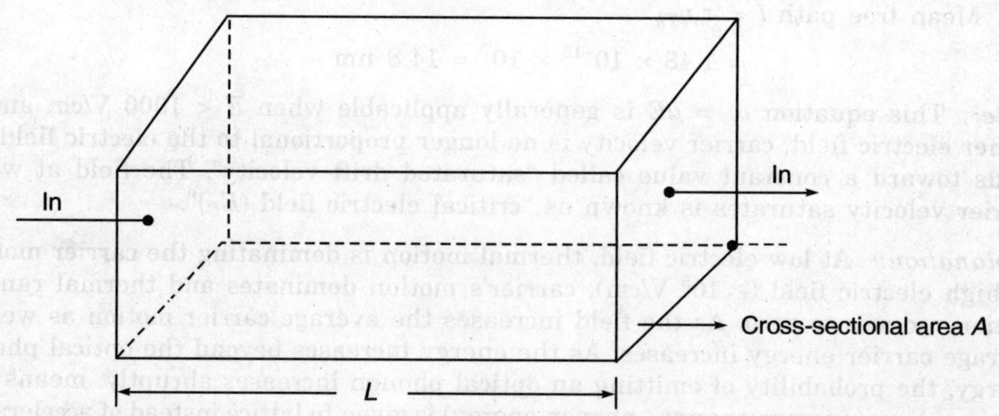

FIGURE 3.5(a) Cross-sectional view of a homogeneous semiconductor material.

The transport of the carriers under the influence of applied electric field E along the length of the bar produces a current, called **drift current**. The current density J_n flowing in the sample is proportional to the number of electrons crossing a unit area per unit time, i.e., $n <v_x>$ and also proportional to the charge on the carrier. In mathematical form, total current density can be expressed as:

$$J_n = \frac{I}{A} = \sum_{i=0} (-q)\, v_i \tag{3.15}$$

where summation is taken over all the electrons.

Equation (3.15) shows that the current density in the sample can be found by summing the product of the electronic charge on each electron times the electron's velocity over all electrons per unit volume.

Equation (3.15) can also be expressed as:

$$J_n = -\, q n v_n \tag{3.16a}$$

If the applied field is low then using $v_n = -\,\mu E$ for electron, Eq. (3.16a) can be rewritten as:

$$J_n = q n \mu_n E \tag{3.16b}$$

Similarly, for holes

$$J_p = q \mu_p p E \tag{3.17}$$

where p is hole density, μ_n and μ_p are electron and hole mobility respectively.

The total current density in the semiconductor bar in presence of low electric field is given by the sum of current densities due to majority as well as minority carriers, i.e.,

$$J = J_n + J_p$$

or

$$J = q(\mu_n n + \mu_p p) E \tag{3.18}$$

Equation (3.18) can be written in the form of Ohm's law as:

$$J = \sigma E \tag{3.19a}$$

where, $\sigma = q(n\mu_n + p\mu_p)$ is called **conductivity** of the semiconductor material and its unit is $(\Omega\text{-cm})^{-1}$ or mho/cm. Conductivity of semiconductor material depends on the electron and hole concentration and also on their mobilities.

The conductivity of a strong n-type semiconductor ($n \gg p$) is given by

$$\sigma_n = q \mu_n n \tag{3.19b}$$

And, for strong p-type semiconductor ($p \gg n$),

$$\sigma_p = q \mu_p p \tag{3.19c}$$

And, hence Eq. (3.18) can also be expressed as:

$$\sigma = \sigma_n + \sigma_p \tag{3.20}$$

The reciprocal of the conductivity is defined as the resistivity which is an intensive material property and is independent of size and shape of the material. Since, resistivity is a function of carrier concentration and mobilities, therefore, for a given concentration the resistivity ($\rho = 1/\sigma$ Ω-cm) of p-type semiconductor material is higher than type semiconductor material and given as:

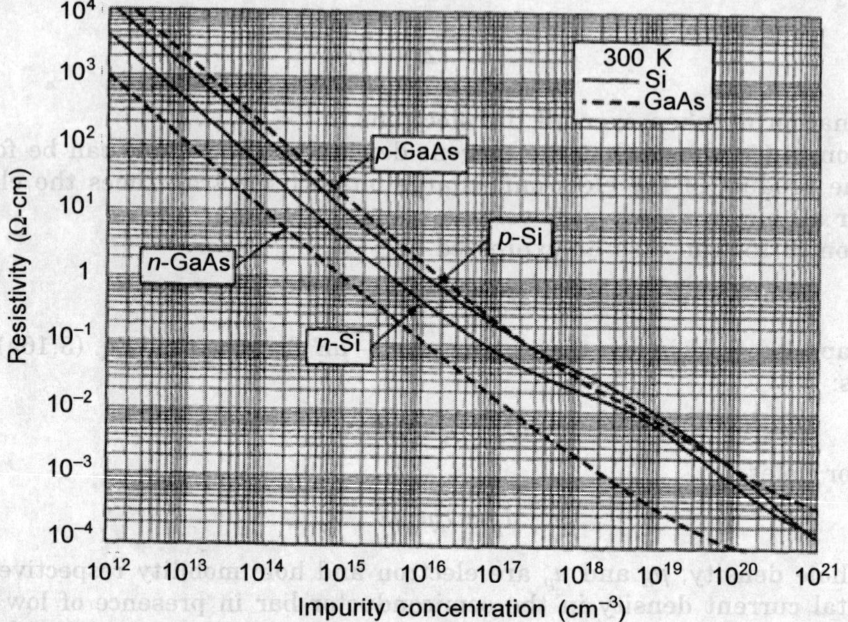

FIGURE 3.5(b)   Plot of resistivity with impurity concentration for Si and GaAs at $T = 300$ K.

$$\rho = \frac{1}{\sigma} = \frac{1}{q(\mu_n n + \mu_p p)} \qquad (3.21)$$

The resistivity vs impurity concentration for n-type and p-type silicon at room temperature is shown in Fig. 3.5(b).

This figure clearly shows that for any impurity concentration, the conductivity due to electron is always higher than hole because of larger electron mobility than hole.

For intrinsic semiconductor material, conductivity is given by

$$\sigma_i = q(\mu_n + \mu_p)\, n_i \qquad (3.22)$$

Note: Drift process is a majority carrier process.

EXAMPLE 3.3

(a) Show that the minimum conductivity of a semiconductor sample occurs when

$$n_0 = n_i \sqrt{\frac{\mu_p}{\mu_n}}.$$

Solution: Using $\sigma = q(\mu_n n + \mu_p p)$

At thermal equilibrium, $\qquad\qquad \sigma = q\left(\mu_n n_0 + \mu_p \dfrac{n_i^2}{n_0^2}\right)$

$$\frac{dv}{dn_0} = q\left(\mu_n - \mu_p \frac{n_i^2}{n_0^2}\right)$$

For minimum conductivity, $\dfrac{dv}{dn_0} = 0$

Hence,
$$\mu_n = \mu_p \frac{n_i^2}{n_0^2} \quad \text{or} \quad n_0^2 = n_s^2 \frac{\mu_p}{\mu_n}$$

or
$$n_0 = n_i \sqrt{\frac{\mu_p}{\mu_n}}$$

Hence proved.

(b) Find the minimum conductivity expression.

$$\sigma_{\min} = q\left(\mu_n n_0 + \mu_p \frac{n_i^2}{n_0}\right)$$

$$= q\left(\mu_n \sqrt{\frac{\mu_p}{\mu_n}} \, n_i + \mu_p n_i^2 \sqrt{\frac{\mu_n}{\mu_p}} \, \frac{1}{n_i}\right)$$

$$= q\left(\sqrt{\mu_n \mu_p} \, n_i + \sqrt{\mu_p \mu_n} \, n_i\right)$$

or
$$\sigma_{\min} = 2q\sqrt{\mu_n \mu_p} \, n_i$$

(c) Compare $\sigma_{\min}$ with intrinsic conductivity.

$$\frac{\sigma_{\min}}{\sigma_i} = \frac{2q\sqrt{\mu_n \mu_p} \, n_i}{q(\mu_n + \mu_p) \, n_i} = \frac{2\sqrt{\mu_n \mu_p}}{(\mu_n + \mu_p)}$$

assuming $\mu_n \simeq 2.5 \, \mu_p$,

$$\sigma_{\min} = 2\frac{\sqrt{\mu_p \times 2.5 \, \mu_p}}{(2.5 + 1)} \simeq 0.903$$

3.3.1 Integrated-Circuit Resistors

Mainly, two types of resistors are used in the integrated circuit: first one is fabricated by the deposition of thin films (such as nichrome or tantalum) on the surface of an insulating layer of SiO_2 on the semiconductor. These materials have sheet resistances of order of 40 to 4000 ohms/square ($\square$).

The other type of resistor is formed by the introduction of impurities into the surface of semiconductor by diffusion or ion implantation techniques. These layers have sheet resistance in range of 1000 ohm/square. The geometry of resistor formed by the latter technique is shown in Fig. 3.6 with a length L, width W and thickness t.

The resistor R of the geometry, as shown in Fig. 3.6,

$$R \propto \text{Length of the bar}$$

$$R \propto \frac{1}{\text{Cross-sectional area of bar}}$$

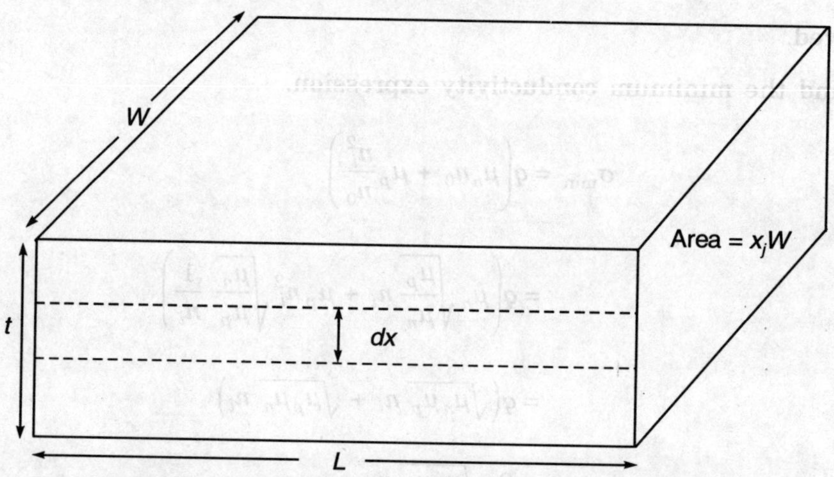

FIGURE 3.6 Geometry of a semiconductor resistor.

Combining both,

$$R \propto \frac{L}{A}$$

$$R = \rho \frac{L}{A} \tag{3.23}$$

where proportionality constant ρ is defined as resistivity of the semiconductor.

The cross-sectional area A of the bar is:

$$A = Wt$$

Hence, Eq. (3.23) can be written as:

$$R = \rho \left(\frac{L}{Wt} \right) \tag{3.24}$$

For $L = W$, the bar reduces to square geometry. Quantity (ρ/t) is defined as the sheet resistance and represented as $R_{\text{sheet}} = R_\square$. In terms of sheet resistance, Eq. (3.24) can be written as:

$$R_\square = R_{\text{sheet}} = \left(\frac{\rho}{t}\right) \text{ ohm}/\square \qquad (3.25)$$

Therefore, for a resistor with length L and width W, the resistance of the slab is given as:

$$R = R_\square (L/W) \qquad (3.26)$$

Where (L/W) gives the number of squares as shown in Fig. 3.7.

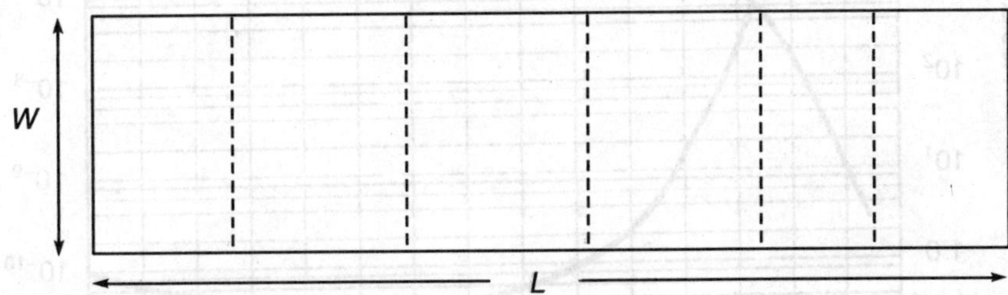

FIGURE 3.7   Representation of number of square in a semiconductor resistor.

In Fig. (3.7), the number of square is 6. Sheet resistivity does not depend on the size of the square. The most common method to measure the sheet resistivity of a thin film is four-probe method.

As the temperature increases, the resistivity of aluminium (Al), which is a metal, increases due to increase in phonon scattering at higher temperatures, whereas for semiconductor, the resistivity decreases. The opposite temperature dependence of the resistivity is one major difference between metals and semiconductors. The temperature dependence curve for Al and p-type Ga is shown in Fig. 3.8.

EXAMPLE 3.4

Consider a silicon bar of uniformly doped with acceptor impurities. The geometry of the bar is same as shown in Fig. 3.6. A current of 2 mA is flowing in the bar when a 5 V battery is connected across the bar. Current density should not be larger than 100 A/cm². Find the cross-sectional area and length.

Take $N_a = 10^{16}/\text{cm}^3$, $\mu_p \simeq 400 \text{ cm}^2/\text{v-s}$.

Solution: $J = \dfrac{I}{A}$ or $A = \dfrac{I}{J} = \dfrac{2\ mA}{100} = 2 \times 10^{-5} \text{ cm}^2$

The resistor of the sheet is given by

$$R = \frac{V}{I} = \frac{5}{2 \times 10^{-3}} = 2.5 \text{ k}\Omega$$

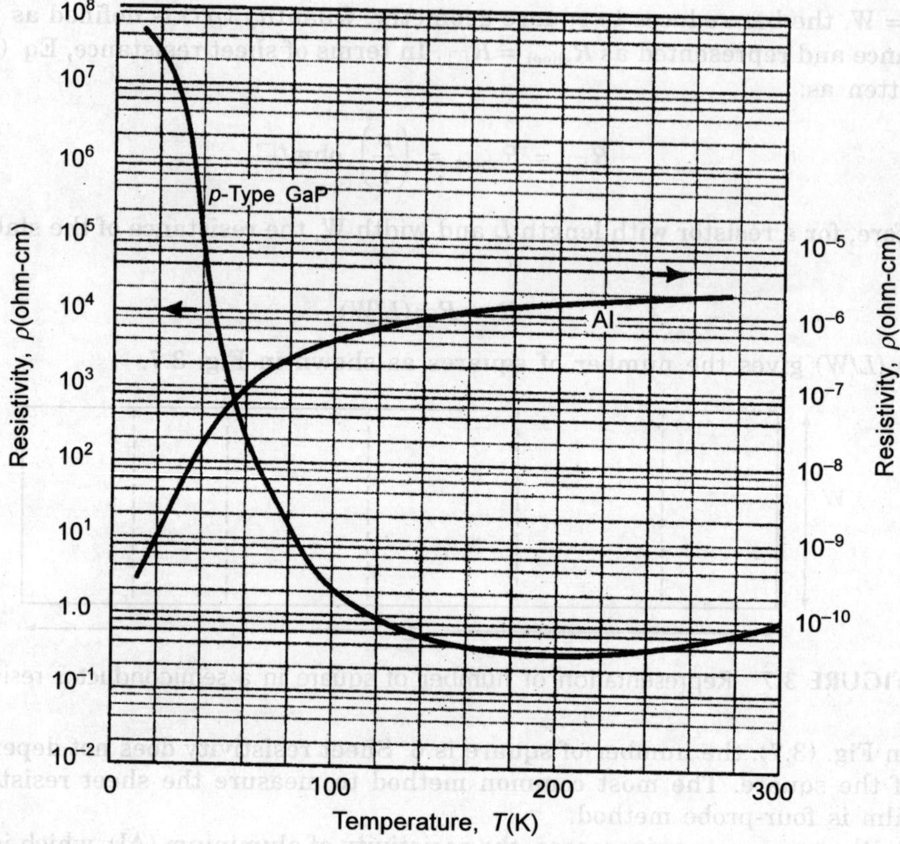

FIGURE 3.8 Plot of resistivity with temperature for metal and semiconductor materials.

From Eq. (3.23),

$$R = \rho \frac{L}{A} \simeq \frac{1}{q\mu_p p}\frac{L}{A}$$

$$= \frac{1}{q\mu_p N_a}\frac{L}{A}$$

or

$$2.5\ \text{k}\Omega = \frac{1}{1.6 \times 10^{-19} \times 400 \times 10^{16}} \times \frac{L}{2 \times 10^{-5}\ \text{cm}^2}$$

or

$$L = 3.2 \times 10^{-2}\ \text{cm}$$

3.4 HALL EFFECT

Hall effect is used to get information about the carriers in semiconductor materials. Using Hall effect, one can determine whether electrons or holes are primarily responsible for charge transport, i.e., whether the semiconductor material is n-type or p-type. It is also used to measure the carrier concentration. Hall effect is one of the convencing method to show the existence of the holes as charge carriers. Figure 3.9 shows a semiconductor of bar of length L, width W and thickness d.

An electric field is applied along the x-axis and magnetic field along the z-direction. Electrons and holes which are moving in the semiconductor bar will experience force as shown in Fig. 3.9. In a p-type material, ($n_0 << p_0$), there will be a positive charge on the surface $y = 0$ and similarly, for n-type semiconductor ($n_0 >> p_0$), negative charge will build up at the surface $y = 0$. This net build up charges induces an electric field along the y-direction as shown in Fig. 3.9.

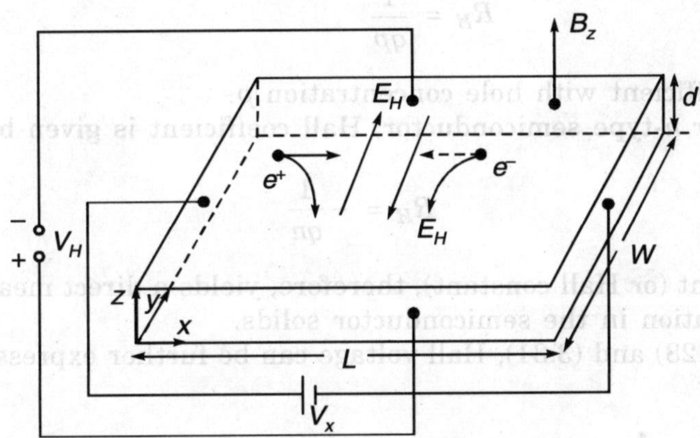

FIGURE 3.9 Hall effect in semiconductors material.

In the steady state, since, there is no net current flow along the y-direction, the induced electric field force will be exactly balanced by the Lorentz force, i.e.,

$$qE_y = - q\mathbf{v} \times \mathbf{B}$$

or

$$qE_y = qv_xB_z \tag{3.27}$$

The induced electric field along the y-direction is called Hall field E_H. The voltage produced by the Hall field in the semiconductor is called **Hall voltage**, and is given by

$$V_H = E_HW \tag{3.28}$$

where electric field E_H is assumed to be positive in y-direction.

In a p-type semiconductor, the Hall voltage will be positive and in n-type semiconductor, Hall voltage will be negative. By knowing the polarity of the Hall voltage, the type of semiconductor material can be easily determined.

Substituting Eq. (3.28) into Eq. (3.27), we have

$$V_H = v_xB_zW \tag{3.29}$$

Let us consider a p-type semiconductor in which the current density is:

$$J_p = q \, v_{dp}p \tag{3.30}$$

Putting Eq. (3.30) in Eq. (3.29), we obtain

$$E_y = E_H = \left(\frac{J_{px}}{qp}\right)B_z$$

$$= \left(\frac{1}{qp}\right)J_{px}B_z$$

$$= R_H J_{px} B_z \tag{3.31}$$

where
$$R_H = \frac{1}{qp} \tag{3.32a}$$

is called Hall coefficient with hole concentration p.

Similarly, for n-type semiconductor, Hall coefficient is given by

$$R_H = -\frac{1}{qn} \tag{3.32b}$$

The **Hall coefficient (or Hall constant)**, therefore, yields a direct measure of Hole and electron concentration in the semiconductor solids.

Using Eq. (3.28) and (3.31), Hall voltage can be further expressed as (for p-type semiconductor):

$$V_H = \left(\frac{J_x}{qp}\right)WB_z = \left(\frac{I_x}{Aqp}\right)WB_z$$

$$= \left(\frac{I_x}{Wdqp}\right)WB_z = \frac{I_x B_z}{qpd} \tag{3.33a}$$

where I_x is the current component in x-direction.

From Eq. (3.33a), hole concentration is determined by

$$p = \left(\frac{I_x B_z}{qdV_H}\right) \tag{3.33b}$$

Similarly, concentration of electrons in n-type semiconductor is:

$$n = -\left(\frac{I_x B_z}{qdV_H}\right) \tag{3.33c}$$

where, all the quantities can be experimentally determined. Since, Hall voltage for n-semiconductor is negative, therefore, the electron concentration obtained by Eq. (3.33c) is a positive quantity. Therefore, from Eqs. (3.33b) and (3.33c) it is clear that the electron concentration in n-type semiconductor and hole concentration in

p-type semiconductor can be determined by measuring the current through the respective bar in x-direction, magnetic field value in z-direction and Hall voltage.

Once, the value of majority carrier concentration has been determined, the low field mobility can be easily calculated.

For a p-type semiconductor material,

$$J_x = \sigma E = q\mu_p p E_x$$

$$\frac{I_x}{A} = q\mu_p p E_x$$

$$\text{or} \qquad \mu_p = \frac{I_x L}{qp V_x W d} \qquad\qquad (3.34)$$

Similarly, one can obtain the mobility of electron in n-type semiconductor.

Therefore, it is concluded that by Hall experiment one can determine the type of semiconductor material (either n-type or p-type), majority carrier concentration and their low field mobilities.

If both types of charges like electrons and holes contribute to the transport process in a semiconductor material, the Hall coefficient of the compensated semiconductor at thermal equilibrium is given by

$$R_H = \frac{1}{q} \frac{[p\mu_p^2 - n\mu_n^2]}{[p\mu_p + n\mu_n]^2}$$

where p and n are concentration of holes and electrons respectively. Hall mobility for compensated semiconductor is

$$\mu_H = \sigma|R_H| = \left| \frac{p\mu_p^2 - \mu_n^2 n}{p\mu_p + n\mu_n} \right|$$

EXAMPLE 3.5

A 10 cm long and 0.2 cm × 0.1 cm cross section sample of germanium has an electron concentration of $10^{17}/cm^3$ at $T = 300$ K. The sample is used in Hall measurement with 64 mA current flowing across the length and a magnetic field of 2×10^3 Gauss applied normal to the 10 cm × 0.1 cm surface. Determine Hall voltage and Hall coefficient. Assume a drift field of 80 meV/cm in the sample.

Solution: Using Eq. (3.32b),

$$R_H = -\frac{1}{qn_0}$$

$$= -\frac{1}{1.6 \times 10^{-19} \times 10^{17}} = -62.5 \text{ cm}^3/\text{C}$$

Using Eq. (3.33c),

$$n_0 = -\frac{I_x B_z}{qV_H \times d}$$

or
$$V_H = -\frac{I_x B_z}{qn_0 d}$$

or
$$V_H = -\frac{64 \times 10^{-3} \times 2 \times 10^{-5}}{1.6 \times 10^{-19} \times 10^{17} \times 0.1}$$

$$= -\frac{128 \times 10^{-8} \times 10^2}{0.16} = -800 \times 10^{-16}$$

$$= -800 \times 10^{-4}$$

$$= -0.8 \text{ mV}$$

3.5 CARRIER DIFFUSION

3.5.1 Diffusion Process

Another important component in semiconductor material, other than drift current, is diffusion current. This current is due to spatial variation of the carrier concentration. Carriers always move from higher concentration to lower concentration.

To understand diffusion process in a semiconductor, consider an electron concentration which varies in one dimension, i.e., along x-direction as shown in Fig. 3.10.

The temperature is assumed to be uniform so that the average thermal velocity of the electrons is independent of x. To calculate the current due to concentration gradient, we will determine the net number of electrons crossing the plane at $x = 0$ per unit time and per unit area. Due to finite temperature electrons move randomly with an average thermal velocity v_{Th}. The average distance an electron travels between collisions is l. This distance is called **mean free path**, and is given by

$$l = \tau v_{Th}$$

where τ is the mean free time.

The electron at $x = -l$ (i.e., one mean free path left to the surface $x = 0$) has same probability of moving right or left. Then, in one mean free time τ, half of them will move across the plane at $x = 0$. In other words, one half of electrons at $x = -l$ will be moving to the right at any instant of time and one half of the electrons at $x = +l$ will be moving to the left at any instant of time.

Therefore, the net rate of flow of electrons F_n in +x-direction at $x = 0$ is given by

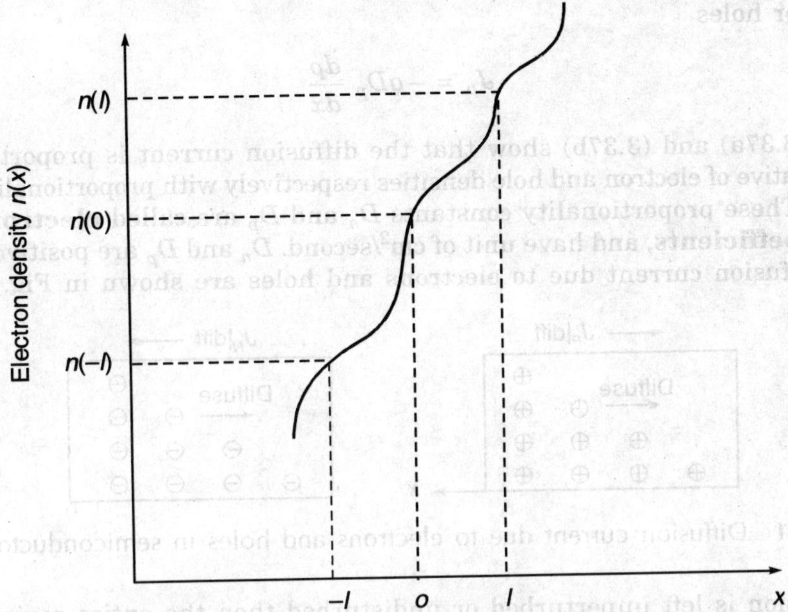

FIGURE 3.10 Electron density variation along x-direction.

$$F_n = \left[\frac{1}{2}\, v_{Th}\, n(-l) - \frac{1}{2}\, v_{Th}\, n(+l) \right] \tag{3.35}$$

Approximating the densities at $x = \pm l$ by the two terms of a Taylor's series expansion, we have

$$F_n = \frac{1}{2}\, v_{Th} \left[\left\{ n(0) - l\,\frac{dn}{dx} \right\} - \left\{ n(0) + l\,\frac{dn}{dx} \right\} \right]$$

or

$$F_n = - v_{Th}\, l\, \frac{dn}{dx}$$

$$= - D_n\, \frac{dn}{dx} \tag{3.36a}$$

where $D_n = v_{Th}\, l$, called the **diffusion coefficient** or **diffusivity**. Equation (3.36a) tells that due to concentration gradient, electrons move from higher concentration to lower concentration and results in diffusion current. This current is given by

$$J_n = -\, qF_n \tag{3.36b}$$

Substituting Eq. (3.36a) into Eq. (3.36b), we get

$$J_n = qD_n\, \frac{dn}{dx} \tag{3.37a}$$

Similarly, for holes

$$J_p = -qD_p \frac{dp}{dx} \tag{3.37b}$$

Equations (3.37a) and (3.37b) show that the diffusion current is proportional to the spatial derivative of electron and hole densities respectively with proportionality constants D_n and D_p. These proportionality constants D_n and D_p are called **electron and hole diffusion coefficients**, and have unit of cm^2/second. D_n and D_p are positive quantities. The diffusion current due to electrons and holes are shown in Fig. 3.11.

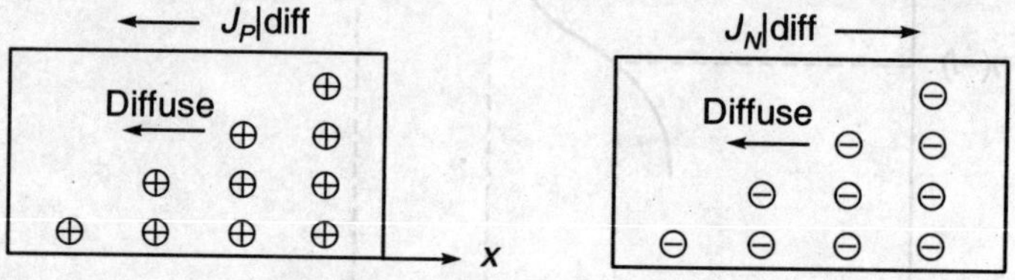

FIGURE 3.11 Diffusion current due to electrons and holes in semiconductor material.

If diffusion is left unperturbed or undisturbed then the entire region will have uniform doping after sometime, and diffusion process will stop.

Diffusion process does not need any external field to apply to the semiconductor. Diffusion is a minority carrier process. The particles moves about using their thermal motion.

EXAMPLE 3.6

Calculate the diffusion current density in an n-type Si semiconductor material at room temperature where electron concentration varies linearly from 1×10^{18}/cm^3 to 7×10^{17}/cm^3 over a distance of 0.1 cm. Assume $D_n = 22.5$ cm^3/s.

Solution: Using Eq. (3.37a), we have

$$J_n = qD_n \frac{dn}{dx}$$

$$= 1.6 \times 10^{-19} \times 22.5 \times \frac{10^{18} - 0.7 \times 10^{18}}{0.1}$$

$$= 10.8 \text{ A/cm}^2$$

3.5.2 Induced Electric Field (Non-homogeneous Semiconductor Material)

Consider a non-uniformly doped n-type semiconductor with donor impurities. The variation of donor impurity concentration with x is shown in Fig. 3.12.

Figure 3.12 shows that the concentration increases along the bar length. This results in a concentration gradient and a diffusion of electrons from the region of higher concentration to lower concentration starts, i.e., in the $-$ve x-direction. The

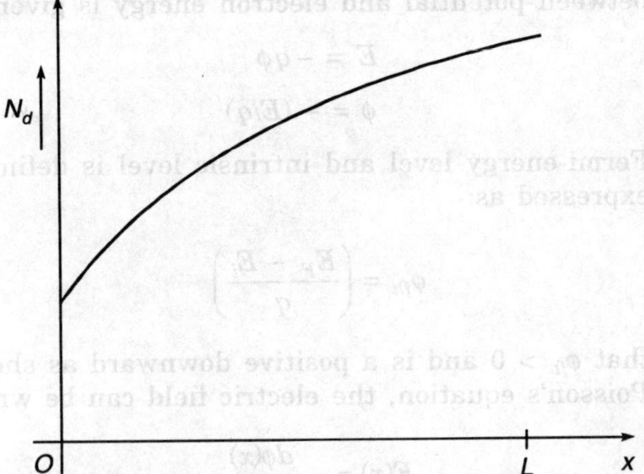

FIGURE 3.12 Variation of donor impurity concentration along x-direction.

movement of electrons from higher concentration to lower concentration will leave behind positively charged immobile donor atoms at the higher concentration side. The separation of negative charge (due to mobile electrons) and positive charge (due to immobile ionized donor atoms) induces an electric field. This induced electric field always opposes the diffusion process and causes drift current.

At thermal equilibrium, the drift current and diffusion currents in the material become equal and the induced electric field prevents any further separation of charges. The energy band diagram for a non-uniformly doped semiconductor material at the thermal equilibrium is shown in Fig. 3.13.

As seen in Fig. 3.13 that fermi level is spatially constant throughout the material because of thermal equilibrium condition.

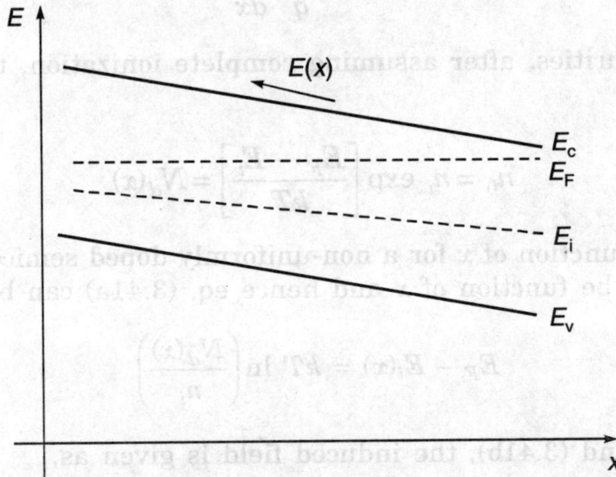

FIGURE 3.13 Band diagram in a non-homogeneous semiconductor material.

The relation between potential and electron energy is given by

$$E = -q\phi \tag{3.38a}$$

or

$$\phi = -(E/q) \tag{3.38b}$$

The difference in Fermi-energy level and intrinsic level is defined as the potential difference, and is expressed as:

$$\phi_{fn} = \left(\frac{E_F - E_i}{q}\right) \tag{3.39}$$

Since, $E_f > E_i$, so that $\phi_{fn} > 0$ and is a positive downward as shown in Fig. 3.13.

According to Poisson's equation, the electric field can be written as:

$$E(x) = -\frac{d\phi(x)}{dx} \tag{3.40a}$$

Therefore, due to non-uniform doping in the semiconductor material, the induced electric field is given (using Eqs. (3.39) and (3.40a)) as:

$$E(x) = -\frac{d}{dx}\left(\frac{E_F - E_i}{q}\right)$$

or

$$E(x) = -\frac{1}{q}\frac{dE_F}{dx} + \frac{1}{q}\frac{dE_i}{dx} \tag{3.40b}$$

The first term of R.H.S. of Eq. (3.40b) is zero because material is at thermal equilibrium, and hence

$$E(x) = \frac{1}{q}\frac{dE_i}{dx} \tag{3.40c}$$

For donor impurities, after assuming complete ionization, the majority carrier concentration is:

$$n_0 = n_i \exp\left[\frac{E_F - E_i}{kT}\right] = N_d(x) \tag{3.41a}$$

Since, N_d is a function of x for a non-uniformly doped semiconductor, therefore, $E_F - E_i$ should also be function of x and hence eq. (3.41a) can be expressed as

$$E_F - E_i(x) = kT \ln\left(\frac{N_d(x)}{n_i}\right) \tag{3.41b}$$

Using Eqs. (3.40c) and (3.41b), the induced field is given as,

$$E(x) = -\left(\frac{kT}{q}\right)\frac{1}{N_d(x)}\frac{d}{dx}[N_d(x)] \tag{3.42}$$

The induced field is also called **Built-in field**, and is important for establishing equilibrium in non-uniformly doped semiconductor material in absence of applied external electric field.

EXAMPLE 3.7

An intrinsic Si sample is doped with donors from one side such that $N_d = N_0 \exp(-bx)$

(a) Find the expression for induced field at equilibrium over the range for which $N_d \gg n_i$.
(b) Find the value of induced field when $a = 1/\mu\text{m}$.
(c) Sketch the band diagram and variation of concentration with x.

Solution: (a) Using Eq. (3.42b), we have

$$E(x) = -\left(\frac{kT}{q}\right)\frac{1}{N_0 e^{-ax}} \times -aN_0 e^{-ax} = \frac{kT}{q}\,a$$

(b)
$$E(x) = 26 \text{ mV} \times 10^4 = 26 \text{ V/cm}$$

(c) The sketches of band diagram and variation of concentration with x are shown in Figs. 3.14(a) and (b).

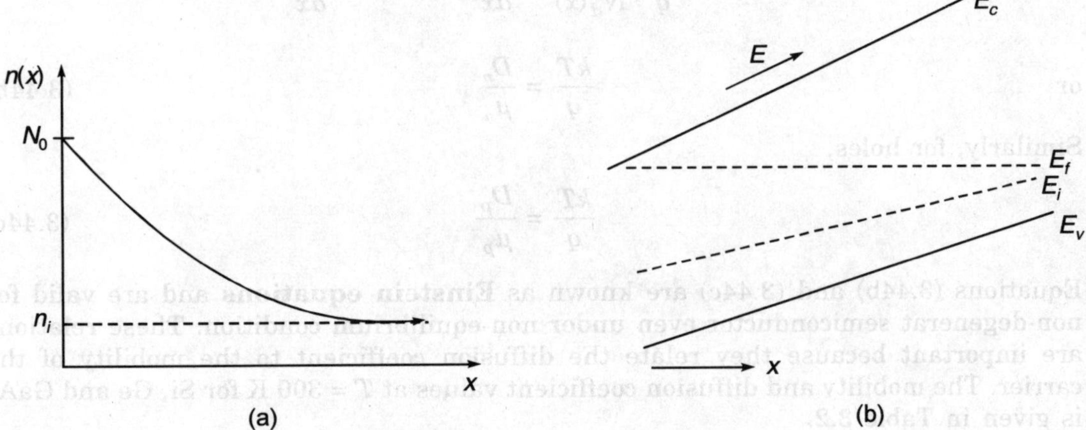

FIGURE 3.14 Sketch of band diagram and variation of concentration with x.

3.5.3 The Einstein Equation

Consider a non-uniformly doped semiconductor material whose band diagram is shown in Fig. 3.13. Let us assume that no external field is applied on the semiconductor. The non-uniform doping or concentration gradient in the material results in diffusion current. Due to the concentration gradient an induced electric field or built-in field will develop in the semiconductor as clear from Eq. (3.42b). This induced electric field causes drift of the carriers even in the absence of external field and gives drift current. In absence of applied external electric field, material acquires thermal equilibrium when diffusion current component is equal to drift current component. At thermal equilibrium, net current flow is zero.

For electron at thermal equilibrium,

$$J_n = \text{Drift component of current plus diffusion component of current}$$
$$= 0$$

or

$$J_n = q\mu_n n_0 \, E(x) + qD_n \frac{dn_0}{dx} = 0$$

or

$$q\mu_n n_0 \, E(x) = -qD_n \frac{dn_0}{dx} \tag{3.43}$$

Assuming quasi-charge neutrality condition, i.e., $n_0 \simeq N_d(x)$, Eq. (3.43) can be further written as

$$\mu_n N_d(x) E(x) = -D_n \frac{dN_d(x)}{dx} \tag{3.44a}$$

Using Eq. [3.42(b)] in Eq. [3.44(a)], we have

$$-\mu_n N_d(x) \frac{kT}{q} \frac{1}{N_d(x)} \frac{dN_d(x)}{dx} = -D_n \frac{dN_d(x)}{dx}$$

or

$$\frac{kT}{q} = \frac{D_n}{\mu_n} \tag{3.44b}$$

Similarly, for holes,

$$\frac{kT}{q} = \frac{D_p}{\mu_p} \tag{3.44c}$$

Equations (3.44b) and (3.44c) are known as **Einstein equations** and are valid for non-degenerat semiconductor even under non-equilibrium condition. These relations are important because they relate the diffusion coefficient to the mobility of the carrier. The mobility and diffusion coefficient values at $T = 300$ K for Si, Ge and GaAs is given in Table 3.2.

Table 3.2 Mobility and Diffusion Coefficient Values

Material	cm²/Vs		cm²/Vs	
	μ_n	μ_p	D_n	D_p
Si	1350	480	35	12.4
Ge	8500	400	220	10.4
GaAs	3900	1900	101	49.2

EXAMPLE 3.8

Minority carrier holes are injected into a homogeneous n-type semiconductor sample at one point. An electric field of 50 V/cm is applied across the sample, and the carriers move 1 cm in 100 μs in the presence of field. Find the drift velocity and diffusivity of the minority carriers.

Solution:
$$v_{dp} = \frac{1 \text{ cm}}{100 \ \mu s} = 10^4 \text{ cm/s}$$

$$\mu_p = \frac{v_p}{E} = \frac{10^4}{50} = 200 \text{ cm}^2/\text{Vs}$$

Using Eq. (3.44c),

$$D_p = \left(\frac{kT}{q}\right) \mu_p = \text{Diffusivity}$$

$$= (26 \text{ mV}) \times 200 = 5.2 \text{ cm}^2/\text{Vs}$$

3.6   CARRIER GENERATION AND RECOMBINATION

The operation of optoelectronic devices like photodiodes, laser etc. depends on the generation-recombination process of carriers. These two processes are due to the interaction of electron with lattice of crystal or with the phonon. The generation and recombination processes involve movement of electrons from the valence band to conduction band or conduction band to valence band. In these processes either electrons gain energy or lose energy.

The generation and recombination processes are complement of each other. When valence electrons in semiconductor material get sufficient larger energy than the forbidden gap of material due to external excitation, then valence electrons jump to conduction band. In this process, free electrons are created in the conduction band and same number of holes are created in the valence band. The creation of electron in the conduction band and hole in the valence band is known as Electron-hole pair (EHP) generation or simply generation. The generation process can be achieved either by thermal excitation, optical excitation, ionization or multiplication. There are actually two dominating generation processes, thermal generation and optical generation (due to incident of light with energy $hv \geq E_g$). The optical generation occurs in non-equilibrium condition, i.e., due to fall of light on the semiconductor material, equilibrium disturbed. The thermal generation can occur in equilibrium as well as non-equilibrium conditions. The generation process is expressed in terms of generation rate which is defined as the number of electron-hole pairs per unit volume per unit time. The thermal generation rate and optical generation rate are represented by G_{Th} and G_L respectively. The unit of generation rate in $\text{cm}^{-3}\text{sec}^{-1}$.

For each generation process, there is recombination process. In recombination process, free electrons fall from conduction band to the valence band and recombine with holes. In this process both carriers annihilate each other. The energy equivalent to the difference between initial and final state of electron is given off.

When an electron jumps directly into the valence band from the conduction band, then the process is known as radiative recombination or direct recombination process. In radiative recombination process the excess energy is emitted in form of photon. The other types of recombination process is shown in Fig. 3.15. In non-radiative recombination, energy is observed. The radiative recombination process is also known as band-to-band recombination. Trap-assisted recombination or non-radiative recombination occurs when an electron falls into a "Trap state". The electron in the trap state can fall into an empty state in valence band in the second attempt; thereby completing the recombination process.

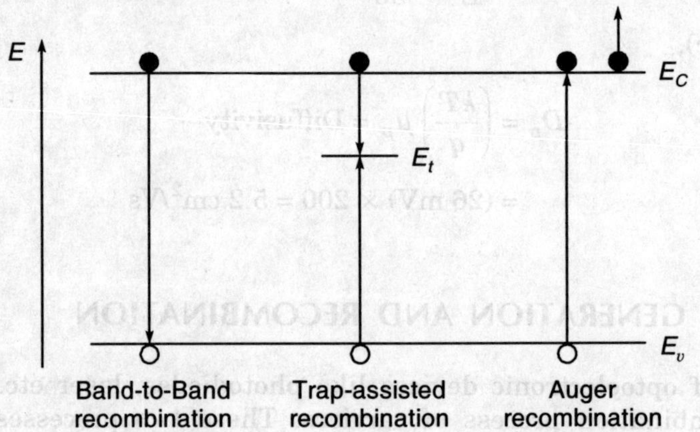

FIGURE 3.15 Representation of carrier recombination processes.

Auger recombination is explained in detail in section 3.6.4.

In direct band-to-band recombination process at thermal equilibrium;

Electron generation rate in conduction band = Hole generation rate in valence band

$$i.e., \quad G_{nTh} = G_{pTh} \tag{3.45a}$$

Similarly, argument is true for recombination rates,

$$i.e., \quad R_{nTh} = R_{pTh} \tag{3.45b}$$

Since, at thermal equilibrium,

$$\text{Generation rate} = \text{Recombination rate}$$

Therefore, $$G_{nTh} = G_{pTh} = R_{nTh} = R_{pTh} \tag{3.45c}$$

At thermal equilibrium,

$$\text{Generation rate} \propto \text{Electron concentration in conduction band}$$

and $$\text{Generate rate} \propto \text{Hole concentration in valence band}$$

Combining these two,

$$G_{Th} \propto n_0 p_0$$

$$= \alpha_r n_0 p_0 = \alpha_r n_i^2(t) \tag{3.45d}$$

This expression shows that generation or recombination rate is the function of temperature, and for a given temperature, both rates are equal to square of the intrinsic carrier constant at that temperature.

3.6.1 Excess Carrier Generation

When a light of energy $h\nu \geq E_g$ falls on the semiconductor material, then electrons are excited from valence band to conduction band. Due to that extra electrons and holes (than their equilibrium value) are created in the conduction band and valence band respectively. This additional electrons in conduction band and holes in valence band are called **excess-electrons** and **excess-holes**. In this case the mass action equation $n_0 p_0 = n_i^2$ does not hold and material is said to be in non-equilibrium condition. The process of introducing excess carriers in the semiconductor is called **carrier injection**. Injection increases the electrons and holes concentration in their respective bands above their thermal equilibrium values.

The excess electron concentration Δn is given by

$$\Delta n = n - n_0 \tag{3.46a}$$

where n is non-equilibrium electron concentration and n_0 is the thermal equilibrium concentration.

Equation (3.46a) can also be written as:

$$n = n_0 + \Delta n \tag{3.46b}$$

Similarly, for holes, the excess concentration Δp is:

$$\Delta p = p - p_0 \tag{3.47a}$$

where p is non-equilibrium hole concentration and p_0 is thermal equilibrium hole concentration. Eq. (3.47a) can be further written as:

$$p = p_0 + \Delta p \tag{3.47b}$$

The magnitude of excess-carrier concentration relative to the thermal equilibrium majority carrier concentration will determine the level of injection. For example, let us assume a n-type semiconductor material doped with 10^{16} donor atoms/cm^3. It is assumed that all the donors are ionized means thermal equilibrium majority carrier (electrons) concentration is 10^{16}/cm^3 and minority carrier hole concentration is 10^4/cm^3.

Let us assume that excess carriers are introduced in the material due to photoexcitation technique. Now, the excess of electron concentration Δn must be equal to excess-hole concentration because electrons and holes are produced in pairs.

i.e.,
$$\Delta n = \Delta p$$

Also, charge neutrality must hold.

If $10^{12}/cm^3$ excess minority carrier holes are introduced into the n-type semiconductor due to photoexcitation then hole concentration at non-equilibrium becomes

$$p = 10^4 + 10^{12} \simeq 10^{12}/cm^3$$

Similarly, non-equilibrium electron concentration is:

$$n = 10^{16} + 10^{12} \simeq 10^{16}/cm^3$$

The percentage change in the majority carrier concentration due to injection of excess electron concentration is about 0.01% which is very small and hence, its contribution for majority carrier is neglected; whereas, for minority carrier concentration, a large increase is observed from their equilibrium concentration.

The condition, in which the excess-carrier concentration is small compared to thermal-equilibrium majority carrier concentration, is called **low-level injection**.

A high-level injection is that in which excess carrier concentration is comparable or larger than majority carrier thermal equilibrium concentration.

From this discussion, it is concluded that in the low-level injection only minority carrier concentration gets changed after the injection whereas majority carrier concentration gets changed in high-level injection.

Whenever, thermal equilibrium condition is disturbed, a process exists to restore the system to equilibrium.

In case of low-level injection, the system restores the equilibrium either by radiative recombination or non-radiative recombination process of excess minority carriers with the majority carriers.

3.6.2 Direct Recombination

Let us consider a direct semiconductor material at thermal equilibrium. As discussed in Section 3.5, due to thermal energy valence electron will be excited to conduction band, leaving one hole in valence band. After finite time, the electron makes a transition from conduction band to valence band, and electron-hole pair is annihilated due to recombination process. Under the thermal equilibrium condition, the generation rate must be equal to recombination rate to maintain the constant carrier concentration so that $n_0 p_0 = n_i^2$.

Therefore, at thermal equilibrium,

$$G_{Th} = R_{Th} \tag{3.48}$$

This process is illustrated in Fig. 3.16.

Now, a light of energy $h\nu \geq E_g$ is incident on the direct semiconductor material as shown in Fig. 3.16. This causes introduction of excess carriers into the material and disturb the equilibrium condition. Due to shine of light on the semiconductor, electron-hole pairs are produced at the rate of G_L as shown in Fig. 3.16. In direct semiconductor material, there is high probability that the excess electrons and holes recombine directly because the bottom of the conduction band and top of the valence band are lined up, and momentum is conserved in the transition. Since the direct recombination is a spontaneous process, therefore, the probability of an electron and hole recombination is constant with time.

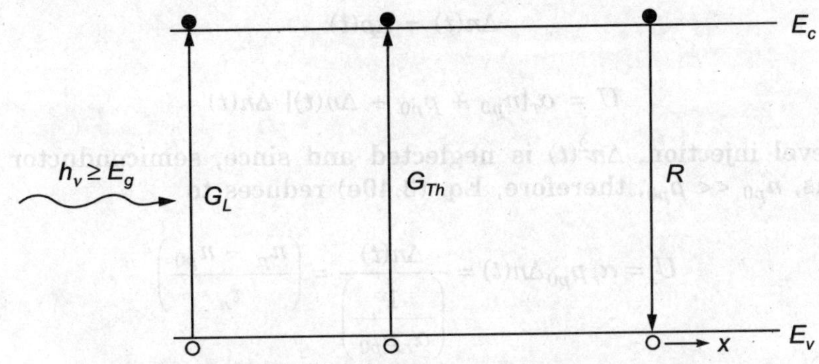

FIGURE 3.16 Representation of direct recombination.

The rate at which electron recombines must be proportional to the electron concentration in conduction band at any time t and hole concentration in the valence band at the same time.

i.e.,

$$R \propto n(t)$$
$$\propto p(t)$$

or

$$R \propto n(t)p(t)$$
$$R = \alpha_r n(t)p(t) \qquad (3.49a)$$

where α_r is proportionality constant.

The net rate of change of electron concentration in a p-type semiconductor is given by

$$\frac{dn_p}{dt} = G - R = \text{Generation rate} - \text{Recombination rate} \qquad (3.49b)$$

where $G = G_L + G_{Th}$

$$= \text{Optical generation rate} + \text{Thermal generation rate} \qquad (3.49c)$$

If the semiconductor material is not exposed in the light then $G_L = 0$.

For steady-state condition, $\dfrac{dn_p}{dt} = 0$, therefore, Eq. (3.49b) reduces to

$$G - R = 0 \quad \text{or} \quad G = R$$

or

$$G_L = R - G_{Th} = U \qquad (3.49d)$$

where U represents the net recombination rate.

Substituting values of R from Eq. (3.49a) and $G_{Th} = \alpha_r n_{p0}p_{p0}$ in Eq. (3.49d), we have

$$U = \text{Net recombination rate} = \alpha_r[n(t)p(t) - n_{p0}p_{p0}]$$
$$= \alpha_r[(n_{p0} + \Delta n(t))(p_{p0} + \Delta p(t)) - n_{p0}p_{p0}]$$
$$= \alpha_r[n_{p0}\Delta p(t) + p_{p0}\Delta n(t) + \Delta n(t)\Delta p(t)]$$

Since $$\Delta n(t) = \Delta p(t)$$
Hence

$$U = \alpha_r[n_{p0} + p_{p0} + \Delta n(t)]\ \Delta n(t) \tag{3.49e}$$

Using low-level injection, $\Delta n^2(t)$ is neglected and since, semiconductor material is p-type means, $n_{p0} \ll p_{p0}$, therefore, Eq. (3.49e) reduces to

$$U = \alpha_r p_{p0} \Delta n(t) = \frac{\Delta n(t)}{\left(\dfrac{1}{\alpha_r p_{p0}}\right)} = \left(\frac{n_p - n_{p0}}{\tau_n}\right) \tag{3.49f}$$

Equation (3.49f) shows that the net recombination rate is proportional to the excess minority carrier concentration and is inversely proportional to minority carrier life time. At thermal equilibrium, net recombination rate $U = 0$.

Similarly, the net recombination rate for holes in n-type semiconductor material is:

$$U = \left(\frac{p_n - p_{n0}}{\tau_p}\right) \tag{3.49g}$$

Therefore, from Eq. (3.49d),

$$G_L = \text{Optical generation rate}$$

$$= \left(\frac{p_n - p_{n0}}{\tau_p}\right) \text{ for holes in } n\text{-type}$$

$$= \left(\frac{n_p - n_{p0}}{\tau_n}\right) \text{ for electrons in } p\text{-type}$$

If at any time $t = 0$, the light is suddenly turned off, means, $G_L = 0$, (called dark case), the net rate of change of electron population in p-type semiconductor material is:

$$\frac{dn(t)}{dt} = G_{Th} - R$$

$$= \alpha_r[n_0 p_0 - n(t)p(t)] \tag{3.50}$$

From Eqs. (3.46b) and (3.47d), Eq. (3.50) becomes

$$\frac{d}{dt}[n_{p0} + \Delta n(t)] = \alpha_r[n_{p0} p_{p0} - (n_{p0} + \Delta n(t))(p_{p0} + \Delta p(t))]$$

or $$\frac{d\Delta n(t)}{dt} = \alpha_r[n_{p0} p_{p0} - n_{p0} p_{p0} - n_{p0}\Delta p(t) - \Delta n(t)\, p_{p0} - \Delta n(t)\, \Delta p(t)]$$

Taking $$\frac{dn p_0}{dt} = 0$$

Using $\Delta n(t) = \Delta p(t)$ to maintain overall charge neutrality,

$$\frac{d\Delta n(t)}{dt} = -\alpha_r[(n_{p0} + p_{p0})\Delta n(t) + \Delta^2 n(t)] \tag{3.51}$$

Equation (3.51) is a non-linear equation and can be solved by imposing the low-level injection condition, i.e., neglecting second term $[\Delta^2 n(t)]$ of right-hand side of Eq. (3.51).

Hence Eq. (3.51) reduces to

$$\frac{d\Delta n(t)}{dt} = -\alpha_r(n_{p0} + p_{p0})\Delta n(t) \tag{3.52}$$

Since in p-type semiconductor material $n_{p0} \ll p_{p0}$, Eq. (3.52) can be written as:

$$\frac{d}{dt}(\delta_n(t)) = \frac{d\Delta n(t)}{dt} = -\alpha_r p_{p0}\Delta n(t) \tag{3.53a}$$

Solution of differential Eq. (3.53a) under suitable boundary conditions is given as:

$$\Delta n(t) = \Delta n(o)e^{-t/\tau_{n0}} \tag{3.53b}$$

where

$$\tau_{n0} = \frac{1}{\alpha_r p_{p0}} \tag{3.53c}$$

is called **excess minority carrier life time** or **recombination life time** of electron in p-type semiconductor. This is constant for the low-level injection. Equation (3.53b) describes the decay of excess minority carrier electrons in p-type semiconductor.

Similarly, the excess minority carrier hole life time in a n-type semiconductor under low-level injection is:

$$\tau_{p0} = \frac{1}{\alpha_r n_{n0}} \tag{3.53d}$$

A general expression for carrier life time is:

$$\tau_n = \frac{1}{\alpha_r(n_0 + p_0)} \tag{3.53e}$$

Equation (3.53e) is valid for n-type or p-type semiconductor if the injection level is low.

Equation (3.53b) can be further written as, using boundary conditions, at $t = 0$ $n_p(t = 0) = n_{p0} + \tau_n G_L$ and $t = \infty$, $n_p(t = \infty) = n_{p0}$.

$$n_p(t) = n_{p0} + \Delta_n(o)\,e^{-t/\tau n}$$

or,

$$n_p(t) = n_{p0} + (\tau_n G_L)\,e^{-t/\tau n} \tag{3.54a}$$

Similarly, for holes in n-type semiconductor material is:

$$p_n(t) = p_{n0} + (\tau_p G_L)\, e^{-t/\tau_p} \qquad (3.54b)$$

The minority carriers recombine with majority carriers and decay exponentially with a time constant τ_p or τ_n.

Figure 3.17 shows the variation of p_n with time.

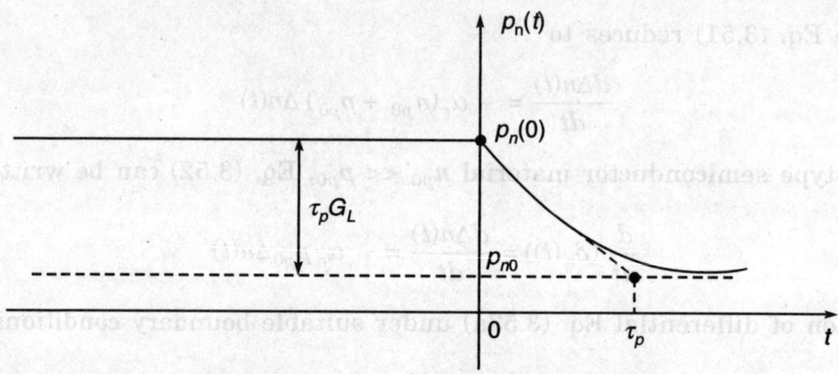

FIGURE 3.17 Variation of p_n with time.

EXAMPLE 3.9

A silicon sample is doped with the donor impurities of $10^{14}/cm^3$, and this sample is illuminated with light. Due to shine of light on the surface, 10^3 electron-hole pairs are created every microsecond. If $\tau_n = i_p = 2\mu s$. Find the minority carrier concentration?

Solution: Since dopant material is donor type, hence resulting silicon sample is n-type. Given $n_{no} = 10^{14}/cm^3$.

Minority carrier hole concentration $p_{n0} = \dfrac{n_i^2}{n_{n0}}$

$$= 2.25 \times 10^6/cm^3$$

Hence, before illumination, $p_{n0} = 2.25 \times 10^6/cm^3$.

After illumination, using Eq. (3.54b),

$$p_n(t) = p_{n0} + (\tau_p G_L)\, e^{-\frac{t}{\tau_p}}$$

Take $t = 0$ sec.,

$$p_n(t) = p_{n0} + \tau_p G_L$$

$$= 2.25 \times 10^6 + 2 \times 10^{-6} \left(\frac{10^{13}}{10^{-6}} \right)$$

$$\cong 2 \times 10^{13}/cm^3$$

3.6.3 Indirect Recombination

For indirect semiconductor materials (e.g., Si and Ge), direct transition from conduction band to valence band is not possible because of violation of momentum conservation. Therefore, the dominant recombination process in such semiconductor is indirect transition via localized energy states or trap states in the forbidden gap. These localized energy states act as a steeping stones between the conduction band and valence band, and is called **recombination centre** or **trapping centre**. The resulting loss in energy by recombining electron is usually given to the lattice in form of heat. Any impurity or crystalline defect in semiconductor will serve as a recombination centre within the forbidden gap if it is capable of receiving a carrier of one type and subsequently capturing the opposite type of carriers and annihilate the pairs. Various transitions that occur in the recombination process through the recombination centre in the indirect semiconductor material is shown in Fig. 3.18. The arrow indicates the transition of electron from one energy level to other.

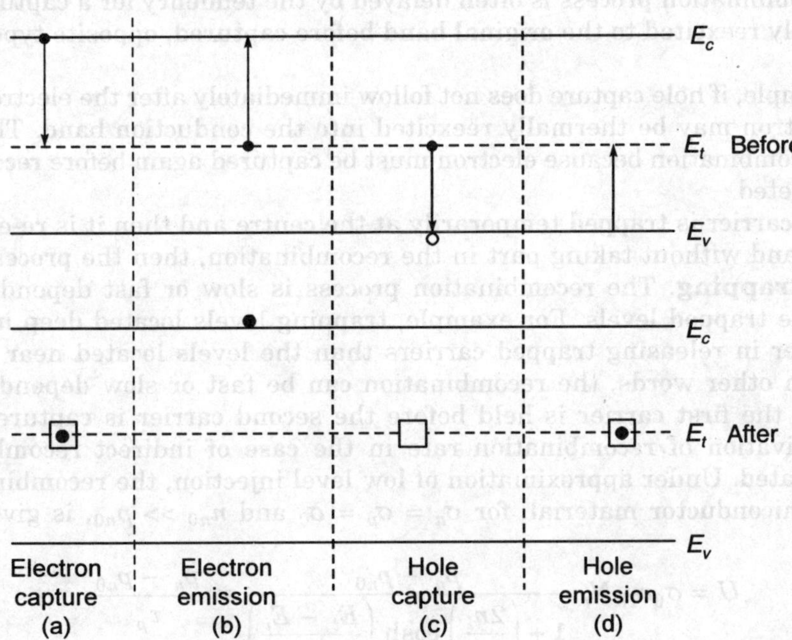

FIGURE 3.18 Indirect generation-recombination processes at thermal equilibrium.

Four basic transition processes are:

- Electron capture
- Electron emission
- Hole capture
- Hole emission

An acceptor type recombination centre is neutral when not occupied by electron and becomes negatively charged when it is occupied by an electron and reverse is true for donor type of recombination centre.

It is assumed that initially recombination centre is neutral because all electrons are occupied by conduction band. When electron jumps from conduction band edge to recombination centre, process is called **electron capture** and centre becomes negatively charged.

In other process, electron makes a transition from recombination centre to conduction band. This process is called **electron emission**, and centre is now neutral.

In third process, electron jumps from recombination centre to valence band leaving behind an empty state in the recombination centre and process is called **hole capture**. In this process, energy is given to the lattice in the form of heat.

Similarly, if an electron moves from valence band edge to unoccupied recombination centre, process is known as **hole emission**.

When both processes like electron capture and hole capture have occurred simultaneously, then electron-hole are annihilated and recombination centre is back to its original state and electron-hole pair is missing in this process. Therefore, after one recombination is over, the centre is ready to participate in another recombination.

The recombination process is often delayed by the tendency for a captured carrier to be thermally reexcited to the original band before captured, opposite type of carrier can occur.

For example, if hole capture does not follow immediately after the electron capture, then the electron may be thermally reexcited into the conduction band. This process delays the recombination because electron must be captured again before recombination can be completed.

When a carrier is trapped temporarily at the centre and then it is re-excited into its original band without taking part in the recombination, then the process is called **temporary trapping**. The recombination process is slow or fast depending on the position of the trapped levels. For example, trapping levels located deep in the band gap are slower in releasing trapped carriers than the levels located near one of the band edge. In other words, the recombination can be fast or slow depending on the average time the first carrier is held before the second carrier is captured.

The derivation of recombination rate in the case of indirect recombination is more complicated. Under approximation of low level injection, the recombination rate in n-type semiconductor material, for $\sigma_n = \sigma_p = \sigma_0$ and $n_{n0} \gg p_{n0}$, is given as:

$$U = \sigma_0 v_{\text{Th}} N_t \frac{p_n - p_{n0}}{1 + \left(\dfrac{2n_i}{n_{n0}}\right) \cosh\left(\dfrac{E_t - E_i}{k_T}\right)} \cong \frac{p_n - p_{n0}}{\tau_p} \tag{3.55}$$

where, v_{Th} is thermal velocity of carriers

σ_n, σ_p are electron and hole cross-sectional area

τ_p depends upon the location of the recombination centres

3.6.4 Auger Recombination

Auger (oh-jay) recombination process results from electron-electron and hole-hole collisions and becomes important in direct band gap materials with higher doping concentration and E_g (band gap) less than about 1.0 eV.

In electron collision process, two electrons collide and one electron will drop to the empty state in the valence band and second electron assumes the energy difference in between the conduction band to valence band. After receiving this energy, second electron becomes an energetic electron and loses its energy to the lattice by scattering events as shown in Fig. 3.19.

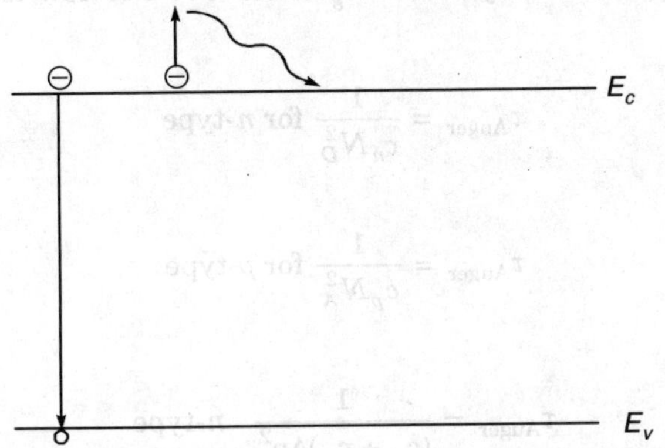

FIGURE 3.19 Auger recombination process involving two electrons and one hole.

In second process, two holes collide in which one hole recombines with electron, coming from conduction band, while second hole is excited into the split-off valence band by the recombination energy equal to forbidden gap E_g as shown in Fig. 3.20.

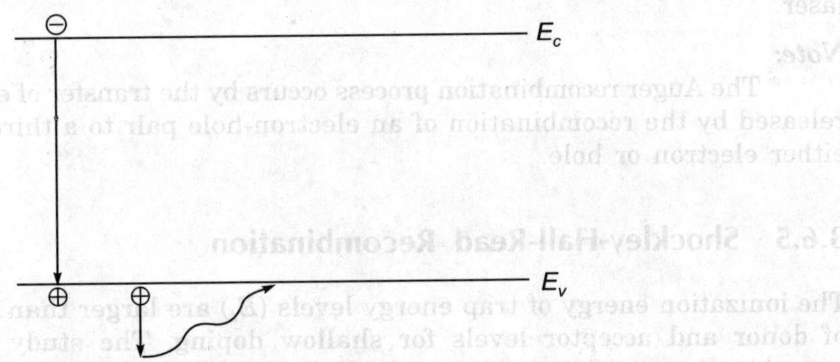

FIGURE 3.20 Auger recombination involving two holes and one electrons.

Therefore, Auger recombination process is three-particle process. The third particle involved in this process will eventually lose its energy to lattice in form of heat.

The process involving two holes and one electron would occur predominantly in heavily doped p-type semiconductor and process involving two electrons and one hole would occur primarily in heavily doped n-type semiconductor.

The total rate of Auger recombination is given by

$$R_{\text{Auger}} = C_n n^2 p + c_p n p^2 \tag{3.56}$$

where c_n/c_p is proportionality constant in unit of cm^6/sec and has a strong temperature dependence. This is called **Auger coefficient**. Auger coefficient depends on the energy gap and becomes important only when $E_g \leq 1.0$ eV. For low-level injection, Auger lifetime is given by

$$\tau_{\text{Auger}} = \frac{1}{c_n N_D^2} \text{ for } n\text{-type} \tag{3.57}$$

$$\tau_{\text{Auger}} = \frac{1}{c_p N_A^2} \text{ for } p\text{-type}$$

Considering high

$$\tau_{\text{Auger}} = \frac{1}{(c_n + c_p)\Delta p^2} \quad n\text{-type}$$

$$\tau_{\text{Auger}} = \frac{1}{(c_n + c_p)\Delta n^2} \quad p\text{-type}$$

where Δn and Δp are excess carrier densities.

Auger recombination is a non-radiative recombination process for semiconductor laser.

Note:

* The Auger recombination process occurs by the transfer of energy and momentum released by the recombination of an electron-hole pair to a third particle that can be either electron or hole.

3.6.5 Shockley-Hall-Read Recombination

The ionization energy of trap energy levels (E_t) are larger than the ionization energy of donor and acceptor levels for shallow doping. The study of the non-radiative recombination was first carried out by Hall, Shockley and Read, and they have developed a theory, called **Shockley-Hall-Read (SHR) recombination theory**. Any energy level having larger ionization-energy than the shallow donors and acceptors, serves as recombination centre, as shown in Fig. 3.21.

In the recombination process, the recombination centre first captures an electron/hole and then eliminate an electron-hole pair by capturing a hole/electron. This is usually non-radiative recombination process.

Consider a p-type semiconductor, for which the non-radiative recombination centre E_t is above the valence band edge with concentration of N_t. A capture cross section σ_n represents the effectiveness of the centre to capture an electron and is measure,

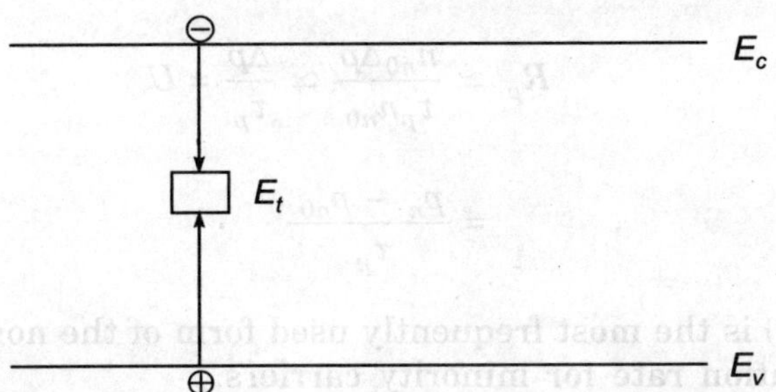

FIGURE 3.21 Shockley-Hall-Read recombination process.

how close the electron has to come to the centre to be captured. The capture rate is proportional to the number of recombination centre N_t. Therefore, the carrier life time for minority carrier electron is given by

$$\tau_{n/p} = \frac{1}{\sigma_{n/p} v_{Th} N_t} \tag{3.58}$$

where v_{Th} is thermal velocity which is equal to 1×10^7 cm/s for most of the semiconductor. τ_n/τ_p are the minority-electrons/holes lifetime, τ_n/σ_p are capture cross-section for electrons/holes. For a non-equilibrium but steady state case, the net rate of capture of electrons must be equal to that of holes. The net non-radiative recombination rate is given as:

$$U = \frac{np - n_i^2}{\tau_p(n + n_1) + \tau_n(p + p_1)} \quad (n_i^2 = n_i p_i) \tag{3.59}$$

where n_1 and p_1 are electron and hole concentration for $E_F = E_t$. These are obtained as:

$$n_1 = n_i e^{(E_t - E_i)/kt} = N_C e^{(-E_C + E_t)/kT} \tag{3.60a}$$

$$p_1 = n_i e^{(E_i - E_t)/kt} = N_v e^{(-E_t + E_v)/kT} \tag{3.60b}$$

For $E_t = E_i$, $n_1 = n_i = p_1$ and recombination rate has its maximum.

For a n-type semiconductor, the net recombination rate for the minority carrier holes is:

$$R_p = \frac{(n_{n0} + \Delta n)(p_{n0} + \Delta p) - n_i^2}{\tau_p(n_{n0} + \Delta n + n_1) + \tau_n(p_{n0} + \Delta n + p_1)}$$

or

$$R_p = \frac{n_{n0}(\Delta p + \Delta n)}{\tau_p(n_{n0} + \Delta n + n_1) + \tau_n(p_{n0} + \Delta n + p_1)} \tag{3.61a}$$

Assuming low-level injection, neglecting $\Delta n \Delta p$, and $n_{n0} \Delta n$, Eq. (3.61a) becomes

$$R_p \simeq \frac{n_{n0}\Delta p}{\tau_p n_{n0}} \simeq \frac{\Delta p}{\tau_p} = U$$

$$= \frac{p_n - p_{n0}}{\tau_p} \qquad (3.61b)$$

Equation (3.61b) is the most frequently used form of the non-radiative Shockley-Read-Hall recombination rate for minority carriers.

Similarly, for a p-type semiconductor material, the minority carrier electron recombination rate is:

$$R_n = \frac{\Delta n}{\tau_n} = U = \frac{n_p - n_{p0}}{\tau_n} \qquad (3.62a)$$

or

$$R_n = \sigma_p(n_p - n_{p0})\, v_{Th} N_t \qquad (3.62b)$$

Therefore, it is clear from Eq. (3.62b) that the recombination rate is proportional to the number of impurity states.

From Eq. (3.58), it is clear that for device to operate efficiently one needs long recombination life time. This is only possible if the recombination centre's concentration must be minimized. On the other hand, for high speed switching operation, short-recombination life time is required which can be obtained by choosing a heavily doped semiconductor with recombination centres.

An impurity level close to the middle of the energy gap serves as the efficient recombination centre. The minority carrier life time can be changed by high energy irridation, which causes the displacement of lattice atoms and introduces energy levels in the energy gap.

3.7 SURFACE RECOMBINATION

Minority carrier lifetime is a basic parameter determining the performance of a large variety of semiconductor devices. It is now well established concept that highly localized deep defects in the bulk and on semiconductor surfaces are extremely efficient recombination centres for the minority carriers. Thus, a recombination via deep levels associated with transition metal impurities can be a dominant mechanism to control the minority carrier lifetime in Si. The issue of recombination via imperfections at semiconductor surfaces and at internal interfaces is becoming critical and dominant issue as size of the semiconductor devices decreasing steadily.

In general, the lifetime of minority carriers is given by

$$\frac{1}{\tau_{\text{minority}}} = \frac{1}{\tau_{\text{radiative}}} + \frac{1}{\tau_{\text{defects}}} + \frac{1}{\tau_{\text{surface}}}$$

Recombination at surfaces and interfaces can have a significant impact on the behaviour of semiconductor devices because surfaces and interfaces typically contain a large number of recombination centres due to the abrupt termination of the semiconductor crystal, which leaves a large number of electrically active states. In addition, the surfaces and interfaces are more likely to contain impurities since they are exposed during the device fabrication process. The net recombination rate due to trap-assisted recombination and generation is given by

$$U_{s,\,SHR} = \frac{pn - n_i^2}{p + n + 2n_i \, \cos h\left(\dfrac{E_i - E_{st}}{kT}\right)} N_{st} v_{Th} \sigma_s$$

This expression is almost identical to that of Shockley–Hall–Read recombination except the recombination is due to a two-dimensional density of traps N_{ts}, as the traps only exist at the surface or interface. Let us assume, an electron in a quasi-neutral p-type region $p >> n$ and $p >> n_i$ so that for $E_i = E_{st}$, and above equation can be simplified as:

$$U_{s,\,n} = R_{s,\,n} - G_{s,n} = v_s(n_p - n_{p0})$$

where v_s is the surface velocity and given as:

$$v_s = N_{st} v_{th} \sigma_s$$

It was observed the semiconductors with inherently large surface recombination velocity are not useful light emitters irrespective of other material parameters.

There are two ways to lower the surface recombination rate. One is the chemical passivation of the surface and other involves the formation of a potential barrier that would prevent minority carriers from reaching the surface. Such a potential can be accomplished by making an appropriate heterojunction with low interface recombination rate (e.g. AlGaAs on GaAs) or by strong doping of the surface region that electrostatically bends the bands in a way that would confine minority carriers in the bulk.

In a surface with no recombination, the movement of carriers towards the surface is zero, and hence the surface recombination velocity is zero. In a surface with infinitely fast recombination, the movement of carriers towards this surface is limited by the maximum velocity they can attain, and for most of the semiconductors it is of the order of 1×10^7 cm/sec.

3.8 CONTINUITY EQUATION

When drift, diffusion and recombination occur simultaneously in a semiconductor material then the governing equation is called **Continuity equation**. The continuity equation allows to calculate the carrier distribution in the presence of generation and recombination, and it applies to both majority and minority carriers. The continuity equation implies the conservation of electrons and hole densities. To drive one dimensional continuity equation for electrons, let us consider a infinitesimal slice of thickness dx located at x as shown in Fig. 3.22. The number of electrons in the slice may increase due to net current flow and net carrier generation in the slice.

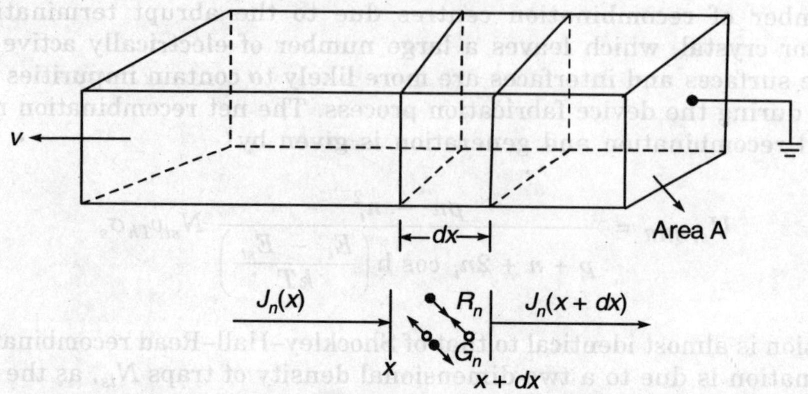

FIGURE 3.22 An infinitesimal semiconductor slice of thickness dx.

The overall rate of electron increase into the slice is the algebraic sum of four components:

* (Number of electrons flowing into the slice at x) – (Number of electrons flowing out at $x + dx$) + (Rate at which electrons are generated – Rate at which they combine with holes in the slice)

or   Net rate of change in the number of electrons into the slice

$$= \frac{1}{-q} \, [J_n(x) - J_n(x + dx)] \, A + [G_n - R_n] \, A \, dx$$

where,

    q is charge on the electron

    A is cross-sectional area of slice

    $A dx$ is volume of slice

In mathematical form, we have

$$\frac{\partial n}{\partial t} \, A dx = \left(-\frac{1}{q}\right) A[J_n(x) - J_n(x + dx)] + [G_n - R_n]Adx \qquad (3.63)$$

where, $J_n(x)$ is current density at x and $J_n(x + dx)$ is current density due to electrons coming out from slice at $(x + dx)$.

Expanding the term $J_n(x + dx)$ in Taylor series, we have

$$J_n(x + dx) = J_n(x) + \frac{\partial J_n(x)}{\partial x} \, dx + \dots \qquad (3.64a)$$

Substituting Eq. (3.64a) into Eq. (3.63), we have

$$\frac{\partial n}{\partial t} \, A \, dx = \left(-\frac{1}{q}\right) A\left[J_n(x) - J_n(x) - \frac{\partial J_n(x)}{\partial x} \, dx\right] + (G_n - R_n) \, A \, dx$$

or
$$\frac{\partial n}{\partial t} = \frac{1}{q}\frac{\partial J_n(x)}{\partial x} + (G_n - R_n) \tag{3.64b}$$

In series expansion of Eq. (3.64a), we have retained only two terms and neglected higher terms because the slice thickness dx is very small.

Equation (3.64b) represents the continuity equation for electrons. Similarly, the continuity equation for holes can be written as:

$$\frac{\partial p}{\partial t} = -\frac{1}{q}\frac{\partial J_p}{\partial x} + (G_p - R_p) \tag{3.64c}$$

The current density consists of both drift and diffusion, given as:

$$J_n = q\mu_n n E + qD_p \frac{dn}{dx}$$

and
$$J_p = q\mu_p p E - qD_p \frac{dp}{dx}$$

Substituting these two expressions in Eqs. (3.64b) and (3.64c) the continuity equation for electron in p-type semiconductor is:

$$\frac{\partial n_p}{\partial t} = \mu_n n_p \frac{\partial E}{\partial x} + \mu_n E \frac{\partial n_p}{\partial x} + D_n \frac{\partial^2 n_p}{\partial x^2} + (G_n - R_n) \tag{3.65a}$$

where n_p is the minority electron concentration in p-type semiconductor material.

Similarly, the continuity equation for minority carrier holes in n-type semiconductor is:

$$\frac{\partial p_n}{\partial t} = -\mu_p p_n \frac{\partial E}{\partial x} - \mu_p E \frac{\partial p_n}{\partial x} + D_p \frac{\partial^2 p_n}{\partial x^2} + (G_p - R_p) \tag{3.65b}$$

where p_n is minority carrier hole concentration in n-type semiconductor.

Now, putting the value of recombination rate as given by Eq. (3.62a), the two one-dimensional continuity Eqs. (3.65a) and (3.65b) can be further written as:

$$\frac{\partial n_p}{\partial t} = -\mu_n n_p \frac{\partial E}{\partial x} + \mu_n E \frac{\partial n_p}{\partial x} + D_n \frac{\partial^2 n_p}{\partial x^2} + G_n - \frac{n_p - n_{p0}}{\tau_n} \tag{3.66a}$$

and

$$\frac{\partial p_n}{\partial t} = -\mu_p p_n \frac{\partial E}{\partial x} - \mu_p E \frac{\partial p_n}{\partial x} + D_p \frac{\partial^2 p_n}{\partial x^2} + G_p - \frac{p_n - p_{n0}}{\tau_p} \tag{3.66b}$$

In Eqs. (3.66a) and (3.66b), it is assumed that the mobility and diffusion coefficient are independent of x. Since, we have already seen that in the semiconductor material, electric field is zero in the bulk, therefore, Eqs. (3.65a) and (3.65b) reduce to

$$\frac{\partial n_p}{\partial t} = D_n \frac{\partial^2 n_p}{\partial x^2} + (G_n - R_n) \tag{3.67a}$$

and

$$\frac{\partial p_n}{\partial t} = D_p \frac{\partial^2 p_n}{\partial x^2} + (G_p - R_p) \tag{3.67b}$$

For steady-state condition,

$$\frac{\partial n_p}{\partial t} = 0 = \frac{\partial p_n}{\partial t}$$

Hence,

$$D_n \frac{\partial^2 n_p}{\partial x^2} + (G_n - R_n) = 0 \tag{3.68a}$$

$$D_p \frac{\partial^2 p_n}{\partial x^2} + (G_p - R_p) = 0 \tag{3.68b}$$

Another frequent application of continuity equation is when distribution of carrier concentration is assumed to be uniform, i.e.,

$$\frac{\partial^2 n_p}{\partial x^2} = 0 = \frac{\partial^2 p_n}{\partial x^2}$$

and, therefore, continuity Eq. (3.67a) and (3.67b) reduce to

$$\frac{\partial n_p}{\partial t} = (G_n - R_n) \tag{3.69a}$$

$$\frac{\partial p_n}{\partial t} = (G_p - R_p) \tag{3.69b}$$

3.9 STEADY-STATE INJECTION FROM ONE SIDE

Let us consider a n-type semiconductor material which is illuminated by light ($h\nu \gg E_g$) from one side as shown in Fig. 3.23(a). Due to illumination and absorption, excess carriers are injected into the material from that side. It is assumed that light penetration is negligibly small.

At steady state and assuming zero field and no generation, the continuity Eq. (3.66b) reduces to

$$0 = D_p \frac{\partial^2 p_n}{\partial x^2} - \left(\frac{p_n - p_{n0}}{\tau_p} \right)$$

or
$$\frac{\partial^2 p_n}{\partial x^2} = \frac{p_n - p_{n0}}{\tau_p D_p} = \frac{p_n - p_{n0}}{L_p^2} \tag{3.70}$$

where $L_p = \sqrt{\tau_p D_p}$ is the diffusion length. Equation (3.70) is a second order differential equation whose solution is given by

$$p_n(x) = c_1 \exp\left(-\frac{x}{L_p}\right) + c_2 \exp\left(\frac{x}{L_p}\right) + p_{n0} \tag{3.71a}$$

where c_1 and c_2 are arbitrary constants and determined by boundary conditions,

$$p_n(x = 0) = p_n(0)$$

and

$$p_n(x = \infty) = p_{n0}$$

After using these two boundary conditions, we have

$$c_1 = p_n(0) - p_{n0}$$

and

$$c_2 = 0$$

Hence, solution becomes

$$p_n(x) = p_{n0} + [p_n(0) - p_{n0}] \exp\left(-\frac{x}{L_p}\right) \tag{3.71b}$$

Figure 3.23(b) shows the variation of minority carrier concentration which decays exponentially with the characteristic length L_p.

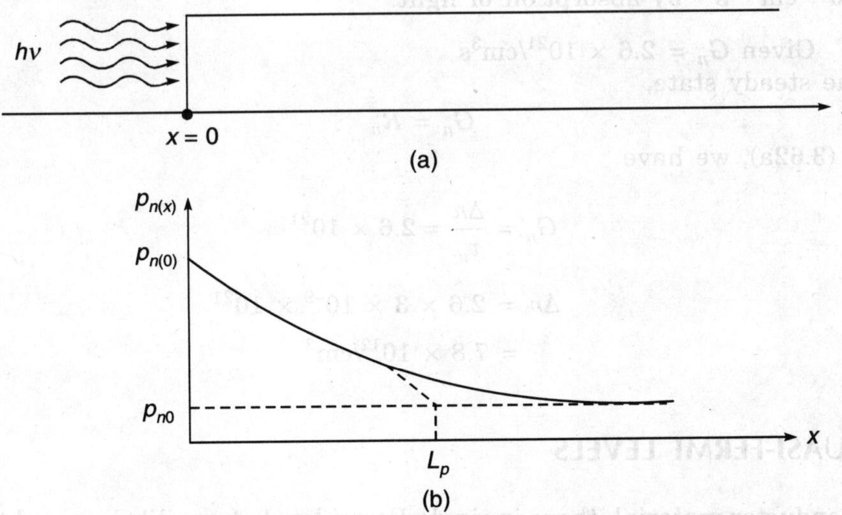

FIGURE 3.23 (a) Semiconductor material which is illuminated by light $hv \geq E_g$ from one side. (b) Variation of minority carrier concentration.

At $x = L_p$, the excess carrier concentration has decayed by exp (–1) of the initial value which is 36.8%.

The current density in the direction of decreasing concentration is:

$$J_p(x) = - qD_p \frac{dp_n(x)}{dx}$$

$$= \frac{qD_p}{L_p} \left[p_n(0) - p_{n0}\right] e^{-x/L_p}$$

$$= \frac{qD_p}{L_p} \left[p_n(x) - p_{n0}\right]$$

In terms of excess carrier density above equation can be further rewritten as:

$$J_p(x) = \frac{qD_p}{L_p} \, \delta p(x) \tag{3.72}$$

The diffusion current, as given by Eq. (3.72), at any x is just proportional to the excess concentration δp at that position.

Since L_p or L_n is much larger than the active dimension of a VLSI device, therefore, in general, generation-recombination plays a very little role in device operation.

EXAMPLE 3.10

Find the steady state minority carrier concentration for a p-type GaAs semiconductor with $N_a^+ = N_a = 3 \times 10^{16}/\text{cm}^3$ and life time $\tau = 3 \times 10^{-9}$ s. The uniform excitation rate is 2.6×10^{21} cm^{-3} s^{-1} by absorption of light.

Solution: Given $G_n = 2.6 \times 10^{21}/\text{cm}^3\text{s}$
      At the steady state,

$$G_n = R_n$$

From Eq. (3.62a), we have

$$G_n = \frac{\Delta n}{\tau_n} = 2.6 \times 10^{21}$$

or

$$\Delta n = 2.6 \times 3 \times 10^{-9} \times 10^{21}$$

$$= 7.8 \times 10^{12}/\text{cm}^3$$

3.10   QUASI-FERMI LEVELS

In a semiconductor material, there is single Fermi level at equilibrium and this Fermi level is spatially constant throughout the material. The position of Fermi level with respect to conduction band edge (CB) or valence band edge (VB) measures the concentration of electrons in conduction band and holes in the valence band respectively. If the semiconductor material is subjected to any external disturbances like illumination, electric field etc., the material is said to be in non-equilibrium and single Fermi level will split into two distinct Fermi levels, one close to conduction band edge and other close to valence band edge. These two Fermi levels are called **quasi-Fermi levels**, and are represented by E_{Fn} and E_{Fp} for electrons and holes respectively. The two Fermi levels are related to the non-equilibrium carrier concentration in the same way as Fermi level E_F is related to the equilibrium carrier concentration. Mathematically,

$$n = n_i \, e^{(E_{Fn} - E_i)/kT} \tag{3.73a}$$

$$p = n_i \, e^{(E_i - E_{Fp})/kT} \tag{3.73b}$$

Therefore,
$$E_{Fn} = E_i + kT \ln\left(\frac{n}{n_i}\right) \tag{3.74a}$$

and
$$E_{Fp} = E_i - kT \ln\left(\frac{p}{n_i}\right) \tag{3.74b}$$

In other way, Eqs. (3.74a) and (3.74b) can also be written as:

$$E_{Fn} = E_c(x) - kT \ln\left(\frac{N_c(x)}{n(x)}\right) \tag{3.75a}$$

and
$$E_{Fp} = E_v(x) + kT \ln\left(\frac{N_v(x)}{p(x)}\right) \tag{3.75b}$$

In general, E_{Fn} and E_{Fp} are the two distinct values as shown in Fig. 3.24.

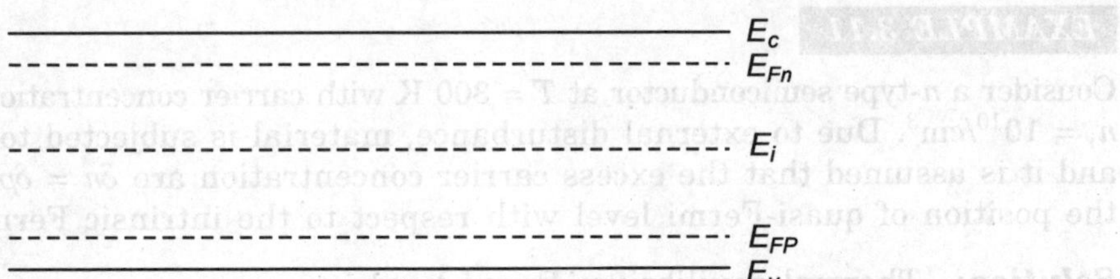

FIGURE 3.24   Representation of quasi-Fermi levels in non-equilibrium semiconductor material.

The separation of quasi-Fermi levels ($E_{Fn} - E_{Fp}$) is a direct measure of the deviation from equilibrium and that will tend back to single Fermi level E_F as the semiconductor goes back to equilibrium.

Quasi-Fermi levels are steady state analogous of the equilibrium Fermi level E_F when excess carriers are present. The deviation of E_{Fn} and E_{Fp} from E_F will indicate how far the electron and the hole concentration are from equilibrium values n_0 and p_0. A given concentration of excess electron-hole pairs causes a large shift in the minority carrier quasi-Fermi level compared to that for majority carriers.

The current density equations for electron and hole in the non-equilibrium condition is given by

$$J_n(x) = \mu_n n(x) \, \Delta E_{Fn} \tag{3.76a}$$

And

$$J_p(x) = \mu_p p(x) \, \Delta E_{Fp} \tag{3.76b}$$

Equations (3.76a) and (3.76b) are extremely useful for analysis of energy bands and current transport in electronic devices.

For one dimensional case,

$$J_n(x) = q\mu_n n(x) \, \frac{d}{dx}\left(\frac{E_{Fn}(x)}{q}\right)$$

$$= \sigma_n(x) \, \frac{d}{dx}\left(\frac{E_{Fn}(x)}{q}\right)$$

$$J_n(x) = \mu_n n(x) \, \frac{d}{dx} E_{Fn}(x) \tag{3.77a}$$

Similarly,

$$J_p(x) = \sigma_p(x) \, \frac{d}{dx}\left(\frac{E_{Fp}(x)}{q}\right)$$

$$= \mu_p p(x) \, \frac{d}{dx} E_{Fp}(x) \tag{3.77b}$$

These two Eqs. (3.77a) and (3.77b) reflect that quasi-Fermi level gradient in an energy-band diagram indicates a current flow.

EXAMPLE 3.11

Consider a n-type semiconductor at $T = 300$ K with carrier concentration of $n_0 = 10^{15}/\text{cm}^3$, $n_i = 10^{10}/\text{cm}^3$. Due to external disturbance, material is subjected to non-equilibrium, and it is assumed that the excess carrier concentration are $\delta n = \delta p = 10^{13}/\text{cm}^3$. Find the position of quasi-Fermi level with respect to the intrinsic Fermi level.

Solution: Thermal equilibrium Fermi level is:

$$E_F - E_i = kT \ln\left(\frac{n_0}{n_i}\right) = 0.2982 \text{ eV}$$

Using Eq. (3.74a),

$$E_{Fn} - E_i = kT \ln\left(\frac{n}{n_i}\right) = kT \ln\left(\frac{n_o + \delta_n}{n_i}\right)$$

$$= 0.2984 \text{ eV}$$

Using (3.74b),

$$E_i = E_{Fp} = kT \ \ln\left(\frac{p_0 + \delta_p}{n_i}\right)$$

$$= kT \ \ln\left(\frac{10^5 + 10^{13}}{10^{10}}\right) = 0.179 \ \text{eV}$$

3.11 THE HAYNES-SHOCKLEY EXPERIMENT

Minority carriers play an important role in device operation. J.R. Haynes and W. Shokley performed the experiment in 1951 to investigate the dynamic behaviour of the minority carriers. In the experiment they have created electron-hole pairs in extrinsic semiconductor. This experiment allows the independent measurement of mobility, life time and diffusion coefficient for minority carriers. The basic experimental setup is shown in Fig. 3.25.

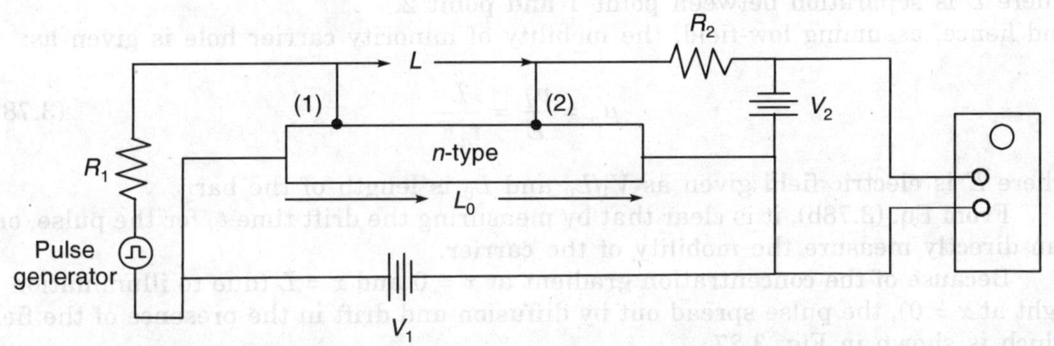

FIGURE 3.25 Experimental setup for Haynes-Shockley experiment.

A voltage V_1 is applied across the n-type semiconductor bar ($n_0 >> p_0$) to establish the electric field in the bar along the length of the bar (i.e., along x-direction). Excess minority carriers are injected into the semiconductor bar the point 1 (i.e. $x = 0$) either by light pulse or pulse generator as shown in Fig. 3.26. A fraction of the excess carriers will be collected at point 2 as they drift through the semiconductor bar in the presence of the field and spreads out by diffusion. The time required for minority carrier holes to drift a given distance in the presence of the field measures the mobility of the carrier and spreading of the pulse during a given time is used to calculate the diffusion coefficient.

Assuming case of low-level injection, the excess hole created in the material will drift in the direction of the electric field and reach at $x = L$. By measuring the drift time t_d from $x = 0$ to $x = L$, (i.e., from position 1 to position 2) the drift velocity can be calculated as:

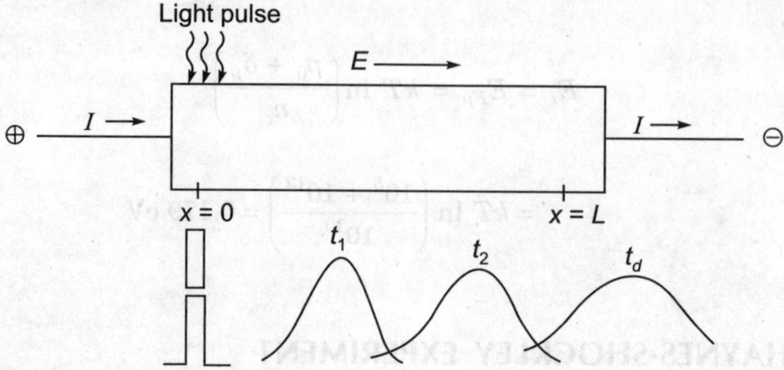

FIGURE 3.26 A pulse of excess carrier generation by a light flash at some point $x = 0$ in the semiconductor bar.

$$v_d = \frac{L}{t_d} \tag{3.78a}$$

where L is separation between point 1 and point 2.
and hence, assuming low-field, the mobility of minority carrier hole is given as:

$$\mu_p = \frac{v_d}{E} = \frac{L}{t_d E} \tag{3.78b}$$

where E is electric field given as V_1/L_0 and L_0 is length of the bar.

From Eq. (3.78b), it is clear that by measuring the drift time t_d for the pulse, one can directly measure the mobility of the carrier.

Because of the concentration gradient at $x = 0$ and $x = L$ (due to illumination of light at $x = 0$), the pulse spread out by diffusion and drift in the presence of the field which is shown in Fig. 3.27.

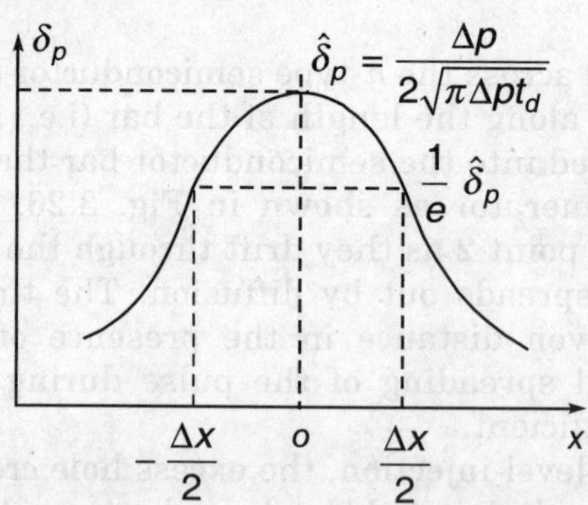

FIGURE 3.27 Spreading of pulse due to diffusion as the pulse drift in the presence of the field.

By measuring the pulse spread, one can calculate D_p using equation at point x,

$$\delta_p(x, t) = \left[\frac{\Delta p}{2\sqrt{\pi D_p t}}\right] e^{-x^2/4 D_p t} \tag{3.79}$$

where Δp is the number of holes per unit area created over a negligibly small distance at $t = 0$.

The peak value of δp is obtained at $x = 0$ and decreases with time. The exponential factor predicts the spread of the pulse in the positive and negative x-direction. The most convenient choice is the point $(\Delta x/2)$, where the δp is down by $(1/e)$ of its peak value (δp). At this point one can write

$$e^{-1}\hat{\delta}_p = \hat{\delta}_p e^{-(\Delta x/2)^2/4 D_p t_d}$$

or

$$D_p = \frac{(\Delta x)^2}{16\, t_d} \tag{3.80a}$$

Since, there is no direct way to find Δx, so it can be found by using

$$\Delta x = \Delta t\, v_d$$

$$= \Delta t\, \frac{L}{t_d} \tag{3.80b}$$

where Δt is the time taken for pulse to move from $x = 0$ to $x = L$.

The Haynes-Shockley experiment also provides the information about recombination process along with diffusion process in the semiconductor material. The recombinate rate can be measured by observing the variation of the area under the trace (Fig. 3.27) as it propagates from the flash point to the contact point 2.

EXAMPLE 3.12

As shown in Fig. 3.25, the silicon n-type semiconductor material is used. The length of the sample is 1 cm and the points 1 and 2 are separated by 0.95 cm.

The applied voltage on the sample is 2 V. A pulse arrives at point 2 at 0.25 ms after injection at point 1. Assume the width of the pulse is 120 μs. Calculate the mobility of minority carrier hole and diffusion coefficient.

Solution:

$$v_d = \frac{0.95}{0.25 \times 10^{-3}} = \frac{95}{25} \times 10^3 \text{ cm/s}$$

For low electric field,

$$\mu_p = \frac{v_d}{E}$$

$$\therefore \qquad E = \frac{2 \text{ volt}}{1 \text{ cm}} = 2 \text{ V/cm}$$

$$\mu_p = \frac{95}{25} \times \frac{1}{2} \times 10^3 = 1900 \text{ cm}^2/\text{V s}$$

Using Eq. (3.80b),

$$\Delta x = \frac{120 \times 10^{-6} \times 0.95}{0.25 \times 10^{-3}} = \frac{120 \times 95}{25} \times 10^{-3}$$

$$= 456 \times 10^{-3} \text{ cm} = 0.456 \text{ cm}$$

Now, using Eq. (3.75a),

$$D_p = \frac{(0.456)^2}{16 \times 0.25} \times 10^3$$

$$= 51.985 \text{ cm}^2/\text{s}$$

3.12 SUMMARY

At thermal equilibrium, no net current flows through semiconductor material due to the random motion of the electrons. For a low electric field, the drift velocity of the carrier is directly proportional to the field and proportionality constant is called **mobility** of the carrier. Mobility measures the responsiveness of the carrier motion within the semiconductor crystal to an electric field. Most of the electronic devices preferred n-type semiconductor material due to larger mobility of electron than the hole.

The two most important scattering processes which limit the mobility of the carriers are: lattice scattering and ionized impurity scattering. Lattice scattering dominates at higher temperature and low doping whereas ionized impurity scattering dominates in heavily doped semiconductor material at low temperature.

The conductivity of a semiconductor material depends on the electron and hole concentration and their respective mobilities. The reciprocal of the conductivity is defined as the resistivity which is an important electrical property of the solid state material. Hall effect is used to determine the type of the carrier in the semiconductor material. With Hall effect one can also measure the carrier concentration of majority carrier. Hall coefficient is positive for hole and negative for electron. Due to non-uniform doping in the semiconductor material, an induced electric field is produced which causes drift of the carriers. This induced field is also called **built-in-field,** and this field is important for restoring equilibrium in the non-uniformly doped semiconductor in the absence of any applied field.

There are basically two types of recombination processes:

1. Direct or radiative recombination
2. Indirect or non-radiative recombination

When a light of energy $h\nu \geq E_g$ falls on the semiconductor material, extra electron hole pairs are created. These extra carriers are called excess carriers. Now, equilibrium is disturbed and mass-action law is now not valid. The process of introducing excess carriers are called **injection**. If majority carrier concentration is changed due to excess carrier generation, injection is called **high-level injection** otherwise low-level injection.

Auger recombination process is a three particle process in which the energy and momentum released by the recombination of electron-hole pair is transferred to third particle. The continuity equation is applied to both majority as well as minority carriers and it allows the calculation of carrier distribution in the presence of generation and recombination.

The Haynes-Shockley experiment allows the independent measurement of mobility, lifetime and diffusion coefficient of minority carriers.

REVIEW QUESTIONS

1. Why no current flows in the absence of applied field at thermal equilibrium in semiconductor material?

2. Why mobility of electron is larger than mobility of hole?

3. Most of the modern electronic devices are preferred n-type semiconductor material. Why?

4. Why mobility is limited by scattering processes?

5. Explain, why lattice scattering is dominant at high temperature and low doping whereas ionized impurity scattering dominates at low temperature and heavily doped semiconductor material.

6. At higher electric field ($E \geq 10^5$ V/cm), the velocity of the carriers saturate. Why?

7. Two semiconductor bars of length L_1 and L_2 are given to you. Assuming both have same area and resistivity. Taking $L_1 = 2L_2$, which semiconductor bar has larger resistance value and why?

8. Conductivity in any semiconductor material is always larger for n-type semiconductor than p-type semiconductor, why?

9. Explain why drift is assumed to be a majority carrier process and diffusion as a minority carrier process.

10. How can one determine the type of the given unknown semiconductor material?

11. Why built-in-field in any semiconductor material is required to establish the thermal equilibrium?

12. Why mobility decreases from its low-field value at very high electric field?

13. If the mobility of the carrier is given in any semiconductor material then the diffusion coefficient of the carrier can be determine. If yes, how?

14. Assume the donor impurity concentration in a semiconductor is given by

$$N_d(x) = 10^{15} \exp\left(-\frac{x}{L_n}\right) \text{ for } x \geq 0 \text{ and } L_n = 10^{-4} \text{ cm}$$

    (a) Is it homogeneous semiconductor material or non-uniformly doped semiconductor?

    (b) Find the induced electric field in the material.

15. A silicon semiconductor sample is doped with the donor impurity of $10^{16}/cm^3$ at room temperature and now this sample is illuminated with a light. Due to illumination 10^{13} electron-hole pairs/cm^3 are created every microsecond. Whether this is case of low level injection or high level injection? Explain.

16. In Problem 15, find out the carrier concentration of majority carrier after illumination.

17. Explain, why mass-action law $n_0 p_0 \neq n_i^2$ does not hold at non-equilibrium.

18. Why for optical excitation of electron from valence band to conduction band, light of energy $h\nu > E_g$ is required?

19. Why Auger recombination process is non-radiative recombination process?

20. Show that the unit for $\dfrac{1}{q\mu_n n_0}$ is ohm-cm.

21. What is the carrier recombination centre?

22. Explain, physically why the carrier life time is much smaller in direct gap materials than indirect.

NUMERICAL PROBLEMS

1. A silicon sample has been doped with 10^{17} As atoms/cm^3. Calculate the conductivity of the sample at $T = 300$ K and $T = 400$ K. Interpret your result.

   The above n-type silicon sample is further doped with 9×10^{16} B atoms/cm^3. Calculate the conductivity of sample at $T = 300$ K and $T = 400$ K. Comment on the result.

2. A sample of germanium has an intrinsic resistivity of 60 Ω-cm. Calculate the value of current density in an applied field of 10 mV/cm; if there are 10^{13} donor/cm^3 and 4×10^{12} acceptors/cm^3. Take $\mu_n = 4200$ cm^2/V-s. and $\mu_p = 2000$ cm^2/V-s. Assume all the impurities are ionized.

3. Electrons in undoped piece of germanium have a mobility of 1400 cm^2/V-s. Calculate the average time between collisions. Calculate the distance travelled between two collisions (i.e., mean free path). Use an average velocity of 10^7 cm/s and $m_n^* = 0.19\, m_0$ where m_0 is free electron mass.

4. A piece of silicon doped with arsenic ($N_d = 10^7/\text{cm}^3$) is 100 μm long, 10 μm wide and 1 μm thick. Calculate the resistance of this sample when contacted each end. Take the mobility of electrons to be 800 cm^2/V-s.

5. The hole density in an n-type silicon wafer ($N_d = 10^{17}/\text{cm}^3$) decreases linearly from $10^{14}/\text{cm}^3$ to $10^{13}/\text{cm}^3$ between $x = 0$ to $x = 1$ μm. Calculate the hole diffusion current density. Take $\mu_p = 300$ cm^2/V-s.

6. Calculate the minimum and maximum resistivity of n-type Si semiconductor at $T = 300$ K. Assume $N_c = N_v = 2.78 \times 10^{13}/\text{cm}^3$ and $n_i = 1.5 \times 10^{10}/\text{cm}^3$, $\mu_n = 1000$ cm^3/V-s.

7. An experiment is carried out at $T = 300$ K on n-type Si doped semiconductor at $N_d = 10^{17}/\text{cm}^3$. The conductivity is found to be 10.0 mho-cm^{-1}. When light with certain intensity shines on the material, the conductivity changes to 11.0 mho-cm^{-1}. The light is turned-off at time $t = 0$ and is found that at time $t = 1$ μs the conductivity is 10.5 mho-cm^{-1}. The light induced excess conductivity is found to decay as

$$\delta\sigma = \delta\sigma(0)\, e^{-t/\tau}$$

where τ is the carrier life time. Calculate:
  (a) What fraction of donors is ionized before light is made to fall on the semiconductor?
  (b) What is the diffusion length of hole in this material (Assume mobility of electron = 1100 cm^2/V-s.)

8. A sample of silicon is doped with 10^{16} phosphorus atoms/cm^3. Find the Hall voltage in a sample with $W = 500$ μm, $A = 2.5 \times 10^{-3}$ cm^2, $I = 1$ mA and $B_z = 10^{-4}$ Wb/cm^2.

9. Consider a p-type Si integrated circuit resistor with a resistivity of 0.1 ohm-cm at $T = 300$ K and 0.2 ohm-cm at $T = 400$ K. The resistor has thickness of 3.0 μm, a width of 4.0 μm and length of 14 μm.
  (a) What is the sheet resistance at $T = 300$ K?
  (b) What is the resistance of the resistor at $T = 300$ K?

10. Find the minority carrier life time for an n-type GaAs with $N_d^+ = 3 \times 10^{16}$. Assume all the donors are ionized. By absorption of light, uniform excitation rate is $2.6 \times 10^{22}/\text{cm}^3$-s. Take the steady-state minority carrier concentration as $8 \times 10^{12}/\text{cm}^3$.

11. For a Si sample at room temperature, the intrinsic Fermi level decreases linearly by 0.2 eV in a distance of 0.15 μm. Find the induced electric field.

12. A current of 5 mA flows through a semiconductor bar of length 100 μm long and the ends of the bar are 10 μm by 5 μm. A battery of 2 V is applied across the bar. Find its resistivity. Also find the type of semiconductor, whether it is Ge or Si.

13. Find the resistivities of intrinsic silicon and intrinsic GaAs at $T = 300$ K. What will happen to resistivity if the temperature increases?

14. Assume that the mobility of the electrons in a Si semiconductor is limited by the lattice scattering at any temperature. The mobility of electron in the material is 1000 cm^2/V-s at $T = 300$ K.

Find the mobility at $T = 200$ K and $T = 400$ K. Interpret your result.

15. Two scattering mechanisms exist in the semiconductor material. The resultant mobility when both the scattering mechanisms exist at the same time is 600 cm^2/V-s at a given temperature $T = 300$ K. At the same temperature the mobility due to presence of first mechanism only is 1500 cm^2/V-s. Find the mobility of carrier in the presence of second scattering mechanisms only.

16. The ionized impurity scattering limited mobility of electrons in a Si-semiconductor material at room temperature for $N_d = 10^{16}$/cm^3 is 1000 cm^2/V-s. Find the mobility at same temperature for $N_d = 10^{14}$/cm^3 and $N_d = 10^{17}$/cm^3. Discuss your result.

17. Consider a compensated n-type silicon at $T = 300$ K with the conductivity of $16(\Omega$-cm$)^{-1}$. If the concentration of acceptor impurity is 10^{17}/cm^3, find the concentration of donor impurity. Also find the electron mobility.

18. A semiconductor is doped with $N_d(N_d \gg n_i)$ and has a resistance of R_i. The same semiconductor is then doped with an unknown amount of acceptors $N_a(N_a \gg N_d)$, yielding a resistance of 0.5 R_1. Find N_a in terms of N_d. Take $(D_n/D_p) = 50$.

19. A silicon sample of unknown doping is used in Hall experiment as shown in Fig. 3.9. Hall measurement provides the following information: $W = 0.05$ cm, $A = 1.6 \times 10^{-3}$ cm^2, $J = 2.5$ mA and magnitude of magnetic field is 30 nT, where 1 J $= 10^{-4}$ Wb/cm^2. If the Hall voltage of 10 mV is measured, find the value of Hall coefficient, majority carrier concentration, conductivity type, resistivity and mobility of the semiconductor sample.

20. An intrinsic Si sample is doped with donor impurities from one side such that $N_d(x) = N_0 \, e^{-ax}$ where N_0 is donor concentration for $x = 0$. Assume all the donors are ionized and take donor concentration at $x = 0$ is 10^{16}/cm^3.

(a) Whether it is homogeneous or non-homogeneous semiconductor material?
(b) Is any electric field exist in the semiconductor? If yes, why and what is the role of this field?
(c) Find the expression for diffusion current density.
(d) Calculate the value of current density at $x = 1$ μm. Take mobility of electron as 1000 cm^2/V.s.

Repeat part(d) for $x = 0$ and $x = \infty$ also.

21. Consider a silicon sample at $T = 300$ K, doped with acceptor impurity of 10^{16}/cm^3, and this gives a resistivity of 0.3 ohm-cm. The thickness of the silicon bar is 3.0 μm and width is 10 μm.

(a) Find the length of the bar which can give the resistance of 2.5 kW.
(b) If a battery of 5 V is applied across the bar, find the current following into the bar? Take $l = 5$ μm.

22. At thermal equilibrium, $n_0 p_0 = n_i^2$ and the resistivity is given by

$$\rho = \frac{1}{q(\mu_n n_0 + \mu_p p_0)}$$

(a) Find the ratio of (n_0/p_0) for maximum resistivity.
(b) Write an expression for maximum resistivity in terms of mobilities and n_i.

23. Find the mean free time τ_n and τ_p in a silicon semiconductor material at room temperature. Assuming μ_n = 1200 cm^2/V-s and μ_p = 400 cm^2/V-s. Take m_n^* = 0.26 m_0 and m_p^* = 0.36 m_0. Also calculate mean free path for electron assuming average thermal velocity as 10^7 cm/s.

24. Consider a silicon sample of length L = 10 μm doped with donor impurity concentration of 10^{16}/cm^3 at T = 300 K. A battery of voltage 50 mV is connected across the silicon sample. Take μ_n = 1000 cm^2/V-s. Calculate the drift current density.

Repeat the calculation for a battery of 10 mV and 60 mV. Interpret these results.

25. The hole concentration in silicon at T = 300 K varies linearly from x = 0 to x = 0.01 cm. The hole diffusion coefficient D_p = 10 cm^2/s; the hole diffusion current density is 20 A/cm^2. The hole concentration at x = 0 is 4×10^{17}/cm^3. Find the hole concentration at x = 0.01 cm.

26. The hole concentration in silicon is given by $p(x) = 10^{15}\, e^{-x/L_p}$/cm^3 for $x \geq 0$. The hole diffusion coefficient is 10 cm^2/s. The value of diffusion current density at x = 0 is 6.4 A/cm^2. What is the value of L_p.

27. The electron concentration in silicon is given by $n(x) = 10^{15}\, e^{-x/L_n}$/cm^3 for $x \geq 0$. The electron diffusion coefficient is 25 cm^2/s. Find the value of x for which diffusion current density is equal to –14.7 A/cm^2. Take L_n = 1 μm.

28. Find the current density for an electric field of 4.5×10^3 V/cm in an n-type semiconductor with $n_o = 1.5 \times 10^{16}$/cm^3 and the mobility of electron is 1200 cm^2/V-s. at T = 300 K.

29. (a) For a n-type Si semiconductor sample at T = 300 K, $n_0 = 2.5 \times 10^{16}$/cm^3. A light of energy $h\nu > E_g$ is allowed to fall on the sample and excess carriers of 10^4/cm^3 are generated. Find the non-equilibrium hole concentration.
(b) If the Si semiconductor material has energy gap of 1.12 eV at T = 300 K then find the wavelength of optical rays to produce excess electron-hole pairs of 10^4/cm^3.

30. Calculate the electron-hole concentration under steady-state illumination in an n-type silicon with $G_L = 10^{16}$/cm^3-s; $N_d = 10^{15}$/cm^3 and $\tau_n = 2p = 10$ μs.

31. Assume that an n-type semiconductor is uniformly illuminated, producing uniform excess generation rate G. Show that in steady state, the change in the semiconductor conductivity is given by

$$\Delta\sigma = q(\mu_n + \mu_p)\, \tau_p G$$

32. For a carrier life time of 1×10^{-8} s; what is the recombination rate for an excess carrier density of 5×10^{13}/cm^3?

33. Consider a Si sample with $N_d^+ = 3.5 \times 10^{16}$/cm^3 and $N_a^- = 1.7 \times 10^{16}$/cm^3 at room temperature. For a uniform excitation, with the generation rate of 3×10^{22}/cm^3-s and a hole life time of 2×10^{-8} s. What is the steady-state hole concentration. When the excitation is turned off, how long does it take for the hole concentration to decay to 10% of its steady-state value?

34. Find the minority carrier diffusion length in a n-type Si semiconductor at room temperature with 5×10^{16}/cm^3 donor impurities and a life time of 5×10^{-8} s. Assume mobility of electron $\mu_n = 1000$ cm^2/V-s.

35. A silicon sample at thermal equilibrium contains 3×10^{16}/cm^3 ionized phosphorus atoms, 2×10^{16}/cm^3 ionized boron atoms and 2×10^{16}/cm^3 ionized arsenic atoms. The minority carrier life time is 6×10^{-8} s.

   (a) Find the majority carrier and its concentration.
   (b) Find the resistivity of the material.
   (c) If light shines on the semiconductor surface and produces 10^5 excess electron-hole pairs then find the enhanced conductivity.
   (d) Compare the resistivity when light is ON to resistivity when light is turned-off. Comment on the result.
   (e) Determine minority carrier diffusion length.
   (f) Find the net recombination and optical generation rate at steady-state.

36. A silicon sample with 10^{15}/cm^3 donors is uniformly optically excited at room temperature such that 10^{18}/cm^3 electron-hole pairs are generated per second. Find:

   (a) The optical generation rate and net recombination rate at steady state.
   (b) Change in conductivity upon shining the light. Take minority carrier life time as 10 μs and $D_p = 12$ cm^2/s.
   (c) The separation between two quasi fermi levels. Also draw the band diagram.

37. For a 2 cm long Si bar, doped with donor impurities of 10^{16}/cm^3 and having cross-sectional area as 0.05 cm^2, find the current if a 10 V battery is connected across it. Find the resistivity of the material also. If 10^{20} electron-hole pairs are generated uniformly in the bar then find out the value of new current. Assume $\tau_n = \tau_p = 10^{-4}$ s; $\mu_n = 1000$ cm^2/V.

38. Let us assume a photodetector in a form of bar of length L and cross-sectional area A. A constant voltage source of value V is applied across it. It is optically illuminated at a rate of $G_L = g_{0p}$ EHP/cm^3. This excitation is assumed to be uniform throughout the sample. If $\mu_n \gg \mu_p$ then it can be assumed that optically induced change in current ΔI is dominated by the mobility μ_n and τ_n

for electrons. Show that $\Delta I = \dfrac{qALg_{op}\tau_n}{\tau_t}$ for this photodetector, where τ_t is the transit time of electrons drifting down the length of the bar.

39. Show that the average distance a hole diffuses into n-type semiconductor material before recombining is given by diffusion length L_p.

40. Show that for a p-type semiconductor, the average time an electron spends in the conduction band is equal to the time constant τ_n.

41. A direct gap semiconductor sample is illuminated at one end with light of $\lambda = 500$ nm with an intensity of 1 mW/cm^2. The area of illuminated surface is 1 cm^2. Assume carrier life time is 10 ns, and sample has forbidden gap of 1.43 eV at $T = 300$ K.

(a) Whether the incident light is able to generate excess electron-hole pairs or not?

(b) Find the number of photons striking the sample per second.

(c) If every photon is absorbed uniformly (with x) within 1 μm of the surface, what are the excess carrier concentration Δn and Δp in this region?

42. Consider a bar of semiconductor illuminated as shown in Fig. 3.28.

(a) Sketch the concentration of electrons and holes as a function of x.

(b) In which directions electron and hole flow?

(c) In what directions do the electron and hole diffusion currents go?

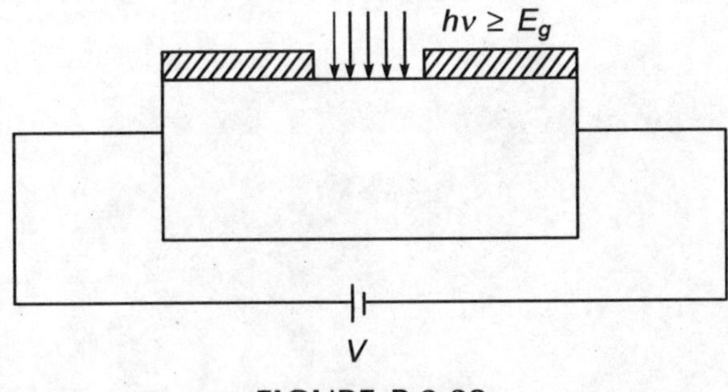

FIGURE P.3.28

43. The total current in a semiconductor is constant and is equal to sum of drift and diffusion currents. The electron concentration is equal to 10^{16} cm^{-3} and assumed to be constant. The hole concentration is given by

$$p(x) = 10^{15}\, e^{-x^2/L_p}/\text{cm}^3 \qquad x \geq 0$$

where L_p is diffusion length = 10 μm. The hole diffusion coefficient is $D_p = 12$ cm^2/s. Take $\mu_n = 1000$ cm^2/V-s. The total current density $J = 4.8$ A/cm^2. Calculate

(a) Hole diffusion current density in terms of x.

(b) Electron current density in terms of x.

(c) The electric field in terms of x. Also plot it with x.

44. Phosphorus is diffused into an intrinsic Si sample. The donor distribution is shown in Fig. 3.29. Sketch the equilibrium band diagram and also show the direction of the resulting electric field.

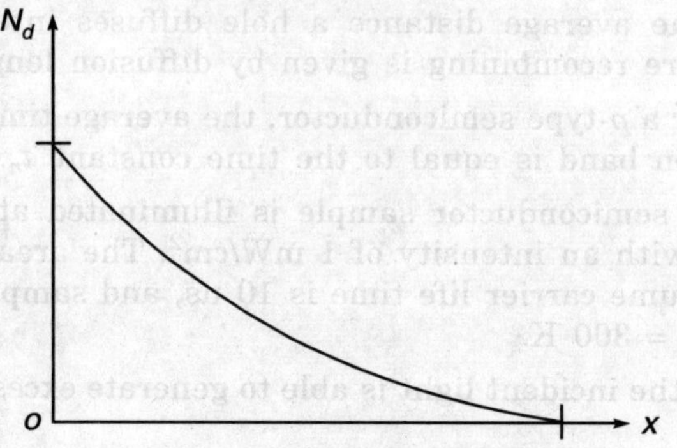

FIGURE P.3.29 Donor distribution.

45. In a Haynes-Shockley experiment, the maximum amplitudes of the minority carriers at $t_1 = 100$ μs and $t_2 = 200$ μs differ by a factor of 5. Calculate the minority carrier life time.

CHAPTER 4

p-n Junction

LEARNING OBJECTIVES

- *p-n* junction diode and its equilibrium band diagram
- Abrupt junction
- One sided *p-n* junction
- Non-uniformly doped *p-n* junction
- Biasing of the *p-n* junction
  - Forward bias
  - Reverse bias
- Current-voltage characteristics of an Ideal and Real *p-n* junction diode
- Capacitance of the *p-n* junction
  - Junction capacitance
  - Diffusion capacitance
- Varactor diode
- Reverse recovery transient time
- *p-n* junction as a rectifier
- Junction breakdown
  - Zener breakdown
  - Avalanche breakdown
- Heterojunction

4.1 INTRODUCTION

In the previous chapters, we have studied the carrier concentrations and transport phenomenon in a homogeneous semiconductor material. In the present chapter, all these concepts will be applied to a single-crystal semiconductor material which contains both *p*-type as well as *n*-type regions to form a junction, called **p-n junction**. The term junction refers to the boundary between two semiconductor regions. *p-n* junctions, is important component of all MOSFET and bipolar devices. Most of the semiconductor devices contain at least one *p-n* junction. Most modern *p-n* junctions are made by planar technology. *p-n* junctions are extensively used in rectification, switching and other operations in electronic circuits. In this chapter, we will also discuss the heterojunction, which is formed between two dissimilar semiconductors. Heterojunction is an important building block for HBJT (Heterojunction BJT), modulation doped field effect transistor, quantum- effect devices and photonic devices.

4.2 *p-n* JUNCTION AND ITS BAND DIAGRAM AT THERMAL EQUILIBRIUM

A *p-n* junction is formed when one region of a single-crystal semiconductor material (substrate) is doped *n*-type (donor impurities) and immediately adjacent region is doped *p*-type (acceptor impurities). The interface separating *n*-type region from *p*-region is referred to as the metallurgical junction. A schematic representation of a *p-n* junction is shown in Fig. 4.1(a).

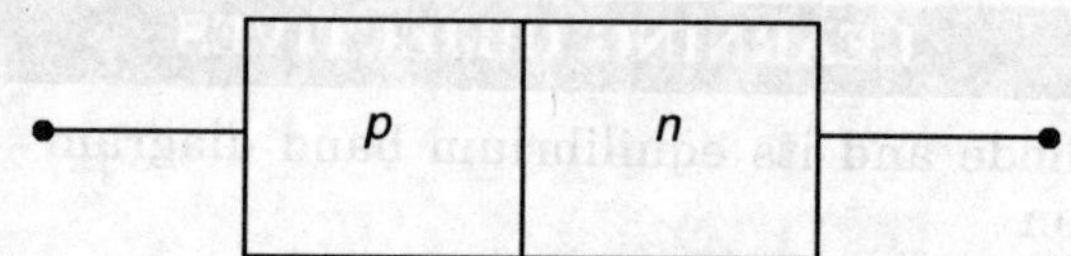

FIGURE 4.1(a) Schematic representation of *p-n* junction.

The symbol of *p-n* junction diode is shown in Fig. 4.1(b).

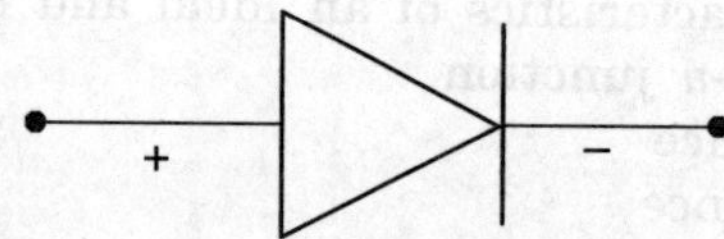

FIGURE 4.1(b) Symbol of *p-n* junction.

Figure 4.1(c) shows the energy band diagrams of an isolated *p*-type and *n*-type silicon semiconductor materials. In *p*-type semiconductor, Fermi level is closer to valence band edge because holes are majority carriers and electrons are minority carriers, i.e., $n_0 \ll p_0$. Similarly, in *n*-type semiconductor Fermi level is closer to conduction band edge due to majority carrier electrons.

When these two isolated materials (*n*-type as well as *p*-type) are joined together then the large carrier concentration gradients at the junction cause diffusion of carriers, i.e., electron will diffuse from *n*-side to *p*-side and recombine with holes and holes will diffuse from *p*-side to *n*-side and recombine with electrons. Since there is no external applied field, this diffusion process cannot continue indefinitely. As electrons diffuse from *n*-region, some of the positive ionized donor atoms (N_d^+) are left uncompensated near the junction. Similarly, as holes diffuse from *p*-region, they uncover negatively charged acceptor atoms (N_a^-). The net positive immobile and negative immobile charges

$n_o \gg p_o$ —————————————— E_C

————————————— E_C

------------------------ E_F

------------------------ E_F

————————————— E_V

------------------------ E_F

————————————— E_V

E_V

p-type   *n*-type

FIGURE 4.1(c)   Band diagram of *p-n* junction before contact.

near the metallurgical junction forms space charge region (transition region). This space charge region induces an electric field that is directed from positive charge towards the negative charge or from the *n* to the *p* region as shown in Fig. 4.2.

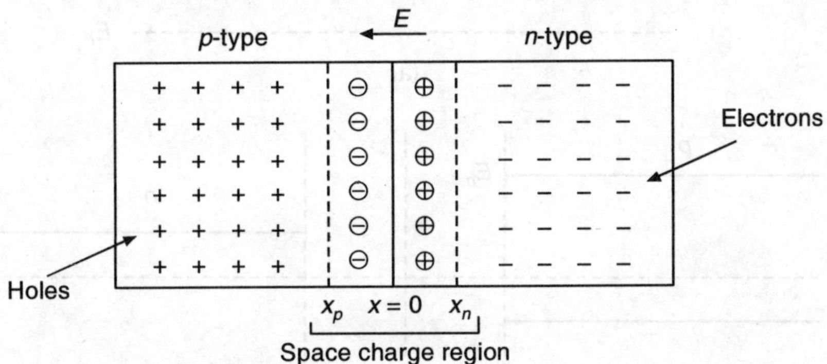

FIGURE 4.2   Formation of space charge region in *p-n* junction at thermal equilibrium.

Due to the induce electric field near the junction, a potential gradient will develop. The potential gradient causes a flow of charge carriers, known as **drift current**, in opposite direction to the diffusion current. Under equilibrium conditions, the diffusion current is exactly balanced by the drift current so that the net current across the *p-n* junction is zero.

In other words, when no external field is applied across the *p-n* junction, the potential gradient due to space charge region forms an energy barrier that prevents further diffusion of charges across the junction as system achieve the thermal equilibrium. The potential gradient across the junction is also called **built-in-potential** which cannot be measured directly with a voltmeter because new potential barrier will be formed between the probes and semiconductor that will cancel built-in potential. The built-in-potential is necessary to establish the equilibrium at the metallurgical junction and also does not imply any external potential.

Since in the space charge region, there is no mobile carriers or the region is depleted of any mobile carriers, therefore, region is also referred to as the depletion region.

The band diagram of a *p-n* junction at thermal equilibrium can be drawn by using the following steps:

(i) At thermal equilibrium, since the Fermi levels in the semiconductor materials must be spatially constant therfore, in *p-n* junction diode, the Fermi levels of the *p*-type and *n*-type semiconductor must be at same level and spatially constant as shown in Fig. 4.3(a).

(ii) The *n* and *p*-regions sufficiently far away from the junction are referred as the bulk quasineutral regions and bands should be essentially flat as shown in Fig. 4.3(b), and respective band characteristic should maintain.

(iii) Requirement of the spatially constant fermi level and continuity of conduction bands and valence bands across the entire structure causes the energy bands of *p*-type region to lie higher than the corresponding energy bands of the *n*-type region and hence results in bending of bands within the space charge region [see Fig. 4.3(c)].

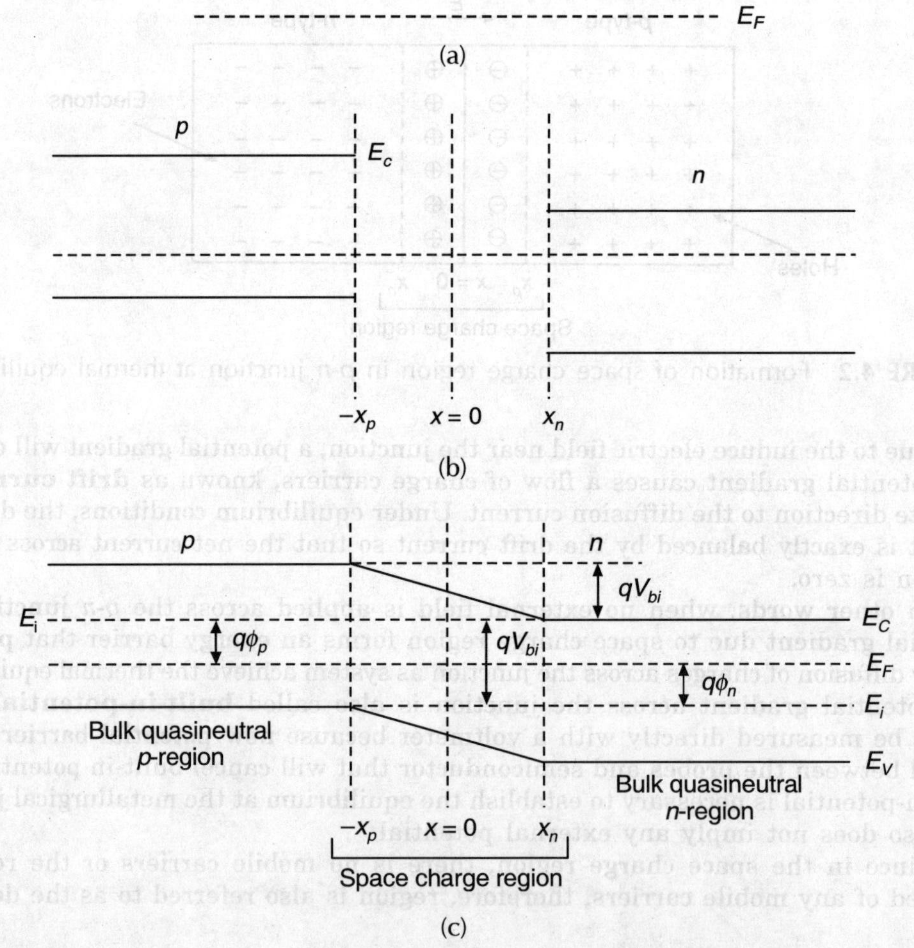

FIGURE 4.3 Band formation in *p-n* junction after contact at thermal equilibrium.

As seen from Fig. 4.3(c), electrons in the conduction band of the *n*-region will observe a potential barrier of height qv_{bi} in trying to move into the conduction band of *p*-region. Similarly, holes in the valence band of *p*-region will experience a potential barrier of height qv_{bi} in moving to valence band of *n*-region. This band bending within space charge region or potential barrier height is responsible for maintaining the thermal equilibrium in a *p-n* junction diode in absence of any external applied voltage. The electric field in the respective bulk quasineutral region is zero. The amount of band bending is the difference in the work function of both regions. The ϕ_p and ϕ_n are defined as the electrostatic potential of *p*-type quasineutral region and *n*-type quasineutral region with respect to the Fermi level, and mathematically expressed as:

$$\phi_p = -\frac{1}{q}(E_i - E_F)\bigg|_{x \le x_p} \tag{4.1a}$$

and

$$\phi_n = -\frac{1}{q}(E_i - E_F)\bigg|_{x \ge x_n} \tag{4.1b}$$

The total electrostatic potential difference between the *p*-side and *n*-side regions at thermal equilibrium, is equal to built-in potential or contact potential, and is given as:

$$V_{bi} = \phi_n - \phi_p \tag{4.2a}$$

Since, the intrinsic Fermi level is equidistant from the conduction band edge through the junction, thus the built-in-potential barrier can be determined as the difference between the intrinsic Fermi levels in *p* and *n*-regions as shown in Fig. 4.3(c).

As clear from Fig. 4.3(c), total depletion width is given by

$$W = |x_n| + |x_p| \tag{4.2b}$$

In depletion region, only ionized donor and acceptor impurities are present in small amount. These ionized impurities are located in the substitutional lattice sites and cannot move in the electric field.

4.2.1 Mathematical Derivation of Built-in-Potential

At thermal equilibrium, in absence of any external excitations at given temperature the net electron or hole current flow is zero. Thus, for each type of carrier the drift current due to induced electric field must be exactly equal to the diffusion current due to concentration gradient and hence these two current components must be in opposite direction of each other

Hence,

$$J_n = J_{n\text{diff.}} + J_{n\text{drift}} = 0$$

$$J_p = J_{p\text{drift.}} + J_{p\text{diff.}} = 0$$

or

$$J_n = q\mu_n\left(n_0 E + \frac{kT}{q}\frac{dn_0}{dx}\right) = 0 \qquad (4.3a)$$

where E is the induced electric field.

$$J_p = q\mu_p\left(p_0 E - \frac{kT}{q}\frac{dp_0}{dx}\right) = 0 \qquad (4.3b)$$

Since,

$$p_0 = n_i\,\exp\frac{[E_i - E_F]}{kT} \qquad (4.3c)$$

Now differentiating Eq. (4.3c) with respect to x, we get

$$\frac{dp_0}{dx} = \frac{p_0}{kT}\left[\frac{dE_i}{dx} - \frac{dE_F}{dx}\right] \qquad (4.3d)$$

Substituting Eq. (4.3d) into (4.3b), we have

$$J_p = q\mu_p\left[p_0 E - p_0 E + \frac{p_0}{q}\frac{dE_F}{dx}\right] = 0$$

which implies

$$\frac{dE_F}{dx} = 0$$

Similarly, for the net electron current density,

$$J_n = \mu_n n_0\frac{dE_F}{dx} = 0$$

Thus, the condition for zero electron or hole current is that the Fermi level should be spatially (i.e., independent of x) constant throughout the sample.

The potential ϕ_p in the p-region, as shown in Fig. 4.3(c), is given as:

$$q\phi_p = E_F - E_i \qquad (4.4a)$$

Therefore, Eq. (4.3c) can be written, after using Eq. (4.4a), as:

$$p_0 = n_i\,\exp\left(\frac{-q\phi_p}{kT}\right) \qquad (4.4b)$$

Similarly, the thermal equilibrium electron concentration n_0 in the n-region in terms of electrostatic potential ϕ_n can be expressed as:

$$n_0 = n_i\,\exp\left(\frac{q\phi_n}{kT}\right) \qquad (4.4c)$$

Setting $n_0 = N_d$ and $p_0 = N_a$ in Eqs. (4.4b) and (4.4c), and after solving, the respective potential can be given as:

$$\phi_n = \frac{kT}{q} \ln\left(\frac{N_d}{n_i}\right) \tag{4.5a}$$

and

$$\phi_p = -\frac{kT}{q} \ln\left(\frac{N_a}{n_i}\right) \tag{4.5b}$$

The built-in-potential across the *p-n* junction diode, using Eqs. (4.2a), (4.5a) and (4.5b), is:

$$qv_{bi} = \phi_n - \phi_p = kT\left[\ln\left(\frac{N_a}{n_i}\right) + \ln\left(\frac{N_a}{n_i}\right)\right]$$

or

$$qv_{bi} = kT\ln\left(\frac{N_d N_a}{n_i^2}\right)$$

or

$$v_{bi} = \frac{kT}{q}\ln\left(\frac{N_d N_a}{n_i^2}\right)$$

$$v_{bi} = V_T\ln\left(\frac{N_d N_a}{n_i^2}\right) \tag{4.6}$$

where, $V_T = kT/q$ is the thermal voltage and it is equal to about 26 mV at the room temperature (300 K). Equation (4.6) gives the value of built-in-potential for abrupt *p-n* junction.

Equation (4.6) tells us that by controlling the concentration of donor and acceptor impurities in the respective regions, one can design a *p-n* junction diode of desired value of built-in-potential. This potential exists at thermal equilibrium and is a direct consequence of the junction between two dissimilar dopant.

4.3 ABRUPT JUNCTIONS

Analysis of a *p-n* junction diode is simpler if the junction is assumed to be abrupt, i.e., the doping impurities are assumed to change abruptly from *p*-type on one side to *n*-type on the other side of the junction, as shown in Fig. 4.4.

The abrupt junction approximation is reasonable for Modern VLSI devices. As seen from Fig. 4.3(c), the *p-n* junction diode is approximated by the three regions. First region with $x < -x_p$ is called **bulk *p*-region** and the second region $x > x_n$ is called **bulk *n*-region**. These two regions are also called **quasineutral regions**. Third region $-x_p < x < x_n$ is called **transition** or **depletion** or **space charge region**. This region

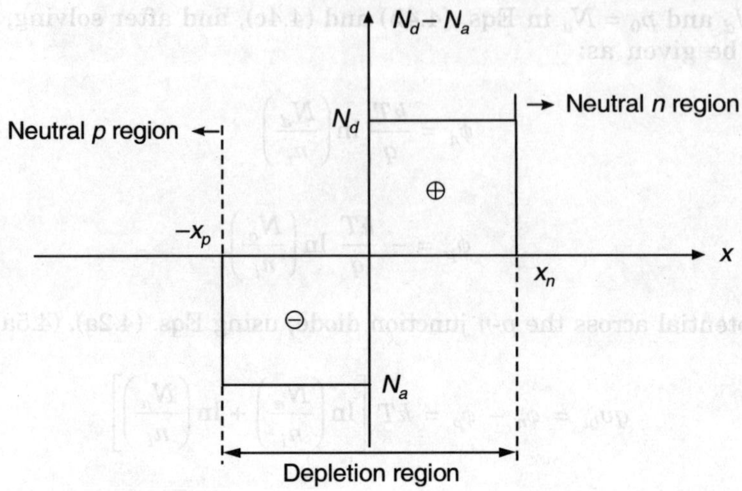

FIGURE 4.4 Distribution of doping impurities in abrupt junctions.

is assumed to be depleted of mobile carriers means there is no mobile carriers in this region, i.e., $n_0 = p_0 = 0$. This approximation is called **depletion approximation**. The analysis of an abrupt junction becomes more simpler by considering depletion approximation. The depletion approximation is quite accurate for all applied voltages except at large forward biases.

Since, there are two sources of charge in the silicon,

(i) Mobile carriers: electrons and holes (n_0, p_0),
(ii) Immobile carriers, or fixed charge: Ionized impurity atoms (N_d^+, N_a^-).

The Poisson's equation,

$$\frac{d^2\psi}{dx^2} = -\frac{q}{\varepsilon_{Si}}\left[p_0(x) - n_0(x) + N_d^+(x) - N_a^-(x)\right]$$

in depletion region, assuming depletion approximation $n_0(x) = p_0(x) = 0$, becomes

$$\frac{d^2\psi}{dx^2} = -\frac{q}{\varepsilon_{Si}}\left[N_d^+(x) - N_a^-(x)\right] \tag{4.7}$$

where ψ is the electrostatic potential, ε_{Si} is silicon permittivity.

For simplicity, it is assumed that all the donors and acceptors within the depletion region are ionized, i.e., $N_d^+(x) = N_d$, $N_a^-(x) = N_a$. We have also assumed that the junction is abrupt and not compensated, i.e., $N_d = 0$ for p-side and $N_a = 0$ for n-side.

Under these approximations (Depletion approximations), Eq. (4.7) can be written as:

$$\frac{d^2\psi}{dx^2} = -\frac{q}{\varepsilon_{Si}}N_d \qquad \text{for } 0 \leq x \leq x_n \tag{4.8a}$$

$$\frac{d^2\psi}{dx^2} = \frac{q}{\varepsilon_{Si}} N_a \qquad \text{for } -x_p \leq x \leq 0 \tag{4.8b}$$

In short, under Depletion approximation

$$\frac{d^2\psi}{dx^2} = \begin{cases} qN_a/\varepsilon_{Si} & \text{for } x_p \leq x \leq 0 \\ 0 & \text{for } x \leq -x_p \text{ and } x \geq x_n \\ -qN_d/\varepsilon_{Si} & \text{for } 0 \leq x \leq x_n \end{cases}$$

and charge density

$$\rho = \begin{cases} -qN_a & \text{for } -x_p \leq x \leq 0 \\ 0 & \text{for } x \leq -x_p \text{ and } x \geq x_n \\ qN_d & \text{for } 0 \leq x \leq x_n \end{cases}$$

Depletion approximation is good description only for a symmetrical *p-n* junction, i.e. $N_a \simeq N_d$ and under high reverse bias. It is not suitable for a single-side (one-sided) abrupt junctions.

The overall charge neutrality of the semiconductor requires that

$$N_a x_p = N_d x_n \tag{4.9}$$

where left hand side of Eq. (4.9) gives the total negative space charge per unit area in the *p*-side and right hand side gives the total positive space charge per unit area in the *n*-side.

The total depletion width is:

$$W = |x_n| + |x_p| \tag{4.10}$$

Integrating Eq. (4.8a) from $x = 0$ to $x = x_n$, and Eq. 4.8(b) from $x = -x_p$ to $x = 0$ and using boundary conditions $\left.\dfrac{d\psi}{dx}\right|_{x=-x_p} = 0 = \left.\dfrac{d\psi}{dx}\right|_{x=x_n}$, we will get the same expression as given by Eq. (4.9). Now, integrating Eq. (4.8a) from $x = 0$ to x_0 we get

$$\int dE = \frac{q}{\varepsilon_{Si}} N_d \int_{x=0}^{x_0} dx$$

or

$$E(x) = \frac{q}{\varepsilon_{Si}} N_d x + C_1 \tag{4.11a}$$

where C_1 is integration constant and can be determining by using $E(x) = 0$ at $x = x_n$, and hence

$$\frac{q}{\varepsilon_{Si}} N_d x_n = -C_1$$

Therefore, Eq. (4.11a) becomes

$$E(x) = \frac{q}{\varepsilon_{Si}} N_d(x - x_n) \qquad \text{for } 0 \leq x \leq x_n \qquad (4.11b)$$

Similarly, integrating Eq. (4.8b) from $x = 0$ to x_0 and using $E(x) = 0$ for $x = -x_p$, the electric field in the p-region is:

$$E(x) = -\frac{qN_a}{\varepsilon_{Si}} (x + x_p) \qquad x_p \leq x \leq 0 \qquad (4.11c)$$

Since, the space charge density in the p-region as well as in the n-region is assumed to be constant (for uniform doping), therefore, the electric field given by Eq. (4.11b) and Eq. (4.11c) must be a linear function of distance as shown in Fig. 4.5(c). Since, the electric field must be **continuous** at the metallurgical junction $x = 0$, therefore, from Eqs. (4.11b) and (4.11c) for $x = 0$, we get Eq. (4.12a).

The maximum electric field is obtained at $x = 0$, and is given as:

$$E_m = \frac{qN_d}{\varepsilon_{Si}} x_n = \frac{qN_a}{\varepsilon_{Si}} x_p \qquad (4.12a)$$

By using Eq. (4.12a), Eqs. (4.11b) and (4.11c) can be further written as:

$$E(x) = -E_m + \left(\frac{qN_d}{\varepsilon_{Si}}\right) x \qquad 0 \leq x \leq x_n \qquad (4.12b)$$

and
$$E(x) = -E_m - \left(\frac{qN_a}{\varepsilon_{Si}}\right) x \qquad -x_p \leq x \leq 0 \qquad (4.12c)$$

Equation (4.12a) shows that the total space charge inside n-region of the depletion region is equal to the total space charge inside p-region of depletion region but in opposite direction. Therefore, the two charge distribution plots have equal area as shown in Fig. 4.5(b).

To get the total potential drop across the p-n junction, namely built-in-potential, one can integrate Eqs. (4.11b) and (4.11c) from $-x_p$ to x_n, i.e.,

$$V_{bi} = -\int_{-x_p}^{x_n} E(x)\, dx = -\left.\int_{-x_p}^{0} E(x)\, dx\right|_{p\text{-side}} - \left.\int_{0}^{x_n} E(x)\, dx\right|_{n\text{-side}}$$

$$= \frac{qN_a x_p^2}{2\varepsilon_{Si}} + \frac{qN_d x_n^2}{2\varepsilon_{Si}}$$

$$= \frac{q}{2\varepsilon_{Si}} [N_d x_n^2 + N_a x_p^2] \qquad (4.12d)$$

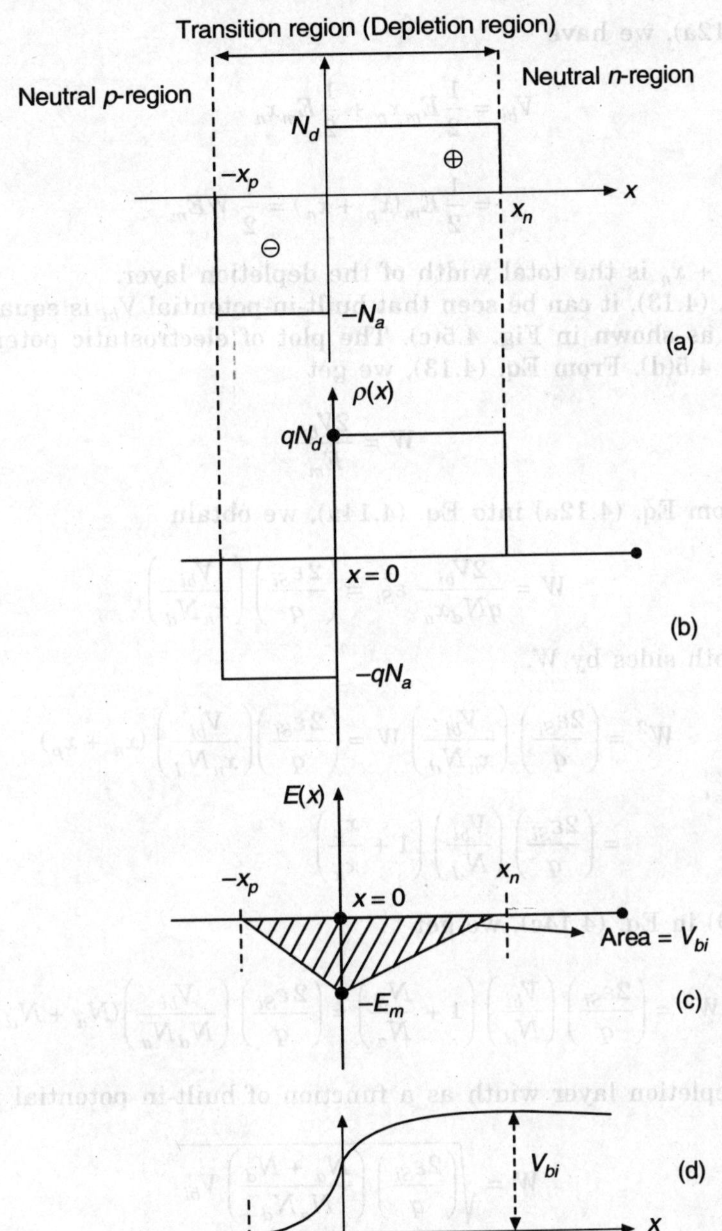

FIGURE 4.5 (a) Dopant distribution in abrupt *p-n* junction, (b) Charge distribution, (c) Electric field near the junction, (d) Plot of electrostatic potential in abrupt *p-n* junction.

Using Eq. (4.12a), we have

$$V_{bi} = \frac{1}{2} E_m x_p + \frac{1}{2} E_m x_n$$

$$= \frac{1}{2} E_m (x_p + x_n) = \frac{1}{2} W E_m \tag{4.13}$$

where $W = x_p + x_n$ is the total width of the depletion layer.

From Eq. (4.13), it can be seen that built-in-potential V_{bi} is equal to the area in $E(x) - x$ plot, as shown in Fig. 4.5(c). The plot of electrostatic potential w.r.t. x is shown in Fig. 4.5(d). From Eq. (4.13), we get

$$W = \frac{2V_{bi}}{E_m} \tag{4.14a}$$

Putting E_m from Eq. (4.12a) into Eq. (4.14a), we obtain

$$W = \frac{2V_{bi}}{qN_d x_n} \varepsilon_{Si} = \left(\frac{2\varepsilon_{Si}}{q}\right)\left(\frac{V_{bi}}{x_n N_d}\right) \tag{4.14b}$$

Multiplying both sides by W,

$$W^2 = \left(\frac{2\varepsilon_{Si}}{q}\right)\left(\frac{V_{bi}}{x_n N_d}\right) W = \left(\frac{2\varepsilon_{Si}}{q}\right)\left(\frac{V_{bi}}{x_n N_d}\right)(x_n + x_p)$$

$$= \left(\frac{2\varepsilon_{Si}}{q}\right)\left(\frac{V_{bi}}{N_d}\right)\left(1 + \frac{x_p}{x_n}\right) \tag{4.14c}$$

Using Eq. (4.9) in Eq. (4.14c), we get

$$W^2 = \left(\frac{2\varepsilon_{Si}}{q}\right)\left(\frac{V_{bi}}{N_d}\right)\left(1 + \frac{N_d}{N_a}\right) = \left(\frac{2\varepsilon_{Si}}{q}\right)\left(\frac{V_{bi}}{N_d N_a}\right)(N_a + N_d)$$

or the total depletion layer width as a function of built-in potential is:

$$W = \sqrt{\left(\frac{2\varepsilon_{Si}}{q}\right)\left(\frac{N_a + N_d}{N_a N_d}\right) V_{bi}} \tag{4.14d}$$

Equation (4.14d) shows the dependence of the total width of the depletion layer on the built-in-potential as well as on the doping concentration of the two sides of the diodes.

Substituting Eq. (4.9) in Eq. (4.12a), and after solving it, the space charge region or depletion width extending in n-region for zero applied voltage is given as:

$$x_n = \left[\frac{2\varepsilon_{Si}}{q} V_{bi}\left(\frac{N_a}{N_d}\right)\left(\frac{1}{N_a + N_d}\right)\right]^{1/2} \tag{4.15a}$$

Similarly, the depletion region extending into *p*-region is:

$$x_p = \left[\left(\frac{2\varepsilon_{Si}}{q}\right)V_{bi}\left(\frac{N_d}{N_a}\right)\left(\frac{1}{N_a + N_d}\right)\right]^{1/2}$$

(4.15b)

Using $W = x_n + x_p$ and putting values of x_n and x_p from Eq. (4.15a) and (4.15b), one can get the same expression as given by Eq. (4.14d).

4.4   ONE-SIDED ABRUPT *p-n* JUNCTION

When the impurity concentration of one side of an abrupt junction is heavily doped compare to the other side then the junction is called **one-sided abrupt junction**, and is represented by either p^+-n or p-n^+ where + sign indicates the heavy doping, i.e., p^+-n means $N_a \gg N_d$ and p-n^+ means $N_d \gg N_a$. The one sided abrupt p-n junction concept is useful in source/drain junction of MOSFET or the emitter-base junction of a bipolar transistor. The doping concentration distribution for one sided p-n junction is shown in Fig. 4.6(b). Since,

$$W = x_n + x_p = x_n\left(1 + \frac{x_p}{x_n}\right)$$

From Eq. (4.9),

$$W = x_n\left(1 + \frac{N_d}{N_a}\right) = x_n\left(\frac{N_a + N_d}{N_a}\right)$$

or

$$x_n = \left(\frac{N_a}{N_a + N_d}\right)W$$

(4.15c)

For n^+-p diode,

$$N_d \gg N_a$$

Hence Eq. (4.15c) reduces to

$$x_n = \left(\frac{N_a}{N_d}\right)W$$

(4.16a)

and

$$x_p \simeq W$$

(4.16b)

From Eqs. (4.16a) and (4.16b), it is clear that whole depletion width will move onto the *p*-side which is lightly doped. Similarly, for p^+-n diode, depletion width will mainly extend into lightly doped *n*-region as shown in Fig. 4.6(a).

Therefore, the characteristics of a one-sided *p-n* diode are primarily determined by the properties of the lightly doped side alone. In this case, practically all the voltage drop occurs across the lightly doped side of the diode.

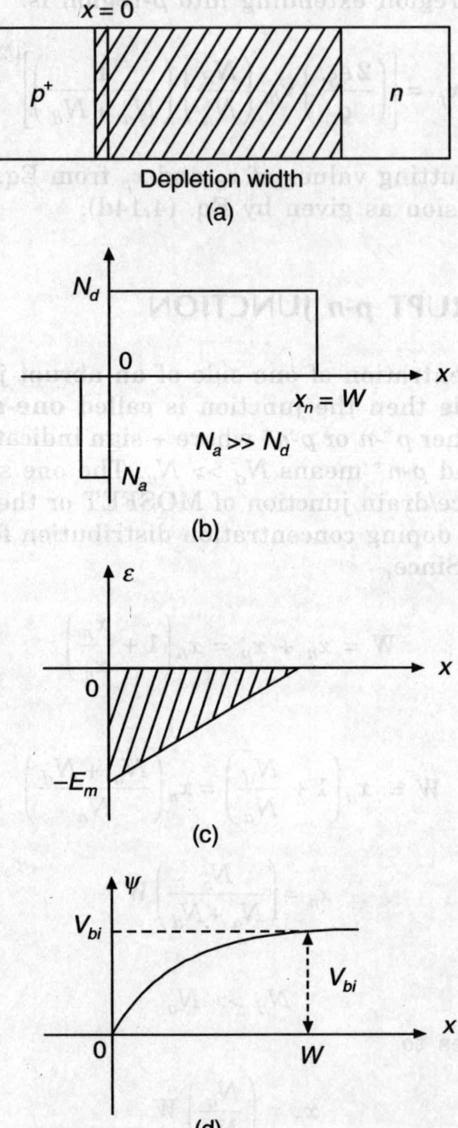

FIGURE 4.6 (a) Schematic representation of one-sided *p-n* junction, (b) Dopant distribution, (c) Field variation with *x*, (d) Plot of electrostatic potential with *x* in one sided *p-n* junction.

Equations 4.15(a) and 4.15(b), for one-sided junction either n^+-p or p^+-n diode is reduced to

$$x_n \simeq \sqrt{\frac{2\,\varepsilon_{Si} V_{bi}}{q N_d}} \quad p^+\text{-}n \text{ diode} \qquad (4.17a)$$

and
$$x_p \simeq \sqrt{\frac{2\varepsilon_{Si}V_{bi}}{qN_a}} \quad n^+\text{-}p \text{ diode} \tag{4.17b}$$

For one-sided p-n junction diode, Eqs. (4.12b) and (4.12c) can be expressed as:

$$E(x) = -E_m \pm \frac{qN_Bx}{\varepsilon_{Si}} \tag{4.18a}$$

where N_B is the lightly doped bulk concentration, i.e., N_d for p^+-n junction and N_a for p-n^+ junction. As clear from Fig. 4.6(c) that field decreases to zero at $x = W$. And hence, at $x = W$, Eq. (4.18a) reduces to

$$E_m = \frac{qN_BW}{\varepsilon_{Si}} \tag{4.18b}$$

Putting the value of Eq. (4.18b) into Eq. (4.18a), we get

$$E(x) = -\frac{qN_BW}{\varepsilon_{Si}} + \frac{qN_Bx}{\varepsilon_{Si}}$$

$$= -\frac{qN_BW}{\varepsilon_{Si}}\left[1 - \frac{x}{W}\right]$$

or
$$E(x) = -E_m\left[1 - \frac{x}{W}\right] \tag{4.18c}$$

Integrating Eq. (4.18c) from $x = 0$ to x and using $\psi(x = 0) = 0$, the potential distribution for one-sided junction is:

$$\psi(x) = \frac{V_{bi}x}{W}\left(2 - \frac{x}{W}\right) \tag{4.18d}$$

The potential distribution is shown in Fig. 4.6(d). It is clear from Eq. (4.18d) that for $x = W$,

$$\psi(W) = V_{bi}$$

i.e., total voltage drop across the depletion region must be equal to built-in-potential.

EXAMPLE 4.1

(a) Find the built-in-potential for a silicon p-n junction where doping concentrations in both regions are:

$$N_a = 10^{18}/\text{cm}^3 \text{ and } N_d = 10^{16}/\text{cm}^3$$

Take $T = 300$ K.

(b) Find the depletion region width.

(c) Draw the depletion width.

(d) Find x_n and x_p and verify the result W.

(e) Find the maximum electric field and plot it.

(f) Find the electrostatic potentials (ϕ_n, ϕ_p) on both sides of the junction and then find built-in-potential.

Solution: (a) Since

$$V_{bi} = \frac{kT}{q} \ln \left[\frac{N_a N_d}{n_i^2} \right]$$

$$= 0.0259 \ln \left[\frac{10^{18} \times 10^{16}}{2.25 \times 10^{20}} \right]$$

$$= 0.0259 \ln \left[\frac{10^{14}}{2.25} \right] = 0.814 \text{ V}$$

(b)

$$W = \sqrt{\frac{2\varepsilon_{Si}}{q} \left(\frac{N_a + N_d}{N_a N_d} \right) V_{bi}}$$

$$= \sqrt{\frac{2 \times 1.04 \times 10^{-12}}{1.6 \times 10^{-19}} \left(\frac{10^{18} + 10^{16}}{10^{34}} \right) \times 0.814}$$

$$= \sqrt{1.3 \times 10^7 \times 10^{-34} \times 101 \times 10^{16} \times 0.814}$$

$$= \sqrt{106.8782 \times 10^{-11}} = 3.269 \times 10^{-5} \text{ cm}$$

$$= 0.33 \ \mu\text{m}$$

(c)

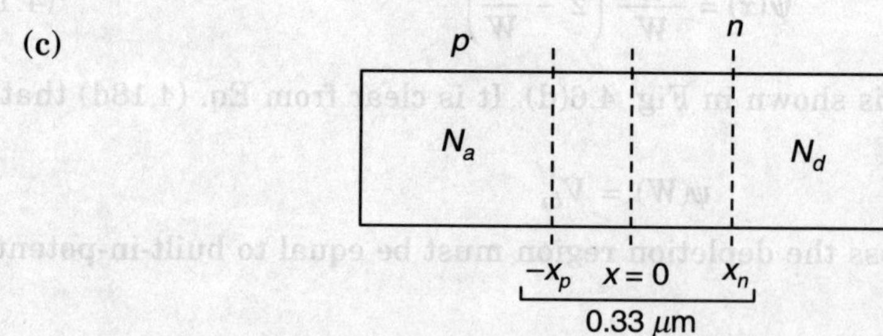

(d) Using Eqs. (4.15a) and (4.15b), we have

$$x_n = \left[1.294 \times 10^7 \times 0.814 \times 10^2 \times \frac{10^{-16}}{10} \right]^{1/2}$$

$$= [0.0104 \times 10^{-7}]^{1/2}$$

$$= 0.322 \ \mu m$$

$$x_p = [0.003 \ \mu m]$$

$$x_n + x_p = W \cong 0.325 \ \mu m$$

(e)
$$E_m = \frac{qNdx_n}{\varepsilon_{Si}}$$

$$= \frac{1.6 \times 10^{-19} \times 10^{16} \times 0.325 \times 10^{-4}}{1.035 \times 10^{-12}}$$

$$= 0.5024 \times 10^5 = 5.024 \times 10^4 \ \text{V/cm}$$

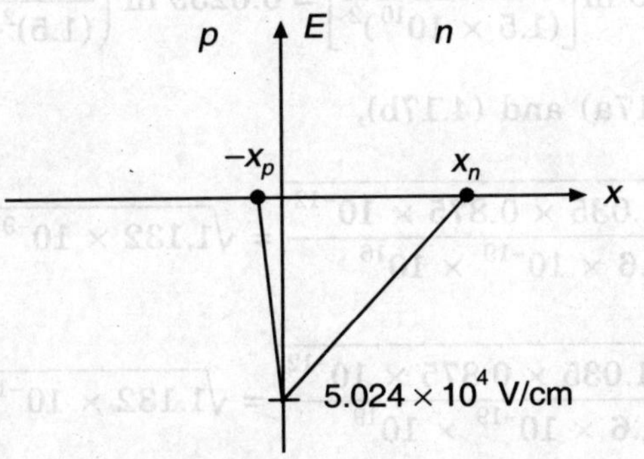

(f)
$$\phi_n = \frac{kT}{q} \ln\left(\frac{N_d}{n_i}\right)$$

$$= 0.0259 \ln\left(\frac{10^{16}}{1.5 \times 10^{10}}\right) = 0.347 \ \text{V}$$

and
$$\phi_p = -\frac{kT}{q} \ln\left(\frac{N_d}{n_i}\right)$$

$$= -0.0259 \ln\left(\frac{10^{18}}{1.5 \times 10^{10}}\right) = -0.467 \ \text{V}$$

Then
$$V_{bi} = |\phi_n| + |\phi_p| = 0.814 \ \text{V}$$

EXAMPLE 4.2

For a silicon one-sided abrupt junction, following parameters are given as:

$$N_a = 10^{19}/\text{cm}^3, \ N_d = 10^{16}/\text{cm}^3, \ T = 300 \text{ K}$$

(a) Calculate the built-in-potential.
(b) Find the depletion layer width along p-side as well as n-side.
(c) Find the depletion layer width.
(d) Find the maximum electric field at zero applied potential.
(e) Find the potential at $x = W$.
(f) Show the distribution of electric field as well as potential with x.

Solution: (a) Using Eq. (4.6), we have

$$V_{bi} = 0.0259 \ln\left[\frac{10^{19} \times 10^{16}}{(1.5 \times 10^{10})^2}\right] = 0.0259 \ln\left(\frac{10^{15}}{(1.5)^2}\right) = 0.873 \text{ V}$$

(b) Using Eqs. (4.17a) and (4.17b),

$$x_n = \sqrt{\frac{2 \times 1.035 \times 0.875 \times 10^{-12}}{1.6 \times 10^{-19} \times 10^{16}}} = \sqrt{1.132 \times 10^{-9}} = 0.336 \ \mu\text{m}$$

$$x_p \simeq \sqrt{\frac{2 \times 1.035 \times 0.875 \times 10^{-12}}{1.6 \times 10^{-19} \times 10^{19}}} = \sqrt{1.132 \times 10^{-12}} = 0.011 \ \mu\text{m}$$

This shows that the depletion region is mainly extending into the lightly doped regions.

(c)
$$W \simeq x_n \simeq 0.34 \ \mu\text{m}$$

(d) From Eq. (4.18b),

$$E_m = \frac{(q \times 0.34 \ \mu\text{m}) \times 10^{16}}{1.035 \times 10^{-12}} = \frac{1.6 \times 10^{-19} \times 0.34 \times 10}{1.035 \times 10^{-12}}$$

$$= 0.526 \times 10^5 \text{ V/cm}$$

(e) From Eq. (4.18d), it is clear that at $x = W$,

$$\psi_s(W) = \text{Potential} = 0.873 \text{ V}$$

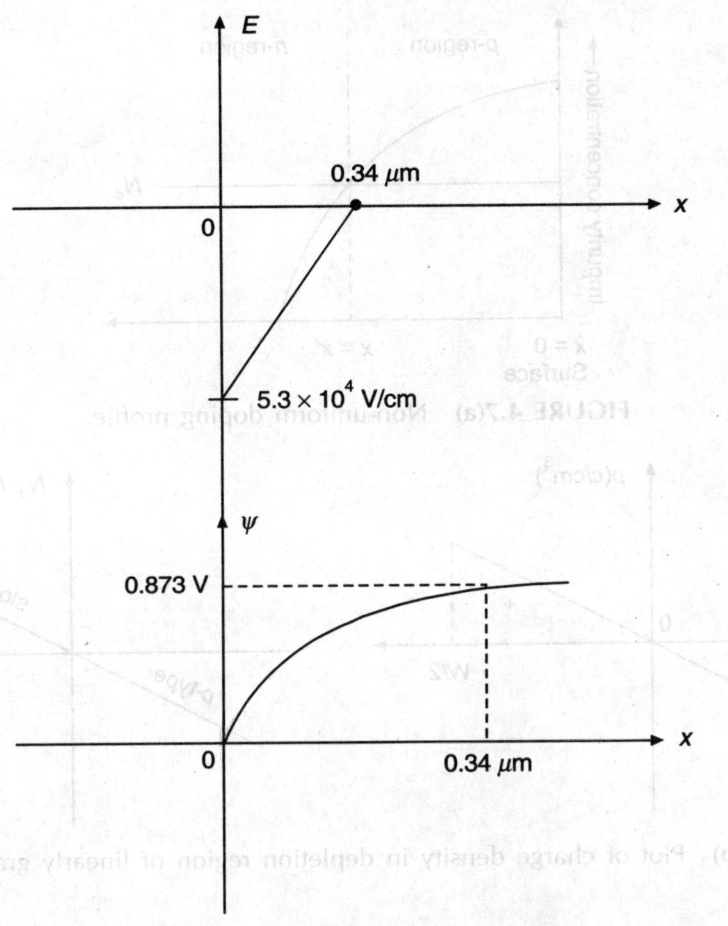

4.5 NON-UNIFORMLY DOPED SEMICONDUCTOR

In many electronic device applications, specific non-uniform doping profile, as shown in Fig. 4.7(a), is used to get some special *p-n* junction capacitance characteristics.

We will consider only linearly graded junctions. In linearly graded junction, the net dopand concentration varies linearly from *p*-type material to *n*-type material as shown in Fig. 4.7(b), i.e., $N(x) = N_d - N_a = ax$. Figure 4.7(b) shows the space charge density in the depletion region of linearly graded junction.

The space charge density for linearly graded junction can be expressed as:

$$\rho(x) = qax \qquad (4.19)$$

where *a* is the impurity gradient, and its unit is /cm^4. The doping concentrations at the edges of the space charge region (−W/2, W/2) are expressed as:

$$N_d(W/2) = aW/2 \qquad (4.20a)$$

$$N_a(-W/2) = aW/2 \qquad (4.20b)$$

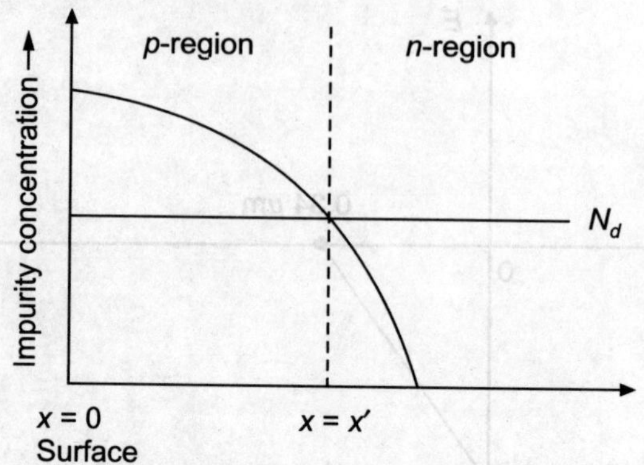

FIGURE 4.7(a) Non-uniform doping profile.

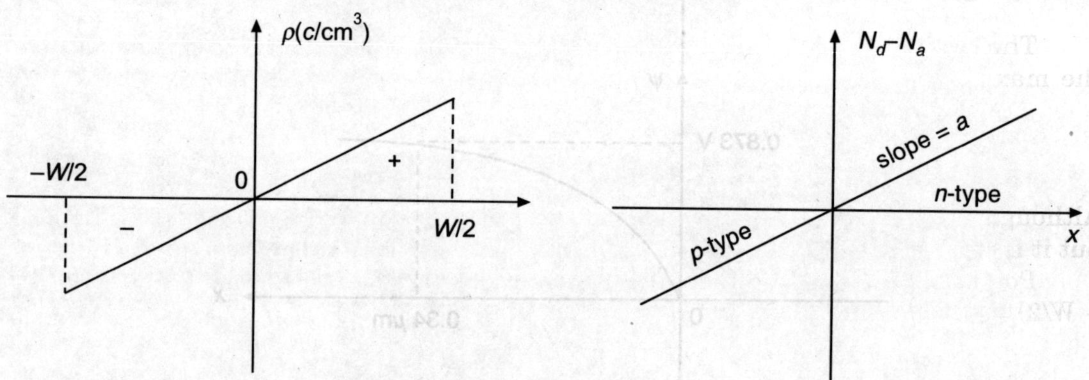

FIGURE 4.7(b) Plot of charge density in depletion region of linearly graded junction.

The Poisson equation for this case is:

$$\frac{d^2\psi}{dx^2} = -\frac{dE(x)}{dx} = -\frac{\rho(x)}{\varepsilon_{Si}} = -\frac{q}{\varepsilon_{Si}}\, ax; \quad -\frac{W}{2} \le x \le \frac{W}{2} \tag{4.21a}$$

and

$$\frac{d^2\psi}{dx^2} = 0 \quad \text{for } x \le -\frac{W}{2} \text{ and } x \ge \frac{W}{2}$$

where, W is the width of the depletion region.

Now, integrating Eq. (4.21a) from $W/2$ to x and using boundary condition that electric field is zero at $\pm W/2$, the electric field is:

$$E(x) = \frac{qa}{2\,\varepsilon_{si}}\left(x^2 - \left(\frac{W}{2}\right)^2\right); \quad -\frac{W}{2} \le x \le \frac{W}{2} \tag{4.21b}$$

And the electric field distribution is shown in Fig. 4.7(c).

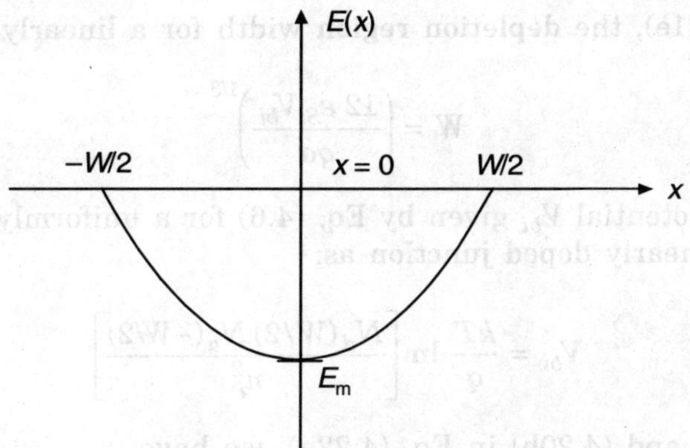

FIGURE 4.7(c)   Electric field distribution in linearly graded junction.

The maximum electric field is obtained at $x = 0$. Now, putting $x = 0$ in Eq. 4.21(b) the maximum field is given as:

$$E_m = -\frac{qa}{8\varepsilon_{Si}} W^2 \tag{4.21c}$$

Although the electric field in a non-homogeneous junction diode is not zero in the bulk but it is very small, so setting $E(x) = 0$ in the bulk region is still a good approximation.

Potential distribution is obtained by integrating Eq. (4.21b) and setting $\phi(x = -W/2) = 0$. The potential across the junction is:

$$\phi(x) = -\frac{qa}{2\varepsilon_{Si}} \left[\frac{x^3}{3} - \left(\frac{W}{2}\right)^2 x \right] + \frac{qa}{3\varepsilon_{Si}} \left(\frac{W}{2}\right)^3 \tag{4.21d}$$

The magnitude of the potential at $x = +(W/2)$ will be equal to the built-in-potential, and hence from Eq. (4.21d),

$$\phi\left(x = \frac{W}{2}\right) = \frac{qaW^3}{12\varepsilon_{Si}} = V_{bi} \tag{4.21e}$$

The potential distribution is shown in Fig. 4.7(d).

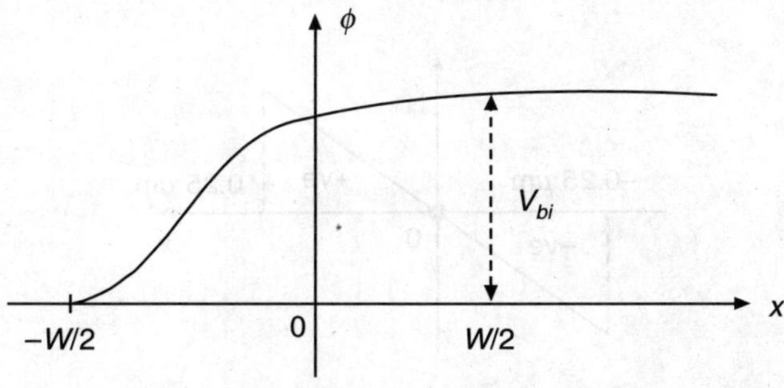

FIGURE 4.7(d)   Variation of electrostatic potential in depletion region of linearly graded junction.

From Eq. (4.21e), the depletion region width for a linearly graded junction is:

$$W = \left(\frac{12\,\varepsilon_{Si}V_{bi}}{qa}\right)^{1/3} \tag{4.21f}$$

The built-in potential V_{bi} given by Eq. (4.6) for a uniformly doped junction can be written for a linearly doped junction as:

$$V_{bi} = \frac{kT}{q}\ln\left[\frac{N_d(W/2)\,N_a(-W/2)}{n_i^2}\right] \tag{4.22a}$$

Using Eqs. (4.20a) and (4.20b) in Eq. (4.22a), we have

$$V_{bi} = \frac{kT}{q}\ln\left[\frac{(aW/2)(aW/2)}{n_i^2}\right]$$

$$V_{bi} = \frac{kT}{q}\ln\left[\frac{(aW)}{2\,n_i}\right]^2 = \frac{2kT}{q}\ln\left[\frac{aW}{2\,n_i}\right] \tag{4.22b}$$

EXAMPLE 4.3

The space charge density for a given junction is given by

$$\rho(x) = q \times 10^{20}x$$

The depletion width is 0.5 μm. Take $T = 300$ K.

(a) Draw the space charge density with x.
(b) Calculate the maximum electric field.
(c) Find the built-in-potential.
(d) Find the concentrations of donor and acceptor at the edges of the space charge regions.

Solution: (a)

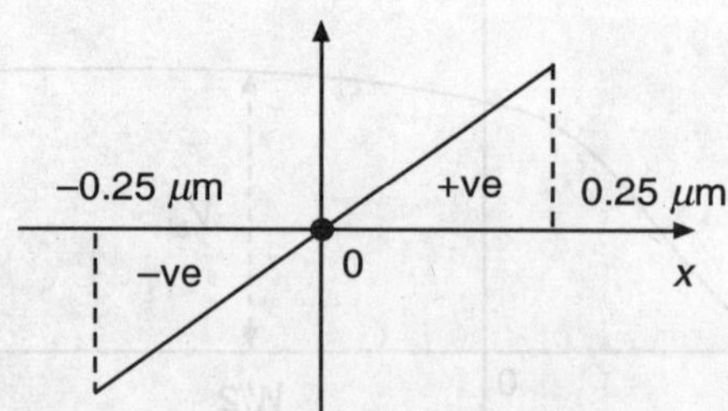

(b) From Eq. (4.21c),

$$|E_m| = \frac{-1.6 \times 10^{-19} \times 10^{20}(0.5 \times 10^{-4})^2}{8 \times 11.9 \times 8.85 \times 10^{-14}}$$

$$= 4.75 \times 10^3 \text{ V/cm}^2$$

(c) Using Eq. (4.22d),

$$V_{bi} = 2 \times \frac{0.0259}{m} \ln\left[\frac{10^{20} \times 0.5 \times 10^{-4}}{2 \times 1.5 \times 10^{10}}\right]$$

$$= 0.052 \ln\left[\frac{0.5 \times 10^{16} \times 10^{-10}}{3}\right] = 0.625 \text{ V}$$

(d)
$$N_d(W/2) = 10^{20} \times 0.25 \times 10^{-4} = 0.25 \times 10^{16}/\text{cm}^3$$
$$N_a(W/2) = 10^{20} \times 0.25 \times 10^{-4} = 0.25 \times 10^{16}/\text{cm}^3$$

4.6 BIASING ON THE UNIFORMLY DOPED *p-n* JUNCTION

At thermal equilibrium, the drift component of the current caused by induced electric field in the depletion region is exactly equal to the diffusion current component caused by the concentration gradients across the metallurgical junction. Therefore, the net current flow is zero in the diode. When an external voltage is applied between the *p* and *n* regions, the device is no longer at the thermal equilibrium and current component balance is disturbed, and a resulting current will flow in the junction diode.

The most interesting and practical properties of a *p-n* junction are observed under non-equilibrium condition. Non-equilibrium condition is due to applied bias voltage or due to illumination. Based on the polarity of applied external bias voltage, a *p-n* junction diode can be biased in two ways.

If we apply a positive potential to the *p*-region and negative potential to the *n*-region, the *p-n* junction is said to be forward biased. Similarly, if *n*-region is connected to the positive voltage and *p*-region with negative voltage, the junction is said to be reverse biased.

We will consider the current flow in the diode for these two biasing conditions separately.

4.6.1 Forward-Biased *p-n* Junction

The forward-biased junction is shown in Fig. 4.8(b) where the positive terminal of the battery (V_f) is connected to *p*-region and the negative terminal to *n*-region. In this arrangement, more free electrons are injected from the *n*-side into the *p*-side, and

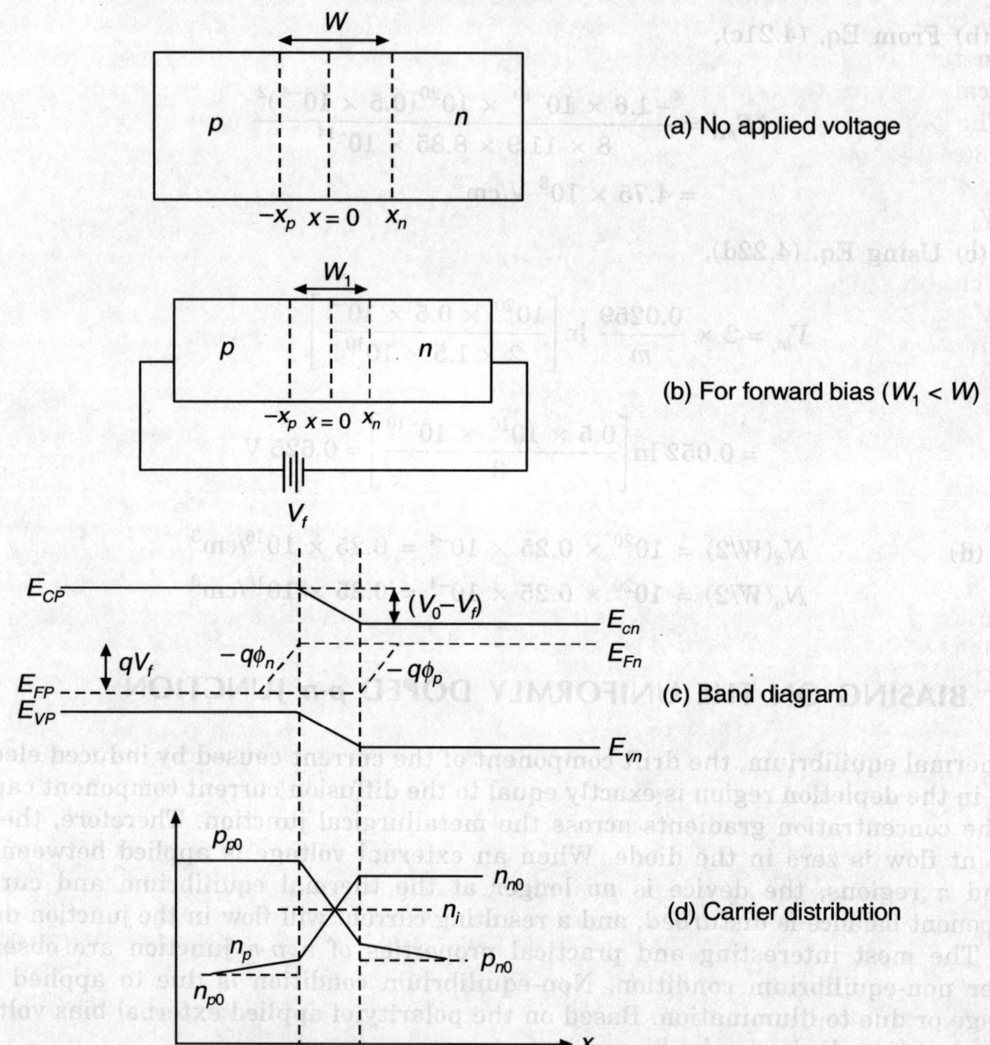

FIGURE 4.8   (a) Schematic representation of *p-n* junction under no applied bias, (b) Schematic representation of *p-n* junction in forward bias condition, (c) Band diagram formation in forward bias condition, (d) Carrier distribution in forward bias *p-n* junction.

more holes are injected from *p*-side into *n*-side. This means, an enhanced hole diffusion from *p* to *n* side and electron diffusion from *n* to *p*-side. As a result, space charge region becomes effectively narrower and energy barrier decreases to an insignificant value. The drift current is reduced in comparison to the diffusion current. The electrostatic potential barrier at the junction is lowered due to forward bias from the equilibrium contact potential V_{bi} to $(V_{bi} - V_f)$. The lowering of the potential barrier occurs because forward bias raises the electrostatic potential on the *p*-side relative to *n*-side. This shows that electrons now receive sufficiently large energy to surmount the junction barrier. The electric field within the space charge region decreases with forward bias. This in turn means the width of the space charge region is reduced because fewer electrical charges are required to maintain this electric field.

The band diagram in the presence of applied forward bias voltage is shown in Fig. 4.8(c). Carrier concentration distribution in forward biased *p-n* junction is shown in Fig. 4.8(d).

W_1 is the forward bias depletion layer width. E_{Fn} and E_{Fp} are called quasi-Fermi levels due to non-equilibrium condition. Therefore, the total voltage drop across the space charge region or transition region or depletion region in forward bias is $(V_{bi} - V_f)$. A large current results due to lowering of potential barrier even at very small increase in forward bias voltage.

4.6.2 Reverse-Biased *p-n* Junction

In reverse bias, the positive terminal of the battery is connected to *n*-type and the negative terminal to the *p*-type material as shown in Fig. 4.9(b), whereas Fig. 4.9(a) shows the unbiased *p-n* junction at equilibrium. In this arrangement, the free electrons in the *n*-type material are attracted towards the positive terminal of the battery, i.e., away from the junction. At the same time, holes from the *p*-type material are attracted towards the negative terminal of the battery, i.e., away from the junction. Due to that space charge region at the junction becomes effectively wider than the equilibrium space charge region width. Hence, the barrier becomes so large $(V_{bi} + V_r)$ that virtually no electron in the conduction band of *n*-region or holes in valence band of the *p*-region get enough energy to surmount this barrier. Therefore, the diffusion current component is usually negligible. The band diagram is shown in Fig. 4.9(c). Figure 4.9(d) shows the carrier distribution in the reverse biased *p-n* junction. Here V_r is applied reverse bias voltage.

The current flow is extremely small and electric field within the transition region increases with increase in applied reverse voltage. Therefore, a small current or zero current flows in the reverse bias *p-n* junction diode. From above discussion it is observed that a large current flows in forward bias *p-n* junction, whereas very small or negligible current flows in reverse bias *p-n* junction. This asymmetry of the current flow makes the *p-n* junction diode as a rectifier, in which the alternating current is converted into direct current. Biased *p-n* junction can be used as a voltage-variable capacitor, photocells, light emitters, etc.

Equations (4.14d), (4.15a), (4.15b) can be used to find out the values of W_1, W_2, x_n and x_p in the presence of applied voltage if V_{bi} is replaced by $(V_{bi} - V_a)$ where $V_a = +V_f$ for forward bias and $V_a = -V_r$ in reverse biased.

Notes: (i) It is assumed that the total applied voltage drop across the space charge or transition region of the junction rather than in the quasineutral *n*- and *p*-regions. Although, some voltage drop occurs in the quasineutral region, if current flows through it. But in most of the *p-n* junction diodes, the length of each quasineutral region is small as compared to its area, and also doping is usually moderate to heavy in these regions. According to relation:

$$R = \frac{\rho L}{A}$$

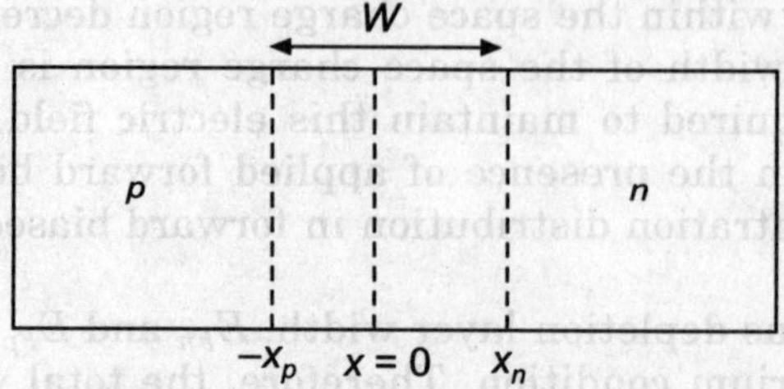

(a) For thermal condition, no applied voltage

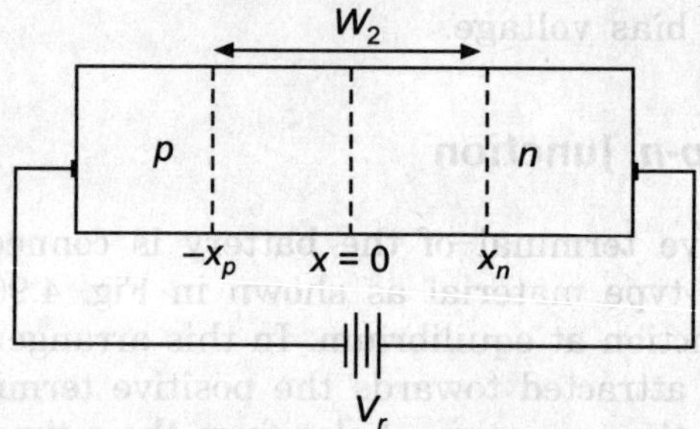

(b) Under reverse bias ($W_2 > W$)

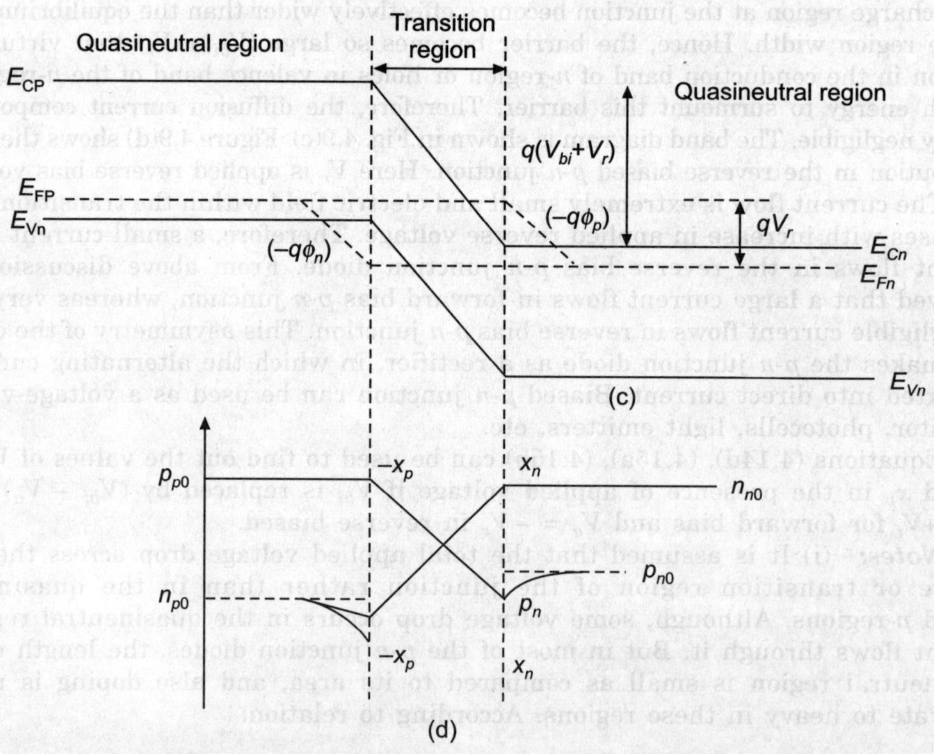

FIGURE 4.9 (a) Schematic representation of *p-n* junction at thermal equilibrium, (b) *p-n* junction under reverse bias condition, (c) Band formation in reverse biased *p-n* junction, (d) Carrier distribution in *p-n* junction when it is reverse biased.

The resistance of each quasineutral region is very small and hence only small voltage drop can occur outside the transition region and for almost all calculation this drop is ignored, and it is assumed that all the bias voltage drops across the transition region.

(ii) As clear from Fig. 4.3(c) that the separation of energy bands is a direct measure of the electrostatic potential barrier at the junction, and the barrier height is given by qV_{bi} where V_{bi} is the height of electrostatic potential or built-in-potential. Thus, the bands (conduction band/valence band) are less separated under forward bias than its equilibrium condition by $q(V_{bi} - V_f)$ as shown in Fig. 4.8(b) and more than equilibrium value under reverse bias by $q(V_{bi} + V_r)$ as shown in Fig. 4.9(c). Whereas in quasineutral regions equilibrium is maintained. The shifting of bands from their equilibrium condition under bias implies a separation of Fermi level on either side of the junction as shown in Figs. 4.8(b) and 4.9(c). Under forward bias condition, the Fermi level E_{Fn} at the n-side is above than the Fermi level E_{Fp} on the p-side by energy qV_f, i.e., $E_{Fn} - E_{Fp} = qV_f$. Similarly, for reverse bias, the Fermi level E_{Fp} on p-side is higher than Fermi level E_{Fn} by qV_r, i.e., $E_{Fp} - E_{Fn} = qV_r$. In other words, we can say that the separation of Fermi levels in these two quasineutral regions will be equal to the applied voltage V_f or V_r.

4.7 CURRENT-VOLTAGE CHARACTERISTICS

4.7.1 Ideal Characteristics

Here, we will consider the current-voltage characteristics based on the following assumptions:

1. Assuming abrupt junction and the depletion approximation are applicable.
2. The carrier concentration at the boundaries of space charge regions are related by the electrostatic potential difference V_{bi} across the junction.
3. The low-level injection maintained in the quasineutral (bulk) region of the device.
4. Assuming no generation and recombination in the depletion region, then hole current leaving the p-side is same as the hole current entering the n-side. Same is also true for electrons. This means, electron and hole currents are constant throughout the depletion region. To avoid the recombination process, it is assumed that the length of the neutral regions is short enough, i.e. short base diode. For this to be valid, the width of p- and n-regions must be substantially smaller than respective diffusion length (L_p and L_n). If this is not the case, the diode becomes of the long-base type.
5. There are no external sources of carrier generation.
6. The steady-state dc solution is required.
7. The bulk regions are uniformly doped and electric field for the minority carriers is zero in the bulk region. Resistance of the neutral region is negligible.

Therefore, to determine the total current flowing in the diode, it is required to determine the hole current entering the n-side and electron current entering the p-side.

Let us assume n_{n0} represents the electron concentration on n-side at thermal equilibrium and n_{p0} is electron concentration on p-side at equilibrium. Similarly, p_{p0} and p_{n0} are the hole concentration on p-side and n-side at thermal equilibrium. Also, assuming at thermal equilibrium,

$$n_{n0} \simeq N_d$$

and

$$p_{p0} \simeq N_a$$

Hence, Eq. (4.6), becomes

$$V_{bi} = \frac{kT}{q} \ln\left[\frac{n_{n0}p_{p0}}{n_i^2}\right]$$

$$= \frac{kT}{q} \ln\left[\frac{n_{n0} \times n_i^2/n_{p0}}{n_i^2}\right]$$

$$= \frac{kT}{q} \ln\left[\frac{n_{n0}}{n_{p0}}\right]$$

or

$$n_{n0} = np_0 e^{qV_{bi}/kT} \tag{4.23a}$$

Similarly, we have

$$p_{p0} = p_{n0} e^{qV_{bi}/kT} \tag{4.23b}$$

Equations (4.23a) and (4.23b) suggest that the electron concentration and hole concentration at the two boundaries of the space charge region are related through the electrostatic or built-in-potential V_{bi} at thermal equilibrium.

These two Eqs. (4.23a) and (4.23b) are also true for unequilibrium condition, i.e., when junction is subjected to forward bias or reverse bias by replacing V_{bi} to $(V_{bi} - V_a)$ where $V_a = V_f$ for forward bias and $V = -V_r$ for reverse bias.

Therefore,

$$n_n = n_p e^{q(V_{bi} - V_a)/kT} \tag{4.24a}$$

and

$$p_p = p_n e^{q(V_{bi} - V_a)/kT} \tag{4.24b}$$

where p_p and p_n are hole concentration in p-region and n-region for the non-equilibrium condition. Similarly, n_p and n_n are the non-equilibrium electron concentration in p-side and n-side respectively.

Dividing, Eqs. (4.23a) and (4.24a), we have

$$\left(\frac{n_{n0}}{n_n}\right) = \frac{n_{p0}}{n_p} \cdot e^{qV_a/kT} \tag{4.25a}$$

Since, for low-level injection, thermal equilibrium majority carrier concentration remain unchanged due to injection means,

$$n_n \simeq n_{n0}$$

Therefore, Eq. (4.25a) becomes

$$n_p(x = -x_p) = n_{p0}\, e^{qV_a/kT} \tag{4.25b}$$

Similarly,

$$p_n(x = x_n) = p_{n0}\, e^{qV_a/kT} \tag{4.25c}$$

These two Eqs. (4.25b) and (4.25c) can be further expressed as:

$$\Delta n_p(-x_p) = n_p - n_{p0} = n_{p0}(e^{qV_a/kT} - 1) \tag{4.26a}$$

and

$$\Delta p_n(x_n) = p_n - p_{n0} = p_{n0}(e^{qV_a/kT} - 1) \tag{4.26b}$$

where Δn_p and Δp_n are the excess minority electron concentration and excess minority hole concentration at the edges of the space charge region, i.e., at $-x_p$ and x_n respectively. Equations (4.25b) and (4.25c) are the two most important boundary conditions governing a *p-n* junction diode. They relate the minority carrier concentrations at the boundaries of the depletion layer with their thermal equilibrium values and to the applied voltage across the junction.

For forward bias, $V_a > 0$,

$$n_p \gg n_{p0} \text{ at } x = -x_p$$
$$p_n \gg p_{n0} \text{ at } x = x_n$$

For reverse bias, $V_a < 0$,

$$n_p \ll n_{p0} \text{ at } x = -x_p$$
$$p_n \ll p_{n0} \text{ at } x = x_n$$

According to fourth assumption, there is no generation or recombination within the depletion region means all currents come from the neutral regions. The steady state continuity equation with no generation for holes on *n*-side neutral region where the electric field is zero (zero drift current), is:

$$\frac{d^2 p_n}{dx^2} - \frac{p_n - p_{n0}}{D_p \tau_p} = 0 \tag{4.27a}$$

The solution of Eq. (4.27a) with the boundary conditions of Eq. (4.26b) and p_n $(x = \infty) = p_{n0}$ gives

$$p_n - p_{n0} = p_{n0}(e^{qV_a/kT} - 1)\, e^{-(x - x_n)/L_p} \quad x \geq x_n \tag{4.27b}$$

where $L_p = \sqrt{D_p \tau_p}$ is called diffusion length of minority carrier holes in the *n*-region.

Since, there is no electric field in the *p*-region, there is no hole drift-current component, only hole diffusion-current component exist at $x = x_n$.

$$J_{pn}(x_n) = -qD_p \left.\frac{dp_n}{dx}\right|_{x=x_n}$$

$$= \frac{qD_p \cdot p_{n0}}{L_p}(e^{qV_a/kT} - 1) \qquad (4.27c)$$

Similarly, for the neutral p-region,

$$n_p - n_{p_0} = n_{p_0}(e^{qV_a/kT} - 1)\, e^{(x+x_p)/L_n} \quad x \le -x_p \qquad (4.28a)$$

and diffusion current due to electron in p-region is

$$J_{n_p}(-x_p) = qD_n \left.\frac{dn_p}{dx}\right|_{x=-x_p} = \frac{qD_n n_{p0}}{L_n}(e^{qV_a/kT} - 1) \qquad (4.28b)$$

where $L_n = \sqrt{D_n \tau_n}$ is the diffusion length of electrons in the p-region.

The total current is constant throughout the device, and is given by the sum of the electron and hole current densities, i.e.,

$$J = J_p(x_n) + J_p(-x_p)$$

From Eqs. (4.27c) and (4.28b), we get

$$J = (e^{qV_a/kT} - 1)\left[\frac{qD_p p_{n0}}{L_p} + \frac{qD_n n_{p0}}{L_n}\right] = qn_i^2\left[\frac{D_p}{L_p N_d} + \frac{D_n}{L_n N_0}\right](e^{qV_a/kT} - 1)$$

This equation is also known as Shockley equation which is based on the number of assumption which might not be valid for real diode.

or $\qquad\qquad J = J_s(e^{qV_a/kT} - 1) \qquad\qquad\qquad (4.29a)$

or $\qquad\qquad I = I_s(e^{qV_a/kT} - 1) \qquad\qquad\qquad (4.29b)$

where

Saturation current density $J_s = \left[\dfrac{qD_p p_{n0}}{L_p} + \dfrac{qD_n n_{p0}}{L_n}\right]$

and Saturation current $I_s = \left[\dfrac{qD_p p_{n0}}{L_p} + \dfrac{qD_n n_{p0}}{L_n}\right] \cdot A$

Equation (4.29) is the ideal diode equation, A is cross-sectional area of the p-n junction diode, I_s is the reverse saturation current V_a is applied bias voltage which is positive for forward bias and negative for reverse biased, T is absolute temperature and k is Boltzmann's constant $= 1.38 \times 10^{-23}$ J/k. The ideal current-voltage characteristics is shown in Fig. 4.10(a) and minority carrier concentration

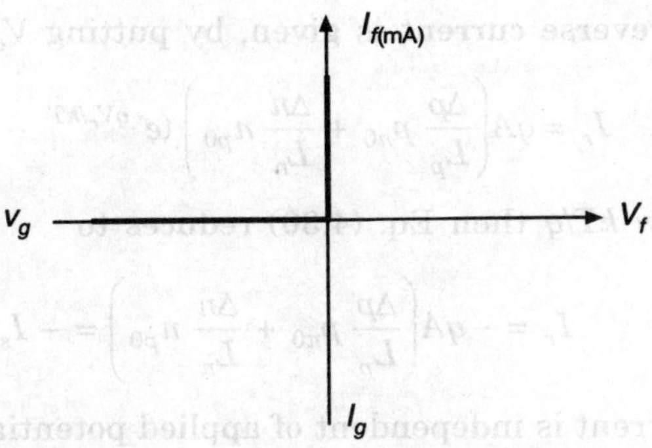

FIGURE 4.10(a) *V-I* characteristics of an ideal diode.

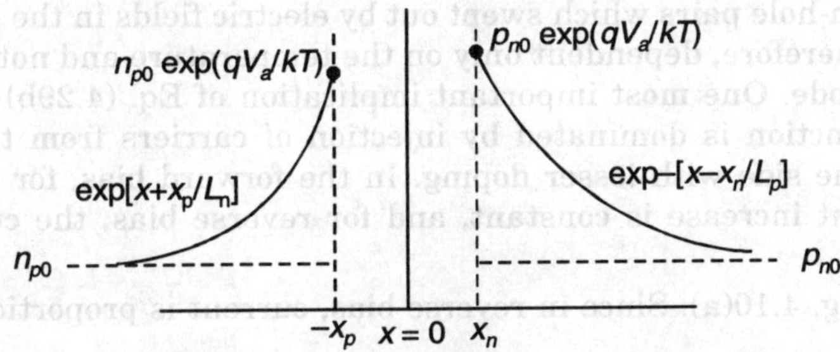

FIGURE 4.10(b) Minority carrier concentration variations in the neutral region of *p-n* junction.

variations in the neutral region of *p-n* junction, as given by Eqs. (4.27b) and (4.28a), are shown in Fig. 4.10(b).

The injected minority carrier distribution and electron and hole currents for forward bias as well as reverse bias is shown in Figs. 4.8(d) and 4.9(d).

The diode equation given by Eq. (4.29b) says nothing about the possibility of reverse bias breakdown.

In forward bias, current becomes very much larger than I_s, hence Eq. (4.29b) reduces to

$$I_{\text{forward}} \cong I_s\, e^{qV_f/kT}$$

This shows that forward current rises exponentially with V_f until the slope becomes more gradual.

The slope of the forward-bias characteristics, as shown in Fig. 4.10(a), at any point gives the resistance of the forward bias junction and is given by

$$\frac{dV_a}{dI} \simeq \frac{kT}{qI_s e^{qV_a/kT}} = \frac{(kT/q)}{I} = r_d$$

and called the **dynamic resistance** of the forward-biased junction. When current is expressed in mA then the dynamic resistance r_d is given as:

$$r_d \simeq \frac{25}{I}$$

Using Eq. (4.29b), the reverse current is given, by putting $V_a = -V_r$, as:

$$I_r = qA\left(\frac{\Delta p}{L_p}\, p_{n0} + \frac{\Delta n}{L_n}\, n_{p0}\right)(e^{-qV_r/kT} - 1) \tag{4.30}$$

If V_r is greater than few kT/q then Eq. (4.30) reduces to

$$I_r = -qA\left(\frac{\Delta p}{L_p}\, p_{n0} + \frac{\Delta n}{L_n}\, n_{p0}\right) = -I_s \tag{4.31}$$

i.e., the total reverse current is independent of applied potential and remain constant at I_s, called **reverse saturation current** which is typically a few μA. This shows that the extremely small reverse curent is due to minority carriers from thermally generated electron-hole pairs which swept out by electric fields in the depletion region. This current is, therefore, dependent only on the temperature and not on the potential barriers of the diode. One most important implication of Eq. (4.29b) is that the total current at the junction is dominated by injection of carriers from the more heavily doped side into the side with lesser doping. In the forward bias, for $V_a = V_f \geq 3kT/q$, the rate of current increase is constant, and for reverse bias, the current saturates

at I_s as seen in Fig. 4.10(a). Since in reverse bias, current is proportional to $\dfrac{n_i^2}{(N_d/N_a)}$ and minority carrier concentration drops to negligible values and hence current saturates. The reverse saturation current is restricted by two factors; the limited availability of minority carriers and the fact that the concentration gradient does not change much once the reverse-bias voltage is sufficiently large.

EXAMPLE 4.4

For the Si *p-n* junction, following parameters are given:

$N_a = 5 \times 10^{16}/cm^3,$ $N_d = 10^{16}/cm^3,$ $n_i = 1.5 \times 10^{10}/cm^3$
$D_n = 21 \ cm^2/s,$ $D_p = 10 \ cm^2/s,$ $\tau_n = 2p = 5 \times 10^{-7}$

Cross-sectional area of junction is $2 \times 10^{-4} \ cm^2$.
Find the reverse saturation current at $T = 300$ K.

Solution: From Eq. (4.31),

$$I_s = qA\left(\frac{D_p}{L_p}\, p_{n0} + \frac{D_n}{L_n}\, n_{p0}\right)$$

or,

$$I_s = qAn_i^2\left[\frac{1}{N_D}\sqrt{\frac{D_p}{\tau_p}} + \frac{1}{N_A}\sqrt{\frac{D_n}{\tau_n}}\right]$$

$$= 1.6 \times 10^{-19} \times 2 \times 10^{-4} \times (1.5)^2 \times 10^{20}\left[\frac{1}{10^{16}}\sqrt{\frac{10}{5 \times 10^{-7}}} + \frac{1}{5 \times 10^{16}}\sqrt{\frac{21}{5 \times 10^{-7}}}\right]$$

$$= 41.53 \ pA$$

EXAMPLE 4.5

For a Si *p-n* junction, the band diagram is shown in Fig. 4.11 under the application of applied voltage V_a. Use same reverse saturation current as given in Example 4.4. Find:

(a) Whether the *p-n* junction is forward or reverse bias
(b) Total current
(c) Applied voltage

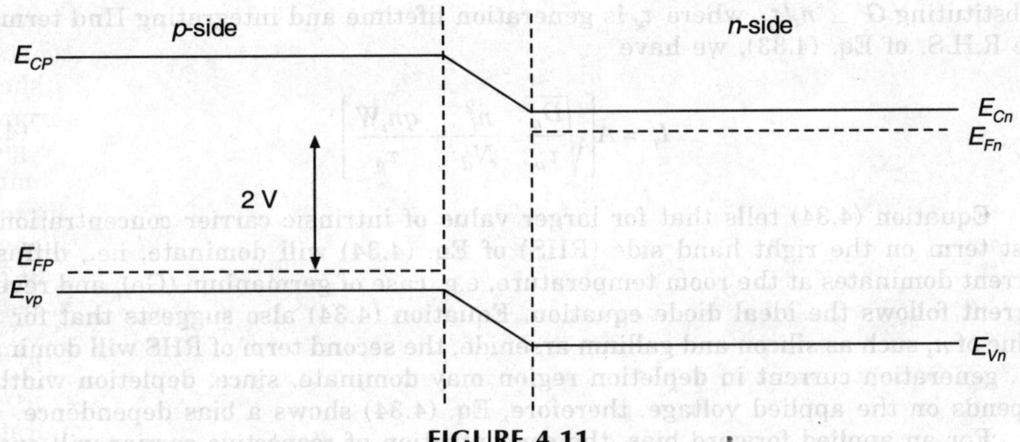

FIGURE 4.11

Solution:
(a) *p-n* junction is forward biased.
(b) $I = I_s \, (e^{qV_a/kT} - 1)$
(c) $V_a = 2$ V

4.7.2 Current-Voltage Characteristics for Real Diode

The ideal diode equation given by Eq. (4.29b) gives only the qualitative agreement with the experimental findings because of the generation or recombination of carriers in the depletion region of a real diode.

Let us first consider reverse-bias diode and under reverse bias, the carrier concentrations fall far below their equilibrium concentrations in the depletion region. The dominant generation-recombination processes are those of electron and hole emissions through band gap generation-recombination centres.

The current due to generation in the depletion region is:

$$I_{\text{generation}} = \int_0^W qAG \, dx \tag{4.32}$$

where W is depletion width, G is the generation rate and A is cross-sectional area.

In most of the practical (real) diodes, one of the sides has a substantially higher doping level and hence produces larger number of minority carriers.

Let us consider a one-sided p-n junction, i.e., p^+-n junction for which $N_a \gg N_d$, and also assume $V_r > 3\,kT/q$. The total reverse current is given by the sum of the diffusion current in the neutral region and generation current in the depletion region, i.e.,

$$I_r = A\left[q\sqrt{\frac{D_p}{\tau_p}} \cdot \frac{n_i^2}{N_d} + \int_0^W qG\,dx \right] \tag{4.33}$$

Substituting $G \simeq n_i/\tau_g$, where τ_g is generation lifetime and integrating IInd terms of the R.H.S. of Eq. (4.33), we have

$$I_r = A\left[\sqrt{\frac{D_p}{\tau_p}} \cdot \frac{n_i^2}{N_d} + \frac{qn_iW}{\tau_g} \right] \tag{4.34}$$

Equation (4.34) tells that for larger value of intrinsic carrier concentration n_i, first term on the right hand side (RHS) of Eq. (4.34) will dominate, i.e., diffusion current dominates at the room temperature, e.g. case of germanium (Ge); and reverse current follows the ideal diode equation. Equation (4.34) also suggests that for low value of n_i such as silicon and gallium arsenide, the second term of RHS will dominate, i.e., generation current in depletion region may dominate. since, depletion width W depends on the applied voltage, therefore, Eq. (4.34) shows a bias dependence.

For an applied forward bias, the concentration of respective carrier will exceed their thermal equilibrium values and carriers will attain their equilibrium value by recombination. Therefore, the dominant generation-recombination processes in the depletion region are the capture processes.

The recombination current is:

$$I_{\text{rec.}} = A\int_0^W qU\,dx \tag{4.35}$$

where U is the recombinate rate and for $V_f > 3\,kT/q$,

$$U \cong \frac{1}{2}\,\sigma_0 v_{Th} N_t n_i e^{qV_f/2kT} \tag{4.36}$$

where σ_0 is the effective cross-section, i.e., $\sigma_p = \sigma_n = \sigma_0$, and N_t is the density of trapped states.

Putting Eq. (4.36) into Eq. (4.35), the recombination current after simplification is given as:

$$I_{\text{rec.}} = \frac{qAWn_i}{2\tau_r}\,e^{qV_f/2kT} \tag{4.37}$$

where τ_r is the recombination lifetime, and is given by

$$\tau_r = \frac{1}{\sigma_0 v_{Th} N_t}$$

For $p_{n0} \gg n_{p0}$ and $V_f > 3kT/q$, the total forward current is given by

$$I_f = A\left[q\sqrt{\frac{D_p}{\tau_p}}\,\frac{n_i^2}{N_d}\,e^{qV_f/kT} + \frac{qWn_i}{2\tau_r}\,e^{qV_f/2kT} \right] \tag{4.38}$$

Again, for low values of n_i, the recombination current will dominate in the *p-n* junction diode.

In general, the experimentally observed current-voltage characteristics for forward bias condition assuming generation and recombination in the space-charge region, can be empirically formulated as:

$$I_f \cong \exp\left(\frac{qV_f}{\eta kT}\right) \tag{4.39}$$

where η is an empirical constant, called **ideality factor**. It is also found that when ideal diffusion current dominates, i.e., for larger n_i, as clear from Eq. (4.38), $\eta = 1$, whereas for low value of n_i where recombination current dominates, $\eta = 2$. When both currents are comparable, η has a value in between 1 and 2. The experimentally measured value of forward current for Si and GaAs is shown in Fig. 4.12. This is clear from Figure that at low current levels (due to low value of n_i) the recombination current dominates and $\eta = 2$ whereas for higher current level (i.e., large value of n_i) diffusion current dominates and η approaches 1.

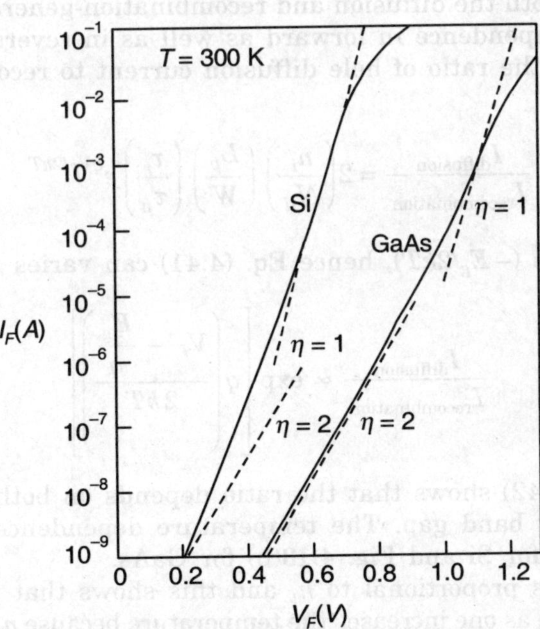

FIGURE 4.12   Plot of experimentally measured value of forward current of Si and GaAs with forward bias potential.

It is still observed that even at higher current levels, the current departs from the ideal case, i.e., $\eta = 1$ and increases more gradually with applied forward bias voltage. This is due to two effects: one is series resistance and other is due to high-level injection. At low and medium current levels, the voltage drop $I_f R$ across the neutral regions is usually small and comparable with kT/q, where, I_f is forward bias current and R is series resistance. This potential drop $I_f R$ in the neutral regions will reduce the bias voltage across the depletion region and hence current becomes

$$I = I_s \exp\left[\frac{q(V_a - I_f R)}{kT}\right]$$

$$= I_s \frac{\exp\left(\dfrac{qV_a}{kT}\right)}{\exp\left(\dfrac{I_f R}{kT}\right)} \tag{4.40}$$

Equation (4.40) shows that the ideal current is reduced by a factor of $\exp(I_f R/kT)$ due to potential drop in the neutral regions when series resistance is taken into account.

Due to high level injection, the injected minority carrier concentration becomes comparable to the majority carrier concentration and it was found that current is now proportional to $\exp\left(\dfrac{qV_f}{2kT}\right)$. Thus, the current increases at a slower rate under the high-level injection. Both the diffusion and recombination-generation currents have a strong temperature dependence in forward as well as in reverse bias.

In forward bias, the ratio of hole diffusion current to recombination current is given by

$$\frac{I_{\text{diffusion}}}{I_{\text{recombination}}} = 2\left(\frac{n_i}{N_d}\right)\left(\frac{L_p}{W}\right)\left(\frac{\tau_r}{\tau_p}\right) e^{qV_f/2kT} \tag{4.41}$$

Since, n_i varies as $\exp(-E_g/2kT)$, hence Eq. (4.41) can varies as:

$$\frac{I_{\text{diffusion}}}{I_{\text{recombination}}} \propto \exp\left[q\left(\frac{V_f - \dfrac{E_g}{q}}{2\,kT}\right)\right] \tag{4.42}$$

Ratio given by Eq. (4.42) shows that the ratio depends on both the temperature as well as on the energy band gap. The temperature dependence for forward bias is shown in Fig. 4.13(a) for Si and Fig. 4.13(b) for GaAs.

Since, the ratio is proportional to n_i and this shows that the diffusion current should tend to dominate as one increases the temperature because n_i is strongly dependent on the temperature. Since, silicon has smaller band gap value than the GaAs, therefore, the ratio given by Eq. (4.42) has smaller value for silicon than GaAs. In the reverse-bias condition for a p^+-n junction, the ratio of diffusion to generation current is:

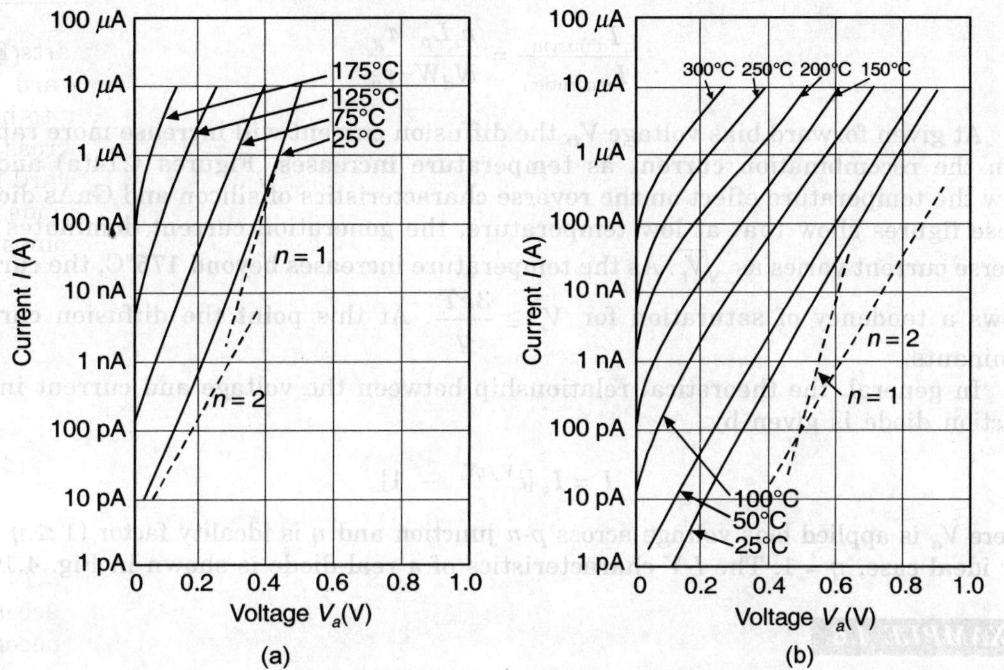

FIGURE 4.13   (a) Plot of forward bias current in Si at different temperature with applied voltage, (b) Plot of forward current with applied potential at different temperature for GaAs.

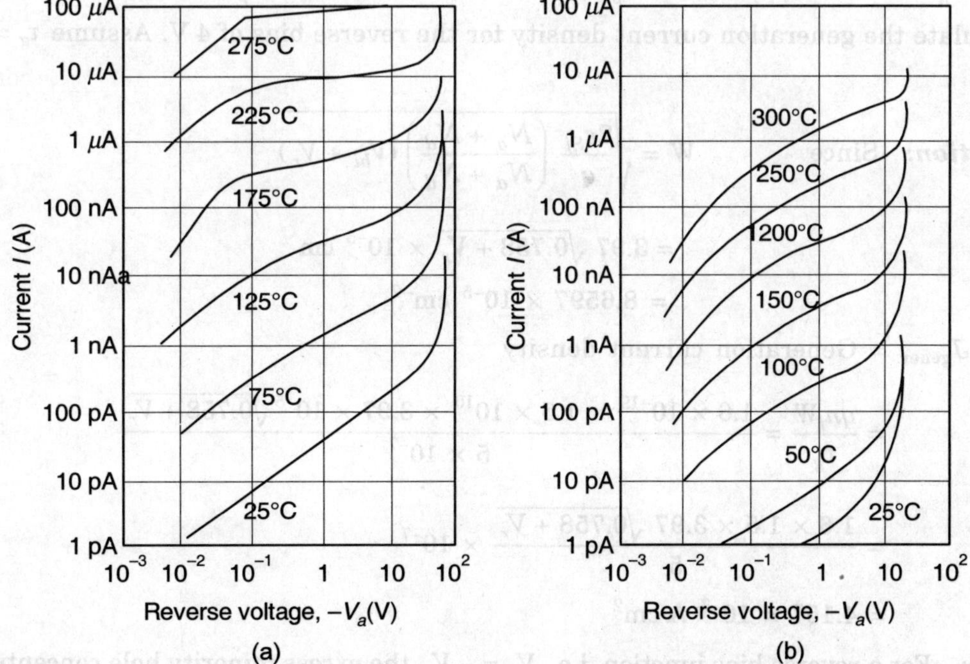

FIGURE 4.14   (a) Plot of current with reverse bias potential at different temperature for Si, (b) Plot of current with reverse bias voltage at different temperature for GaAs.

$$\frac{I_{\text{diffusion}}}{I_{\text{generation}}} = \frac{n_i L_p}{N_d W} \frac{\tau_g}{\tau_p} \tag{4.43}$$

At given forward bias voltage V_f, the diffusion current will increase more rapidly than the recombination current as temperature increases. Figures 4.14(a) and (b) show the temperature effect on the reverse characteristics of silicon and GaAs diodes. These figures show that at low temperature, the generation current dominates and reverse current varies as $\sqrt{V_r}$. As the temperature increases beyond 175°C, the current shows a tendency of saturation for $V_r \geq \dfrac{3kT}{q}$. At this point the diffusion current dominants.

In general, the theoretical relationship between the voltage and current in p-n junction diode is given by

$$I = I_s \left[e^{V_a / \eta V_t} - 1 \right]$$

where V_a is applied bias voltage across p-n junction and η is ideality factor $(1 \leq \eta \leq 2)$. For ideal case, $\eta = 1$. The *I-V* characteristics of a real diode is shown in Fig. 4.10(a).

EXAMPLE 4.6

Consider the Si p-n junction diode for which the following parameters are given,

$$N_a = 5 \times 10^{16}/\text{cm}^3, \qquad N_d = 10^{16}/\text{cm}^3, \qquad n_i = 1.5 \times 10^{10}/\text{cm}^3$$
$$D_n = 21 \text{ cm}^2/\text{s}, \qquad D_p = 10 \text{ cm}^2/\text{s}, \qquad \tau_n = \tau_p = 5 \times 10^{-7} \text{ s}$$

Calculate the generation current density for the reverse bias of 4 V. Assume $\tau_g = 5 \ ns = \tau_p$.

Solution: Since

$$W = \sqrt{\frac{2\varepsilon_{Si}}{q} \left(\frac{N_a + N_d}{N_a + N_d} \right) (V_{bi} + V_r)}$$

$$= 3.97 \sqrt{0.758 + V_r} \times 10^{-5} \text{ cm}$$

$$= 8.6597 \times 10^{-5} \text{ cm}^{-2}$$

$J_{\text{gener.}}$ = Generation current density

$$= \frac{q n_i W}{\tau_g} = \frac{1.6 \times 10^{-19} \times 1.5 \times 10^{10} \times 3.97 \times 10^{-5} \sqrt{0.758 + V_r}}{5 \times 10^{-7}}$$

$$= \frac{1.6 \times 1.5 \times 3.97 \sqrt{0.758 + V_r}}{5} \times 10^{-7}$$

$$= 4.157 \times 10^{-7} \text{ A/cm}^2$$

Note: For a reverse bias junction, i.e., $V_a = -V_r$, the excess minority hole concentration at the edge $(x = -x_p)$ of the junction, from Eq. (4.26b) becomes

$$\Delta p_n = p_{n0}(e^{-qV_r/kT} - 1) \tag{4.44}$$

If $V_r \gg kT/q$, Eq. (4.44) can be reduced to

$$\Delta p_n \simeq - p_{n0}$$

or
$$p_n - p_{n0} = - p_{n0}$$

or
$$p_n = 0 \tag{4.45}$$

This shows that for reverse bias of more than few tenths of a volt, the hole minority carrier at the edge ($x = x_n$) approaches to zero. This is analogous to extraction of minority carriers whereas in forward bias there is injection of minority carriers.

Near the junction (just outside the depletion region) the majority carrier concentration changes by exactly same amount as the minority carrier concentration to maintain the space charge neutrality. Very far away, i.e., 3 to 5 diffusion lengths, the majority carrier concentration is independent of space and total current is due to drift of majority carriers whereas near the junction majority carrier concentration changes spatially and majority carrier current is both due to diffusion as well as drift. The drift current due to majority carriers always dominates except for a high level doping.

So, one important conclusion is that the electric field in the neutral regions cannot be zero otherwise there is no drift component of current. Thus, the assumption that all the applied voltage drop across the depletion region is not correct. But since, in the neutral regions majority carrier concentration is very large in comparison to minority carrier concentration, so why a small electric field is needed for the majority carrier drift current. Therefore, in most of the cases, our assumption is almost true.

EXAMPLE 4.7

Physically explain, what is the meaning of extraction of minority carriers in the reverse bias?

Solution: Physically, extraction occurs because the minority carriers at the edges of the space charge region are swept down the barrier at the junction to the other side, and are not replaced by an opposing diffusion of carriers.

4.8 MINORITY-CARRIER STORAGE

In forward bias, electrons are injected from the n-region into p-region and holes are injected from p-region into n-region. Once injected across the junction, minority carriers recombine with majority carriers and decay exponentially with distance. These minority-carrier distributions lead to current flow and charge storage in the p-n junction.

The charge of injected minority carriers per unit area stored in the neutral n-region can be obtained by integrating the excess holes in the neutral region, i.e.,

$$Q_p = q \int_{x_n}^{\infty} (p_n - p_{n0})\, dx$$

Using Eq. (4.27b), we have

$$Q_p = q \int_{x_n}^{\infty} p_{n0}(e^{qV_a/kT} - 1)\, e^{-(x-x_n)/L_p}\, dx$$

and after simplification, we have

$$Q_p = qL_p \cdot p_{n0}(e^{qV_a/kT} - 1) \tag{4.46}$$

Equation (4.46) shows that the number of storage minority carriers depends on both the diffusion length and the charge density at the boundary of the depletion region. Dividing Eq. (4.46) and (4.27c), we get

$$\frac{Q_p}{J_{pn}(x_n)} = \frac{q/L_p p_{n0} L_p}{q D_p p_{n0}} = \frac{L^2 p}{D_p}$$

or

$$Q_p = \frac{L^2 p}{D_p} J_{pn}(x_n)$$

$$= \tau_p \cdot J_{pn}(x_n) \tag{4.47}$$

Equation (4.47) states that by taking the product of current density and minority carrier's lifetime, one can get the stored minority carrier charge per unit area. This is because the injected hole diffuse farther into the *n*-region before recombining with electron if their lifetime is longer, thus more holes are stored. Similar argument is true for stored electron charge in the neutral region of *p*-side.

4.9 CAPACITANCE OF *p-n* JUNCTIONS

It is also essential to understand the response of the device to change in its bias conditions. This is known as transient or dynamic behaviour of the device which determines transient or dynamic behaviour of the device which determines the maximum speed at which the device can be operated. Since the operation mode of the diode is the function of the charges present in neutral and space charge region. This results in capacitive behaviour of the diode and the dynamic behaviour of the diode is strongly determined by how fast the charges can be moved around. There are basically two types of capacitance associated with the *p-n* junction:

  (i) The depletion or junction capacitance
  (ii) The diffusion capacitance

4.9.1 Depletion Capacitance or Junction Capacitance

Since, depletion region contains only few mobile carriers (in real case) or almost zero mobile carriers (in ideal case), hence it can be conceived as an insulator with dielectric constant of ε_{Si} (permittivity of silicon). The n- and p-region of the p-n junction act as the two capacitor plates. This structure is similar to parallel plate capacitor. A small change in applied voltage dV causes a change in the space charge $d\theta$ and hence the junction capacitance or depletion-layer capacitance is defined as:

$$C_j = \frac{dQ}{dV} \tag{4.48}$$

Since the charge Q on both sides of the transition region varies non-linearly with the applied voltage, this capacitance is dominant under the reverse-bias condition. Here, we will derive the capacitance for an abrupt junction due to space-charge depletion region. To find the depletion capacitance, as clear from Eq. (4.48), it is necessary to obtain an expression for the charge in terms of the voltage.

Since we are dealing with the unequilibrium case, therefore, the depletion width can be given by $(V_{bi} - V_a)$ in Eq. (4.14d), i.e.,

$$W = \left[\left(\frac{2\varepsilon_{Si}}{q} \right) \left(\frac{N_a + N_d}{N_a N_d} \right) (V_{bi} - V_a) \right]^{1/2} \tag{4.49}$$

where V_a is applied voltage. It takes +ve value for forward bias and –ve value for reverse bias. Since, the uncompensated charge Q on each side of the junction related to the width W of the transition region, therefore, variation in W with applied voltage causes the variation in Q also and results in capacitive behaviour of the junction. The total positive and negative space charge Q must be equal, and can be written in terms of doping concentration and transition region width on each side of the junction as:

$$|Q| = qAx_n\, N_d^+ = |-qN_a^- x_p A| \tag{4.50}$$

where A is the cross-sectional area and q is electronic charge.

The relations between the total width W and individual widths x_n and x_p are given by Eqs. (4.15c) and (4.16b). Substituting these values in Eq. (4.50), we have

$$|Q| = qA\, \frac{N_d N_a}{N_a + N_d}\, W \tag{4.51a}$$

Assuming complete ionization, i.e., $N_d^+ = N_d$; $N_a^- = N_a$. Using Eq. (4.49) in Eq. (4.51a), we have

$$|Q| = qA\, \frac{N_d N_a}{N_a + N_d} \left[\frac{2\varepsilon_{Si}}{q} \left(\frac{N_a + N_d}{N_a N_d} \right) (V_{bi} - V_a) \right]^{1/2}$$

or $$|Q| = A\left[\frac{2\varepsilon_{Si} q}{q} \left(\frac{N_a N_d}{N_a + N_d} \right) (V_{bi} - V_a) \right]^{1/2} \tag{4.51b}$$

Equation (4.51b) shows a non-linear relation of charge $|Q|$ with the applied voltage V_a. Since, the voltage that varies the charge in the transition region is the barrier height $(V_{bi} - V_a)$, therefore, the junction capacitance can be obtained by taking derivative of Eq. (4.51b) w.r.t. $(V_{bi} - V_a)$

i.e.,

$$C_j = \left| \frac{dQ}{d(V_{bi} - V_a)} \right|$$

Using Eq. (4.51b) one gets

$$C_j = \frac{A}{2} \left[\frac{2q\varepsilon_{Si}}{(V_{bi} - V_a)} \cdot \frac{N_d N_a}{N_d + N_a} \right]^{1/2} \tag{4.52}$$

or it can be written as;

$$C_j = \frac{C_{jo}}{\sqrt{1 - V_a/V_{bi}}} \quad \text{where } C_{jo} = A \sqrt{\frac{\varepsilon_{Si} q}{2} \frac{N_a N_d}{N_a + N_d} \cdot \frac{1}{V_{bi}}}$$

is capacitance under zero-bias condition and it is only function of physical parameter of the device.

Typically, in design A is omitted from expressions and C_j and C_{jo} are expressed in capacitance per unit area.

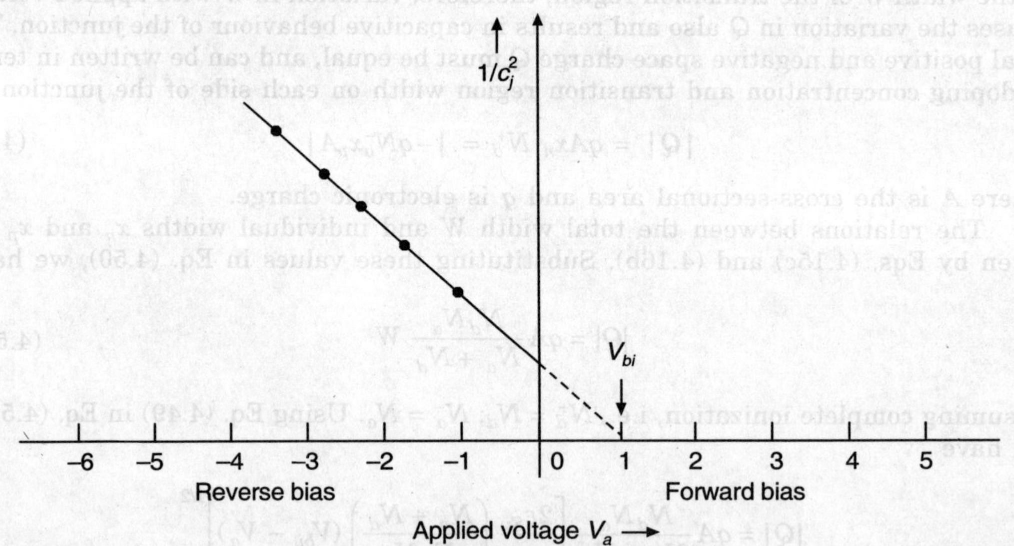

FIGURE 4.15 Variation of $(1/C_j^2)$ with applied voltage in reverse and forward bias conditions.

Equation (4.52) predicts that the junction capacitance is a voltage-variable capacitor. It is also clear from Eq. (4.52) that a plot of $1/C_j^2$ versus V_a produces a straight line for a one-sided abrupt junction. The slope gives the impurity concentration and the intercept (at $1/C_j^2 = 0$) gives V_{bi} as shown in Fig. 4.15.

Equation (4.52) is only valid under the approximation of abrupt *p-n* junction.

There are several important applications for variable capacitors like in tuned circuits and varactor diode.

Using Eq. (4.49) in Eq. (4.52), we can further write

$$C_j = \frac{\varepsilon_{Si} A}{W} \qquad (4.53)$$

In analogy with the parallel plate capacitor, the space charge region width W corresponds with the plate separation of the conventional capacitor.

Consider abrupt one-sided junction n^+-p, where $N_d \gg N_a$ and hence, Eq. (4.52) can be simplified as:

$$C_j = \frac{A}{2} \left[\frac{2q\varepsilon_{Si}}{(V_{bi} - V_a)} N_a \right]^{1/2} \qquad (4.54a)$$

or

$$\frac{1}{C_j^2} = \frac{2(V_{bi} - V_a)}{A^2 q\varepsilon_{Si}} \cdot \frac{1}{N_a} \qquad (4.54b)$$

This shows that junction capacitor for one-sided junction is mainly determined by the concentration of lightly doped side. If the cross-sectional area A is known then N_a can be obtained from Eq. (4.54a). V_{bi} is obtained from $(1/C_j^2)$ plot as shown in Fig. 4.15. These all equations are applicable to a sharp step junction and certain modification must be made in the case of graded junction.

A more generic expression for the junction capacitance can be given as

$$C_j = \frac{C_{j0}}{[1 - V_a/V_{bi}]^m}$$

where m is grading coefficient and equal to $(1/2)$ for the abrupt junction and $1/3$ for linear or graded junction.

Junction capacitor is actually a small-signal parameter whose values vary over the bias points. Since, in digital circuits, the operating bias voltages vary rapidly over the wide ranges. Therefore, under this condition, it is good approximation to replace C_j (voltage-dependent non-linear junction capacitance) by C_{eq} which is defined as voltage varies from V_{low} to V_{high}.

EXAMPLE 4.8

For a silicon one-sided abrupt junction with $N_a = 2 \times 10^{19}/\text{cm}^3$, $N_d = 8 \times 10^{15}/\text{cm}^3$. Calculate the junction capacitance per unit area at zero bias and reverse bias of 4. Assume the room temperature $T = 300$ K.

Solution: In equation,

$$V_{bi} = \frac{kT}{q} \ln \frac{N_a N_d}{n_i^2}$$

$$= 0.0259 \ln \left[\frac{16 \times 10^{34} \times 10^{-20}}{2.25} \right]$$

$$= 0.0259 \ln \left[\frac{16 \times 10^{16}}{225} \right] = 0.886 \text{ V}$$

and $W/V_a = 0$ (for $N_d << N_a$) $= \sqrt{\dfrac{2\varepsilon_{Si}}{q} \dfrac{V_{bi}}{N_d}}$ V   [from Eq. (4.49)]

$$= \sqrt{\frac{2 \times 11.9 \times 8.854 \times 10^{-14} \times 0.886}{1.6 \times 10^{-19} \times 8 \times 10^{15}}} = \sqrt{\frac{186.702 \times 10^{-10}}{12.8}} = 0.382$$

Therefore, from Eq. (4.53),

$$(C_j/A) = \frac{\varepsilon_{Si}}{W}$$

$$= \frac{11.9 \times 10^{-14} \times 8.854}{0.382 \times 10^{-4}} = 2.76 \times 10^{-8} \text{ F/cm}^2$$

When $V_a = -4$ V, Using Eq. (4.52)

$$(C_j/A) = \frac{1}{2} \left[\frac{2 \times 1.6 \times 10^{-19} \times 11.9 \times 8.854 \times 10^{-14}}{(0.886 + 4)} \times N_d \right]^{1/2}$$

$$= \frac{1}{2} \left[\frac{674.321 \times 10^{-18} \times 4}{4.886} \right]^{1/2} = 1.175 \times 10^{-8} \text{ F/cm}$$

4.9.2 Diffusion Capacitance

When the junction is forward biased, it exhibits another capacitive behaviour which is much larger than junction capacitance and known as **diffusion capacitance** or **storage capacitance** and denoted either by C_d or C_s. This extra capacitive effect is due to the excess minority carrier charge stored at the boundaries of the depletion region. The storage capacitance due to holes on the n-side of the junction is given as:

$$C_d = C_s = \left| \frac{dQ_p}{dV_a} \right| \tag{4.55}$$

where Q_p is charge due to excess holes, and is given by

$$Q_p = qA \int_{x_n}^{\infty} [p_n(x) - p_{n0}] \, dx \tag{4.56a}$$

where A is cross-sectional area and $p_n(x)$ is given by Eq. (4.27b). Substituting Eq. (4.27b) in Eq. [4.56(a)], we have

$$Q_p = qA \int\limits_{x_n}^{\infty} \left\{ p_{n0}\left[\exp\left(\frac{qV_a}{kT}\right) - 1 \right] \exp\left(-\frac{x - x_n}{L_p} + p_{n0} - p_{n0} \right) \right\} dx \qquad (4.56b)$$

Integrating Eq. (4.56b) and after solving

$$Q_p = qA\, p_{n0} L_p \left[\exp\left(\frac{qV_a}{kT}\right) - 1 \right] \qquad (4.56c)$$

Using $p_{n0} = \dfrac{n_i^2}{N_d}$ and $L_p^2 = D_p \tau_n$, Eq. (4.56c) can be written in another form as:

$$Q_p = \frac{qA\, D_p n_i^2}{L_p N_d}\, \tau_p \left[\exp\left(\frac{qV_a}{kT}\right) - 1 \right] \qquad (4.56d)$$

Equation (4.56d) shows the storage minority carrier charge is proportional to n_i^2, and inversely proportional to majority carrier concentration. One can also increase the storage charge either by increasing the life time of minority carrier or reducing the mean free path. In terms of forward current density $J_{pn}(x_n)$, the Eq. (4.56d) can be expressed as:

$$Q_p = J_{pn}(x_n)\, \tau_p A = I_d \tau_p \qquad (4.56e)$$

where the diffusion current $I_d = J_{pn}(x_n)A$. Equation (4.56e) demonstrates that the stored charge is the product of diffusion current and the minority carrier life time.
Taking derivative of Eq. (4.56c) w.r.t. V_a, the diffusion capacitance is given as:

$$C_{dp} = C_{sp} = \frac{dQ_p}{dV_a} = \frac{q^2 A p_{n0} L_p}{kT}\, \exp\left(\frac{qV_a}{kT}\right) \qquad (4.57)$$

Similarly, the minority-carrier electrons storage charge on the p-side is:

$$Q_n = qA n_{p0} \left[\exp\left(\frac{qV_a}{kT}\right) - 1 \right] \qquad (4.58a)$$

and diffusion capacitance on n-side is:

$$c_{sn} = c_{dn} = \frac{dQ_n}{dV_a} = \frac{q^2 A n_{p0}}{kT}\, \exp\left(\frac{qV_a}{kT}\right) \qquad (4.58b)$$

The total diffusion capacitance is the sum of the contributions from both the p- and n-sides when N_a and N_d are not greatly different. C_s or C_d is a small-signal capacitance and is only valid around the given bias voltage. For reverse bias condition, it is fair to assume that C_d is negligible because of negligible minority carriers. For a p^+-n junction,

$$P_{n0} \gg n_{p0}$$

and hence contribution to C_d of the stored electrons becomes insignificant because from Eqs. (4.58a) and (4.56c)

$$Q_n << Q_p$$

Therefore, total diffusion capacitance is given by

$$C_d = C_{dp}$$

In many applications, in addition to diffusion capacitance and depletion capacitance, we must include the conductance in the *p-n* junction to account for the current through the device. The conductance is defined as:

$$G = \frac{A_d J}{dV_a} \tag{4.59}$$

Using Eqs. (4.59) and (4.29a), for ideal *p-n* junction,

$$G = \frac{AJ_s}{kT} \times q[e^{qV_a/kT}] = \frac{qA}{kT}(J + J_s) \simeq \frac{qI}{kT} \tag{4.60}$$

The diode equivalent circuit is shown in Fig. 4.16.

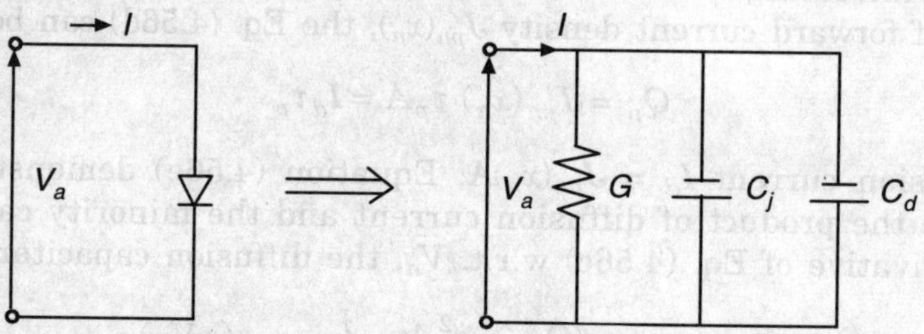

FIGURE 4.16 Diode equivalent circuit.

EXAMPLE 4.9

For an ideal abrupt silicon p^+-n junction with $N_d = 8 \times 10^{15}/\text{cm}^3$, calculate the stored minority carriers per unit area in the neutral region when a forward bias of 1 V is applied. The diffusion length of the hole is 5 μm. Also find the diffusion capacitance.

Solution: From Eq. (4.56c),

$$Q_p = 1.6 \times 10^{-19} \times A \times \left(\frac{n_i^2}{N_d}\right) \times 5 \times 10^{-4}[e^{1/0.0259} - 1]$$

$$\frac{Q_p}{A} = 1.6 \times 10^{-19} \times \left(\frac{2.25 \times 10^{20}}{8 \times 10^{15}}\right) \times 5 \times 10^{-4}[5.863 \times 10^{16}]$$

$$= 13.192 \times 10^{-2} = 0.132 \text{ C/cm}^2$$

From Eq. (4.57),

$$C_d = \left(\frac{C_{dp}}{A}\right) = \frac{q^2 \times p_{n0}L_p}{kT}\exp\left(\frac{qV_a}{kT}\right)$$

$$= \frac{p_{n0} \times L_p}{(kT/q)}\, q \exp\left(\frac{qV_a}{kT}\right)$$

$$= \frac{2.25 \times 10^{20} \times 5 \times 10^{-4} \times 1.6 \times 10^{-19}}{8 \times 10^{15} \times 0.0259} \times e^{1/0.0259}$$

$$= \frac{1.0553}{0.2072} = 5.093 \text{ F/cm}^2$$

4.9.3 Transient Behaviour

For switching applications, very fast time is required for the transition from forward-bias to reverse bias, i.e., transient time should be short.

Let us consider the circuit as shown in Fig. 4.17. Initially switch is at position A, so diode D is in forward bias and a forward current I_f flows in the circuit. Suddenly at time $t = 0$, the switch S is thrown to position B and now diode is in reverse bias and initial reverse current $I_R = V_R/R$ flows in the circuit. The transient time t_{off} is defined as the time required for the current to reach 10% of its initial value (V_R/R).

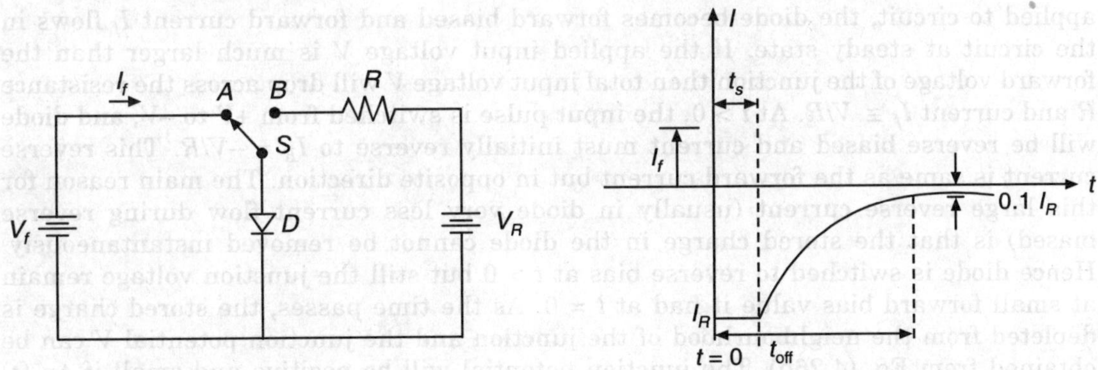

FIGURE 4.17 Transient circuit and transient behaviour of *p-n* junction.

Under the forward bias condition, the stored minority carriers in the n-region of p^+-n junction is given by Eq. (4.56c), i.e.,

$$Q_p = \tau_p J_{pn}(x_n)A = I_f\tau_p$$

where I_f is total forward current and A is area of the device. The turn-off time t_{off} is the time required to remove the total stored charge Q_p, and is given as:

$$t_{\text{off}} \simeq \frac{Q_p}{(I_R)_{aV}} = \tau_p\left(\frac{I_f}{(I_R)_{aV}}\right) \qquad (4.61)$$

where $(I_R)_{aV}$ is the average current flowing during the turn-off period.

Therefore, from Eq. (4.61), it is concluded that turn-off time depends on both the life time of minority carrier as well as the ratio of forward to reverse current. Hence, for fast switching device, it is required to reduce the lifetime of the minority carriers that can be achieved by introducing recombination-generation centres near the middle of the energy gap. Problem of stored charge can also be reduced by designing a p^+-n diode with very narrow n region. If n region is shorter than hole diffusion length very little charge is stored. Thus, little time is required to switch the diode ON and OFF. This type of structure is called **Narrow Base diode**.

4.9.4 Reverse-Recovery Transient

The transient time is associated with the diode when the diode switches from reverse bias to forward bias, i.e., from OFF to ON. When diode switches from reverse bias to forward bias, some finite time is required to reach the steady state forward voltage. This delay in achieving steady state forward voltage is due to the fact that first some time is required to discharge the junction capacitance and then inject the carriers to set up the steady state electron distribution.

Let us consider a circuit as shown in Fig. 4.18(a) where a p^+-n junction diode is used and the circuit is driven from a rectangular input voltage waveform that periodically switches from $+V$ to $-V$ volt as shown in Fig. 4.18(b). When the input voltage $+V$ is applied to circuit, the diode becomes forward biased and forward current I_f flows in the circuit at steady state. If the applied input voltage V is much larger than the forward voltage of the junction then total input voltage V will drop across the resistance R and current $I_f \cong V/R$. At $t > 0$, the input pulse is switched from $+V$ to $-V$, and diode will be reverse biased and current must initially reverse to $I_R = -V/R$. This reverse current is same as the forward current but in opposite direction. The main reason for this large reverse current (usually in diode very less current flow during reverse biased) is that the stored charge in the diode cannot be removed instantaneously. Hence diode is switched to reverse bias at $t > 0$ but still the junction voltage remain at small forward bias value it had at $t = 0$. As the time passes, the stored charge is depleted from the neighbourhood of the junction and the junction potential V can be obtained from Eq. (4.26b). The junction potential will be positive and small if $\Delta p_n(t)$ remains positive and current I in the circuit is equal to $(-V/R)$ until Δp_n goes to zero.

If the total initial stored charged is depleted and Δp_n becomes negative then according to Eq. (4.26b), the junction voltage becomes negative. As time goes on, the magnitude of the reverse current will become smaller because more voltage will drop across the reverse biased junction. This process is continue until the reverse current approaches to reverse saturation current. The time required to make the stored charge equal to zero is called **storage delay time**, represented by t_{sd} and is given by

$$t_{sd} = \tau_p\left[erf^{-1}\left(\frac{I_f}{I_f + I_R}\right)\right]^2 \qquad (4.62)$$

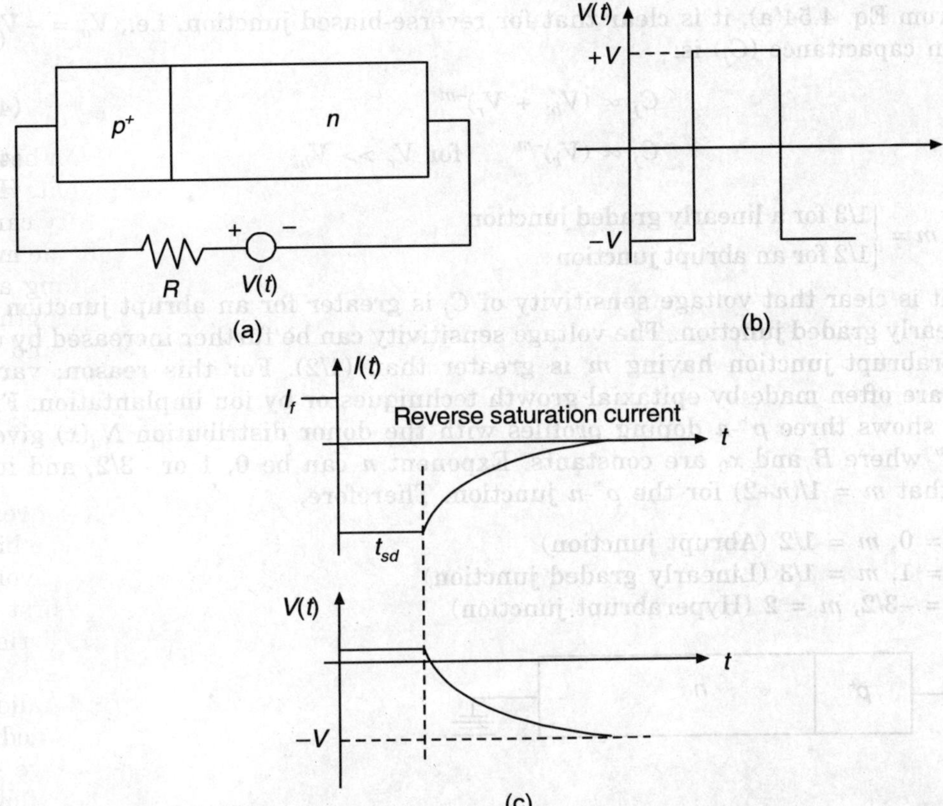

FIGURE 4.18 (a) Transient circuit for a one sided *p-n* junction, (b) Rectangular input voltage waveform, (c) Transient current and voltage response.

where, *erf* is error function and I_R is reverse biased current. This time delay is an important parameter in evaluating diode's switching speed. Generally, it is required that t_{sd} should be small compared with switching time. As clear from Eq. (4.62), the critical parameter which decides the value of t_{sd} is the minority carrier lifetime τ_p. Other application of Eq. (4.62) is that if t_{sd} is determined experimentally then one can easily determine the lifetime of the minority carriers.

The storage delay time can be reduced by introducing recombination centres into the diode material or by utilizing the narrow base diode configuration.

The switching speed of a diode depends upon its construction and fabrication. In general, the smaller the chip, the faster it switches, keeping other parameters same. Switching speed is also limited by chip geometry, doping and temperature. The reverse recovery time t_{sd} is also a limiting parameter for diode design.

4.9.5 Varactor Diode

Many circuit applications employ the voltage-variable properties of reverse biased *p-n* junctions. A *p-n* junction designed specifically for such a purpose is called a varactor which is a shortened form of variable reactor.

From Eq. 4.54(a), it is clear that for reverse-biased junction, i.e., $V_a = -V_r$, the junction capacitance (C_j) is:

$$C_j \propto (V_{bi} + V_r)^{-m} \tag{4.63a}$$

or $\qquad C_j \propto (V_r)^{-m} \quad$ for $V_r \gg V_{bi}$ $\tag{4.63b}$

where $m = \begin{cases} 1/3 \text{ for a linearly graded junction} \\ 1/2 \text{ for an abrupt junction} \end{cases}$

Thus, it is clear that voltage sensitivity of C_j is greater for an abrupt junction than the linearly graded junction. The voltage sensitivity can be further increased by using a hyperabrupt junction having m is greater than $(1/2)$. For this reason, varactor diodes are often made by epitaxial growth techniques or by ion implantation. Figure 4.19(a) shows three p^+-n doping profiles with the donor distribution $N_d(x)$ given as $B(x/x_0)^m$ where B and x_0 are constants. Exponent n can be 0, 1 or $-3/2$, and it was found that $m = 1/(n+2)$ for the p^+-n junction. Therefore,

For $n = 0$, $m = 1/2$ (Abrupt junction)
$\quad\;\; n = 1$, $m = 1/3$ (Linearly graded junction)
$\quad\;\; n = -3/2$, $m = 2$ (Hyperabrupt junction)

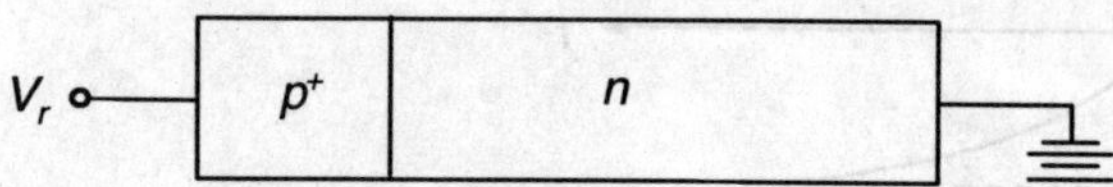

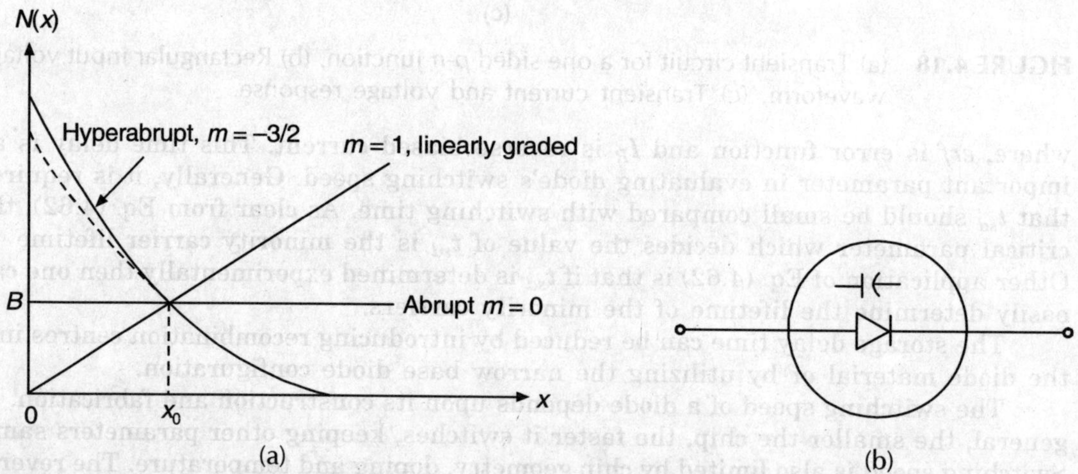

FIGURE 4.19 (a) Three doping profile for one-sided *p*-*n* junction, (b) Symbol for VARACTOR diode.

The hyperabrupt junction is particularly interesting for certain varactor applications where, $C_j \propto V_r^{-2}$. When such capacitor is connected to the induction L in a resonant circuit, the resonant frequency varies linearly with the applied voltage to the varactor, as;

$$\omega_r = \frac{1}{\sqrt{LC_j}} \propto \frac{1}{\sqrt{V_r^{-2}}} \propto V_r \quad \text{(for constant } L\text{)}$$

By choosing proper doping profile, varactor diodes can be designed for specific applications, e.g. for higher frequency applications, varactors can be designed to exploit the forward-bias charge storage capacitance in the short diodes.

The varactor diode symbol is shown in Fig. 4.19(b).

The non-linearity variation in the junction capacitance with the applied reverse bias voltage allows the varactor diode to be used as a harmonic generator.

Major varactor considerations are:

 (i) Capacitance value
 (ii) Voltage applied
 (iii) Variation in capacitance with applied reverse biased voltage
 (iv) Maximum working voltage
 (v) Leakage current

Varactor diodes are not used for rectification but V mainly used to tuning circuits, frequency modulating radio signals, frequency conversion and parametric amplification.

For varactor diodes, one region of junction should be heavily doped so that all the depletion region should move into the lightly doped region and must be operated in reverse biased.

4.9.6 Rectifiers

The diode has a unique property of unidirectional, i.e., current flows only in one direction. Diode will conduct when it is forward biased and there is no conduction when it is reverse biased. An ideal diode acts like a switch. The switch is closed when current flows and switch is open when there is no current. In other way ideal diode acts as a short circuit in forward bias and open circuit in reverse bias, as clear from ideal voltage-current characteristics of a diode shown in Fig. 4.10(a).

The voltage-current characteristics of real diode is shown in Fig. 4.20. The piece-wise linear approximations for junction diode is shown in Fig. 4.21.

As seen from Fig. 4.20, a real diode conducts a small current in the forward direction up to a voltage called **cut-in-voltage** or **knee voltage** V_r or built-in-potential and this voltage is

$$V_r = \begin{cases} 0.3 \text{ V for Ge} \\ 0.7 \text{ V for Si} \end{cases}$$

and for ideal diode $V_r = 0$.

After knee voltage diode conducts heavily in forward direction. If the applied voltage V_a is less than V_r, diode will act like an OFF switch. The slope of the forward characteristic is equal to forward resistance of the diode.

For lineal piece-wise circuit model.

$V_1 = V_r$ for forward bias diode.

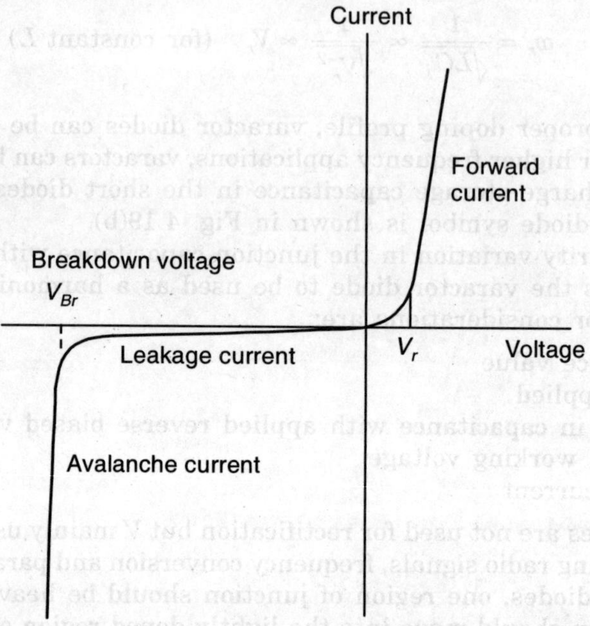

FIGURE 4.20 Current-voltage characteristics for a real *p-n* junction.

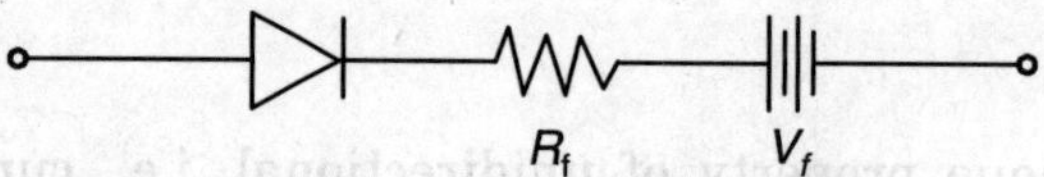

FIGURE 4.21 Piece-wise linear circuit model for real diode in forward bias.

$R_1 = R_f$ for real diode (of order of few ohm)
0 for ideal diode forward bias
R_r (for reverse bias) = ∞ ohm for ideal
= mega ohm for real

For the ideal diode, the reverse-biased current, also called **leakage current** is zero but for real diode there is small leakage current till the reverse break-down voltage. This leakage current is undesirable, and is specified at a voltage less than the breakdown voltage. It is always intended to operate the diode below their breakdown voltage.

The current rating of a diode is determined primarily by the size of the diode chip, and the material and configuration of the package. In diode, average current is used instead of rms current. A larger chip and package of high thermal conductivity will result in a higher current rating.

A diode can be placed in the series with an ac voltage source to provide rectification of the signal. In the rectification process, an alternating signal ac is converted into a unidirectional signal dc. Since, in diode, current will flow only in forward direction, means for first half of signal, i.e., from a to c as shown in Fig. 4.22, diode conducts and for negative half-cycle c-d, diode does not conduct. So, we will get only half positive

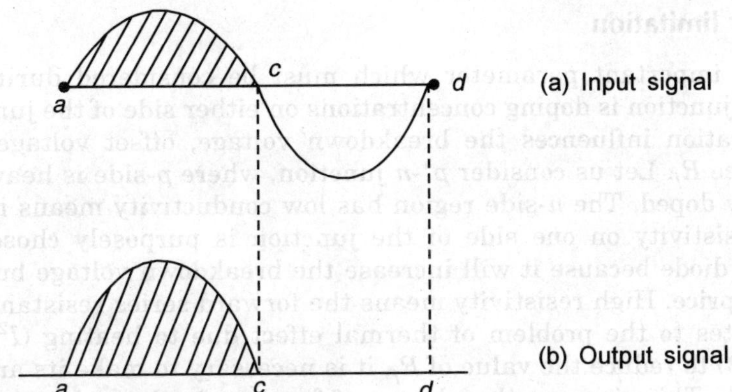

FIGURE 4.22 (a) Input Sinusoidal waveform, (b) The dc output of a diode.

cycle. This rectification is called **half-wave rectification**. The output waveform has positive average value and contains dc component.

The junction diode designed for use of rectifier should have closed current-voltage characteristics of ideal diode. Following requirements must be met in designing a diode required for rectification purpose:

- Should be negligible leakage current.
- Offset voltage should be either zero or as low as possible.
- The forward resistance should be negligible.
- The reverse breakdown voltage should be large.

But unfortunately, all these requirements cannot be met in a single device, so as a designer we have to compromise somewhere to provide best junction diode for rectification applications.

When whole ac signal, i.e. from a–d is converted into dc then rectification process is known as full-wave rectification. Full-wave rectification is obtained either using two junction diode in such a way that only one diode conducts during half of the cycle or using four diodes in form of bridge called bridge rectifier.

Material required for designing a rectifier diode

Since, reverse current decreases with increasing energy bandgap, therefore, a rectifier diode should be made with a wide bandgap material. These devices can also be operated at very high temperature because of the reduction of thermally generated EHPs. Such temperature effects are critical parameters in rectifier, because in forward direction a large current flows in the diode and according to relation $I_f^2 R$, this forward current will generate appreciable heating. On the other hand, built-in potential V_r increases with bandgap. So, a designer should choose in between low leakage current and large value of V_r. But the disadvantage of large V_r is outweighed by the advantage of less leakage current.

Silicon semiconductor material is generally chosen for designing a rectifier diode over Ge for power rectifiers because of its large bandgap, lower leakage current and higher breakdown voltage. Main advantage of using Si is its easy fabrication process.

Another limitation

Another important parameter which must be considered during the fabrication of rectifier junction is doping concentrations on either side of the junction because doping concentration influences the breakdown voltage, offset voltage and series forward resistance R_f. Let us consider p^+-n junction, where p-side is heavily doped and n-side is lightly doped. The n-side region has low conductivity means its resistivity is high. High resistivity on one side of the junction is purposely chosen for designing the rectifier diode because it will increase the breakdown voltage but for that one has to pay the price. High resistivity means the forward series resistance increases and this contributes to the problem of thermal effect due to heating ($I^2 R_f$). So, according to Eq. (3.23) to reduce the value of R_f, it is necessary to make its area larger and reduce its length. This gives another degree of freedom for designing a rectifier, i.e., physical geometry of diode. Limitation on the practical area for diode include problems of obtaining uniform starting material and junction processing over large areas. Any localized flaws in junction uniformity can cause premature breakdown in small region of the device. Similarly, lightly doped region of the junction can not be made arbitrary short due to punch-through problem. In punch-through problem, entire lightly doped region will be depleted. The result of the punch through means breakdown is taking place at a lower voltage than the expected value.

The premature breakdown across the edge of the sample at high reverse bias can be reduced by beveling the edge or by diffusing a guard ring to isolate the junction from the edge of the sample.

EXAMPLE 4.10

How can one avoid the premature breakdown across the sample's edge by beveling the edge or by diffusing a guard ring to isolate the junction from the edge of the sample.

Solution: The electric field is lower at the beveled edge of the sample as shown in Fig. 4.23(a) than the main body of the device. Similarly, the junction at the lightly doped guard ring [as shown in Fig. 4.23(b)] breaks down at higher voltage than p^+-n junction. Since the depletion region is wider in the p-ring than in the p^+-region, the average electric field is smaller at the ring.

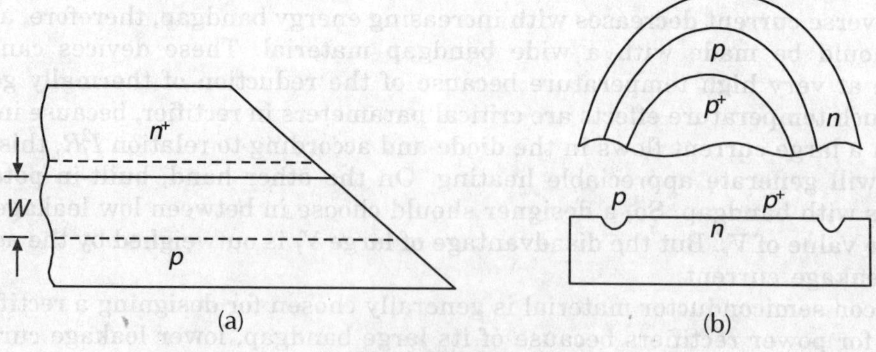

FIGURE 4.23

EXAMPLE 4.11

How one can reduce the forward resistance R_f in the case of diode for high-current device?

Solution: During fabrication of one sided junction p^+-n or n^+-p it is common practice to terminate the lightly doped region with a heavily doped region of same type as shown in Figs. 4.24(a) and 4.24(b). This results in a formation of p^+-n-n^+ and p^+-p-n^+ in which active regions are p^+-n layer and p-n^+ layer. If the lightly doped centre regions are short comparing to the minority carrier diffusion length then the injected

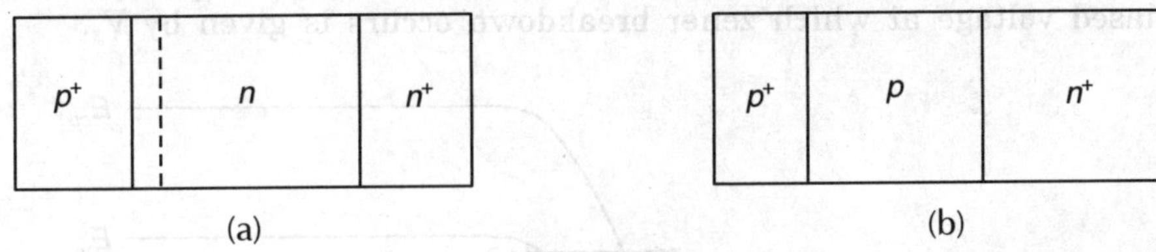

(a)   (b)

FIGURE 4.24

minority carriers in forward bias will enhance the conductivity of the region significantly because the probability of recombination is reduced. This type of increase in conductivity is called **conductivity modulation**. The conductivity modulation reduces the resistivity which results in reduction of forward diode resistance R_f. But for that we have to pay the price like punch-through problem.

4.10 JUNCTION BREAKDOWN

As we have already seen that when a diode is reverse biased, very little current flows, and the diode acts to a first order approximate, as an open circuit. When the reverse voltage is increased further, a point is reached where there is a dramatic increase in current in reverse direction as shown in Fig. 4.21(a). It is equivalent to saying that there is a dramatic reduction in the dynamic resistance. This condition is called **junction breakdown** and the reverse voltage at which breakdown takes place, is called **breakdown voltage** V_{Br}. The reverse breakdown voltage V_{Br} is independent of the reverse current. Although the breakdown process is not inherently destructive but the maximum reverse current should be limited by an external circuit element to avoid excessive junction heating. There are two important breakdown mechanisms in the junction diode, namely,

1. Zener breakdown (tunnelling) for an abrupt junction
2. Avalanche breakdown (Impact Ionization) for graded junction

The avalanche breakdown imposes an upper limit on the reverse bias voltage for most diodes. Avalanche breakdown also limits the collector voltage of a bipolar transistor and drain voltage of the MOSFET.

4.10.1 Zener Breakdown

Zener breakdown occurs in a heavily doped *p-n* junction. The heavy doping in the two regions of junction makes depletion width extremely thin so that the conduction band of *n*-side and valence band of *p*-side of the junction are sufficiently close and the applied high electric field (due to narrow width) across the region in the reverse direction allows electrons from the valence band of *p*-side to tunnel through the energy gap into the conduction band of *n*-side as shown in Fig. 4.25. This process is called tunnelling. Due to the tunnelling of carriers through the energy gap, a large current flow in the reverse direction which results in **zener breakdown** and the reverse biased voltage at which zener breakdown occurs is given by V_z.

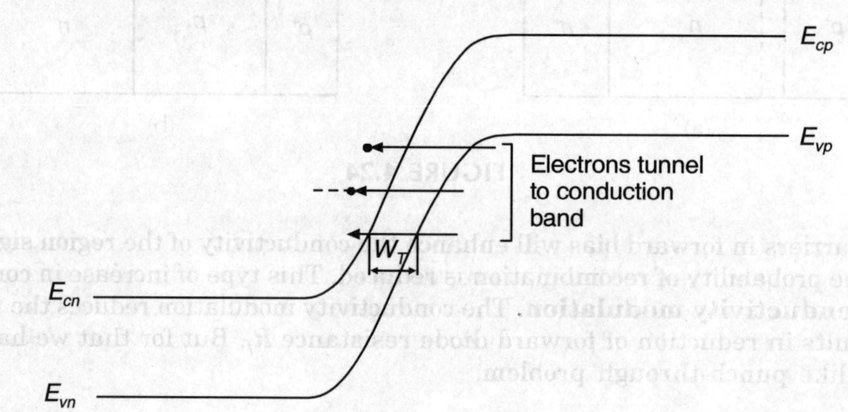

FIGURE 4.25 Tunnelling of electrons in a heavily doped *p-n* junction.

The transmission probability for quantum mechanical tunnelling generally depends exponentially on the depletion width. At a depletion width of order of less than 50 Å, the tunnelling can become a significant contributor to conduction mechanism. The typical field required for tunnelling is 10^6 V/cm or higher for silicon and gallium arsenide (GaAs) and doping concentration for both *p*- and *n*-regions must be of order of 5×10^{17} cm^{-3} or more.

The tunnelling probability T for an electron, having energy E and incident normally to the forbidden gap, is:

$$T = \exp\left[-\pi W_T \sqrt{m^* E_g / 2^{3/2} h}\right] \tag{4.64}$$

where W_T is the tunnelling distance. Tunnelling distance depends on the bandgap and slope of the bandgap edges and given as $W_T = E_g/qE$. Equation (4.64) tells that the tunnelling probability as well as tunnelling current depends strongly on the effective mass, bandgap, doping level and applied voltage.

If the breakdown voltage is less than about $4E_g/q$ where, E_g is forbidden gap, then in silicon and gallium arsenide, breakdown mechanism is dominated by Zener breakdown or tunnelling.

For tunnelling breakdown, the reverse voltage for a fixed current becomes smaller as the temperature is increased.

Zener diode

The zener diode is a heavily doped *p-n* junction which operates in reverse breakdown region to make the use of zener effect. Zener effect means a diode operates with a predetermined breakdown voltage V_z. This voltage is known as Zener breakdown voltage and remains constant for a particular Zener diode. When zener diode operates in the forward bias, it acts like an ordinary diode. The symbol for zener diode is shown in Fig. 4.26.

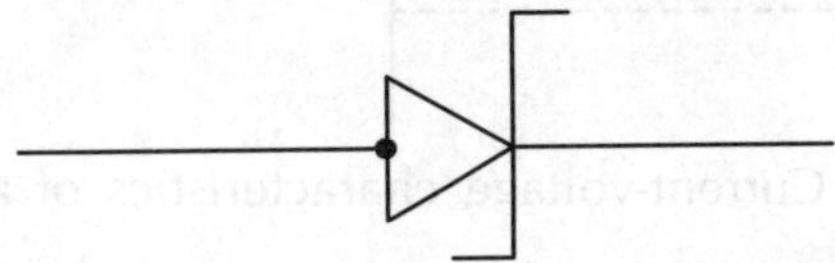

FIGURE 4.26 Symbol for zener diode.

The zener diode maintains a constant voltage across its ends regardless of the input voltage and process is known as **voltage regulation.** The voltage regulator circuit is shown in Fig. 4.27.

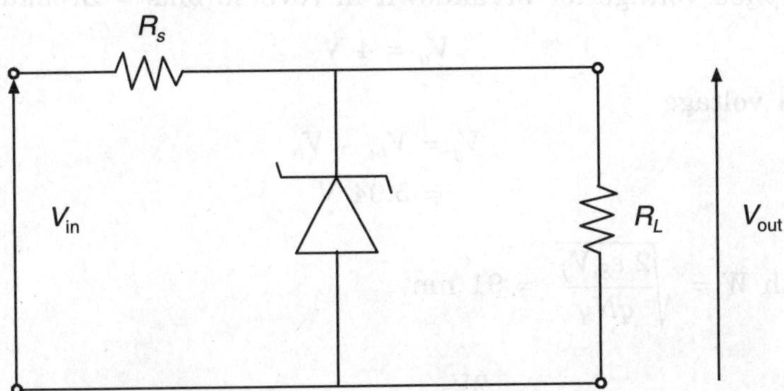

FIGURE 4.27 Voltage regulator circuit.

In zener diode either one or both breakdown mechanisms may be present. At low doping levels and higher voltages, the avalanche breakdown dominates while at heavy doping levels and lower voltages zener mechanism dominates. At certain doping level and around 6 V for Si, both mechanisms are present with temperature coefficients that just cancel.

Typical zener diodes have breakdown voltages anywhere in between 2 V and 200 V depending on the application.

Zener diodes can withstand at a relatively large reverse current without being damaged. The reverse bias voltage leading to zener breakdown is adjustable during manufacture of the device. The current characteristic for a zener diode is shown in Fig. 4.28.

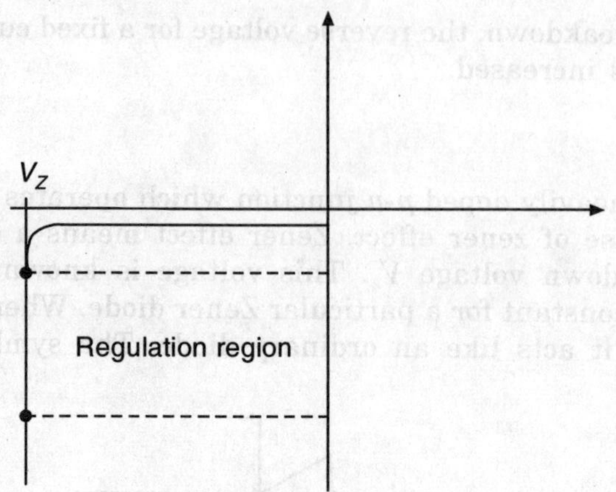

FIGURE 4.28 Current-voltage characteristics of a zener diode.

EXAMPLE 4.12

Estimate the tunnelling distance for appreciable tunnel current. Consider a p^+-n junction with $N_d = 8 \times 10^{17}$. Given built-in-potential = 1.04 V and breakdown voltage = 4 V. Take silicon material for p^+-n junction.

Solution: Applied voltage for breakdown in reverse bias = Breakdown voltage

i.e.
$$V_a = 4 \text{ V}$$

Hence junction voltage
$$V_j = V_{bi} - V_a$$
$$= 5.04 \text{ V}$$

Depletion width $W = \sqrt{\dfrac{2\,\varepsilon_{Si}V_j}{qN_d}} = 91 \text{ nm}$

Therefore,
$$\varepsilon_{\text{max}} = \frac{2V_j}{W} = 1.1 \times 10^8 \text{ V/cm}$$

Using $W_T = \dfrac{E_g}{q\varepsilon_{\text{max}}}$, we have

$$W_T = \frac{1.12 \text{ eV}}{q \times 1.1 \times 10^8} = 10 \text{ nm}$$

4.10.2 Avalanche Breakdown

Avalanche breakdown is caused by an impact ionization of electron-hole pairs at high electric fields. Due to high electric field in reverse bias, the charge carriers (electrons/

holes), which drift across the space charge region, acquire enough kinetic energy to break covalent bonds upon collision with the semiconductor lattice and create an electron-hole pair by ionizing the lattice atom or breaking covalent bond. These carriers can participate in a similar process, which leads to a multiplication of carriers in the space charge region. For example, if the field is sufficiently large across the space charge region, a thermally generated electron A gains sufficient kinetic energy from the field, and collides with the impurity atoms. If the energy of electron A is greater than the binding energy of the respective atom then electron A can break the lattice covalent bonds, creating an electron-hole pair B and B' through a process called **impact ionization**.

Now, there are three carriers A, B, B'. These three carriers acquire sufficient kinetic energy from the field and create additional electron-hole pairs (C and C' etc.). Therefore, the number of carriers increased up to nine, i.e., six electrons and three holes. These all carriers in turn continue the process, creating other electron-hole pairs; leading to avalanche process because we started with one electron and through a succession of impacts create more and more carriers. The avalanche process is schematically shown in Fig. 4.29. The tilted energy bands in Fig. 4.29 represents the application of high electric field (E) to semiconductor.

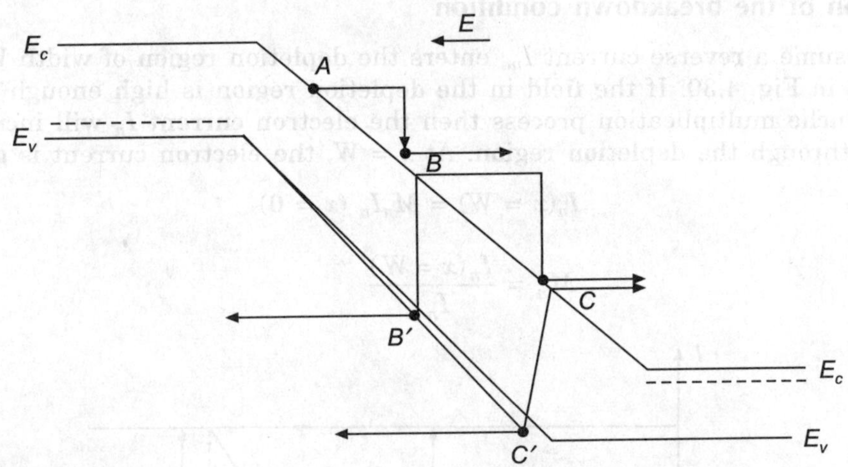

FIGURE 4.29 Schematic representation of avalanche process.

The avalanche breakdown process results in the upper limit on the reverse voltage which may be applied on the *p-n* junction or on the base-collector junction of the bipolar transistor or the drain voltage of the MOSFETs. The avalanche breakdown is similar to situation to an avalanche where a small disturbance causes a whole mountain side of snow to come crashing down.

Avalanche breakdown occurs in the lightly doped *p-n* junctions where the depletion width is comparatively large. The doping density controls the breakdown voltage. The temperature coefficient of the avalanche breakdown is positive, i.e., as the temperature increases the reverse bias breakdown voltage increases. The magnitude of the temperature coefficient also increases with increasing breakdown voltage. The avalanche breakdown results in a junction diode if the breakdown voltage in excess of $(6E_g/q)$. For the breakdown voltage between $(4E_g/q)$ and $(6Eg/q)$, both avalanche as well as zener breakdown occur. A *p-n* junction that experience breakdown above 5 V is caused by avalanche effect whereas any junctions breakdown around 5 V are usually caused by combination of two effects.

The efficiency of an avalanche breakdown effect is characterized by multiplication factor M that depends on the reverse voltage, and given by emperical relation as:

$$M = \frac{1}{1 - \left(\dfrac{V_r}{V_{Br}}\right)^n} \tag{4.65}$$

where n is in range of 2–6, V_r is applied reverse voltage, and V_{Br} is breakdown voltage. When $V_r \to V_{Br}$, $M \to \infty$; this means under this condition, reverse current increass sharply in the reverse direction. An avalanche breakdown initiated by free electrons is known as electron assisted breakdown whereas avalanche breakdown caused by hole is known as hole assisted breakdown.

Derivation of the breakdown condition

Let us assume a reverse current I_{no} enters the depletion region of width W at $x = 0$ as shown in Fig. 4.30. If the field in the depletion region is high enough to initiate the avalanche multiplication process then the electron current I_n will increase with distance through the depletion region. At $x = W$, the electron current is given by

$$I_n(x = W) = M_n I_n \ (x = 0) \tag{4.66a}$$

or

$$M_n = \frac{I_n(x = W)}{I_{no}}$$

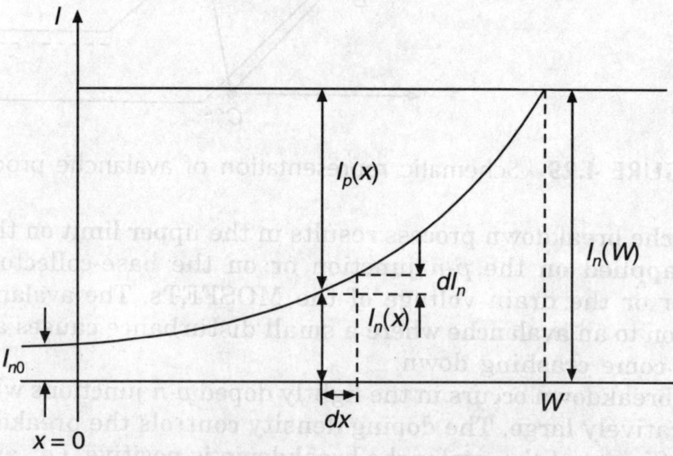

FIGURE 4.30 Electron and hole current components through the space charge region during avalanche multiplication.

where M_n is multiplication factor. Similarly, since hole moves opposite to electron, therefore, hole current I_p increases from $x = W$ to $x = 0$, and reaches its maximum value at $x = 0$. The total current I is given by sum of electron current and hole current, i.e., $I = I_n + I_p$. This total current is constant at steady state. The incremental electron current at some point x equals the number of electron-hole pairs generated per second in the distance dx, i.e.,

$$dI_n(x) = I_n(x)\alpha_n dx + I_p(x)\,\alpha_p\,dx$$

or
$$dI_n(x) = I_n(x)\alpha_n dx + (I - I_n(x))\,\alpha_p\,dx$$

or
$$dI_n(x) = I_n(x)(\alpha_n - \alpha_p)dx + I\alpha_p\,dx$$

or
$$dI_n(x) - (\alpha_n - \alpha_p)\,I_n(x)\,dx = I\alpha_p\,dx \qquad (4.66b)$$

where α_n and α_p are the electron and hole ionization constants respectively.

Let us assume $\alpha_n = \alpha_p = \alpha$, for simplicity, then Eq. [4.66(b)] can be written as:

$$dI_n(x) = I\alpha\,dx \qquad (4.66c)$$

Integrating Eq. (4.66c) through space charge region, i.e. from $x = 0$ to $x = W$, and using the boundary condition $I_n(x = W) = I_n(W)$ and $I_n(x = 0) = I_{n0}$, the solution of differential Eq. (4.66c) can be given as:

$$\frac{I_n(W) - I_{n0}}{I} = \int_0^W \alpha\,dx \qquad (4.66d)$$

Putting Eq. (4.65) into Eq. (4.66 d), we have

$$\frac{M_n I_{n0} - I_{n0}}{I} = \int_0^W \alpha\,dx$$

or
$$\frac{I_{n0}}{I}(M_n - 1) = \int_0^W \alpha\,dx \qquad (4.66e)$$

Since $M_n I_{n0} \simeq I$, Eq. (4.66e) reduces to

$$\frac{1}{M_n}(M_n - 1) = \int_0^W \alpha\,dx$$

or
$$\left(1 - \frac{1}{M_n}\right) = \int_0^W \alpha\,dx \qquad (4.66f)$$

The avalanche breakdown voltage is defined as the voltage where multiplication factor M_n approaches to infinity. Hence, putting $M_n \to \infty$ in Eq. (4.66f), we get

$$\int_0^W \alpha \, dx = 1 \tag{4.67}$$

Since, the ionization constants are strong function of electric field and electric field is not constant through the space charge region, therefore, ionization constant is represented in the simpler form as:

$$\alpha = A \, \exp\{[-B/E(x)^m]\}$$

where A, B and m are empirical fitting parameters. Let us consider a one-sided abrupt junction, i.e., p^+-n junction for which the maximum field is given by

$$E_m = \frac{qN_d x_n}{\varepsilon_{Si}} \tag{4.68a}$$

and depletion width along lightly doped region is:

$$x_n \simeq \left(\frac{2\,\varepsilon_{Si}V_r}{q} \cdot \frac{1}{N_d}\right)^{1/2} \tag{4.68b}$$

where V_r is applied reverse bias voltage. The maximum electric field at which breakdown occurs is defined as the critical electric field. At breakdown condition, replacing

$$E_m \rightarrow E_c$$
$$V_r \rightarrow V_{Br}$$
$$N_d \rightarrow N_B$$

where N_B is the semiconductor doping in the low-doped region of one-sided abrupt junction, we have (after using eqs. (4.68a) and (4.68b)

$$V_{Br} = \frac{\varepsilon_{Si}E_c^2}{2qN_B} = \frac{E_c \times x_n}{2} \simeq \frac{E_c \cdot W}{2} \tag{4.69}$$

where $x_n \simeq W$ for lightly doped n-region

Figure 4.31 shows the variation of critical electric field required for avalanche breakdown with the background doping of the lightly doped region for Si and GaAs at room temperature. From figure, it is clear that for low doping concentration, the breakdown always occurs at slightly higher critical field for GaAs than Si; but at very high doping concentration ($N_B \geq 10^{18}/\text{cm}^3$), breakdown occurs at same electric field for both Si as well as GaAs.

If the depletion layer width $W < W_m$ (maximum depletion width), the device will punched through prior to breakdown. Then, it is required to increase reverse bias further for breakdown to occur. Essentially the critical field is same and the breakdown voltage for the punch-through diode is V_B'.

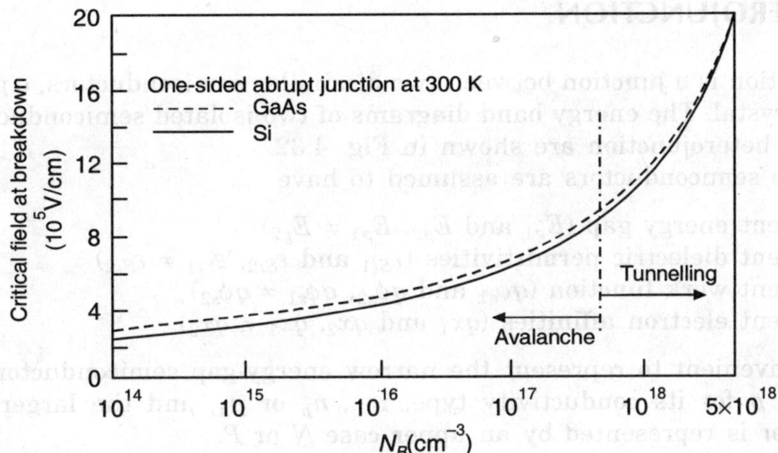

FIGURE 4.31 Variation of critical electric field with background doping of lightly doped region required for breakdown in case of Si and GaAs.

Therefore,

$$\frac{V'_B}{V_B} = \left(\frac{W}{W_m}\right)\left(2 - \frac{W}{W_m}\right) \tag{4.70}$$

Punch-through occurs when doping concentration N_B will become sufficiently low as in a $p^+ = \pi = n^+$ or $p^+ = v = n^+$ diode, where, π stands for lightly doped p-type and v stands for lightly doped n-type semiconductor.

Applications of Breakdown

Mainly in

(i) Reference voltage diode (Zener diode)
(ii) Avalanche photodiode (APD)
(iii) IMPATT diode
(iv) Tunnel diode or Esaki diode

EXAMPLE 4.13

Calculate the breakdown voltage for Si one sided abrupt p^+-n junction with $N_d = 5 \times 10^{16}/cm^3$.

Solution: From Fig. 4.31, $\varepsilon_c = 5.7 \times 10^5$ V/cm for $N_d = 5 \times 10^{16}/cm^3$
Using Eq. (4.69),

$$V_B = \frac{\varepsilon_c^2 \varepsilon_{Si}}{2qN_d} = \frac{11.9 \times 8.8521 \times 10^{-4} \times 5.7 \times 10^5}{2 \times 1.6 \times 10^{-19} \times 5 \times 10^{16}}$$

$$= 21.4 \text{ V}$$

4.11   HETEROJUNCTION

A heterojunction is a junction between two dissimilar semiconductors, e.g., Si and Ge in a single crystal. The energy band diagrams of two isolated semiconductors prior to formation of heterojunction are shown in Fig. 4.32.

The two semiconductors are assumed to have

- different energy gap (E_{g1} and E_{g2}, $E_{g1} \neq E_{g2}$)
- different dielectric permittivities (ε_{Si1} and ε_{Si2}, $\varepsilon_{Si1} \neq \varepsilon_{Si2}$)
- different work function ($q\phi_{s1}$ and $q\phi_{s2}$, $q\phi_{s1} \neq q\phi_{s2}$)
- different electron affinities (qx_1 and qx_2, $qx_1 \neq qx_2$)

It is convenient to represent the narrow energy-gap semiconductor by a lower case of n or p for its conductivity type, i.e., n_p or p_n, and the larger energy-gap semiconductor is represented by an upper case N or P.

From Fig. 4.32, it is clear that

$$qx_1 = E_{\text{vacc}} - E_{c1}$$

$$qx_2 = E_{\text{vacc}} - E_{c2}$$

$$q\phi_{s1} = E_{\text{vacc}} - E_{F1}$$

$$q\phi_{s2} = E_{\text{vacc}} - E_{F2}$$

and
$$\Delta E_c = E_{c1} - E_{c2} = (E_{\text{vacc}} - qx_1) - (E_{\text{vacc}} - qx_2)$$

$$= q(x_2 - x_1) \tag{4.71a}$$

is defined as the difference in the energy of the conduction band edge in the two semiconductors,

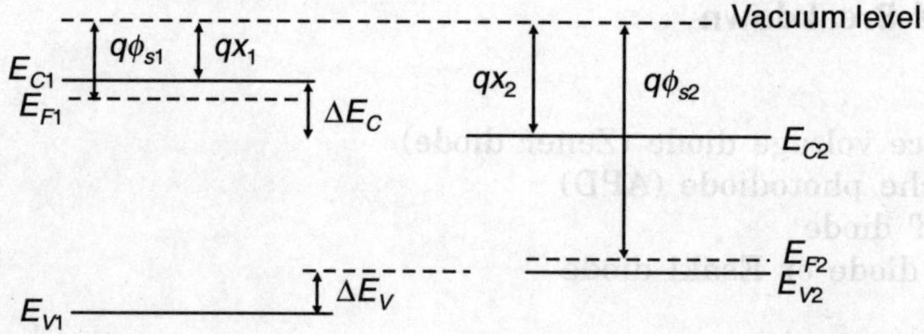

FIGURE 4.32   Band diagram of two isolated dissimilar semiconductor materials.

Similarly,

$$\Delta E_v = E_{v1} - E_{v2} \tag{4.71b}$$

or
$$\Delta E_v = E_{g1} + qx_1 - (E_{g2} + qx_2)$$

$$= \Delta E_g + q(x_1 - x_2) = \Delta E_g - \Delta E_c \tag{4.71c}$$

is the difference in the energy of the valence band edge of the two semiconductors, where $\Delta E_g = E_{g1} - E_{g2}$ is the energy gap difference. When these two semiconductor materials are brought together, a heterojunction *N-p* is formed between them. Let us assume that there is negligible number of traps or generation-recombination centres at the interface of the two dissimilar semiconductors, i.e., it is assumed to be ideal abrupt heterojunction. This assumption is true only when heterojunctions are formed between two semiconductors which have closely matched lattice constants.

To draw the energy-band diagram of the heterojunction at thermal equilibrium, we must meet two basic requirements:

1. Fermi level should be spatially constant on both sides of the interface.
2. Vacuum level must be continuous and parallel to the band edges.

One can draw the thermal equilibrium energy band diagram as:

1. First, see the position of Fermi level in an isolated atom. Since, the Fermi level of first material is above the second material, so electron will jump from the conduction band of first material to conduction band of second material near the junction. This shows the conduction band of first material near the junction will move away from resultant Fermi level whereas the conduction band of second material will move towards the Fermi level and shown as point *a–d* in Fig. 4.34.
2. Similarly trend will observe in the valence band.
3. To keep the electron affinity and work function of the material constant, the vacuum level should also bend, called **local vacuum level**.

Steps:
1. First draw constant Fermi level as:

$$------------------------------- E_F$$

2. Draw conduction band edge and valence band edge of the respective materials in the bulk as shown in Fig. 4.33.

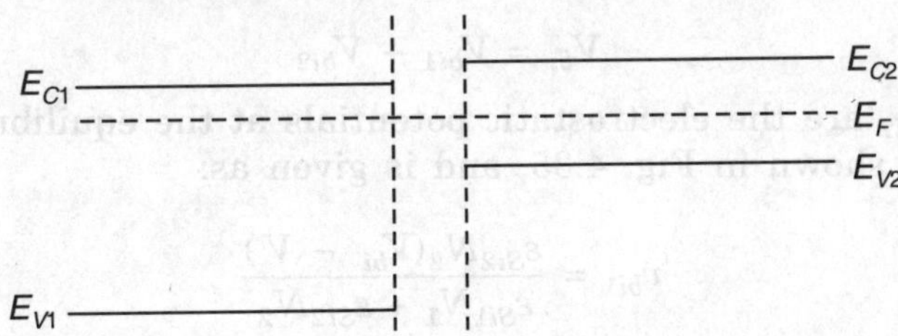

FIGURE 4.33   Conduction band and valence band of two dissimilar materials in the bulk.

3. As discussed, near the junction E_{c1}, moves away from E_F and E_{c2} moves toward E_F. [Refer Fig. 4.34].

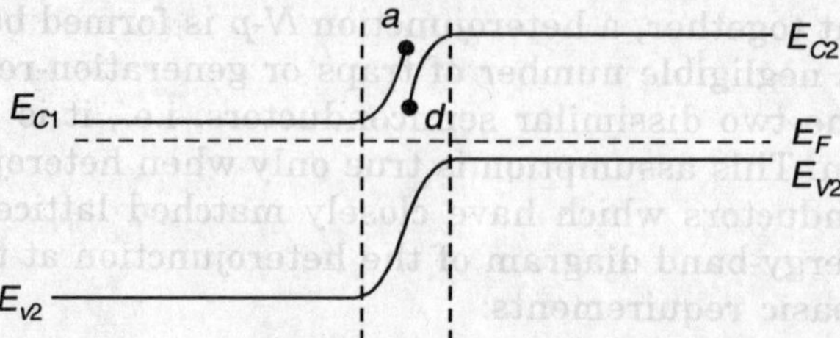

FIGURE 4.34   Bending of conduction band when these two materials are joined.

4. Due to this movement of the conduction band on either side will result in discontinuity as shown in Fig. 4.35.

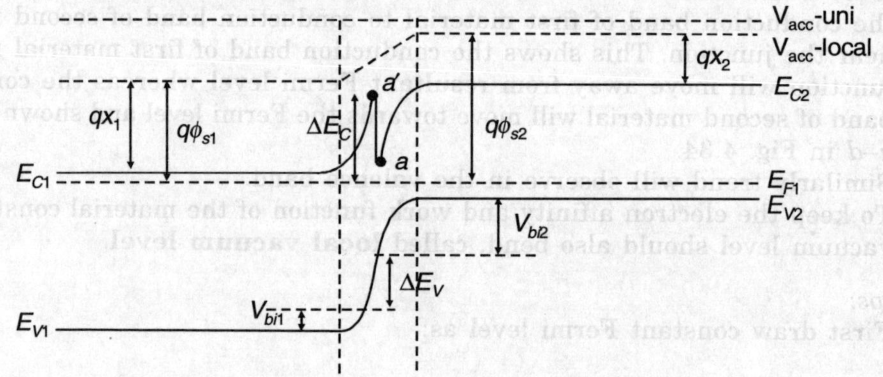

FIGURE 4.35   Band diagram of heterojunction.

where ΔE_c and ΔE_v give the discontinuity in conduction band edges and valence band edges. The total built-in-potential is:

$$V_{bi} = V_{bi1} + V_{bi2} \qquad (4.72)$$

where V_{bi1} and V_{bi2} are the electrostatic potentials at the equilibrium for respective semiconductors as shown in Fig. 4.35, and is given as:

$$v_{bi1} = \frac{\varepsilon_{Si2}N_2(V_{bi} - V)}{\varepsilon_{Si1}N_1 + \varepsilon_{Si2}N_2} \qquad (4.73a)$$

and

$$v_{bi2} = \frac{\varepsilon_{Si1}N_1(V_{bi} - V)}{\varepsilon_{Si1}N_1 + \varepsilon_{Si2}N_2} \qquad (4.73b)$$

where N_1 and N_2 are doping concentrations in respective semiconductor, and V is the applied voltage.

4.12 SUMMARY

Diodes are the simplest of the solid-state devices; consists of a piece of *p*-type material fused to a piece of *n*-type material. The most common forms of diodes are constructed from silicon because germanium is less stable at high temperatures than silicon. At thermal equilibrium, in absence of any external excitation forces, no current flows in the diode. The built-in-potential is responsible for maintaining the equilibrium in absence of any applied voltage and built-in-potential cannot be measured. A diode (*p-n* junction) is a unidirectional device, means current flows when it is forward bias and negligible current in reverse direction. It is used extensively in rectification, switching applications. Analysis of *p-n* junction can be simplified by considering abrupt junction and depletion approximation. The field in the bulk regions of *p*-side and *n*-side is zero. The depletion width entend extirely into the lightly doped region of *p-n* junction.

The ideal *I-V* characteristics of a *p-n* junction shows that current increses rapidly in forward direction with applied forward bias voltage and remain constant (almost zero) in the reverse direction. For ideal diode, built-in-potential as well as forward resistance are zero whereas reverse resistance (R_r) is infinity. However, practical *p-n* junction departs from the ideal characteristic because of generation and recombination in the depletion region. Also other effects which are responsible for deviation from ideal condition are series-resistance effects and high-injection under forward bias.

Two main capacitors associated with the *p-n* junction are junction capacitor which dominates in the reverse-bias and diffusion capacitor, which dominates in the forward-bias. A limiting factor in the operation of *p-n* junction is junction breakdown. Two main breakdown mechanisms in the *p-n* junction are zener breakdown and avalanche breakdown. Zener breakdown, also called tunnelling (which is a quantum mechanical effect) is more visible in abrupt junction where both regions are heavily doped. The typical field required for tunnelling is 10^6 V/cm. Zener breakdown voltage has negative temperature coefficient. The avalanche breakdown imposes an upper limit on the reverse bias for most diodes. Avalanche breakdown is caused by impact ionization and it is also called impact ionization breakdown. The avalanche breakdown mainly dominant in lightly doped semiconductor and avalanche breakdown voltage has positive temperature coefficient. For the breakdown voltage between ($4E_g/q$) and ($6E_g/q$), both breakdown, i.e., avalanche as well as zener breakdown occur. Both breakdowns are occur in reverse biased *p-n* junction. Zener diode is mainly used for voltage regulator because zener diode maintains a constant voltage across its ends regardless of the input voltage. A heterojunction is formed between two dissimilar semiconductors.

REVIEW QUESTIONS

1. Why in absence of external applied field, no current flows in the *p-n* junction at thermal equilibrium?

2. Why built-in-potential cannot be measured?

3. Why *p-n* junction diode is called unidirectional device?

4. For an abrupt Si *p-n* junction, why the built-in-potential become larger at low temperature? Assume that $p = N_a^-$ and $n = N_d^+$ remain constant with temperature.

5. What is meant by a quasi Fermi level?

6. Calculate the value of the built-in-potential for a Si *p-n* junction in which *n*-region is uniformly doped with 10^{16} donors per cm^3 and the *p*-region has a uniform acceptor concentration of $10^{15}/cm^3$.

7. Why Si semiconductor is preferred over the Ge for designing a *p-n* junction diode?

8. Why depletion region extends into lightly doped region in the *p-n* junction? Explain qualitatively.

9. In a p^+-*n* junction the *n*-doping (N_d) is doubled. How do the following change:
   (a) Junction capacitance
   (b) Built-in-potential
   (c) Breakdown voltage
   Assuming other parameters remain constant.

10. Explain qualitatively, why forward bias current increases rapidly with applied forward bias voltage.

11. Why does generation current dominant over the recombination current under reverse bias in *p-n* junction but recombination current dominates over generation current under forward bias?

12. At a given forward bias voltage, as the temperature increases the diffusion current will increase more rapidly than the recombination current. Why?

13. Show the temperature dependence of saturation current and energy gap. Explain the result qualitatively.

14. The band diagram of a *p-n* junction is shown in the Fig. 4.36.
   (a) Whether the junction is at thermal equilibrium or not? Explain.
   (b) What is the type of biasing and find its value?
   (c) How the junction will achieve its equilibrium?
   (d) If at any temperature, the intrinsic carrier concentration becomes 1.5 times than its earlier value. How the reverse saturation current gets affected?

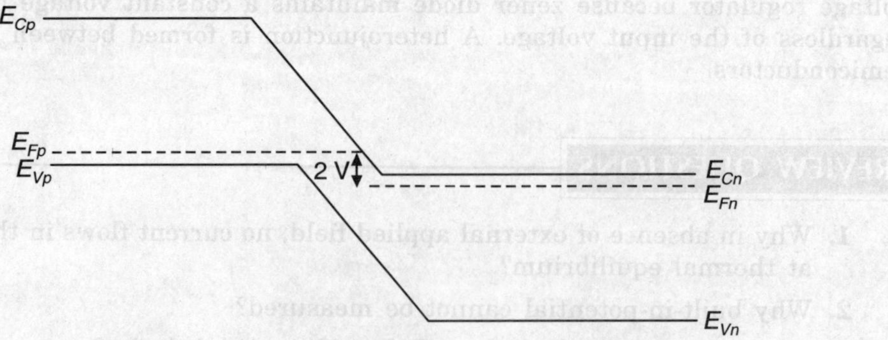

FIGURE 4.36 Band diagram of *p-n* junction.

15. The variation of the junction capacitance with applied voltage is shown in Fig. 4.37 from this graph, is it possible to determine the concentration of both sides of the junction? Explain why $(1/C_j^2)$ is constant after a particular applied voltage.

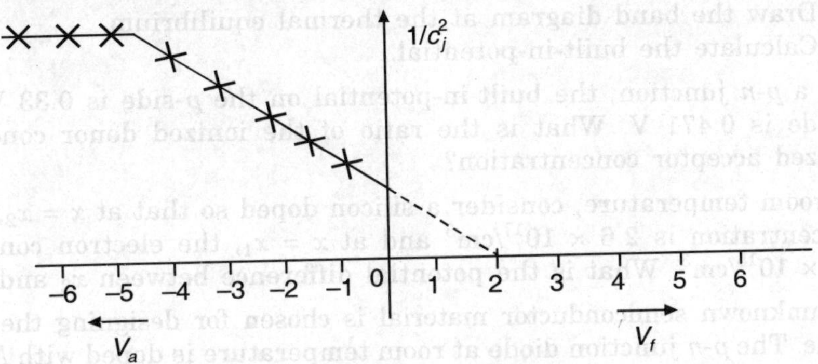

FIGURE 4.37 Variation of junction capacitance with applied voltage.

16. Sketch the variation of
  (a) Carrier distribution
  (b) Electron-and-hole currents in forward-biased long base *p-n* junction diode.

17. Why in many application, in addition to diffusion capacitance and depletion capacitance, one must include the conductance in the diode equivalent circuit?

18. If the forward bias voltage becomes double, find the change in diffusion capacitance.

19. Why voltage sensitivity of junction capacitor is greater for an abrupt junction than the linearly graded junction? For designing a varactor diode whether you will prefer abrupt junction or linearly graded junction. Why?

20. Why varactor diodes are not used for rectification?

21. What is rectification? As an engineer, what criteria will you consider for designing a rectifier?

22. Write the basic differences between zener and avalanche breakdown.

23. Why zener breakdown voltage has negative temperature coefficient whereas avalanche breakdown voltage has positive temperature coefficient?

24. Explain, whether the avalanche breakdown voltage of one sided *p-n* junction increases or decreases with increase in depletion width. Assume critical field remains constant.

25. Explain how a tunnel diode may be thought as a heavily doped zener diode. Does a tunnel diode have a rectifying?

NUMERICAL PROBLEMS

1. For a silicon p-n junction at room temperature, the workfunction on the p-side is 4.90 eV and on the n-side is 4.20 eV.

   (a) Draw the band diagram at the thermal equilibrium.
   (b) Calculate the built-in-potential.

2. For a p-n junction, the built-in-potential on the p-side is 0.33 V and on the n-side is 0.471 V. What is the ratio of the ionized donor concentration to ionized acceptor concentration?

3. At room temperature, consider a silicon doped so that at $x = x_2$, the electron concentration is $2.6 \times 10^{17}/cm^3$ and at $x = x_1$, the electron concentration is $6.3 \times 10^{15}/cm^3$. What is the potential difference between x_2 and x_1?

4. An unknown semiconductor material is chosen for designing the p-n junction diode. The p-n junction diode at room temperature is doped with $N_d = 10^{18}/cm^3$ and $N_a = 10^{16}/cm^3$, and built-in-potential was obtained to be $V_{bi} = 0.814$ V. Find the type of semiconductor material.

5. An abrupt Si p-n junction has $N_a = 10^{18}/cm^3$ on one side and $N_d = 5 \times 10^{15}/cm^3$ on the other side.

   (a) Draw the band diagram at thermal equilibrium.
   (b) Find the position of fermi level in the respective regions.
   (c) Calculate the built-in-potential and compare the result with the value obtained from energy band diagram.
   (d) Find the depletion width value.
   (e) Calculate the maximum field.
   (f) Plot the charge density, potential and electric field with x.

6. Show that for a linearly graded junction, the depletion layer capacitance is given by

$$C_j = \left(\frac{qa\varepsilon_s^2}{12\,(V_{bi} - V_a)} \right)^{1/3} F/cm^2$$

   where V_a is applied potential.

7. Consider the $(1/C_j^2)$ V_s applied voltage (V_a) plot of a silicon p^+-n-n^+ junction as shown in Fig. 4.38. The diode has a junction area of $10^{-3}/cm^2$. The n-region is grown on the n^+ substrate and has thickness W_1.

   (a) Find built-in-potential.
   (b) Find N_a and N_d.
   (c) Calculate W_1.

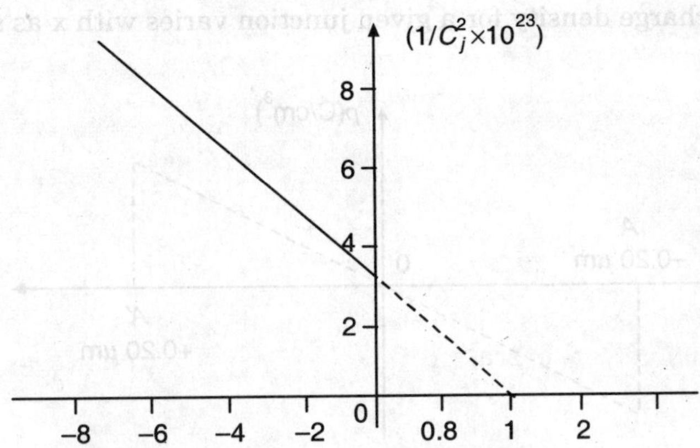

FIGURE 4.38 Plot of $1/C_j^2$ vs applied voltage.

8. A lightly doped n-type sample of Si has a resistivity of 4 ohm-cm. It is used to make a p-n junction with a p-type region in which dopant density is 1000 times higher than in the n-region (Assuming ideal diode case).

    (a) Sketch the transition region with $x = 0$ defined as the metallurgical junction.

    (b) The junction potential is modified to 0.5 V due to applied voltage across the junction. Find

        (i) Whether the junction is forward or reverse biased

        (ii) Applied voltage

        (iii) Current will flow

        (iv) Electron-hole product deviation from its equilibrium value in presence of the field

9. An abrupt junction (Si p-n junction) at $T = 300$ K has an area of 1 mm^2. The measured junction capacitance C_j as a function of bias is given by the relation:

$$(1/C_j^2) = (5/10) \times 10^6 \times (3 - 5\,V_a)$$

where C_j is expressed in microfarad and V_a is the applied voltage in V.

    (a) Find built-in potential.

    (b) How the junction capacitance will vary if one increases the applied voltage? Explain briefly.

10. Show that the maximum electric field for p^+-n junction due to applied voltage V_a is given by

$$\varepsilon_{\max} = \left[\frac{2qN_d}{\varepsilon_{Si}}(V_{bi} - V_a)\right]^{1/2}$$

11. When the reverse bias of an abrupt p^+-n junction diode is increased to -2.55 V, the capacitance becomes one-half of the value at zero bias. What is the value of built-in-potential?

12. The space charge density for a given junction varies with x as shown in Fig. 4.39.

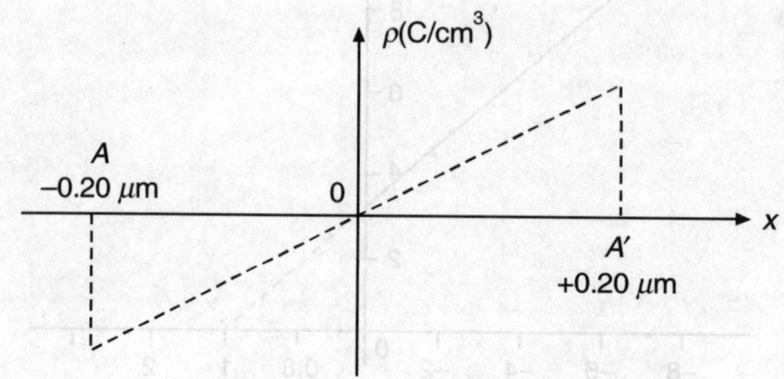

FIGURE 4.39 Variation of space charge density with x.

At $T = 300$ K,

(a) Express the charge density in terms of x.
(b) Find the ratio of donor and acceptor impurity concentration at A and A'.
(c) The variation of potential for such junction is shown in Fig. 4.40. Find:

    (i) Built-in-potential
    (ii) Impurity gradient
    (iii) Maximum electric field and sketch the electric field with x
    (iv) Type of semiconductor material by which the junction is formed.

13. Determine the n-type doping concentration to meet the following specifications for a Si p-n junction.

$$N_a = 10^{18}/\text{cm}^3, \ \varepsilon_{\max} = 4 \times 10^5 \ \text{V/cm at } V_R = 30 \ \text{V and } T = 300 \ \text{K}.$$

14. For a silicon linearly graded junction as shown in Fig. 4.40 with an impurity gradient of $10^{20}/\text{cm}^4$, calculate the junction capacitance at reverse bias of 4 V (Take $T = 30$). Find the percentage change in the value of junction capacitance as the reverse bias voltage is changed to 2 V. Comment on the result.

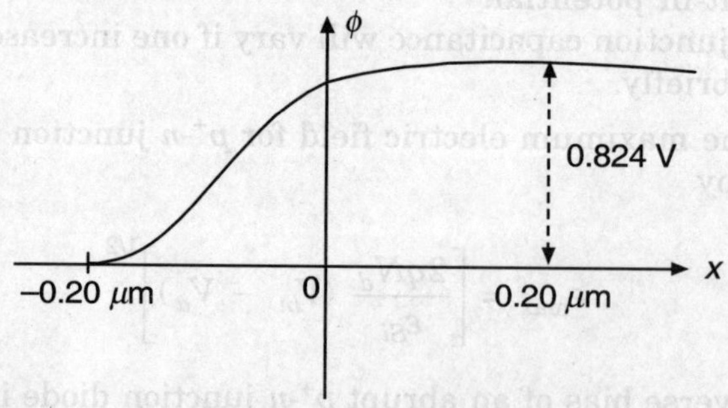

FIGURE 4.40 Silicon linearly graded junction.

15. Consider a Si *p-n* junction with *n*-type doping concentration of 10^{16}/cm^3 and built-in-potential is 0.814 V. The junction is subjected to forward bias of 0.8 V at T = 300 K. Find the minority carriers concentration at the respective edges of the space charge region. Also draw the band diagram for forward bias.

16. Calculate the applied reverse-bias voltage at which ideal reverse current in a *p-n* junction diode at T = 300 K reaches 95% of its reverse saturation current value.

17. Calculate the generation current for a Si *p-n* junction under reverse bias conditions. Take

$$N_a = 10^{17}/\text{cm}^3, \ N_d = 10^{17}/\text{cm}^3$$

$$\tau_n = i_p \cong 6 \times 10^{-6} \ \text{second}, \ (V_{bi} - V_a) = 5 \ \text{V}$$

$$\Delta_n = 20 \ \text{cm}^2/\text{s} \qquad \Delta_p = 11 \ \text{cm}^2/\text{s}$$

Also calculate the diffusion current and compare these two results? What will happen, if a Ge *p-n* junction is used instead of Si *p-n* junction. Interpret your result.

18. Calculate the recombination current for the Si *p-n* junction and compare with the generation current obtained in Problem 17. Assume applied forward bias voltage is 4.1 and take other parameters as given in Problem 17. Explain your findings qualitatively also.

19. Design an abrupt Si p^+-*n* junction diode that has a reverse breakdown voltage of 130 V and has a forward bias current of 2.2 mA at V_f = 0.7 V. Assume τ_{p0} = 10^{-7} sec.

20. An ideal silicon *p-n* junction has $N_d = 10^{18}$/cm^3, $N_a = 10^{16}$/cm^2, $\tau_p = z_n = 10^{-6}$ s, device area of 1.2×10^{-5} cm^2.

(a) Calculate the theoretical saturation current at T = 300 K.
(b) Calculate forward and reverse currents at ± 0.7 V. Compare these two results and discuss qualitatively.

21. For an ideal abrupt Si p^+-*n* junction, $N_d = 10^{16}$/cm^3. Find the stored minority carriers per unit area in the bulk region for an applied forward bias of 1 V. The length of the neutral region is 1 μm and diffusion length of hole is 5 μm.

22. Calculate the turn-off-time for the Si *p-n* junction. Take all the parameters of Problem 20.

23. Show that for a linearly graded junction, the avalanche breakdown voltage is given by

$$V_B = \frac{4 \, \varepsilon_c^{3/2}}{3} \left(\frac{2\varepsilon_{Si}}{q} \right) (a)^{-1/2}$$

where ε_c is the critical field and a is impurity gradient.

Consider an Si p^+-*n* junction at T = 300 K having a linearly varying profile from $N_a = 10^{18}$/cm^3 to $N_d = 10^{18}$/cm^3 over a distance of 2 μm. Calculate the breakdown voltage.

24. Consider an ideal abrupt hetrojunction with a built-in-potential of 1.6 V. The impurity concentration in semiconductors 1 and 2 are 1×10^{16} donors/cm^3 and 3×10^{19} acceptor/cm^3 and the dielectric constants are 12 and 13 respectively. Find the electrostatic potential and depletion width in each material at equilibrium, and draw the band diagram also.

25. The doping profile of a p-n junction is shown in Fig. 4.41.

   (a) Can you use this device for the varactor diode application? Justify your answer.

   (b) How the behaviour of this device will change for varactor diode application if doping level of p-side is made exactly equal to that of n-side?

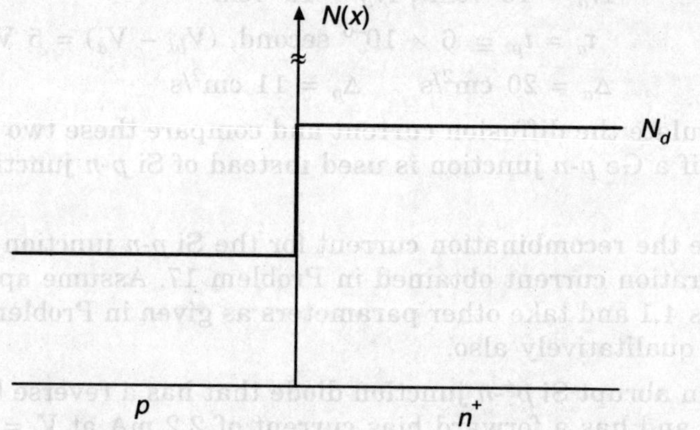

FIGURE 4.41 Doping profile of p-n junction.

CHAPTER

5

Metal-semiconductor Contact

LEARNING OBJECTIVES

- Schottky barrier diodes
  - Ideal *I-V* characteristics
  - Applications
- Ohmic contacts
- Energy band diagrams

5.1 INTRODUCTION

Metal-to-semiconductor contacts are of great importance since they are present in every semiconductor devices either as Schottky barrier (non-ohmic) or as an ohmic contact. The Schottky behaviour or ohmic behaviour of metal-semiconductor contacts depend on the characteristics of the interface.

The electrical characteristics of a metal-semiconductor contact, also called **M-S contact**, are a strong function of doping concentration. A metal contact with lightly doped semiconductor is called **Schottky barrier diode**, whereas a heavily doped semiconductor contact with metal exhibits low-resistance contact and hence this contact is referred as **ohmic-contact**. All semiconductor devices as well as integrated circuits need ohmic contact for connection with other devices in an electronic system. In this chapter, we will consider the energy-band diagram and current-voltage characteristics of both Schottky as well as ohmic contacts.

5.2 SCHOTTKY BARRIER DIODES

In 1938, Walter Schottky suggested that the rectification between a metal and a semiconductor could arise as a result of space charge in the semiconductor known as Schottky effect, i.e., when negative charges are brought near the metal surface, positive

209

(image) charges are induced in the metal. When this image force is combined with an applied electric field, the effective work done is somewhat reduced. Such barrier lowering is called Schottky effect. The structure, as shown in Fig. 5.1, is known as **Schottky diode**. This barrier diodes are mainly used in microwave detectors and as rectifiers in high current application. The Schottky diode is a majority carrier device whose switching speed is not limited by minority carriers. Many metals can create a Schottky barrier on either silicon or GaAs semiconductors.

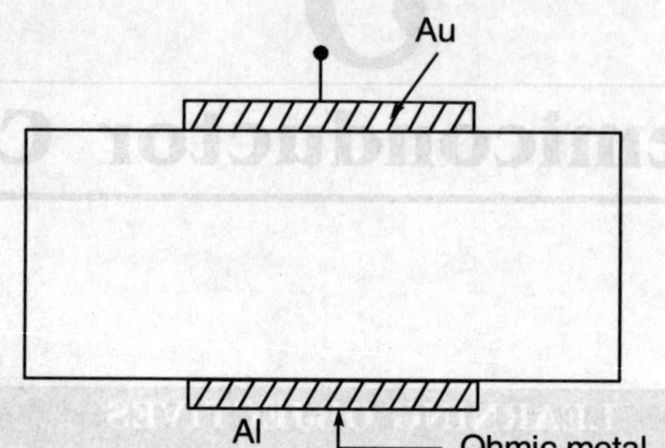

FIGURE 5.1 Structure of Schottky barrier diode.

5.2.1 Energy-band Diagram

The energy band diagram of an isolated metal and an isolated semiconductor at thermal equilibrium is shown in Fig. 5.2(a). The vacuum-level is defined as a level where electron is not bounded by any force. The metal work function is defined as the energy required to move electron from Fermi level E_f to vacuum level E_0. It is represented by $q\phi_m$ and measured in electron-volt. The workfunction of metal is generally different from semiconductor. The workfunction of the semiconductor material, depends on the doping concentration and electron affinity. Electron affinity is constant for a particular semiconductor and must remain constant throughout. However, the Fermi level of the semiconductor and hence workfunction can vary with doping. Therefore, semiconductor workfunction ϕ_s is given as:

$$q\phi_s = q\chi + (E_c - E_f) \tag{5.1}$$

A Schottky or rectifying contact between metal and semiconductor is formed when

$$q\phi_m > q\phi_s \text{ for } n\text{-type S.C.}$$

$$q\phi_m < q\phi_s \text{ for } p\text{-type S.C.}$$

From Fig. 5.2(a) it is clear that $q\phi_m > q\phi_s$. This indicates that metal has larger workfunction than the semiconductor and hence Fermi level at semiconductor side lies above the metal side. Therefore, electrons in the semiconductor have a higher potential energy than electrons in the metal. When metal makes intimate contact with the semiconductor, electron will instantaneously transfer from semiconductor to metal to establish thermal equilibrium. Transfer of electrons will leave positive ionized donor atoms near the junction and these transferred electrons accumulate on the surface of the metal and create a negative surface charge. The positively charged region near the junction of semiconductor is called a depletion region (W). The transfer of carriers result in band bending near the contact. Since, the charge neutrality must be maintained after the contact, therefore, excess charges inside the semiconductor and that inside the metal must be equal to each other in magnitude and opposite in sign. Since, metal is assumed to be sea of free electrons, therefore, the width over which these electrons (excess) move into the metal is negligibly thin in compare to the width inside the semiconductor. This results that the built-in-electric field and band bending are primarily present inside the semiconductor. The positive charge on the semiconductor creates negative field and lowers the band near the junction of semiconductor side; (Schottky effect) as shown in Fig. 5.2(b). The shape of energy diagram of the M-S contact junction is governed by the following three rules:

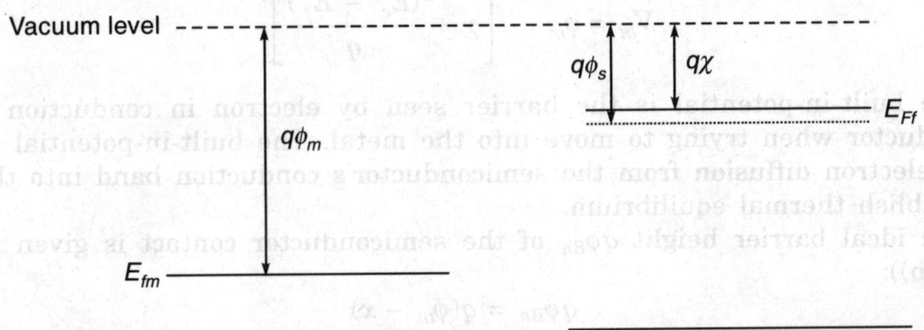

FIGURE 5.2(a) Band diagram of metal and *n*-semiconductor before contact.

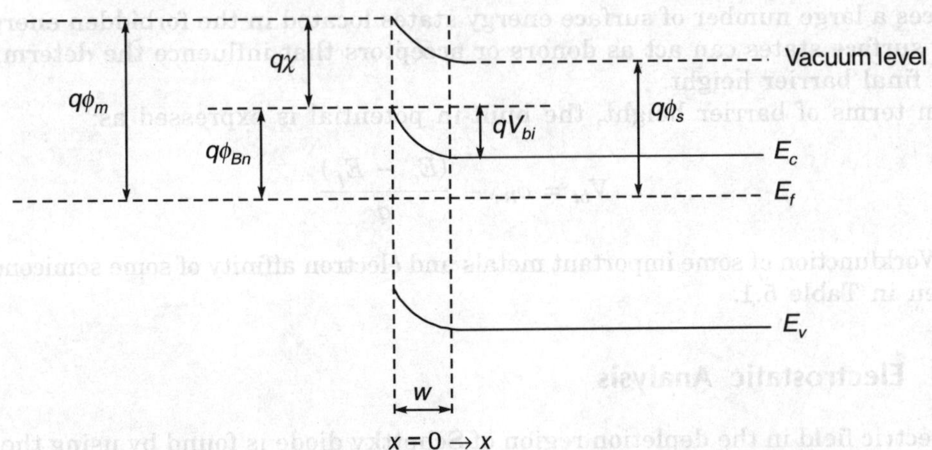

FIGURE 5.2(b) Band diagram of resulting M-S contact.

1. At thermal equilibrium, the Fermi levels for both semiconductor as well as metal must be spatially constant and equal throughout the system.
2. The electron affinity of the semiconductor material must be constant.
3. The vacuum level must be continuous.

To satisfy all the three requirements simultaneously, the valence band and conductor band of the semiconductor are forced to bend at the junction as shown in Fig. 5.2(b). The ideal metal-semiconductor theory applies when both materials are perfectly pure and there is no interaction between two materials and also in between their interfacial layer. For the ideal case, the built-in-potential for Schottky barrier diode is given by the difference between the workfunctions

i.e.
$$V_{bi} = (\phi_m - \phi_s) \tag{5.2}$$

Using Eq. (5.1) in Eq. (5.2), we have

$$V_{bi} = \phi_m - \left[\chi + \frac{(E_c - E_f)}{q} \right] \tag{5.3}$$

The built-in-potential is the barrier seen by electron in conduction band of semiconductor when trying to move into the metal. The built-in-potential prevents further electron diffusion from the semiconductor's conduction band into the metal and establish thermal equilibrium.

The ideal barrier height $q\phi_{Bn}$ of the semiconductor contact is given as (from Fig. 5.2(b)):

$$q\phi_{Bn} = q(\phi_m - x) \tag{5.4}$$

This is the potential barrier seen by electrons in the metal trying to move into semiconductor. This barrier is known as **Schottky barrier**.

It was found experimentally that the barrier height increases with increasing $q\phi_m$, but dependence is not as strong as predicted by Eq. (5.4). This is because in real Schottky diodes, the disruption of the crystal lattice at the semiconductor surface produces a large number of surface energy states located in the forbidden energy gap. These surface states can act as donors or acceptors that influence the determination of the final barrier height.

In terms of barrier height, the built-in potential is expressed as:

$$V_{bi} = \phi_{Bn} - \frac{(E_c - E_f)}{q} \tag{5.5}$$

Workfunction of some important metals and electron affinity of some semiconductors is given in Table 5.1.

5.2.2 Electrostatic Analysis

The electric field in the depletion region of Schottky diode is found by using the Gauss law to the region.

Table 5.1 Workfunctions of Metals and Electron Affinity of Semiconductors

Element	Workfunction ϕ_m(volt)
Ag, silver	4.26
Al, aluminium	4.28
Au, gold	5.1
Cr, chromium	4.5
Mo, molybdenum	4.6
Ni, nickel	5.15
Pd, palladium	5.12
Pt, platinum	5.65
Ti, titanium	4.33
W, tungsten	4.55

Electron affinity of some semiconductors

Element	Electron affinity χ (volt)
Ge, germanium	4.13
Si, silicon	4.01
GaAs, gallium arsenide	4.07
AlAs, aluminium arsenide	3.5

The electric field $E(x)$ is related to the surface charge $\rho(x)$ by one dimensional Gauss's law:

$$\frac{dE(x)}{dx} = \frac{\rho}{\varepsilon_{Si}} = \frac{qN_d}{\varepsilon_{Si}} \tag{5.6}$$

where,

N_d is the donor impurity concentration and assumed to be uniform ε_{Si} in silicon permittivity. Using depletion approximation and assuming field is maximum at the junction and must be zero at the edge of the depletion region, integrating Eq. (5.6) from $x = 0$ to $x = W$, we have

$$E(x) = E_{\max}\left(1 - \frac{x}{W}\right) \tag{5.7}$$

where $E_{\max}$ is the maximum electric field and is given as:

$$E_{\max} = \left(-q\frac{N_d}{\varepsilon_{Si}}\right)W \tag{5.8a}$$

The charge density, electric field and electrostatic potential variation with the distance from contact x is shown in Fig. 5.3 at equilibrium. As seen from Fig. 5.3, the electric field is a simple triangular function and can be easily integrated to give the built-in potential. The built-in-potential is given as:

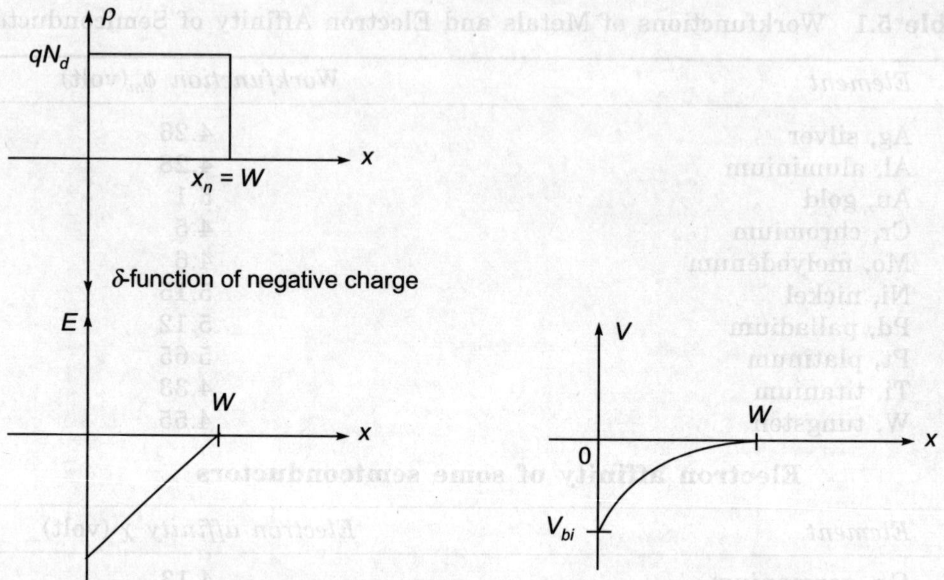

FIGURE 5.3   Charge distribution, electric field variation and electrostatic potential variation with the distance from contact at equilibrium in Schottky diode.

$$V_{bi} = \frac{qN_dW^2}{2\varepsilon_{Si}} \tag{5.8b}$$

where W is the resulting depletion width and given as:

$$W = \sqrt{\left(\frac{2\varepsilon_{si}}{qN_d}\right)V_{bi}} \tag{5.9}$$

By knowing the value of donor concentration and depletion width, the depletion charge can be obtained as:

$$Q_J = qN_d \cdot AW$$

$$Q_J = A\sqrt{2qV_{bi}\varepsilon_{Si}N_d} \tag{5.10}$$

where A is the area of the junction.

5.3   IDEAL *I-V* CHARACTERISTICS OF SCHOTTKY BARRIER DIODE

The current transport in a metal-semiconductor junction is mainly due to majority carrier in contrast to a *p-n* junction where current is mainly due to transport of minority carriers. If a Schottky diode is operated at room temperature (T = 300 K), the dominate transport mechanism is due to the thermionic emission of majority carriers from the semiconductor over the potential barrier into the metal.

At thermal equilibrium, the current is balanced by two equal and opposite flows of carriers and hence there is no current flow. Electrons in the semiconductor tend to flow in metal, as discussed in Section 5.1.1, and there is an equal number of flow of electrons into semiconductor from metal.

At the semiconductor surface, electron can be thermionically emitted into the metal if its energy is more than the barrier height $(q\phi_{Bn})$.

When forward bias is applied to the Schottky-barrier diode as shown in Fig. 5.4, the electron energy is increased relative to barrier height and hence electron emission from the semiconductor into the metal is increased, whereas the current component in opposite direction remains constant. In other words, when a forward bias V_f is applied to contact, the electrostatic potential difference across the barrier is reduced and result in the increased electron density at the surface which is given as:

$$n_2 = N_c \exp\left[-q\left(\frac{\phi_{Bn} - V_f}{kT}\right)\right] \tag{5.11}$$

The semiconductor workfunction is replaced by $q\phi_{Bn}$,
Here, k is Boltzmann's constant (1.37×10^{-23} J/K) and T is absolute temperature.

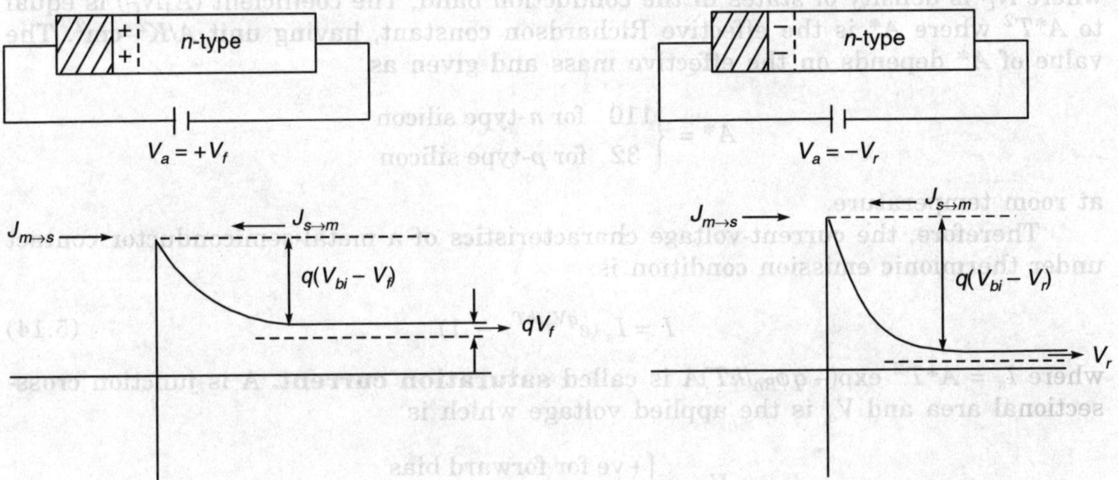

FIGURE 5.4 Band structure in Schottky diode when it is forward and reverse biased.

However, the flux of electrons from metal to semiconductor remains constant because the barrier $(q\phi_B)$ remains at its equilibrium value which results in constant $J_{m \to s}$.

The current results from the electron flow out of the semiconductor is represented by $J_{s \to m}$. In the forward bias, this current density must increase by the same factor as the carrier density given by Eq. (5.11).

The net current density under forward bias is given by the difference between $J_{m \to s}$ and $J_{s \to m}$, i.e., one from metal to semiconductor and other from semiconductor to metal. Therefore,

$$J = J_{s \to m} - J_{m \to s} \tag{5.12a}$$

where

$$J_{s \to m} = A_1 N_c \, \exp\!\left[\frac{-q(\phi_{Bn} - V_f)}{kT}\right] \tag{5.12b}$$

and

$$J_{m \to s} = A_1 N_c \, \exp\!\left[\frac{-q\phi_{Bn}}{kT}\right] \tag{5.12c}$$

Putting values from Eqs. (5.12b) and (5.12c) in Eq. (5.12a), we have

$$J = A_1 N_c \left[e^{-q\phi_{Bn}/kT} \left\{ e^{qV_f/kT} - 1 \right\} \right]$$

or

$$J = (A_1 N_c) \left[e^{-q\phi_{Bn}/kT} \left\{ e^{qV_f/kT} - 1 \right\} \right] \tag{5.13}$$

where N_C is density of states in the conduction band. The coefficient $(A_1 N_C)$ is equal to $A^* T^2$ where A^* is the effective Richardson constant, having unit $A/K^2\text{–cm}^2$. The value of A^* depends on the effective mass and given as

$$A^* = \begin{cases} 110 & \text{for } n\text{-type silicon} \\ 32 & \text{for } p\text{-type silicon} \end{cases}$$

at room temperature.

Therefore, the current-voltage characteristics of a metal-semiconductor contact under thermionic emission condition is:

$$I = I_s \left(e^{qV_a/kT} - 1 \right) \tag{5.14}$$

where $I_s = A^* T^2 \exp(-q\phi_{Bn}/kT)A$ is called **saturation current.** A is junction cross-sectional area and V_a is the applied voltage which is

$$V_a = \begin{cases} +\text{ve for forward bias} \\ -\text{ve for reverse bias} \end{cases}$$

Equation (5.14) is called the **ideal diode equation**. In order to compensate the non-ideal behaviour, Eq. (5.14) is modified to

$$I = I_s \left[e^{qV_a/\eta kT} - 1 \right] \tag{5.15}$$

where η is ideality factor and generally its value is in between 1.05 and 1.25.

In addition to majority carrier (electron in n-type S.C.) current, a current due to minority carrier (due to holes) also exists in a metal and n-type semiconductor contact because of hole injection from metal to semiconductor, and is given as:

$$I_P = I_{PO} \left(e^{qV_a/kT} - 1 \right)$$

where
$$I_{PO} = \frac{qD_p n_i^2}{L_p N_d}$$

Under normal condition, $I_P <<$ current due to electron. Therefore, in Schottky barrier diode, current is mainly due to majority carriers and hence it is a unipolar device. Under the equilibrium condition,

$$|J_{m \to s}| = |J_{s \to m}|$$

When a forward bias is applied to the Schottky barrier, the potential barrier height is reduced by an amount V_f, i.e., from V_{bi} to $(V_{bi} - V_f)$ whereas in the reverse bias, it increases. Therefore, there is easy current flow in the forward bias and little current in the reverse bias. Hence, Schottky contact is also referred as the *rectifying contact*.

5.3.1 Deviation from Ideal Case

Real Schottky diodes do not follow the ideal *I-V* characteristics equation as given by Eq. (5.14) because of the introduction of imperfection at the junction during the time of fabrication and other factors which are not included in the expression. The major limitations are:

Schottky barrier lowering

In Section 5.2, it was assumed that the Schottky barrier height $(\phi_m - x)$ remains unaffected by the bias voltage. But in real case, the barrier height varies with the applied voltage because the conduction electrons in the semiconductor experience a force from their image charges at the metal surface. This force attracts the electrons towards the metal surface from semiconductor and effectively lowered the barrier and hence, the current-voltage relation deviate from ideal one. Theoretically, it was found that due to *image force*, reverse current is not constant with the bias voltage, but varies as $(V_r)^4$. This effect is not usually observed because at very high reverse bias voltage, the carrier generation in the depletion region and tunnelling effects dominate the reverse leakage current. A forward bias voltage approximately above than 0.1 V, causes ideality factor η to deviate slightly from unity. For a Schottky diode, assuming barrier lowering effect is dominating the ideality factor is given by

$$\eta = \frac{1}{\left[1 - \dfrac{dV_{bi}}{dV}\right]} \tag{5.16}$$

where $\dfrac{dV_{bi}}{dV}$ is variation of barriers height with applied voltage, and is given as:

$$\frac{dV_{bi}}{dV} = \frac{1}{4}\frac{q^3 N_d}{8\pi^2 \varepsilon_{Si}^3}\left[V_{bi} - V - \phi_{fc} - \frac{kT}{q}\right]^{-3/4} \tag{5.17}$$

where ϕ_{fc} is the potential difference between the Fermi level and conduction band edge.

Since, this image-induced barrier lowering increases the current density for a Schottky diode, the reverse saturation current becomes

$$I_s = A^*T^2 \ e^{-[(\phi_B - \Delta\phi_B)/kT]} \tag{5.18}$$

where $\Delta\phi_B$ is change in barrier height due to image effect

$$\Delta\phi_B = q\left[\frac{q\,|E_{\max}|}{4\pi\varepsilon_{Si}}\right]^{1/2} \tag{5.19a}$$

and

$$|E_{\max}| = \left[\frac{2qN_d(V_{bi} - V_a)}{\varepsilon_{Si}}\right]^{1/2} \tag{5.19b}$$

Surface imperfections

The semiconductor surface must be extremely clean in order to realize *I-V* characteristics of the Schottky diode. However, inspite of taking utmost care in fabrication, the contact experiences at least a small amount of contamination due to impurities. If diode is subjected to high temperature, undesired chemical compounds will form between the metal and the semiconductor near the junction. During fabrication metal-semiconductor junction subjected to high temperature. The main effect of surface imperfection is to increase the ideality factor as well as reverse current. Surface imperfections are probably the major cause of non-ideal behaviour in Schottky diodes.

Tunnelling

Thermal emission is not only one mechanism by which electrons can cross the barrier at the junction but electrons can also reach to other sides through quantum mechanical tunnelling. This tunnelling has also significant effect on the *I-V* characteristic of the diode. Tunnelling is often responsible for *soft I-V* characteristics at low currents. The tunnelling effect is more dominant in those devices which are used in cryogenic operation. Tunnelling also increase the noise temperature of the diode.

Series resistance

Schottky junction generally requires lightly doped semiconductor with high bulk resistivity. A highly doped substrate would not be practical for diode fabrication because of high series resistance and poor ohmic contact. Practical diodes are, therefore, fabricated on a lightly doped, thin epitaxial layer that is grown on heavily doped, low resistance substance. This structure allows the lightly doped region to be used for the junction and the heavily doped region to minimize series resistance.

Series resistance often creates a lower limit to the diode size because it is strongly dependent on the diode structure and also creates power losses.

Edge effect

In deriving the current-volt equation for Schottky diode, it was assumed that the electric field is perpendicular to the junction over its entire area. But practical diode is formed with a small anode on a large semiconductor surface. This causes fringing effect and due to fringing effect, the electric field near the edge of the metal is greater than in the centre. This results in a large current at the edge of the junction than at the centre.

For this reason, small diodes are sometimes fabricated with metal geometrics to increase the periphery to reduce the series resistance and to reduce the area in order to minimize capacitance.

5.4   APPLICATIONS

Since, the Schottky-barrier diode operates as a majority-carrier device, therefore, there is no minority-carrier storage and the absence of minority-carrier storage results in a faster response than for a p-n junction diode. Schottky-barrier diode is mainly used in high-speed switching applications and as a detector at microwave frequencies. The saturation current I_s of a Schottky diode is several order larger than the p-n junction. This makes Schottky-barrier diode useful for power rectifier. At a given voltage, the current through a Schottky-barrier diode is much larger than the p-n junction.

EXAMPLE 5.1

(a) Find the depletion width of a Schottky diode for a Au/n-Si semiconductor with $n = 10^{15}$/cm^3. What is the polarity connected to the Si for reverse bias, and what is the depletion width for a reverse bias of 10 V? Take

$$\phi_m = 4.75 \text{ V} \quad \text{and} \quad x = 4.05 \text{ V}.$$

  (b) Find the workfunction of semiconductor.
  (c) Find the charge per unit area when current is in reverse bias of 10 V.
  (d) What is the barrier height? Assume ideal case.
  (e) Find the reverse saturation current density. Take $A^* = 110$.
  (f) Find current density for $V_a = -10$ V.

Solution: (a) Using the following relation:

$$E_C - E_F = kT \ln\left(\frac{N_c}{N_d}\right)$$

or

$$E_c - E_f = -kT \ln\left(\frac{n}{N_c}\right)$$

Substituting $N_c = 2.84 \times 10^{19}/\text{cm}^3$ in Eq. (5.3), we get

$$V_{bi} = 4.75 - \left[4.05 + (-0.026) \ln \left\{ \frac{10^{15}}{2.84 \times 10^{19}} \right\} \right]$$

$$= 0.433 \text{ V}$$

Using Eq. (5.9),

$$W = \sqrt{\frac{2 \times 11.7 \times 8.854 \times 10^{-14} \times 0.433}{1.6 \times 10^{-19} \times 10^{15}}} = 0.749 \text{ μm}$$

For $V_r = -10$ V,

$$W = \sqrt{\frac{2 \times 11.7 \times 8.854 \times 10^{-14} \times (0.433 + 10)}{1.6 \times 10^{-19} \times 10^{15}}} \simeq 3.675 \text{ μm}$$

Positive terminal of battery is connected to n-type semiconductor.

(b) Using Eq. (5.2),

$$\phi_s = \phi_m - V_{bi} = 4.75 - 0.433$$

$$= 4.417 \text{ V}$$

(c)
$$\frac{Q_J}{A} = \sqrt{2qV_{bi}\,\varepsilon_{Si}\,N_d} \qquad [\text{Using Eq. (5.10)}]$$

$$= \sqrt{2 \times 1.6 \times 10^{-19} \times 0.433 \times 11.7 \times 8.854 \times 10^{-14} \times 10^{15}}$$

$$= 11.98 \times 10^{-9} \text{ coulomb/cm}^2$$

(d)
$$q\phi_{Bn} = q(\phi_m - x)$$

$$= q(4.75 - 4.05) = 0.70 \text{ eV}$$

(e)
$$J_s = 110 \times (300)^2 \times e^{-0.70/26 \text{ mV}}$$

$$= 1.83 \times 10^{-5} \text{ A/cm}^2$$

(f) For $V_a = -10$ V, the current density

$$J = 1.83 \times 10^{-5} \text{ A/cm}^2$$

5.5 DEPLETION CAPACITANCE/JUNCTION CAPACITANCE

The depletion region of the Schottky diode acts as an insulator which separates metal layer and doped semiconductor layer and acts like a parallel-plate capacitor. This capacitance is determined by the physical dimension of the junction as well as the doping profile of the semiconductor layer.

The expressions for charge in presence of the applied voltage V is given as;

$$Q_J(V) = A\sqrt{2q\varepsilon_{Si}N_d\,(V_{bi} - V)} \qquad (5.20a)$$

The differential capacitance C_d is given as;

$$C_d(V) = \frac{dQ_j(V)}{dV}$$

Now differentiating (5.20a) with respect to V, we get

$$C_d(V) = \frac{-A\sqrt{\dfrac{q\varepsilon_{Si}}{2}N_d}}{\sqrt{V_{bi} - V}} \qquad (5.20b)$$

and for $V = 0$, this reduces to;

$$C_{d0} = \frac{-A\sqrt{\dfrac{q\varepsilon_{Si}}{2}N_d}}{\sqrt{V_{bi}}} \qquad (5.20c)$$

From (5.20b) and (5.20c), it is clear that

$$C_d(V) = \frac{C_{d0}}{\sqrt{1 - \dfrac{V}{V_{bi}}}} \qquad (5.21)$$

Equation (5.21) is most useful expression for circuit analysis. C_{d0} is the depletion or junction capacitance at zero bias. The exponent (1/2) is introduced in the denominator of Eq. (5.21) due to fact that the doping concentration N_d is constant throughout the semiconductor. In real case, N_d may not be uniform, thus changing the exponent, e.g., Mott diode. In Mott diode, the capacitance has relatively weak dependence on the bias voltage.

The variation of $(1/C_d^2)$ with applied voltage V is shown in Fig. 5.5.

EXAMPLE 5.2

Find the donor concentration and the barrier height for the tungsten/n-Si Schottky-barrier diode.

Solution: Take $V_{bi} = 0.42$ V, $1/(C_d/A)^2 = 1.8 \times 10^{15}(\text{cm}^2/\text{F})^2$ at $V = 0$.

Using Eq. (5.20), we have

$$C^2 d = A^2\left[q\varepsilon_{Si}\,\frac{N_d}{2}\,V_{bi}\right] \quad \text{for } V = 0$$

Putting all these values,

$$N_d = 2.8 \times 10^{15}/\text{cm}^3$$

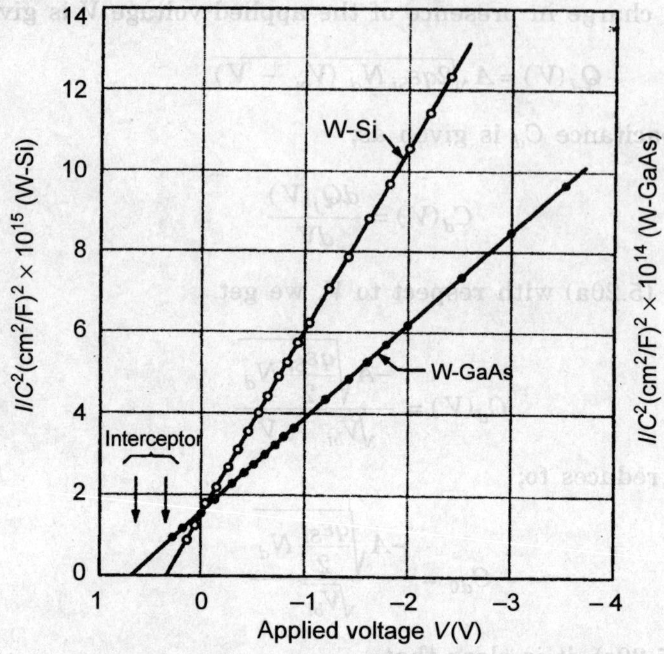

FIGURE 5.5   Variation of $(1/C_d^2)$ with applied voltage (V).

and

$$\phi_{Bn} = V_{bi} + \frac{(E_C - E_F)}{q}$$

$$= 0.42 + (-0.026)\ln\left[\frac{2.8 \times 10^{15}}{2.84 \times 10^{19}}\right]$$

$$= 0.66 \text{ V}$$

5.6   OHMIC CONTACTS

Consider a metal-semiconductor (M-S) contact for $q\phi_m < q\phi_s(n\text{-type})$. In this contact, electrons are extracted from the metal into semiconductor. The resulting contact is known as ohmic contact. To get ohmic contact between metal and p-type semiconductor, the essential condition is $q\phi_m > q\phi_s$. For an ohmic contact, its contact resistance is negligibly small compared to bulk-or series resistance of the semiconductor. A perfect ohmic contact should not significantly degrade the device performance and can pass the required current with a voltage drop that is small compared to voltage drop across the active region of the device. A figure of merit for ohmic contact is defined as the specific contact resistance, which is given as:

$$R_c = \left[\frac{\partial J}{\partial V}\right]_{V=0}^{-1} \Omega\text{-cm}^2 \tag{5.22}$$

As we know that for low doped metal-semiconductor contact, the thermionic emission current dominates and current is given by Eq. (5.14). From Eq. (5.14),

$$\frac{\partial J}{\partial V_a} = Js \cdot \frac{q}{kT} \left\{ e^{qV_a/kT} \right\}$$

$$\left. \frac{\partial J}{\partial V_a} \right|_{V_a = 0} = Js \cdot \frac{q}{kT} \Rightarrow A^* T^2 \, e^{-q\phi_B/kT} \, \frac{q}{kT} \tag{5.23}$$

Using Eqs. (5.22) and (5.23), we have

$$R_c = \frac{k}{q} \frac{e^{q\phi_B/kT}}{A^* T} \tag{5.24}$$

Equation (5.24) does not result in a negligible contact resistance. To reduce the R_c one should reduce the barrier height. Therefore, to obtain ohmic contacts, choose, the semiconductor material which is generally heavily doped to give n^+ region on n-type semiconductor or a p^+ region on a p-type semiconductor. Due to very high doping concentration, the barrier height width becomes very narrow and the tunnelling current becomes dominate. The tunnelling current for n-type Si may be written as:

$$I_T \sim \exp\left[\frac{-2\sqrt{\varepsilon_{Si} m^* n}}{\hbar} \left(\frac{\phi_{Bn} - V_a}{\sqrt{N_d}} \right) \right] \tag{5.25}$$

and the contact resistance for a heavily doping concentration is:

$$R_c \sim \exp\left[\frac{2\sqrt{\varepsilon_{Si} m^* n}}{\hbar} \left(\frac{\phi_{Bn}}{\sqrt{N_d}} \right) \right] \tag{5.26}$$

Equation (5.26) shows that in the tunnelling region the specific contact resistance depends strongly on the impurity concentration and varies exponentially with factor $(\phi_{Bn}/\sqrt{N_d})$. It is observed that for $N_d \geq 10^{19}/\text{cm}^3$, R_c is dominated by tunnelling process and decreases rapidly with increasing doping. On the other hand, if $N_d \leq 10^{17}/\text{cm}^3$, the current is dominated by the thermionic emission and the specific contact resistance is independent of the doping.

For making ohmic contact, it is common practice with silicon to evaporate a metal such as platinum, which will react with Si to form the silicide PtSi when heated to 500°C. Other silicides are Pd_2Si, NiS_2 and TiS_2. Current-voltage characteristics of a Schottky barrier diode and of an ohmic contact are compared and shown in Fig. 5.6. As clear from characteristics that for a good ohmic contact, the current-voltage characteristic should be linear and have very small resistance compared to the resistance of the active region of the semiconductor.

5.6.1 Band Diagram of Ohmic Contact

Let us consider a contact between a metal and n-type semiconductor material for which $\phi_m < \phi_s$. The isolated band diagram is shown in Fig. 5.7(a).

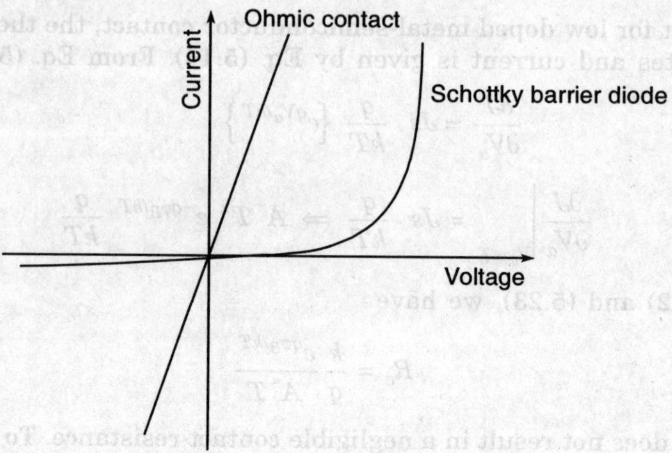

FIGURE 5.6 *I-V* characteristics of ohmic contact and Schottky diode.

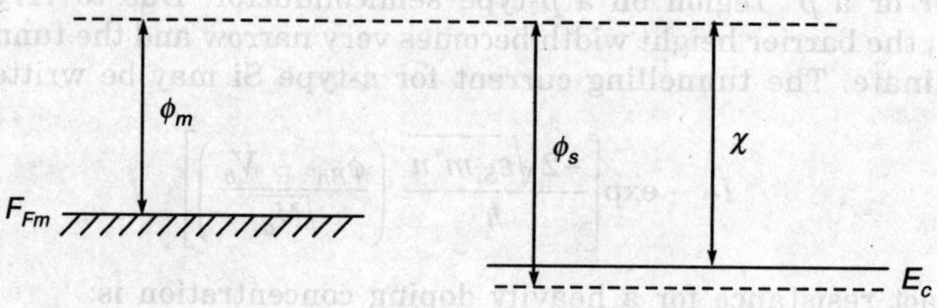

FIGURE 5.7(a) Isolated band diagram of ohmic contact.

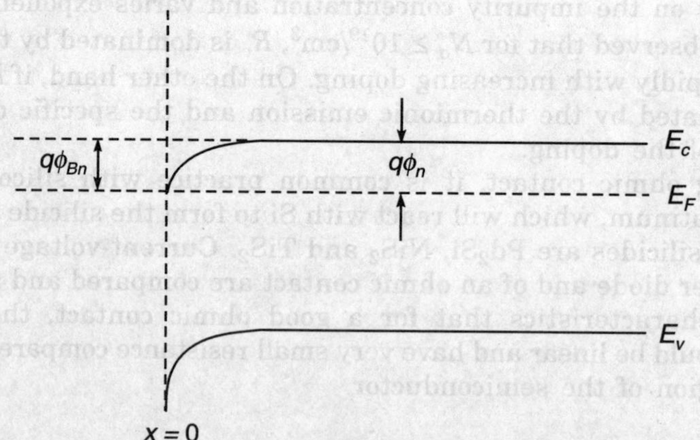

FIGURE 5.7(b) Band diagram of ohmic contact at equilibrium.

As seen from Fig. 5.7(a), the metal's Fermi level is above the semiconductor's Fermi level and hence to align the Fermi levels at thermal equilibrium, electrons must transferred from metal to semiconductor. This raises the semiconductor electron energies relative to the metal at equilibrium. Due to accumulation of electrons near the junction in the semiconductor side, conduction band will move towards the Fermi level as shown in Fig. 5.7(b). The valence level should also follow similar trends to keep the energy gap constant. In this case, the barrier to electron flow between the metal and the semiconductor is small and easily overcome by a small applied voltage. In the resulting contact, no depletion region formed in the semiconductor. The electrostatic potential difference required to align the Fermi levels at equilibrium supports accumulation of majority carriers in the semiconductor instead of depletion of mobile carriers.

One practical method for forming ohmic contact is by high doping near the metal-semiconductor interface. Due to heavy doping depletion width is very narrow, and allows carriers to tunnel through the barrier.

Ideal energy-band diagram of an ohmic contact under forward bias and reverse bias is shown in Figs. 5.8(a) and 5.8(b).

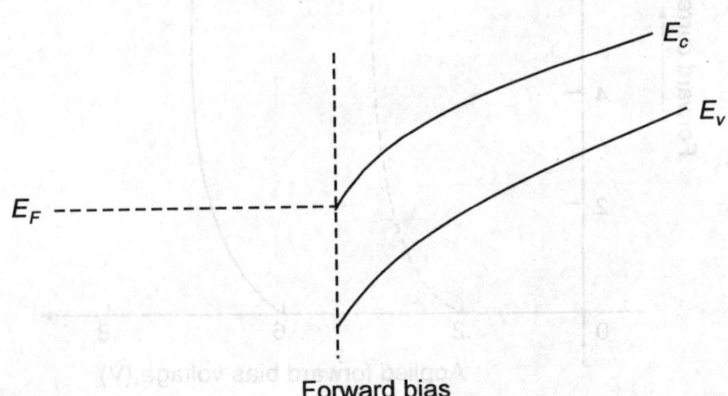

FIGURE 5.8(a)   Band diagram of ohmic contact in forward bias.

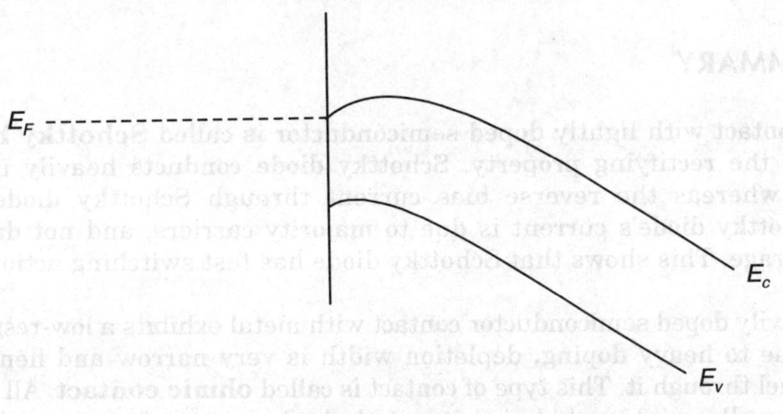

FIGURE 5.8(b)   Band diagram of ohmic contact in reverse bias.

5.7 COMPARISON OF SCHOTTKY DIODE AND THE *p-n* JUNCTION DIODE

It is clear from Eq. (5.14) that Schottky diode has similar ideal current-voltage characteristic equation as *p-n* junction diode. But the magnitude of the reverse saturation current is typically several orders of magnitude larger in the Schottky barrier diode than *p-n* junction diode. The forward bias characteristic is also different for both as shown in Fig. 5.9. The forward bias voltage required to achieve a given current is less in a Schottky diode than in a *p-n* junction. In other words, we can say that effective-turn-on voltage is less in Schottky diode than the *p-n* junction.

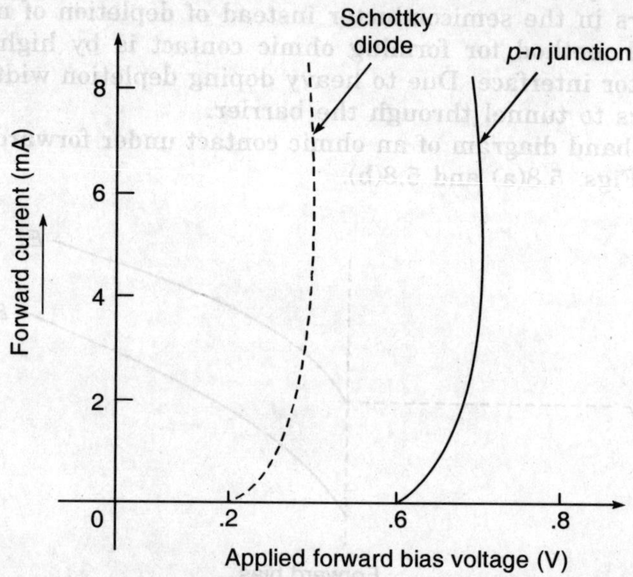

FIGURE 5.9 Variation of forward current with applied forward bias voltage for Schottky diode and *p-n* junction diode.

5.8 SUMMARY

A metal contact with lightly doped semiconductor is called **Schottky barrier diode** which has the rectifying property. Schottky diode conducts heavily in the forward direction, whereas the reverse bias current through Schottky diode is very-very small. Schottky diode's current is due to majority carriers, and not due to minority carrier storage. This shows that Schottky diode has fast switching action than normal diode.

A heavily doped semiconductor contact with metal exhibits a low-resistance contact because due to heavy doping, depletion width is very narrow and hence carrier can easily tunnel through it. This type of contact is called **ohmic contact**. All semiconductor devices as well as integrated circuits need ohmic contact for connection with other devices.

For

$\phi_m > \phi_s$ (*n*-type semiconductor), contact is Schottky
$\phi_m < \phi_s$ (*p*-type semiconductor), contact is Schottky
$\phi_m < \phi_s$ (*n*-type semiconductor), contact is ohmic
$\phi_m > \phi_s$ (*p*-type semiconductor), contact is ohmic

The *I-V* characteristic of Schottky diode is same as the *p-n* junction diode, whereas *I-V* characteristic of ohmic contact is linear one.

REVIEW QUESTIONS

1. Explain why Schottky barrier diode has rectifying property.

2. Explain, how would you use a Schottky-barrier diode to determine whether the semiconductor is *n*-type or *p*-type.

3. What is Schottky effect?

4. What is the basic difference between Schottky barrier diode and normal diode?

5. What are assumptions for the ideal *I-V* characteristics of Schottky diode?

6. Explain Schottky barrier lowering.

7. What is the effect of surface imperfections on the *I-V* characteristics of Schottky diode?

8. Why most of the electronic devices need ohmic contact for connection?

9. What are basic differences between ohmic and Schottky contact?

10. A metal-semiconductor contact is given to you and it is required to know whether contact is ohmic or Schottky. Can you determine the type of contact? Explain.

11. Draw the energy band diagram at thermal equilibrium for a metal and *p*-type semiconductor contact where $\phi_m < \phi_s$. Is contact Schottky or ohmic in nature? Explain your answer.

12. Draw the energy band diagram at thermal equilibrium for a metal and *p*-type semiconductor contact where $\phi_m > \phi_s$. Discuss the nature of the resulting contact.

13. Explain why ohmic contact is also called non-rectifying contact?

14. Why ohmic contact has linear *I-V* characteristics?

NUMERICAL PROBLEMS

1. A gold contact is made to an *n*-silicon semiconductor. The Si has a doping concentration of $N_d = 1.8 \times 10^{16}/cm^3$. Assume the work function for the gold is 4.75 eV and the electron affinity of Si is 4.05 eV. Find

    (a) Whether contact is Schottky or ohmic

    (b) Built-in-potential

    (c) Ideal barrier height seen by electron in the metal trying to move into semiconductor

    (d) Position of conduction band edge w.r.t. Fermi level

    (e) Depletion width

    Whether the depletion width is semiconductor side or metal side? Explain your answer. Also draw the charge density with x near the contact.

2. The accompanying data were obtained on the metal contacts to silicon of equal area (as shown in Fig. 5.10). If Schottky theory applied, which metal probably has the higher work function? Which data were taken on 1 Ω-cm silicon and which on 5 Ω-cm silicon? Justify your answers and explain the word "probably"?

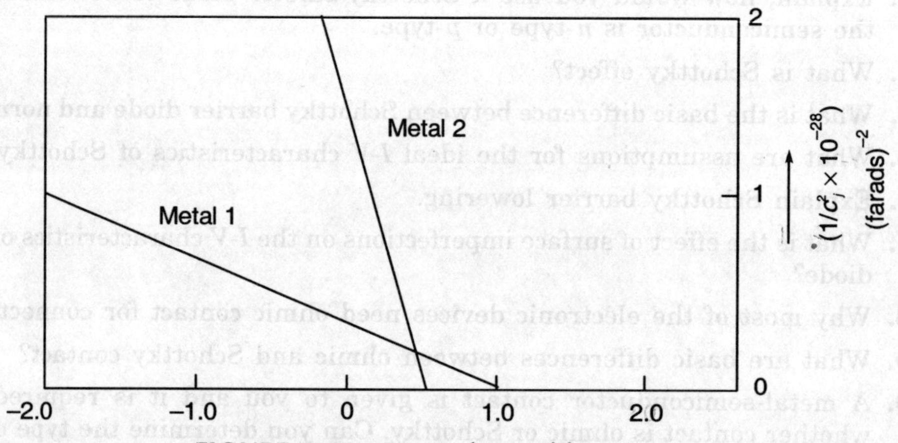

FIGURE 5.10 Figure for problem 2.

3. A Schottky barrier is formed between a metal having a workfunction of 4.3 eV and semiconductor of electron affinity 4 eV. The acceptor doping in the Silicon is 10^{17}/cm^3. Find the ideal current for the Schottky barrier diode in the presence of forward bias voltage of 0.3 V. Also draw the band diagram. What will happen if a reverse bias voltage 3 V is applied. Discuss in terms of energy band diagram. Take A* for Si = 32 at T = 300 K.

4. Consider an ideal Au/n-Si Schottky barrier diode with $n = N_d^+ = 1 \times 10^{16}$/cm^3.

    (a) What is the built-in-potential?

    (b) What is the depletion distance?

    (c) Sketch the thermal equilibrium energy band diagram.

    (d) Sketch the graph potential vs distance.

    (e) What is zero capacitance if the area of Au dot 2×10^{-4} cm^2.

5. A Si Schottky barrier diodes was determined by C-V necesssary to have a built-in potential of 0.310 V and carrier concentration $n = 5.0 \times 10^{16}$/cm^3.

    (a) Find the barrier height.

    (b) Find the electron concentration in the Si at the metal/Si interface.

6. A metal with workfunction ϕ_m = 4.2 V is deposited on an *n*-type silicon semiconductor with χ_s = 4.0 V. Assume the ideal situation. Take T = 300 K.

   (a) Sketch the thermal equilibrium energy band diagram.
   (b) Find the donor concentration.
   (c) What is the potential barrier height seen by electrons in the metal moving into the semiconductor.

7. The energy band diagrams for a metal and a semiconductor are sketched to the same scale as shown in Fig. 5.11(a). Three possible band diagrams for a contact between these materials are sketched in Figs. 5.11(b) to (d). Assume thermal equilibrium. Explain briefly whether each figure (from b to d) is correct or not.

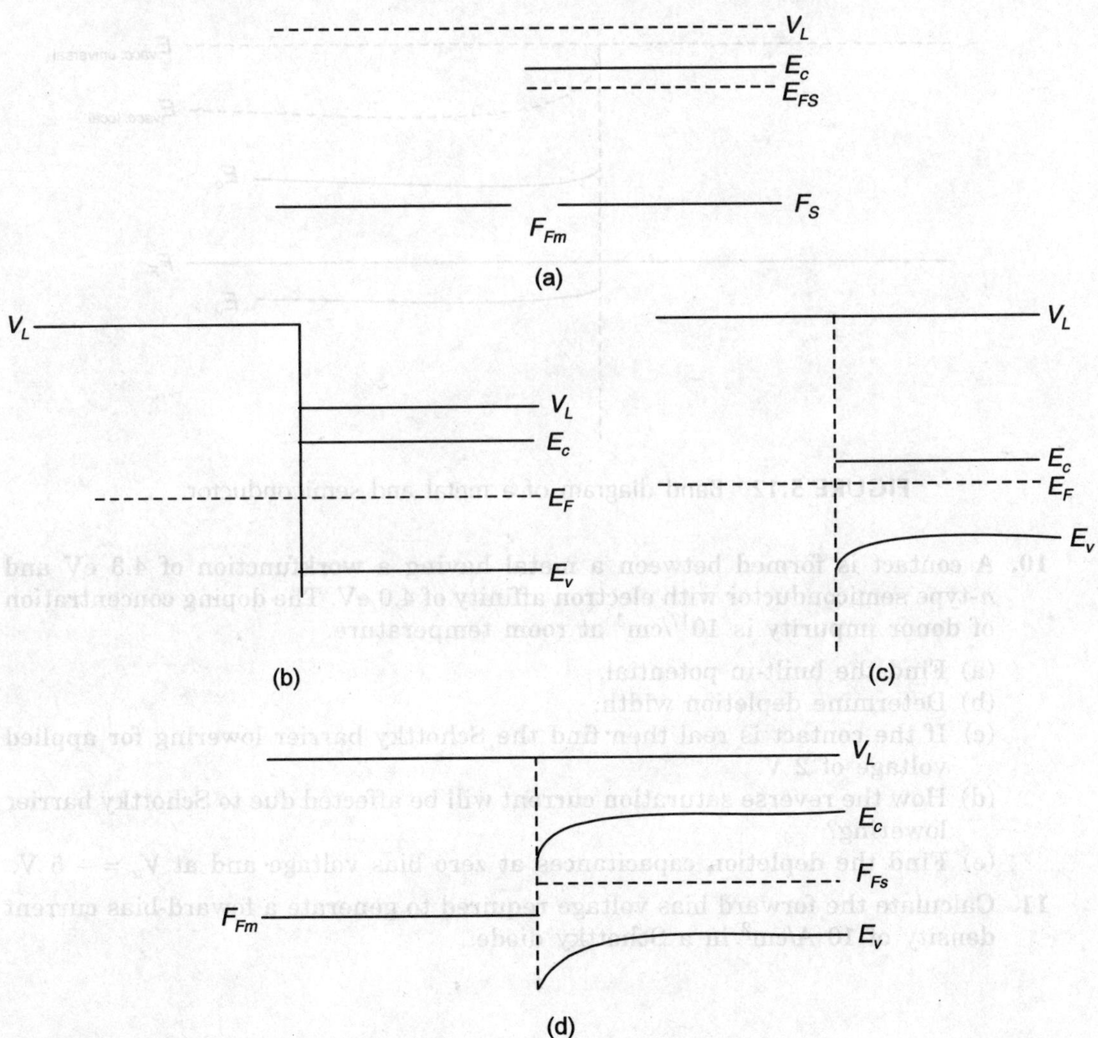

FIGURE 5.11 Band diagrams.

8. Consider an n-GaAs Schottky barrier diode of area 10 μm^2 and potential barrier height at zero bias is 1 eV.

   (a) Find the value of I_S at room temperature.
   (b) Plot the forward I-V characteristics for a diode having $\eta = 1.3$.

   Take $A^* = 8$ A/k^{2}cm^2.

9. The band diagram of a metal and semiconductor after contact at thermal equilibrium is shown in Fig. 5.12.

   (a) Whether the semiconductor material is n-type or p-type?
   (b) Will this junction be ohmic or rectifying? Explain your answer briefly.
   (c) What will be barrier height? Discuss qualitatively only.
   (d) Draw its energy band diagram under forward and reverse bias conditions.

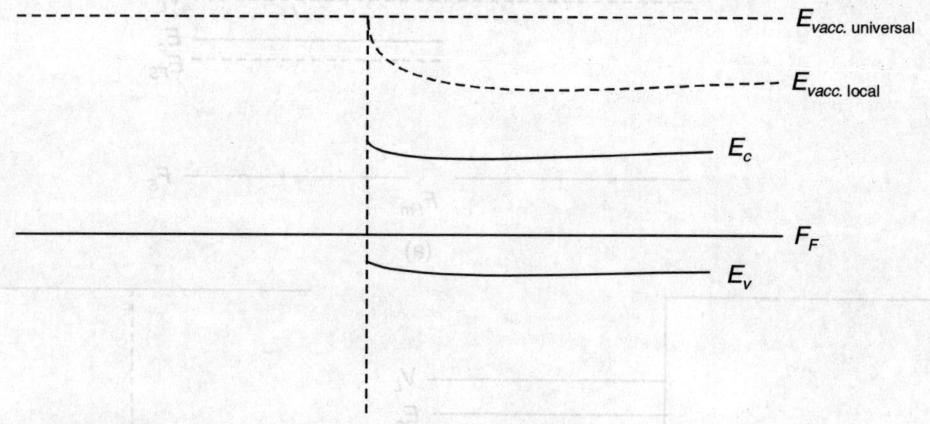

FIGURE 5.12   Band diagram of a metal and semiconductor.

10. A contact is formed between a metal having a workfunction of 4.3 eV and n-type semiconductor with electron affinity of 4.0 eV. The doping concentration of donor impurity is 10^{17}/cm^3 at room temperature.

    (a) Find the built-in potential.
    (b) Determine depletion width.
    (c) If the contact is real then find the Schottky barrier lowering for applied voltage of 2 V.
    (d) How the reverse saturation current will be affected due to Schottky barrier lowering?
    (e) Find the depletion capacitances at zero bias voltage and at $V_a = -5$ V.

11. Calculate the forward bias voltage required to generate a foward-bias current density of 10 A/cm^2 in a Schottky diode.

CHAPTER
6

Field Effect Transistor (FET)

- Field effect transistors
- Metal semiconductor field effect transistors
- Metal oxide semiconductor field effect transistor
- Modulation-doped field effect transistor

6.1 INTRODUCTION

The bipolar junction transistor was the key device which led to revolution in solid state electronics. But the major disadvantages with this transistor was its low input resistance. Another device which achieved the transistor action, with input diode junction reverse biased, is called a **Field effect transistor (FET)**. The field effect transistor is a transistor that relies on an electric field to control the conductivity of the *channel* in the semiconductor materials. The input impedance (resistance) of the FET is very high due to reverse bias of input junction. The field effect transistor is a voltage controlled current source device and sometimes used as a voltage-controlled resistors. FETs are mainly unipolar device where only majority carriers dominate the current.

A FET has three terminals which are known as the **gate (G)**, **drain (D)** and **source (S)**. The voltage applied between the gate and source terminals modulates the current between the source and drain terminals. In FET, drain and source are interchangeable because it is a symmetrical device. The transistor is basically used as amplifier in analog electronics and as a switch in digital electronics. The circuit symbol achieves this designation by placing gate closer to source than to the drain as shown in Fig. 6.1.

The FET is simpler in concept than the Bipolar junction transistor (BJT) and can be constructed from a wide range of materials. The different types of FET can

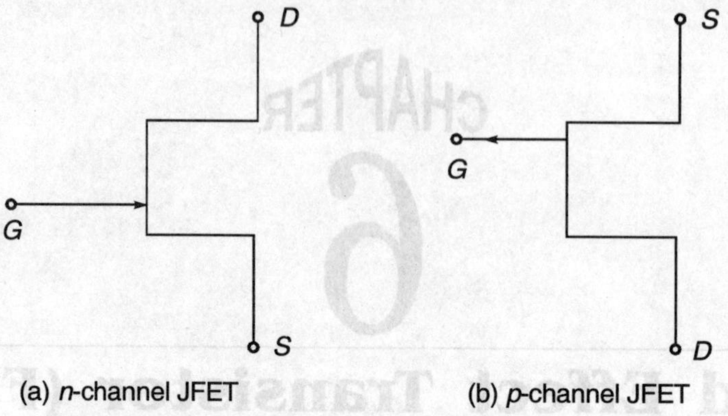

FIGURE 6.1 Symbol representation of transistors.

be distinguished by the method of isolation between channel and the gate. Various types of FETs are:

- Junction field effect transistor (JFET)
- Metal-semiconductor field effect transistor (MESFET)
- Metal-oxide-semiconductor field effect transistor (MOSFET)
- Modulation-doped field effect transistor (MODFET) or high electron mobility transistor (HEMT)

Most FETs are made with conventional bulk semiconductor processing techniques, using the single-crystal semiconductor wafer as the active region or channel.

6.2 JUNCTION FIELD EFFECT TRANSISTOR [JFET]

This FET uses a *p-n* junction diode to isolate the channel from the gate. No inversion layer is formed in the device. The channel is depleted due to reverse bias in between the channel and the gate.

Junction field effect transistor (JFET) is one of the simplest available transistor and available in two polarities like *n*-channel and *p*-channel JFETs. Their circuit symbols are shown in Fig. 6.1. In this present topic, we will discuss the construction and working principle of *n*-channel JFET because it is more widely used transistor due to high mobility of electron.

The arrow in Fig. 6.1 only indicates the direction of current flow. A simplified structure of *n*-channel JFET is shown in Fig. 6.2. It consists of a lightly doped *n*-type silicon slab with heavily doped *p*-regions on its two sides. These two heavily doped *p*-regions (p^+) are electrically connected together and form the gate. The lightly doped *n*-region serves as a channel and at the two ends of the channel, we will make ohmic contacts. These two ends are known as the **source (S)** and **drain (D)**. The whole region between source and drain, known as **channel**, is occupied by the electrons and hence, the device is called *n***-channel JFET**. The gate-source and gate-drain form a *p-n* junctions and act like a one-sided *p-n* junction (p^+-*n*).

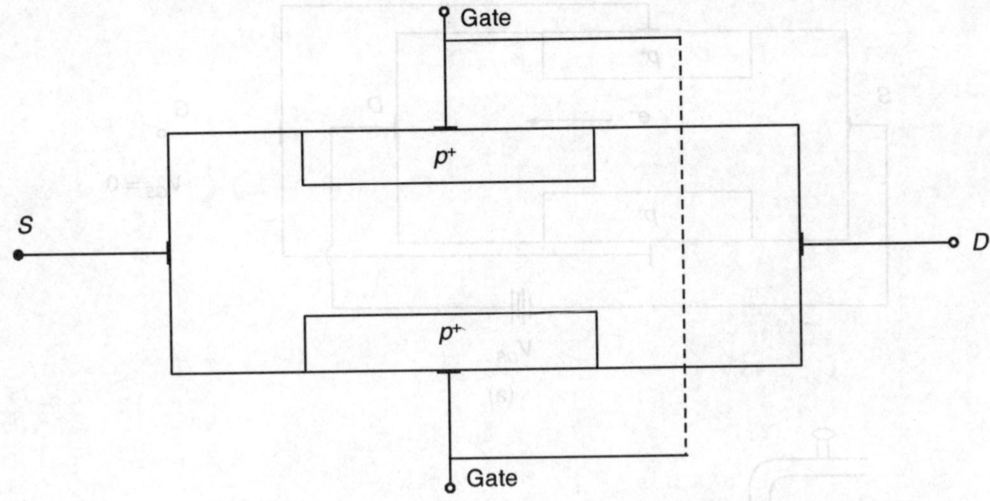

FIGURE 6.2 Simplified structure of *n*-channel JFET.

The operation of the device is based on reverse-biasing of the *p-n* junction between the gate and channel. Indeed, it is the reverse bias on the gate source junction which modifies the width of the channel and hence control the conductivity of the channel. Therefore, *p-n* junction plays an important role in the operation of the transistor, and that is why it is called **junction field effect transistor** (JFET).

The *p*-channel JFET is complementary of *n*-channel JFET and hence it can be constructed by simply reversing all the semiconductor types, i.e., *p*-type semiconductor for channel and heavily doped *n*-type regions for gate.

6.2.1 Operation of JFET

Let us assume that no voltage is applied in between gate and source, i.e., $V_{GS} = 0$ then the application of even a small voltage in between drain to source (V_{DS}) causes a current to flow from the drain to source as shown in Fig. 6.3(a). The end of the channel from which electrons flow is called **source (S)** and the end toward which they flow is called **drain (D)**. Let us consider a hose pipe, whose one end is connected to the tape and other end is connected with the drainage as shown in Fig. 6.3(b). The end which is connected to the tape or which supplies the water acts as a source and the end which drain out water acts as a drain. The hose pipe is similar to channel.

When a negative potential is applied on the gate w.r.t. source, i.e., $V_{GS} = -$ve, the source-gate junction becomes reverse bias and a depletion region will form in between the gate and channel as shown in Fig. 6.3(c). Since the gate region is heavily doped and hence entire depletion region will move into the channel.

If applied drain to source (V_{DS}) potential is small then the depletion region will be symmetrical about source and drain regions, i.e., channel is almost of uniform width. From Fig. 6.3(c), it is clear that due to reverse bias of gate junction, the channel becomes narrower and hence channel resistance increases and current from

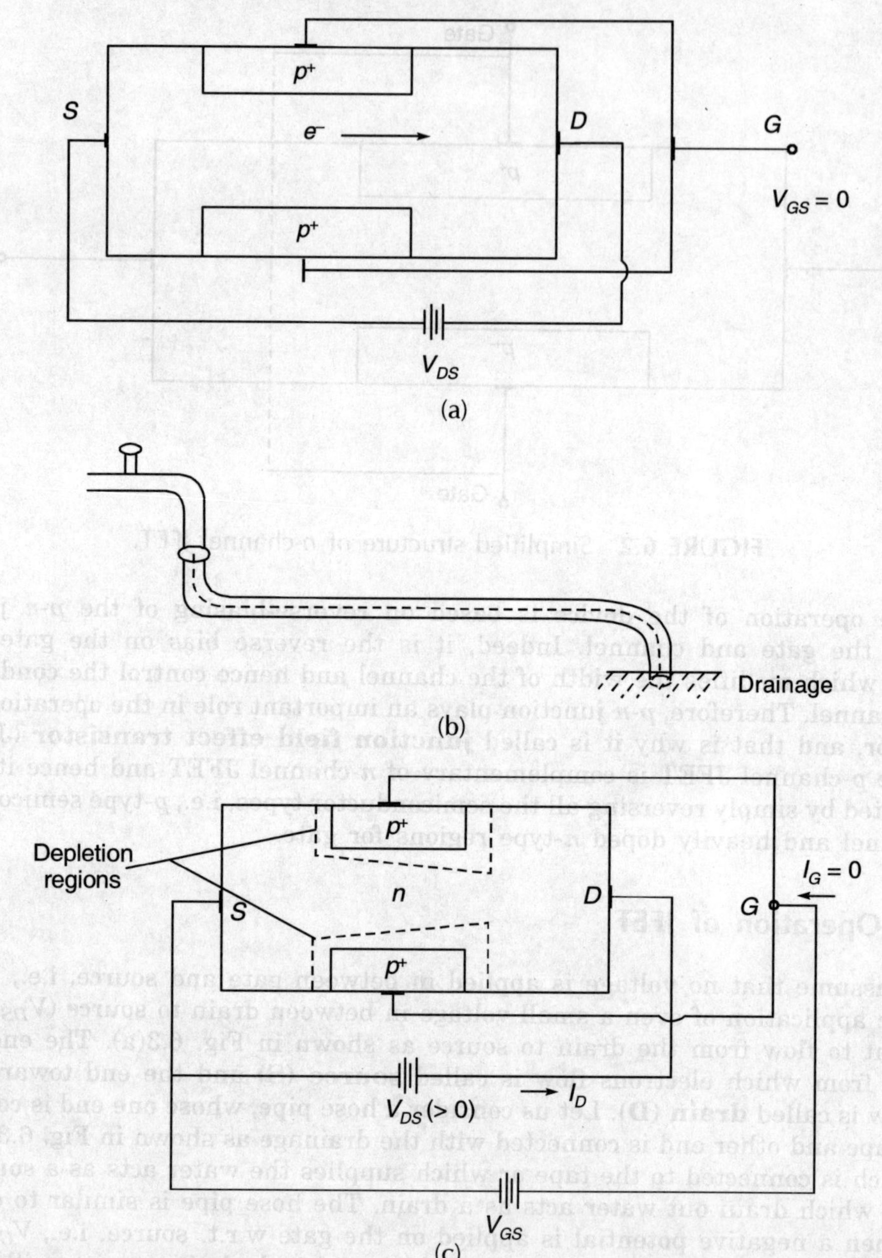

FIGURE 6.3(a)–(c) Current flow in *n*-channel JFET under different bias conditions.

source to drain, called **drain current** or **drain to source current** represented by I_D or I_{DS}, decreases.

For small V_{DS}, the JFET is simply operating as a resistance whose value is controlled by V_{GS}. This region of operation, for small V_{DS}, is known as **linear region** or ohmic region where ohm's law is valid.

If V_{DS} is not small then the shape of depletion region in the channel is asymmetrical as shown in Fig. 6.3(c). The depletion region is wider towards the drain end then the source end because gate-drain junction is more reverse bias than the source-gate junction.

If we keep increasing V_{GS} in the negative direction, the depletion region becomes more and more wider because of increasing reverse bias and at a particular value of V_{GS}, the two depletion regions (due to two gates) just touch to each other and block the channel, i.e., depletion region occupies the entire channel as shown in Fig. 6.3(d). We can say that at this particular gate voltage, the channel is completely depleted of the mobile charge carriers (i.e., electrons), or in other words channel has in effect disappeared.

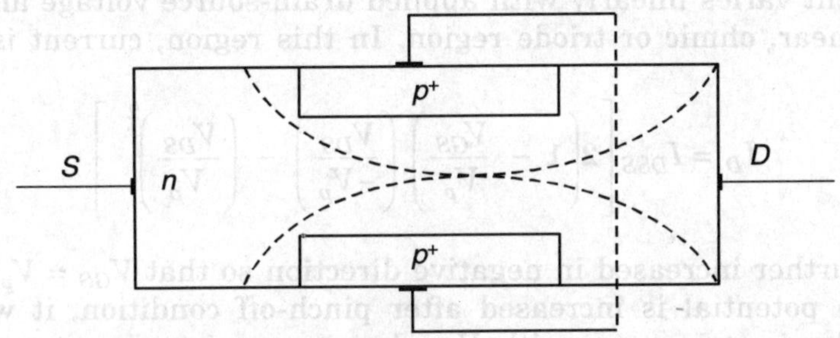

FIGURE 6.3(d)   Pinch-off condition in JFET.

The condition, where channel has disappeared, is called **pinch-off condition** and the gate-source potential at which pinch-off occurs is called **pinch-off voltage**. This voltage is represented by V_P.

Now, let us consider V_{GS} to constant and larger than the pinch-off voltage V_P and V_{DS} is increased. The drain-source voltage V_{DS} appears as the voltage drop across the channel length. This drop increases as one moves from source end to drain end. Drop is minimum or zero at the source end, i.e., $V(x = 0) = 0$ and maximum at the drain end, i.e., $V(x = L) = V_{DS}$. This shows the reverse-bias voltage between gate and channel varies at different points along the channel. Thus, channel appears as a tapered as shown in Fig. 6.3(c) and current-voltage relation is not linear, but it will become non-linear when the reverse bias at the drain end (V_{GD}) becomes smaller than the pinch-off voltage V_p. The channel is pinched off at the drain end and drain current saturates means the drain current becomes independent of drain-source voltage.

If we attempt to apply positive potential on the gate of the n-channel JFET, the gate-channel p-n junction becomes forward biased and the gate ceases to control the channel. From above discussion it is clear that increasing reverse potential on the gate means channel becomes narrower and current decreases. So, we can say that the current in the channel is controlled by applying field on the gate, therefore, such device is called **voltage controlled current device**. Since, in the channel current is only due to majority carriers (electrons for n-channel), therefore, JFET is a unipolar device also.

6.2.2 Current-Voltage Characteristics

We have seen that the pinch-off voltage for n-channel JFET is negative voltage. If a forward bias voltage is applied on the gate, JFET will not act like a transistor and current is zero.

i.e.,
$$\text{for } V_{GS} \leq V_p \text{ and } V_{DS} = 0,\ I_D = 0 \tag{6.1}$$

and transistor is said to be in cut-off region.

For small V_{DS}, and applied gate voltage is larger than V_p,

i.e.,
$$V_p \leq V_{GS} \leq 0 \quad \text{and} \quad V_{DS} \leq (V_{GS} - V_p)$$

The drain current varies linearly with applied drain-source voltage and transistor is said to be in linear, ohmic or triode region. In this region, current is given as:

$$I_D = I_{DSS}\left[2\left(1 - \frac{V_{GS}}{V_p}\right)\left(\frac{V_{DS}}{-V_p}\right) - \left(\frac{V_{DS}}{V_p}\right)^2 \right] \tag{6.2}$$

If V_{GS} is further increased in negative direction so that $V_{GS} = V_p$ then pinch-off occurs. If drain potential- is increased after pinch-off condition, it was found that drain current does not increase with V_{DS}, but it remains almost constant. This is called **Saturation region** and is defined as:

$$V_{GS} \geq V_p \quad \text{and} \quad V_{DS} \geq V_{GS} - V_p$$

and the saturation current is given as;

$$I_D = I_{DSS}\left(1 - \frac{V_{GS}}{V_p}\right)^2 \tag{6.3a}$$

This is a current-voltage equation for a ideal-case. Equation (6.3a) shows that drain current is independent of the drain voltage and relation is parabolic in nature. But in real JFET, due to finite channel resistance, current is not constant, but varies slightly and equation (6.3a) is modified as:

$$I_D = I_{DSS}\left(1 - \frac{V_{GS}}{V_p}\right)^2 (1 + \lambda V_{DS}) \tag{6.3b}$$

where I_{DSS} is maximum drain current for $V_{GS} = 0$ and λ is inverse of early voltage. In the case of finite channel resistance, $\lambda = 1/V_A$, where V_A is early voltage. λ and V_A are positive for n-channel JFET. Since, the gate-channel junction is always reverse-biased and hence only a small-leakage current (of order of pA) will flow through the gate terminal. In most of the calculation purpose, gate current I_G is always assumed to be zero. This gate leakage current strongly depends on the temperature and every 10°C rise in temperature double the leakage current. Due to reverse bias of gate-

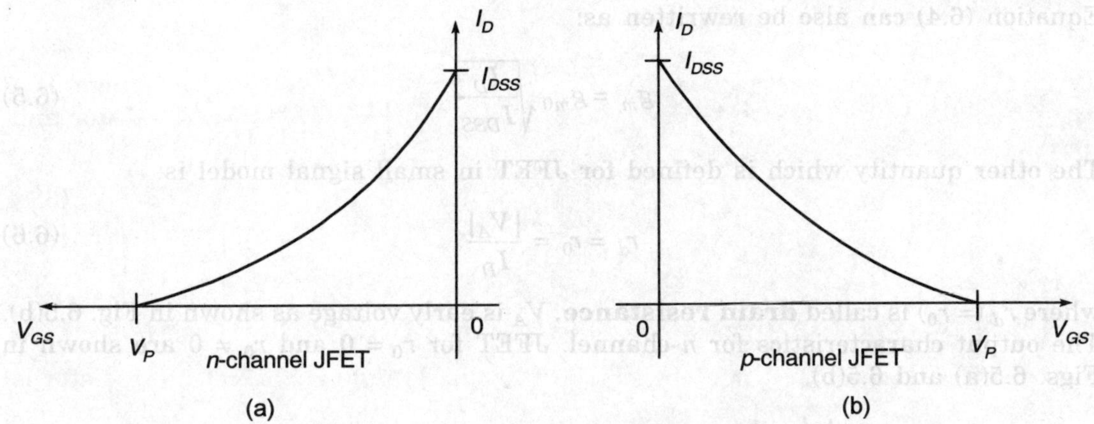

FIGURE 6.4 Input characteristics of *n*-channel and *p*-channel JFETs.

channel junction a very high-input resistance, of order of MΩ, is obtained for JFET. The input characteristics of the *n*-channel JFET as well as *p*-channel JFET are shown in Figs. 6.4(a) and 6.4(b).

Note:

- For proper operation of *p*-channel JFET, apply positive potential on the gate. The pinch-off voltage of *p*-channel JFET is positive.
- The input resistance of the JFET is extremely high, i.e., of order of Megaohm.
- If the channel has a uniform concentration of impurities and the gate is placed in the middle of the channel, the FET is said to be symmetrical. In this case, the source and drain are interchangeable.
- But many FETs are deliberately constructed to be asymmetrical, to enhance the certain parameters and behaviours. In these FET, performance will be impaired if source and drain are interchangeable.

Another important property of JFET is its transconductance. This parameter is generally considered for small signal model JFET and is determined by the saturation current, the pinch-off voltage and DC bias, i.e.,

$$g_m = \frac{\partial I_{D(\text{sat.})}}{\partial V_{GS}}$$

and from Eq. (6.3a),

$$g_m = \frac{2I_{DSS}}{|V_p|}\left[1 - \frac{V_{GS}}{V_p}\right]$$

or

$$g_m = g_{m0}\left[1 - \frac{V_{GS}}{V_p}\right] \tag{6.4}$$

where g_m is called **mutual transconductance**, with units (*A/V*) or siemens or mho.

Equation (6.4) can also be rewritten as:

$$g_m = g_{m0} \sqrt{\frac{I_D}{I_{DSS}}} \tag{6.5}$$

The other quantity which is defined for JFET in small signal model is:

$$r_d = r_0 = \frac{|V_A|}{I_D} \tag{6.6}$$

where r_d (= r_0) is called **drain resistance**, V_A is early voltage as shown in Fig. 6.5(b). The output characteristics for *n*-channel. JFET for $r_0 = 0$ and $r_0 \neq 0$ are shown in Figs. 6.5(a) and 6.5(b).

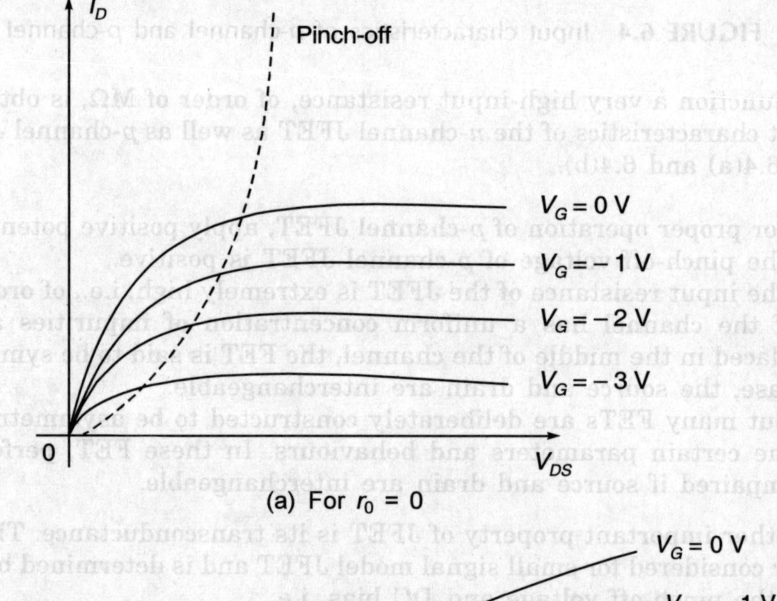

(a) For $r_0 = 0$

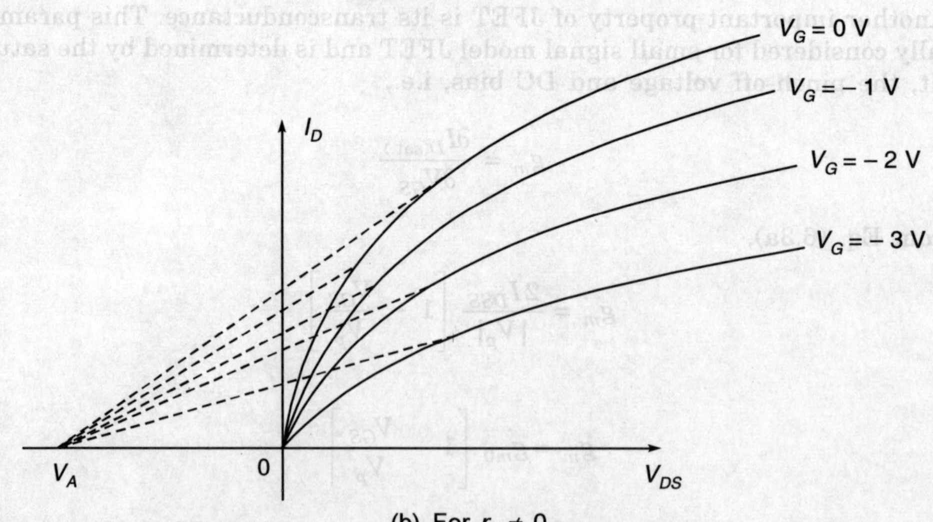

(b) For $r_0 \neq 0$

FIGURE 6.5 Output characteristics of *n*-channel JFET for $r_0 = 0$ and $r_0 \neq 0$.

6.2.3 Calculation of Pinch-Off Voltage

Let us consider a case in which the source and drain regions of JFET separated by large distance, i.e. long channel device. In this case, we can neglect the voltage drop between the source and drain electrodes and the respective ends of the channel. It is a good approximation for such JFET that the total applied voltage at the drain will drop at the drain-channel end, i.e.,

$$V(x = L) = V(\text{drain end}) = V_{DS}$$

and there is a little resistance between the ends of the channel and the electrodes.

Assuming channel is symmetrical, total channel is divided into two equal halves and effect of gates are the same in each half of the channel, as shown in Fig. 6.6. We will consider only one half of the channel, and $h(x)$ will give the channel half-width. Let a = metallurgical half width of the channel, neglecting depletion width. For simplicity, it is assumed that channel width at the drain end decreases uniformly as the reverse gate bias increases to pinch-off. Since, the gate-channel junction can be treated as one sided p-n junction, i.e., p^+-n and pinch-off is generally occurs at the drain end, therefore, depletion width at $x = L$ (drain end) is given as:

$$W(x = L) = \left[\frac{2\varepsilon_{Si}V_a}{qN_d}\right]^{1/2}$$

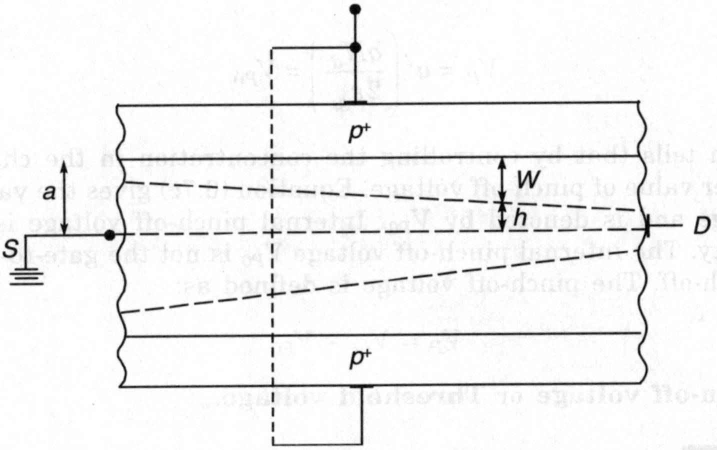

FIGURE 6.6 Structure of symmetrical channel under same gate bias.

Assuming $V_a = -V_{GD}$,

$$W(x = L) = \left[\frac{2\varepsilon_{Si}(-V_{GD})}{qN_d}\right]^{1/2} \tag{6.7a}$$

Neglecting the built-in-potential in comparison to V_{GD}. Since, pinch-off means, two depletion regions should merge with each other. In other words, channel should be disappeared at the drain end. This will occur when

$$h(x) = 0 \qquad (6.7b)$$

Since

$$h(x) = a - W(x = L)$$

Then Eq. (6.7b) becomes

$$a - W(x = L) = 0$$

or

$$W(x = L) = a$$

Substituting value of W from Eq. (6.7a), we have

$$\left[\frac{2\varepsilon_{Si}(-V_{GD})}{qN_d}\right]^{1/2} = a$$

At pinch off,

$$V_{GD} = -V_P = V_G - V_D$$

Hence

$$\frac{2\varepsilon_{Si}(V_P)}{qN_d} = a^2$$

or

$$V_P = a^2\left(\frac{qN_d}{2\varepsilon_{Si}}\right) = V_{P0} \qquad (6.7c)$$

This expression tells that by controlling the concentration in the channel one can select the proper value of pinch-off voltage. Equation (6.7c) gives the value of internal pinch-off voltage and is denoted by V_{P0}. Internal pinch-off voltage is defined as a positive quantity. The internal pinch-off voltage V_{P0} is not the gate-to-source voltage to achieve pinch-off. The pinch-off voltage is defined as:

$$V_P = V_{bi} - V_{P0} \qquad (6.7d)$$

Also called **turn-off voltage** or **Threshold voltage**.

EXAMPLE 6.1

An n-channel JFET at $T = 300$ K has doping concentration of $N_a = 10^{18}/cm^3$, $N_d = 10^{16}/cm^3$. Assume the metallurgical junction channel thickness $a = 0.75$ μm.

(a) Find the internal pinch-off voltage.
(b) Determine built-in-potential.
(c) Calculate turn-off voltage.

Solution:

(a)
$$V_{P0} = \frac{qa^2 N_d}{2\varepsilon_{Si}} = 4.35 \text{ V}$$

(b)
$$V_{bi} = V_{Th}\ln\frac{N_a N_d}{n_i^2} = 0.814 \text{ V}$$

(c)
$$V_{P} = 0.814 - 4.35 = -3.54 \text{ V}$$

EXAMPLE 6.2

Consider a silicon *p*-channel JFET at $T = 300$ K. Given parameters are $N_d = 10^{18}/\text{cm}^3$. Find channel doping and channel thickness for pinch-off voltage of 2.25 V. Take built-in potential = 0.832 V.

Solution:

$$0.832 = V_{Th}\ln\frac{N_a N_d}{n_i^2}$$

or

$$N_a = 2 \times 10^{16}/\text{cm}^3$$

$$V_{P0} = 2.25 \text{ V} + 0.832 = 3.08 \text{ V}$$

$$a = \left(\frac{2\varepsilon_{Si}V_{P0}}{qN_a}\right)^{1/2} = 0.446 \text{ μm}$$

Note: For *n*-channel JFET,

$$V_{DS} \text{ (sat.)} = V_{P0} - (V_{bi} - V_{GS})$$

For *p*-channel JFET,

$$V_{DS} \text{ (sat.)} = V_{P0} - (V_{bi} - V_{GS})$$

6.3 GALLIUM ARSENIDE (GaAs) DEVICES: THE MESFET

The word MESFET stands for Metal-semiconductor field-effect transistor. It was proposed in 1966 by Mead. It is basically a voltage controlled resistor and has three metal contacts: Schottky contact for gate electrode, Two ohmic contacts for source and drain electrodes. In the MESFET, the conduction process involves predominately by one kind of carrier, namely majority-carrier and, therefore, it is called a **unipolar transistor**. The schematic representation of the MESFET is shown in Fig. 6.7. MESFET was fabricated by Hooper and Lehrer by using an *n*-type GaAs epitaxial layer on the semi-insulating substrate. The gate depletion region and the semi-insulating substrate form the boundaries of the conducting channel. Semi-insulating materials have been

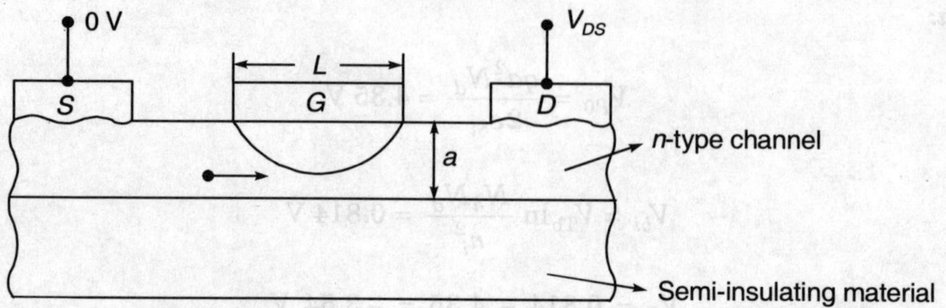

FIGURE 6.7   Schematic representation of MESFET.

chosen only to reduce the parasitic capacitance. The basic device parameters include gate length L, the gate width Z and thickness of the epitaxial layer a. A MESFET is often described in terms of the gate dimensions. If the gate length L is 0.5 μm and gate width Z is 300 μm then the device is referred to as a 0.5×300 μm^2 device. The current handling capability of a MESFET is directly proportional to the gate width.

6.3.1   Principles of Operation

Figure 6.7, shows an ***n*-channel MESFET** which is commonly used due to high electron mobility. Under normal operating conditions, the gate voltage should be zero or reverse biased and drain voltage is zero or forward biased.

If $V_{GS} = 0$ and a small drain voltage is applied then a small drain current I_D will flow in the channel; and this current is given by $I_D = V_{DS}/R$, where R is the channel resistance. This shows that current varies linearly with voltage and region of operation is called **linear** or **ohmic** or **triode region**. Due to finite resistance of the channel, the voltage along the channel varies from zero at the source end to V_{DS} at the drain end for the applied drain voltage V_{DS}. This drop of voltage along the channel makes Schottky barrier more reverse bias as one moves from source to drain. As V_{DS} increases, depletion width becomes more wider and hence width of the conducting channel reduces. This causes an increase in channel resistance and as a result current increases at slower rate as shown in Fig. 6.8(a).

A further increase in drain voltage causes gate-channel junction more reverse bias towards the drain end and depletion width becomes more wider at the drain end than the source end. At certain drain voltage, the depletion region touches the semi-insulating material substrate as shown in Fig. 6.8(b) and the source and drain are pinched-off or completely separated by reverse-bias depletion region. This pinch-off happens when $W = a$ and the current after pinch-off points saturates. The drain voltage at which current saturates is called **saturation voltage** $V_{D\,sat}$; and is given by

$$V_{D\,sat} = \frac{qN_d a^2}{2\varepsilon_{GaAs}} - V_{bi} \qquad \text{for } V_{GS} = 0 \tag{6.8}$$

The current at the pinch-off voltage is called **saturation current** $I_{D\,sat}$.

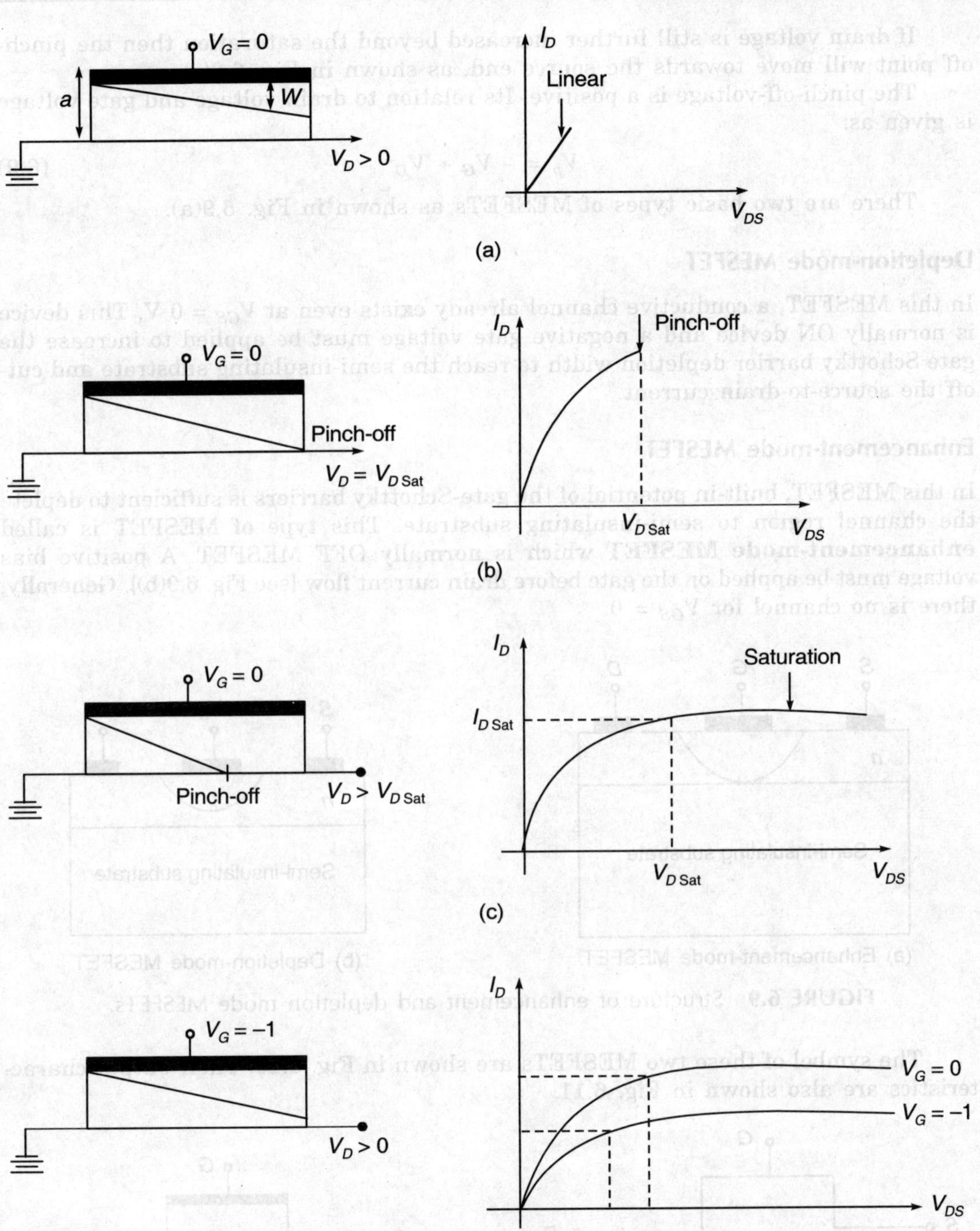

FIGURE 6.8 Channel conditions and *I-V* characteristics at different bias voltages.

If drain voltage is still further increased beyond the saturation then the pinch-off point will move towards the source end, as shown in Fig. 6.8(c).

The pinch-off-voltage is a positive. Its relation to drain voltage and gate voltage is given as:

$$V_p = -V_G + V_D \qquad (6.9)$$

There are two basic types of MESFETs as shown in Fig. 6.9(a).

Depletion-mode MESFET

In this MESFET, a conductive channel already exists even at $V_{GS} = 0$ V. This device is normally ON device and a negative gate voltage must be applied to increase the gate Schottky barrier depletion width to reach the semi-insulating substrate and cut-off the source-to-drain current.

Enhancement-mode MESFET

In this MESFET, built-in potential of the gate-Schottky barriers is sufficient to deplete the channel region to semi-insulating substrate. This type of MESFET is called **enhancement-mode MESFET** which is normally OFF MESFET. A positive bias voltage must be applied on the gate before drain current flow [see Fig. 6.9(b)]. Generally, there is no channel for $V_{GS} = 0$.

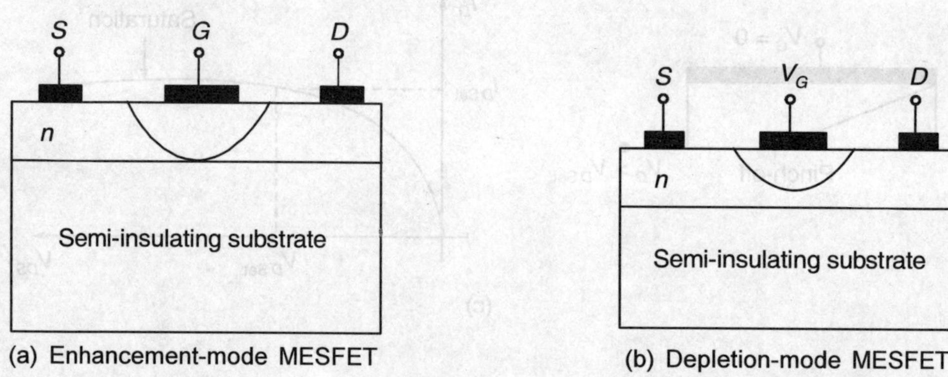

FIGURE 6.9   Structure of enhancement and depletion mode MESFETs.

The symbol of these two MESFETs are shown in Fig. 6.10. Their output characteristics are also shown in Fig. 6.11.

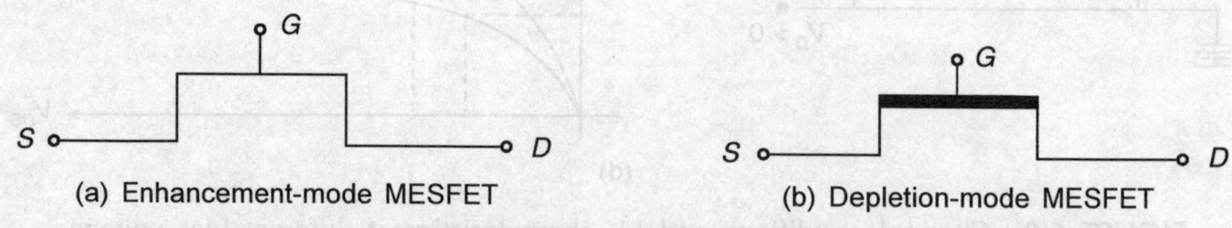

FIGURE 6.10   Symbol of enhancement and depletion mode MESFETs.

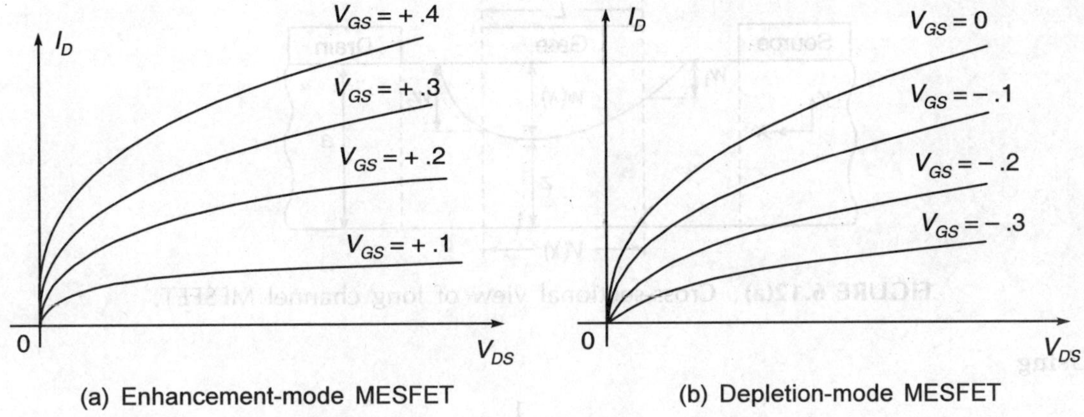

(a) Enhancement-mode MESFET   (b) Depletion-mode MESFET

FIGURE 6.11 *I-V* characteristics of MESFETs.

For high-speed, low-power applications the normally OFF device (enhancement-mode device) is preferred.

6.3.2 Threshold-Voltage

The two MESFETs (Enhancement-mode, depletion-mode) are distinguished by their different threshold voltages V_T which is the gate voltage required to fully deplete the doped channel layer or to enhance the depletion layer. The threshold voltage is positive for enhancement-mode MESFET and negative for the depletion-mode MESFET. For a normally OFF MESFET, a positive bias must be applied to the gate before channel current begins to flow. The required voltage, called the **threshold voltage** V_T, is given by

$$V_T = V_{bi} - V_P \tag{6.10}$$

where V_P is pinch-off voltage and is defined as:

$$V_P = \frac{qN_d a^2}{2\varepsilon_{GaAs}} \tag{6.11}$$

6.3.3 *I-V* Characteristics

The cross-sectional view of a long channel MESFET is shown in Fig. 6.12(a). In Figure, L is the channel length, $W(x)$ is the depletion width, a is epi-layer thickness, A is corss-sectional area and Z is channel width. Consider a small elemental section of thickness dx of the channel and voltage drop across this section dx is dV which is given as:

$$dV = I_d dR = I_d \rho \frac{dx}{A} \tag{6.12a}$$

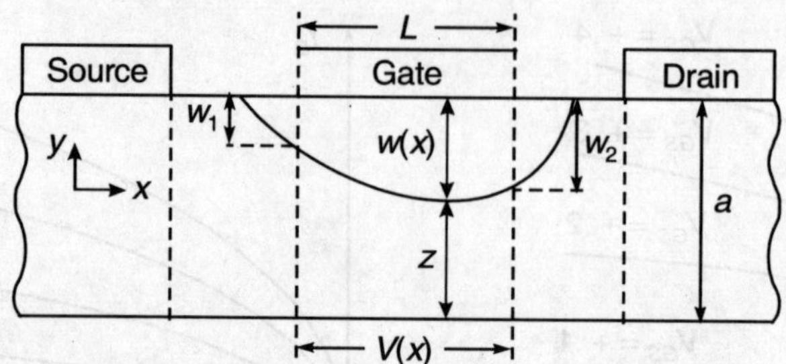

FIGURE 6.12(a) Cross-sectional view of long channel MESFET.

Using

$$\rho = \frac{1}{q\mu_n N_d}$$

and

$$A = Z[a - W(x)]$$

Equation (6.12a) can be further expressed as:

$$dV = I_d \frac{1}{q\mu_n N_d} \cdot \frac{dx}{Z[a - W(x)]} \tag{6.12b}$$

where N_d is donor concentration and μ_n is mobility of electron.

or
$$I_d \cdot dx = q\mu_n N_d \, Z[a - W(x)] \, dV \tag{6.12c}$$

where $W(x)$ is the depletion-layer width at a distance x from the source, and is given by

$$W(x) = \sqrt{\frac{2\varepsilon_{GaAs}[V(x) - V_G + V_{bi}]}{qN_d}} \tag{6.13a}$$

Squaring on both sides, we have

$$W^2(x) = \frac{2\varepsilon_{GaAs}}{qN_d} \, [V(x) - V_G + V_{bi}]$$

or

$$V(x) = \left(\frac{qN_d}{2\varepsilon_{GaAs}}\right) W^2(x) + V_G - V_{bi}$$

Differentiating w.r.t. $W(x)$, we have

$$dV(x) = \left(\frac{qN_d}{2\varepsilon_{GaAs}}\right) [2W(x)] \, dW$$

$$= \frac{qN_d}{\varepsilon_{GaAs}} \, W(x) \, dW \tag{6.13b}$$

Substituting Eq. (6.13b) into Eq. (6.12c), we have

$$I_d \, dx = q\mu_n N_d \, Z[a - W(x)] \, q \frac{N_d}{\varepsilon_{GaAs}} \, W(x) \, dW$$

$$= \frac{q^2 N_d^2 \mu_n Z}{\varepsilon_{GaAs}} \, [a - W(x)] \, W(x) \, dW \tag{6.13c}$$

Integrating left side from $x = 0$ to $x = L$, and right side from W_1 to W_2, we have

$$\int_0^L I_d \, dx = \frac{q^2 N_d^2 \mu_n Z}{\varepsilon_{GaAs}} \int_{W_1}^{W_2} [a - W(x)] \, W(x) \, dW$$

or
$$I_d = \frac{Z}{L} \frac{\mu_n q^2 N_d^2}{2 \, \varepsilon_{GaAs}} \left[a(W_2^2 - W_1^2) - \frac{2}{3} (W_2^3 - W_1^3) \right]$$

From Eq. (6.13a),

$$W_2^2 = \frac{2 \, \varepsilon_{GaAs}[V_D - V_G + V_{bi}]}{qN_d} \quad \text{at the drain end}$$

$$W_1^2 = 2 \, \varepsilon_{GaAs} \left[\frac{-V_G + V_{bi}}{qN_d} \right] \quad \text{at the source end}$$

Therefore,

$$W_2^2 - W_1^2 = a^2 \left[\frac{V_D}{V_P} \right]$$

and
$$W_2^3 - W_1^3 = \frac{a^3}{V_p^{3/2}} [(V_D - V_G + V_{bi})^{3/2} - (-V_G + V_{bi})^{3/2}]$$

Substituting these values in Eq. (6.13d), and representing

$$I_P = \frac{Z}{L} \frac{\mu_n q^2 N_d^2}{\varepsilon_{GaAs}} \, a^3 \qquad \text{(known as conductance of channel for } V_G = 0\text{)}$$

We have

$$I_D = I_P \left[\frac{V_D}{V_P} - \frac{2}{3} \left\{ \left(\frac{V_D - V_G + V_{bi}}{V_P} \right)^{3/2} - \left(\frac{-V_G + V_{bi}}{V_P} \right)^{3/2} \right\} \right]$$

or
$$I_D = \frac{I_P}{V_P}\left[V_D - \frac{2}{3}\left\{\frac{(V_D - V_G + V_{bi})^{3/2}}{\sqrt{V_P}} - \frac{(-V_G + V_{bi})^{3/2}}{\sqrt{V_P}}\right\}\right] \quad (6.14)$$

Equation (6.14) gives the $I_D - V_{DS}$ characteristics of a MESFET. The expression clearly suggests that for a given gate voltage V_G, the drain current I_D increases with V_{DS} until a maximum is reached which is the limit of the validity for this analysis This maximum drain current is called **saturation drain current** $I_{D\,sat}$ and the drain voltage at which saturation takes place is called **drain saturation voltage** $V_{D\,sat}$ as shown in Fig. 6.12. The current for $V_D > V_{D\,sat}$ is taken as a constant. The saturation voltage is also defined as the voltage where the conducting channel is completely depleted near the drain, i.e.

$$V_{D\,sat} = V_P + V_G - V_{bi}$$

or
$$V_{D\,sat} = V_G - V_T \quad (6.15a)$$

The whole current-voltage characteristics of a MESFET is generally divided into three regions.

Region I: When drain-voltage is small, i.e., V_D is small, the drain-current is linearly proportional to drain-voltage and *I-V* characteristic is linear. This region of operation is called **linear region**.

Region II: For $V_D \geq V_{D\,sat}$, the current saturates and this region of operation is called **saturation region**.

Region III: If drain voltage is further increased beyond the saturation voltage, avalanche-breakdown of the gate-to-channel diode occurs, and the drain current suddenly increases drastically. This region of operation is called **breakdown region**.

The complete *I-V* characteristics of a MESFET is shown in Fig. 6.12(b).

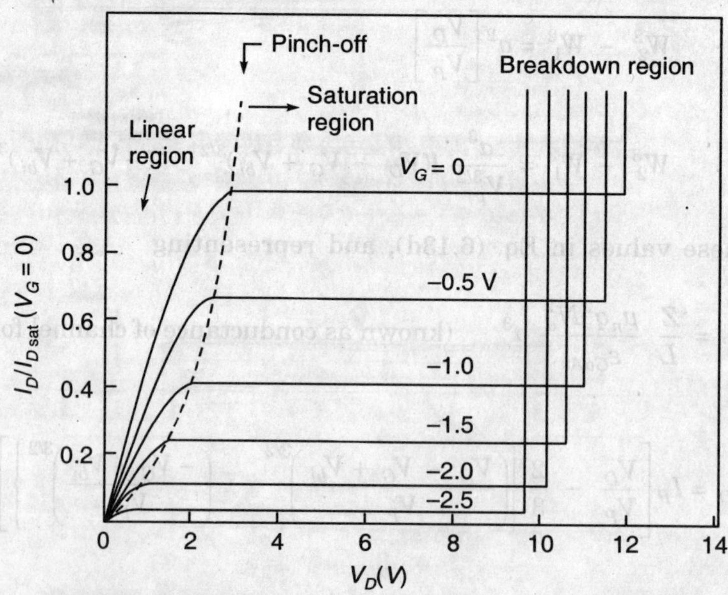

FIGURE 6.12(b) Complete *I-V* characteristics of MESFETs.

For the linear region, $V_D \leq V_G - V_T$ applying Binomial approximation Eq. (6.14) reduces to

$$I_D = I_{D\,\text{linear}} = \frac{I_P}{V_P} \left[1 - \sqrt{\frac{-V_G + V_{bi}}{V_P}} \right] V_D \qquad (6.15b)$$

The saturation current can be evaluated at the pinch-off condition, i.e., $V_P = -V_G + V_D + V_{bi}$ and hence Eq. (6.14) reduces to

$$I_D = I_{D\,\text{sat}} = I_P \left[\frac{1}{3} - \left(\frac{-V_G + V_{bi}}{V_P} \right) + \frac{2}{3} \left(\frac{-V_G + V_{bi}}{V_P} \right)^{3/2} \right] \qquad (6.15c)$$

Schockley also proposed an approximate form of Eq. (6.15c), which is called MESFET square law for the drain current and given as

$$I_{D\,\text{sat}} \cong I_{DSS} \left[1 - \frac{V_{GS}}{V_P} \right]^2 \qquad (6.15d)$$

where, I_{DSS} = Insat for $V_{GS} = 0$.

An important parameter of a MESFET is the transconductance parameter, represented by g_m and defined as the change of drain current with the gate voltage at a given drain voltage, i.e.,

$$g_m = \left(\frac{\partial I_D}{\partial V_G} \right)_{V_D} \qquad (6.16)$$

Therefore, the transconductance in the saturation region can be given as:

$$g_m = \frac{I_P}{V_P} \left[1 - \left(\frac{-V_G + V_{bi}}{V_P} \right)^{1/2} \right] \qquad (6.17)$$

Breakdown occurs generally at the drain end of the channel, where the reverse bias voltage is maximum. The breakdown voltage is given as:

$$\text{Breakdown voltage } V_B = V_D + |V_G| \qquad (6.18)$$

For high-speed low-power applications, normally enhancement-mode MESFET device is preferred. This device does not have conducting channel at zero applied gate voltage. For an enhancement MESFET, a positive gate potential must be applied, before channel current begins to flow. This minimum required gate voltage for flow of current is called **threshold voltage** V_T.

For enhancement-mode MESFET, the saturated drain current is given as:

$$I_{D\,\text{sat}} \cong \frac{Z \mu_n \varepsilon_{GaAs}}{2aL} (V_G - V_T)^2 \qquad (6.19)$$

In deriving Eq. (6.19), we have used the negative sign for V_G to account for its polarity.

Note: Normally, the output characteristic of ON and OFF MESFET devices are same as shown in Fig. 6.11 except the applied gate voltage polarity.

The input characteristics of two modes devices, i.e., ON mode and OFF mode are shown in Fig. 6.13.

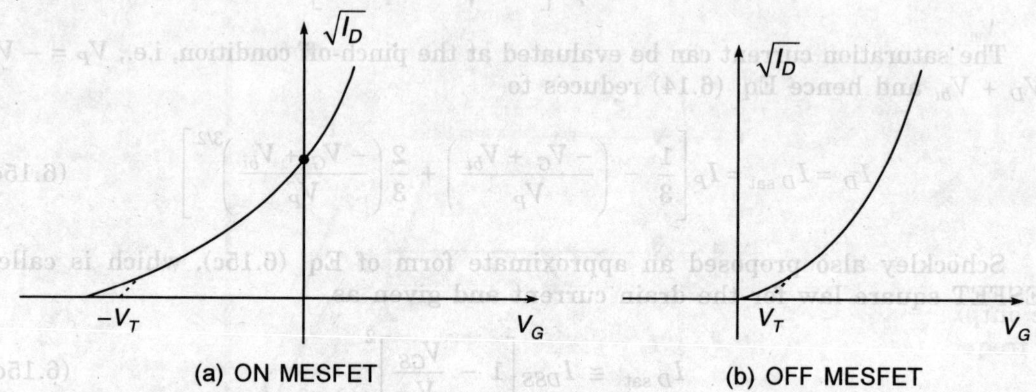

(a) ON MESFET (b) OFF MESFET

FIGURE 6.13 Input characteristics of two mode MESFETs.

6.3.4 Small Signal AC Model

The low frequency as well as high-frequency small signal models of MESFET are shown in Figs. 6.14(a) and 6.14(b) respectively.

In high-frequency model, the capacitors C_{GS}, C_{GD} and C_{DS} are called **parasitic capacitances**, which are dominating at higher frequency, whereas their effects are negligible at lower frequency.

For high frequency applications of MESFET, an important figure of merit is the cut-off frequency f_T. The cut-off frequency is defined as the frequency at which the MESFET can no longer amplify the input signal or it is the frequency at which

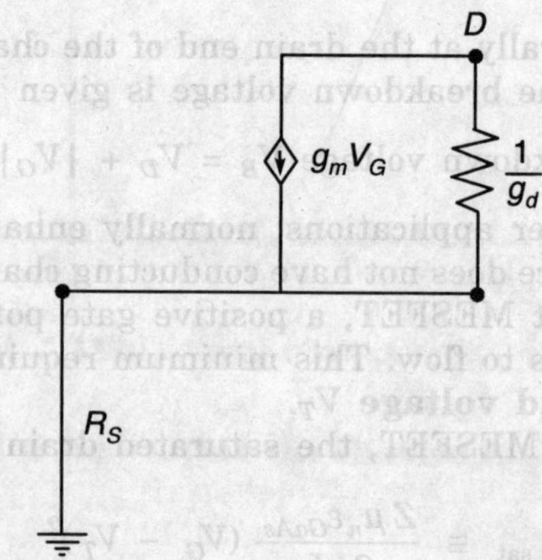

FIGURE 6.14(a) Low-frequency model of MESFET.

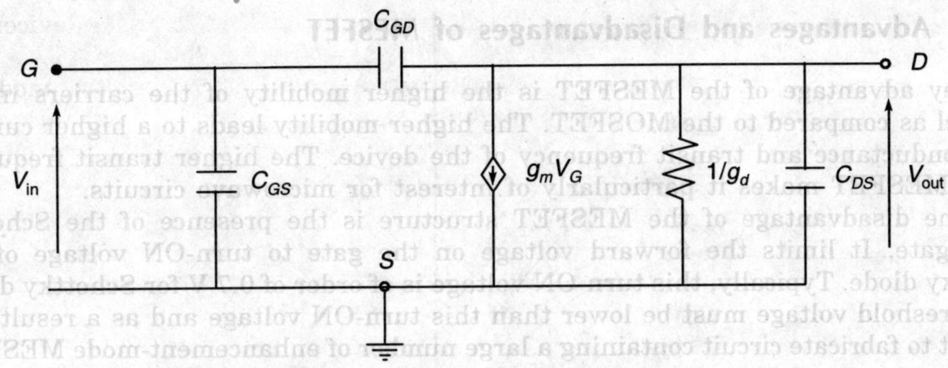

FIGURE 6.14(b)   High-frequency model of MESFET.

the output current equals the input current with the output short circuited. The cut-off frequency is given as:

$$f_T = \frac{g_m}{2\pi C_G} \qquad (6.20)$$

where $C_G = C_{GS} + C_{GD}$
Equation (6.20) can be also expressed as:

$$f_T = \frac{\mu_n q N_d a^2}{2\pi \varepsilon_{GaAs} L^2} \qquad (6.21a)$$

This derivation is based on the fact that the carrier mobility in the channel is constant and independent of the applied field. But, for a very-high frequency operations, the longitudinal field is sufficiently high that the carrier travel at the saturated velocity, i.e.,

$$V = V_s$$

and saturated current is:

$$I_{D\,sat} = Z(a - W)\, q N_d\, V_s \qquad (6.21b)$$

and

$$g_m = \frac{Z V_s \varepsilon_{GaAs}}{W} \qquad (6.21c)$$

and hence, cut-off frequency under velocity saturation condition is:

$$f_T = \frac{V_s}{2\pi L} \qquad (6.21d)$$

Equation (6.21d) clearly shows that if one wants to increase the cut-off frequency then either L should be reduced (short circuit) or carrier velocity should be increased.

6.3.5 Advantages and Disadvantages of MESFET

The key advantage of the MESFET is the higher mobility of the carriers in the channel as compared to the MOSFET. The higher mobility leads to a higher current transconductance and transit frequency of the device. The higher transit frequency of the MESFET makes it particularly of interest for microwave circuits.

The disadvantage of the MESFET structure is the presence of the Schottky metal gate. It limits the forward voltage on the gate to turn-ON voltage of the Schottky diode. Typically, this turn-ON voltage is of order of 0.7 V for Schottky diode. The threshold voltage must be lower than this turn-ON voltage and as a result it is difficult to fabricate circuit containing a large number of enhancement-mode MESFET.

Note: A MESFET can also be constructed using silicon semiconductor material. All the derived expressions for GaAs are also true for silicon MESFET after replacing ε_{GaAs} by ε_{Si}. The use of GaAs rather than silicon provides higher room temperature mobility (i.e. $\mu_{GaAs} \simeq 5\mu_{Si}$) and make it possible to fabricate semi-insulating (SI) GaAs substrate to remove the problem of absorbing microwave processor.

EXAMPLE 6.3

Consider a n-channel MESFET at T = 300 K with a gold contact. Assume the barrier height is 0.89 V. Design a device for which the pinch-off voltage is 0.53 V and channel doping is 2×10^{15}/cm^3. The dielectric constant of GaAs is 12.4. Also, find out built-in-potential and threshold voltage of the GaAs MESFET.

Solution: Using the following equation:

$$v_p = \frac{qN_d}{2\varepsilon_{GaAs}} \cdot a^2$$

$$a = \left(\frac{2v_p\, \varepsilon_{GaAs}}{qN_d} \right)^{1/2}$$

$$= \left[\frac{2 \times 0.53 \times 12.4 \times 8.854 \times 10^{-14}}{1.6 \times 10^{-19} \times 2 \times 10^{15}} \right]^{1/2}$$

$$= 0.6 \ \mu m$$

$$V_n = \frac{kT}{q} \ln\left(\frac{N_C}{N_d} \right) = 0.14 \ V$$

Using $N_c = 4.7 \times 10^{17}$/cm, hence

Built-in-potential = V_{bi} = Barrier height = $-$ 0.14 V = 0.75 V

Now,
$$V_T = V_{bi} - V_P$$
$$= 0.75 - 0.53$$
$$= 0.22 \text{ V}$$

EXAMPLE 6.4

Find the drain saturated current for an enhancement GaAs MESFET of gate length 2 μm and gate width of 10 μm. Other parameters are same as in Example 6.3. Take $V_G = 2$ V and $\mu_n = 1000$ cm^2/Vsec.

Solution: Using Eq. (6.19), we have

$$I_{D \text{ sat}} \simeq \frac{(10) \times (1000) \times (12.4 \times 8.854 \times 10^{-14})}{2 \times (0.6)(2)} \times (2 - 0.22)^2$$

$$\simeq \frac{12.4 \times 8.854 \times 10^{-10}(2 - 0.22)^2}{1.2 \times 2} \times 10^4$$

$$\simeq \frac{12.4 \times 8.854 \times 10^{-6} \times (2 - 0.22)^2}{1.2 \times 2}$$

$$\simeq \frac{347.857}{2.4} \times 10^{-6} \simeq 0.145 \text{ mA}$$

EXAMPLE 6.5

Consider the GaAs MESFET as discussed in Examples 6.3 and 6.4 operated at a very high frequency. Find its cut-off frequency? Take $W = 0.2$ μm

Solution: Using Eq. (6.21d),

$$f_T = \frac{v_s}{2\pi L}$$

Using Eq. (6.21b),

$$I_{D \text{ sat}} = Z(a - W)\, qN_d v_s$$

or

$$v_s \simeq \frac{I_{D \text{ sat}}}{Z(a - W)qN_d} \simeq \frac{0.145 \text{ mA}}{(10 \text{ μm})(0.6 - 0.2)\text{μm } qN_a}$$

$$\simeq \frac{0.145 \text{ mA}}{(10 \text{ μm})(0.4 \text{ μm}) \times 1.6 \times 10^{-19} \times 2 \times 10^{15}}$$

$$\simeq \frac{0.145 \times 10^{-3}}{4 \times 10^{-8} \times 1.6 \times 10^{-4} \times 2}$$

$$\simeq \frac{0.145 \times 10^9}{12.8} \simeq 11.33 \times 10^6 \text{ cm/sec}$$

$$\therefore \qquad f_T = \frac{11.33 \times 10^6}{6.283 \times 2 \times 10^{-4}}$$

$$\simeq 9.02 \times 10^9 \text{ Hz}$$

6.4 MODFET

MODFET stands for modulation-doped field effect transistor. Other common name used for this device is High-electron mobility transistor (HEMT) or two-dimensional electron-gas field effect transistor (TEGFET) or selectively doped heterostructure transistor (SDHT). This field-effect transistor incorporates a junction between two materials of different band gaps as the channel instead of doped region. A commonly used material combination is GaAs with AlGaAs. Depending on the applications various combinations are possible. The two different materials used for heterojunction must have the lattice constant. In semiconductors, these discontinuities form deep-level traps and greatly reduce the device performance. The structure of conventional MODFET is shown in Fig. 6.15.

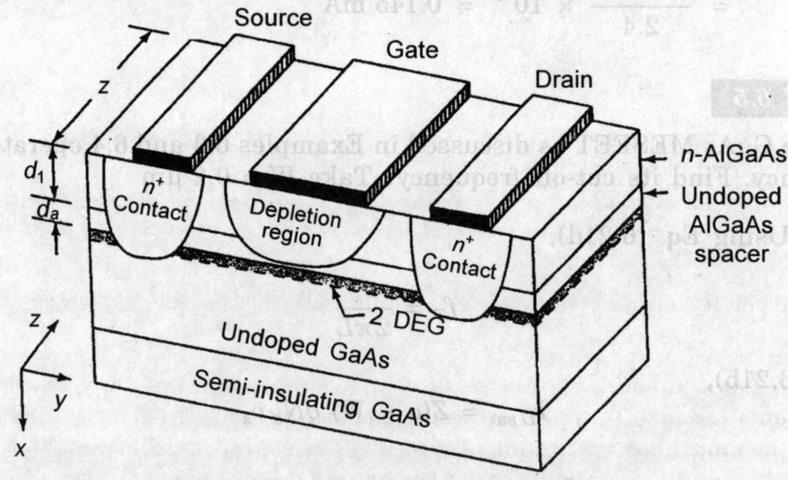

FIGURE 6.15 Structure of MODFET.

MODFET uses a Schottky barrier diode to isolate the channel from the gate. The special feature of MODFET are its heterojunction structure under the gate and the modulation doped layers.

The simplified view of modulation doping (showing only the conduction band) is shown in Fig. 6.16(a) which is similar to potential well structure.

As clear from Fig. 6.16(a), electrons from AlGaAs will diffuse into the undoped GaAs. This diffusion of electrons into the GaAs will increase the concentration of the conduction channel because conduction channel is formed at the surface of GaAs. Since, concentration of carriers in the channel are increasing, and hence process is known as modulation doping. These diffused electrons will confine into the GaAs due to the potential barrier height at AlGaAs–GaAs interface. Since the GaAs is undoped, therefore, scattering probability from the ionized impurity atoms are negligible and this will enhance the mobility of carrier. Therefore, the device is also called **High electron mobility transistor (HEMT)**. The increase in mobility of the electron are higher at low temperature because at low temperature lattice (phonon) scattering is also low.

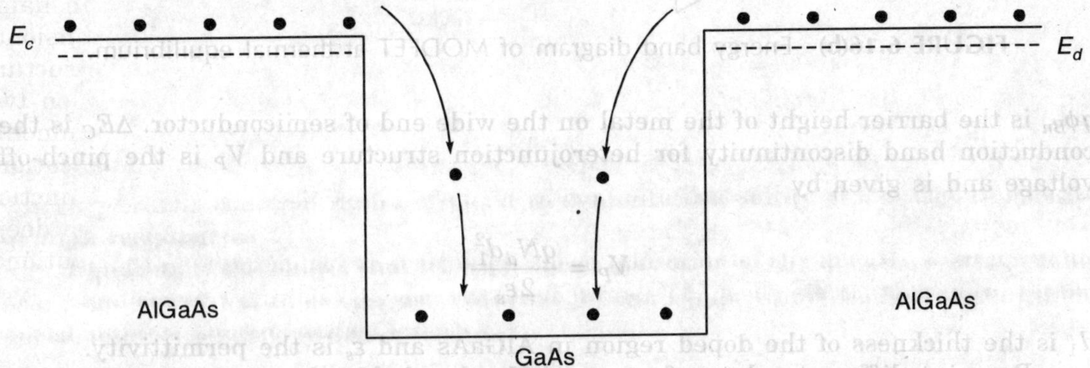

FIGURE 6.16(a) Simplified view of modulation doping.

The energy band diagram of MODFET at thermal equilibrium is shown in Fig. 6.16(b). From previous discussion, it is clear that electrons will accumulate at the corners of the well due to band bending at the AlGaAs/GaAs interface. In fact only one heterojunction (AlGaAs/GaAs) is required to trap electrons as shown in Fig. 6.16(b). The trapping of electrons at the AlGaAs/GaAs junction appears as two-dimensional electron gas and hence device is also called **two-dimensional electron-gas FET**. As seen from Fig. 6.16(b), the spacer moves ionized impurities away from the 2-DEG, reducing scattering probability. Typical spacer is ~10A° for MODFETs. 2-DEG electron layer is either inversion or accumulation layer. As clear from Fig. 6.16(b), the electrons are confined in the z-direction by a triangular well of very small dimensions, thus the allowed electron energies are quantized.

In a properly designed structures, the electron transport approaches that of bulk GaAs with no impurities, so that mobility is limited by lattice scattering.

A key parameter for the operation of a MODFET is the threshold voltage V_T which is defined as the gate voltage at which the channel starts to form between source and drain and is given as:

$$V_T = \phi_{Bn} - \frac{\Delta E_C}{q} - V_P \tag{6.22}$$

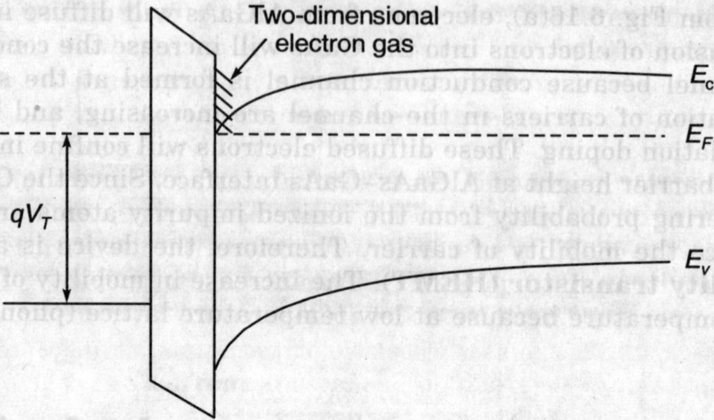

FIGURE 6.16(b)   Energy band diagram of MODFET at thermal equilibrium.

$q\phi_{Bn}$ is the barrier height of the metal on the wide end of semiconductor. ΔE_C is the conduction band discontinuity for heterojunction structure and V_P is the pinch-off voltage and is given by

$$V_P = \frac{qN_d d_1^2}{2\varepsilon_s}$$

d_1 is the thickness of the doped region in AlGaAs and ε_s is the permittivity.

By using different values of $q\phi_{Bn}$ and V_P, threshold voltage of a MODFET can be adjusted.

MODFET is also classified as:

- Enhancement-mode MODFET which is normally OFF device and has positive threshold voltage
- Depletion-mode MODFET which is normally ON device and for which threshold voltage is negative. Since in the MODFET, AlGaAs is doped semiconductor whereas GaAs is undoped semiconductor, therefore, the device is also called **separately doped FET** (SEDFET).

6.4.1 *I-V* Characteristics of MODFET

By using the Gradual channel approximation (GCA), the current at any point along the channel is given by

$$I_D = Zq \ \mu n \ n_s \ E_\chi \tag{6.23}$$

where n_s is density of electrons in the quantized levels at the interface and is given as

$$n_s = - \ \frac{C_i}{q} \ [(V_G - V_T) - V_x] \tag{6.24a}$$

Using Eq. (6.24a) in Eq. (6.23), we have

$$I_D = Z\mu_n C_i [V_G - V_T - V_x] \frac{dV(x)}{dx} \qquad (6.24b)$$

Integrating Eq. (6.24b) from $x = 0$ to $x = L$, assuming drain current is constant, and using the boundary conditions,

$$V(x = 0) = 0$$

and

$$V(x = L) = V_{DS}$$

The current is given by

$$I_D = \frac{Z}{L}\,\mu_n C_i \left[(V_G - V_T)V_{DS} - \frac{V_{DS}^2}{2} \right] \qquad (6.25)$$

where Z is the width of the MODFET device. For linear region,

$$V_{DS} << (V_G - V_T)$$

Equation (6.25) reduces to

$$I_D = \frac{Z}{L}\,\mu_n C_i (V_G - V_T)\, V_{DS} \qquad (6.26)$$

For very large drain-to-source voltage V_{DS}, the charge density reduced to zero at the drain end and this condition is referred as a pinch-off condition.

The saturated drain-source voltage is defined as the drain-source voltage for which $ns(x = L) = 0$ and is given by

$$V_{D\,sat} = V_G - V_T \qquad (6.27)$$

and hence saturated current is given as:

$$I_{D\,sat} = \frac{Z\mu_n C_i}{2L} (V_G - V_T)^2 \qquad (6.28)$$

where

$$C_i = \frac{\varepsilon_s}{d_1 + d_0 + \Delta d} \qquad (6.29)$$

d_1 and d_0 are doped and undoped AlGaAs thickness and Δd is the thickness of the inversion layer (~ 8 nm). The I-V characteristics of a MODFET is similar to MESFET.

At very high-electric field (longitudinal E_x), the velocity of carriers become saturated which will result in saturation current and hence

$$I_{D\,sat} \text{ (at high field)} = Z\, qn_s\, v_s \qquad (6.30a)$$

where v_s is the saturated velocity.

Hence,

$$g_m = \frac{\partial I_{D\,sat}}{\partial V_{DS}} = Z v_s C_i \qquad (6.30b)$$

From Eqs. (6.30a) and (6.30b), it is clear that the drain current is independent of channel length L at high electric field whereas transconductance g_m is independent of both channel length L as well as gate voltage.

The speed of a MODFET is measured by the cut-off frequency, which is given by

$$f_T = \frac{g_m}{2\pi(\text{Total capacitance})}$$

$$= \frac{v_s}{2\pi(L + C_p/ZC_i)} \qquad (6.31)$$

where C_p is total parasitic capacitance.

Note:
- Other alternative structure of a MODFET is δ-doped MODFET.
- GaAs MESFET has an f_T about three times higher than the silicon MOSFET.
- Other MODFETs are:
  * Conventional GaAs MODFET
  * Pseudomorphic SiGe MODFET
  * SiGe MODFET
  * AlGaN/GaN MODFET

6.4.2 Primary Advantage of MODFET

- Higher mobility will give higher velocity and overshoot in higher f_T.
- Higher velocity and overshoot result in higher f_T.
- Lower scattering and buried channel will give lower noise figure.
- Higher forward gate bias will result improved digital signal margin.
- Thinner and fixed channel charge-gate distance will give higher f_T.

Note: SiGe MODFETs are attractive because they can be processed in a silicon fabrication facility.

- The advantages of the HEMT over Si MOSFET are: (a) The higher mobility and maximum electron velocity in GaAs compared with silicon and (b) Smoother interfaces possible with an AlGaAs/GaAs heterojunction compared with the Si/SiO$_2$ interface. The higher performance of the HEMT results in extremely high cut-off frequency and device with fast access times.
- The reason for avoiding Al$_x$Ga$_{1-x}$ As is the presence of the deep level defect called the Δx centre for $x > 0.2$ which traps electrons and impairs the HEMT operation.
- It is observed that the noise figure changes slightly with bias current for MODFET. Lower noise figure and easier design have led to widespread application of MODFET in cellular base station receivers.

EXAMPLE 6.6

Consider an abrupt AlGaAs/GaAs heterojunction with n-AlGaAs doped with 3×10^{18}/cm^3. For such structure Schottky barrier is 0.89 V and discontinuity in the conduction band is 0.23 eV. To achieve the threshold voltage of -0.5 V$_1$, find the thickness of the doped AlGaAs layer. Take $\varepsilon_{sAlGaAs} = 12.3$.

Solution: Using Eq. (6.22),

$$- 0.5 \text{ V} = 0.89 \text{ V} - 0.23 - \left(q \frac{N_d}{2\varepsilon_{AlGaAs}} \right) d_1^2$$

$$-0.5 - 0.89 + 0.23 = - \frac{1.6 \times 10^{-19} \times 3 \times 10^{18}}{2 \times 8.854 \times 12.3 \times 10^{-14}} d_1^2$$

$$-1.16 = - \frac{4.8 \times 10^{-1}}{217.8084} \times 10^{14} d_1^2$$

$$d_1^2 = 52.64 \times 10^{-14}$$

or

$$d_1 = 7.25 \times 10^{-7} = 0.0073 \ \mu m$$

EXAMPLE 6.7

Find the saturated current for a AlGaAs/GaAs MODFET at a gate voltage of -1.5 V, with a 10 nm undoped AlGaAs spacer. Take channel width of 8 nm. Assume $Z = 5L$, and $\mu_n = 1200$ cm^2 V/sec.

Solution: Using Eqs. (6.28) and (6.29),

$$C_i = \frac{12.3 \times 8.854 \times 10^{-14}}{7.25 \times 10^{-7} + 10 \times 10^{-7} + 8 \times 10^{-7}} = \frac{108.904 \times 10^{-7}}{25.25}$$

$$= 4.313 \times 10^{-7} \text{ F}$$

$$I_{d \text{ sat}} = \frac{5 \times 1200 \times 4.313}{2} [-1.15 + 0.5]^2 \times 10^{-7}$$

$$= 1.294 \text{ MA}$$

6.5 SUMMARY

Field effect transistors are unipolar devices and current is only due to majority carriers. Various types of FETs are JFET, MESFET, MODFET and MOSFET. Most of these FETs are made with conventional semiconductor processing techniques.

Junction field effect transistor (JFET) uses a *p-n* junction diode to isolate the channel from the gate. JFET is one of the simplest available transistor and always available in two polarities like *n*-channel JFET and *p*-channel JFET. The operation of JFET is based on the reverse bias of the *p-n* junction between the gate and the channel. The conductivity of the channel can be controlled by controlling the reverse-bias voltage on the gate and hence it is called **junction-field effect transistor**. The current in the channel is saturated after the pinch-off voltage which is −ve for *n*-channel JFET and +ve for *p*-channel JFET.

Metal-Semiconductor contacts are building blocks for MESFET and MODFET devices. MESFET device is proposed by Mead in 1966. It is basically a voltage controlled resistor and has three metal contacts: Schottky barrier as a gate electrode and two ohmic contacts as the source and drain electrodes. This device is important for high-frequency applications. Most MESFETs are made with *n*-type III-V compound semiconductors because of their high Mobilities. There are two types of MESFETs: enhancement-mode and depletion-mode: For enhancement-mode, threshold voltage is positive and it is negative for depletion-mode *n*-type MESFET. Enhancement-mode MESFET is **normally OFF MESFET** whereas depletion mode MESFET is **normally ON MESFET**.

For high-speed, low-power applications, generally enhancement-mode MESFET device is preferred.

The MODFET is a device with enhanced high-frequency performance. This device is similar to MESFET except there is a heterojunction under the gate. A two-dimensional electron-gas, which serves as channel, is formed at the heterojunction interface and electron mobility is increased due to undoped channel region.

The output characteristics of all FETs are similar. They all behave as voltage-controlled-current device and have linear region at low-drain biases.

As the bias voltage increases, the output current increases and at certain bias voltage current saturates. If still bias voltage increases beyond saturation, the breakdown occurs at the drain side.

The cut-off frequency f_T of any FET is a figure of merit for the high-frequency performance. The conventional GaAs MODFET and pseudomorphic SiGe MODFET have an f_T about 30% higher than that of GaAs MESFET.

REVIEW QUESTIONS

1. Why FETs are called majority carrier device?
2. FETs are voltage controlled current device. Explain.
3. Explain basic differences between JFET and MESFET.
4. What is pinch-off in the JFET and why current saturate after pinch-off?
5. Why pinch-off voltage can be controlled by controlling doping of the channel in FETs?
6. What is MESFET?
7. What are normally ON MESET and normally OFF MESFET?

8. Define threshold voltage in MESFET.

9. For high-speed, low-power applications, normally enhancement-mode MESFET device is preferred, why? Explain qualitatively.

10. Give some disadvantages of the MESFET.

11. Use of GaAs rather than Si MESFET provides higher electron mobility at room temperature. Explain.

12. What is the main purpose of fabricating a semi-insulating (SI) GaAs substrate in MESFET?

13. How the electron mobility is increased in MODFET?

14. Why MODFET is called modulation-doped FET?

15. List the advantages of HEMT over the Si MODFET.

16. Out of the two available FETs like MESFET and MODFET, which has greater higher cut-off frequency and why?

17. Why junction field effect transistor (JFET) is not classified as enhancement JFET and depletion JFET?

18. Why source and drain are interchangeable in JFET?

19. Explain the principle of operation of JFET.

20. Why the input impedance of the JFET is extremely high?

NUMERICAL PROBLEMS

1. Consider the JFET as shown in Fig. 6.17. The two gate regions are doped with acceptor impurities of 10^{18}/cm^3 and channel is doped with a donor impurity of 10^{16}/cm^3. If the channel half width is 1 μm then find:

    (i) Type of JFET
    (ii) Depletion width at pinch-off
    (iii) Pinch-off voltage (internal)
    (iv) Built-in potential
    (v) Drain-voltage at pinch off if $V_G = -2$ V
    (vi) Turn-off voltage
    (vii) Saturation drain-voltage for $V_G = -3$ V

    Compare built-in potential with pinch-off voltage also.

2. The output characteristics of a n-channel JFET is shown in Fig. 6.17.

    (i) Find the drain-current at $V_G = -1$ V.
    (ii) If applied gate voltage $V_G = 1$ V, find the current I_D.
    (iii) Find the transconductance at $V_G = -1$ V.
    (iv) What is the doping of the channel if the depletion width of the channel is 0.75 μm?

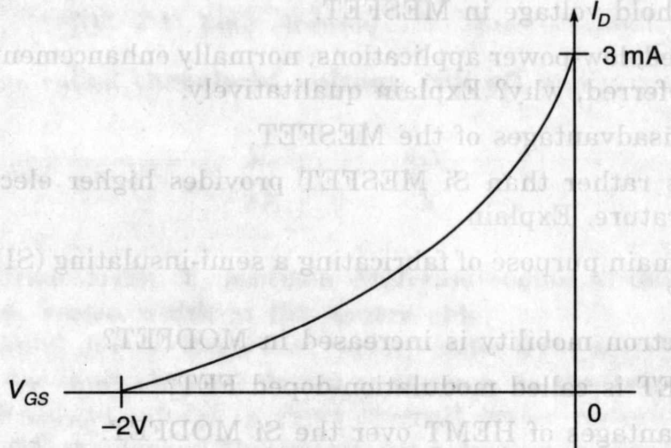

FIGURE 6.17

(v) Calculate the required gate-to-drain voltage for the pinch-off.

(vi) What value of gate voltage turns the device off such that channel conductance goes to zero?

(vii) Discuss the effect of the finite drain resistance on the current.

3. A *n*-channel JFET at T = 300 K has a channel doping of 10^{16}/cm^3 and the pinch-off voltage is − 4 V.

   (i) Calculate the doping concentration of respective gates to achieve the built-in pot-of 0.814 V.

   (ii) Find the channel half width and channel-depletion width.

   (iii) Compare these two at the pinch-off.

   (iv) Find the maximum drain-current I_{DSS} for a drain current of 1 mA at V_{GS} = 3 V.

   (v) Plot the input characteristic curve.

   (vi) If the small signal drain resistance is 50 Ω, find the early-voltage for the same current as given in (iv).

4. Assume the doping in GaAs MESFET is N_d = 7×10^{16}/cm^3 and dimensions are a = 0.03 µm, L = 1.5 µm and Z = 5 µm. Take μn = 4500 cm^2/V sec and the barrier height is 0.891.

   (i) Find the pinch-off-voltage

   (ii) Determine built-in-potential

   (iii) Calculate threshold-voltage of the device

   (iv) Whether it is depletion-mode or enhancement-mode MESFET. Explain.

   (v) Find the drain-source-saturated voltage at V_{GS} = 0 V.

   (vi) Calculate drain current in the linear region as well as in saturation region and compare these two for V_G = 0 V.

   (vii) Calculate the ideal value of g_m for V_D = 1 V, V_G = 0 in

   (a) Linear region

   (b) Saturation region

   (viii) Calculate breakdown voltage at $V_D = 10$ V and $V_G = 0$.

   (ix) Find the higher cut-off frequency.

5. The barrier height for two GaAs n-MESFETs are same and equal to 0.85 V. The channel doping of the first device is 4.7×10^{16}/cm^3 and of the second device is 4.7×10^{17}/cm^3. Determine the channel thickness required in each device such that the threshold voltage is zero for each device. Compare their pinch-off voltages and discuss the result qualitatively.

6. Find the thickness of the undoped spacer layer such that the two-dimensional electron-gas concentration of an AlGaAs/GaAs heterojunction is 1.25×10^{12}/cm^3 at zero gate bias. Assume that the n-AlGaAs is doped to 10^{18}/cm^3 and has a thickness of 50 nm (d_i), the Schottky barrier height is 0.89 V and $(\Delta E_c/q) = 0.23$ V. Given $\varepsilon_{sAlGaAs} = 12.3$.

7. Take the same parameter as given in Problem 6, find

   (i) Pinch-off voltage

   (ii) Threshold voltage

   (iii) Whether the MODFET is in depletion-mode or Enhancement-mode.

8. Consider a AlGaAs/GaAs HFET with a 50 nm n-AlGaAs and 10 nm undoped AlGaAs spacer. The channel width is 8 nm and permittivity of AlGaAs is 12.3. Assume that the n-AlGaAs is doped to 10^{18}/cm^3 and $\Delta E_C = 0.25$ eV.

   (i) Calculate the barrier height for threshold voltage of -1.3 V.

   (ii) Find the intrinsic capacitor value

   (iii) Determine the required drain voltage w.r.t. source for saturation at $|V_G| = 1\ V$.

   (iv) Calculate the saturation current for $Z/L = 5$ and $\mu n = 4200$ cm^2/V for $|V_G| = 1\ V$.

9. Consider an n-AlGaAs–intrinsic GaAs abrupt heterojunction. Assume that the AlGaAs is doped to $N_D = 3 \times 10^{18}$/cm^3 and has a thickness of 35 nm. Assume there is no spacer. Let $\phi_{Bn} = 0.89$ V and $(\Delta E_C/q) = 0.24$ V, $\varepsilon_{sAlGaAs} = 12$. Calculate

   (i) V_P

   (ii) n_c for $V_G = 0$

   (iii) Saturation velocity at very high field for $I_{D\ sat} = 2$ mA and $Z = 5\ \mu m$

10. Consider an AlGaAs/GaAs heterojunction with n-AlGaAs doped to 10^{18}/cm^3 and a thickness of 30 nm. Assume the thickness of the undoped spacer is 2 nm and barrier height is 0.85 V. Take $\Delta E_C = 0.23$ eV, $\varepsilon_{sAlGaAs} = 12.3$. Calculate the electron gas concentration for such heterojunction at $V_G = 0$.

     Drain current in the linear region for $(Z/L) = 5\ \mu n = 4200$ cm^2/V sec and applied drain-voltage w.r. to source is 2 V.

     Also, find saturation current and compare the two results.

Metal Oxide Semiconductor Field Effect Transistor (MOSFET)

- The MOS diode: Ideal MOS capacitor
- Electrostatic potential and charge distribution in silicon
- Non-ideal oxide capacitance-voltage behaviour
- Long channel MOSFET
- Subthreshold characteristics
- Substrate bias effect
- MOSFET channel mobility
- Small signal Model of MOSFET
- Short channel MOS devices
- Narrow width effects
- Gate induced drain leakage (GIDL)
- Source-drain series resistances

7.1 INTRODUCTION

In the present chapter, we discuss the basic operation of the one of the most widely used electronic device, particularly in digital integrated circuits. Because of its relatively small size, millions of devices can be fabricated in a single integrated circuit. The device has four terminals: source, drain, gate and bulk (body or substrate). Current in the channel can be controlled by applying voltage on the gate electrode which is isolated from the channel by an insulating material. The resulting device is called either **Metal-insulator-semiconductor (MIS) FET** or **Insulated-gate-field-effect transistor (IGFET)**. Since most of such devices are fabricated using silicon semiconductor and SiO_2 (silicon dioxide) as an insulator and hence the term *metal-*

oxide semiconductor field-effect transistor (MOSFET) is commonly used. MOSFET is composed of an MOS diode and two *p-n* junctions placed immediately adjacent to the MOS diode. The first MOSFET on silicon substrate (starting material for fabrication), using SiO_2 as the gate insulating material, was fabricated in 1960 by Kohang and Atalla. Although the MOSFET devices are slower than the bipolar transistors, but occupy a relatively smaller area on the chip and their fabrication uses fewer processing steps. The technological advantage together with relative simplicity of MOSFET operation, have helped to make the MOS transistor the most widely used switching device in LSI and VLSI circuits. However, just like the bipolar circuits, single polarity MOSFET circuit suffered from a large standby power dissipation and hence were limited in the level of integration on a chip. In this chapter, we will examine the basic operation, principle and characteristic of MOSFET devices.

7.2 THE MOS CAPACITOR

In practical application, the MOS diode is the heart of the MOSFET which is the most important device for advanced integrated circuits. MOS diode can also be used as a storage capacitor in integrated circuits and forms the basic building block for charge-coupled devices (CCD).

A perspective view of MOS diode is shown in Fig. 7.1. The structure consists of three layers: the metal gate electrode, the insulating silicon dioxide (SiO_2) layer and *p* bulk semiconductor (for *n*-type MOSFET), called **substrate** or **body**. This structure is similar to parallel plate capacitor, where two plates (one of metal and other of semiconductor) are separated by silicon dioxide insulating material of thickness t_{ox}. Such MOS structure is, therefore, also called as **MOS capacitor**. The carrier concentration and its local distribution within the semiconductor substrate can be manipulated by the external voltage applied on the gate and the substrate terminals. MOS capacitor is the most important concept for investigating nearly all the electrical properties of the MOS system.

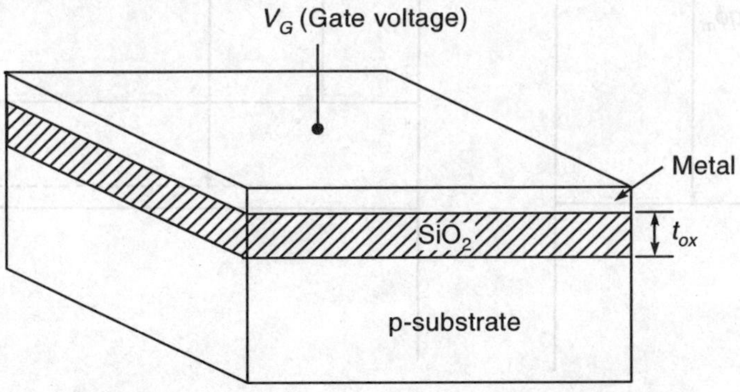

FIGURE 7.1 Perspective view of MOS diode.

Advantages of MOS capacitor

By using MOS capacitor concept, following properties of the MOS system can be obtained.

- (i) Surface band bending and depletion layer width as a function of gate voltage in the silicon.
- (ii) Doping profile of the silicon material.
- (iii) Interface trap density
- (iv) Surface recombination velocity
- (v) Oxide thickness
- (vi) Gate oxide capacitance
- (vii) Type of conductivity

Apart from these listed electrical properties, the MOS capacitor can also be used in determining the ways to control oxide fixed charge and interface trapped densities and so on.

7.2.1 The Ideal MOS Capacitor

Thermal-equilibrium energy-band diagram

Figure 7.2 shows the energy-band diagram of an ideal MOS capacitor. The workfunction of metal (Al) is defined as the energy difference between the vacuum level or free electron energy E_0 and the metal's Fermi level (E_{fm}) and is represented by ϕ_m, i.e.,

$$q\phi_m = E_0 - E_{fm} \tag{7.1}$$

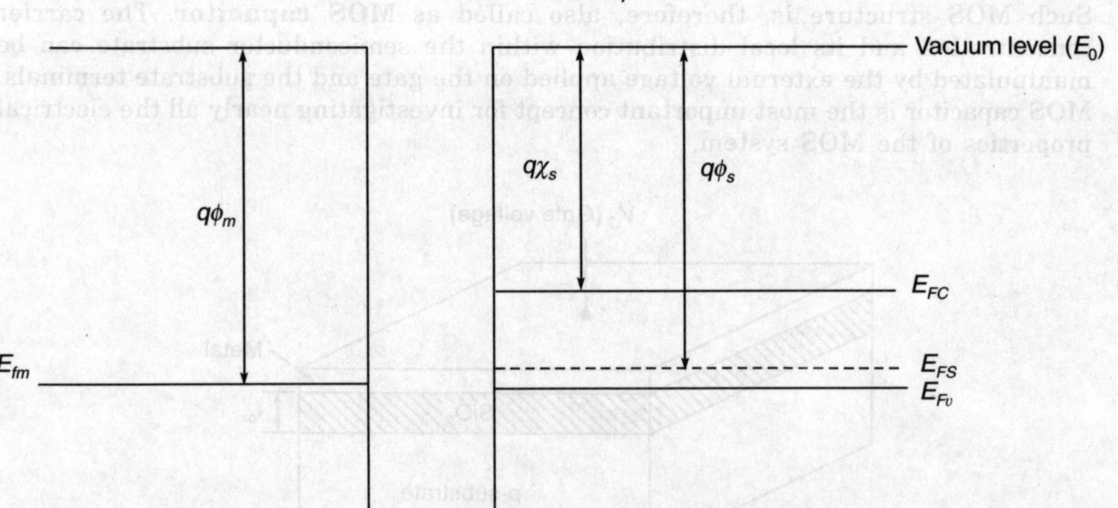

FIGURE 7.2 Band diagram of an isolated ideal *n*MOSFET.

The electron affinity χ_s of the semiconductor is the energy difference between vacuum level to conduction band edge E_c, i.e.,

$$q\chi_s = E_0 - E_c \tag{7.2}$$

The silicon workfunction $q\phi_s$ is defined as the energy difference between the vacuum level E_0 to semiconductor's Fermi level, i.e.,

$$q\phi_s = E_0 - Ef_s$$

or

$$q\phi_s = (E_0 - E_c) + (E_c - Ef_s)$$

$$= q\chi_s + (E_c - Ef_s)$$

or

$$q\phi_s = q\chi_s + (E_c - E_v) + (E_v - Ef_s)$$

$$= q\chi_s + E_g + (E_v - Ef_s)$$

or

$$q\phi_s = q\chi_s + E_g - (Ef_s - E_v) \tag{7.3}$$

For an ideal case, the interface and the oxide are considered to be free from any charges. For an ideal MOS capacitor,

(a) The workfunctions of metal and semiconductor are assumed to be equal in absence of applied bias, i.e., $\phi_m = \phi_s$

or

$$q\phi_{ms} = 0$$

or from Eq. (7.3), we have

$$q\phi_m - [q\chi_s + E_g - (Ef_s - E_v)] = 0 \tag{7.4}$$

In other words, $q\phi_m = q\phi_s$, is possible only if Fermi levels of metal as well as of semiconductor are at same point. Therefore, after the contact, there is no transfer of electrons either from metal or semiconductor side to achieve the equilibrium and hence there is no band-bending. This condition is called **flat-band condition** (in the absence of applied bias).

(b) The only charges that exist in the diode under any biasing conditions are those in the semiconductor and those on the metal surface adjacent to the oxide. The charges on metal surface and semiconductor are equal and opposite to each other.

(c) There is no carrier transport through the oxide under dc biasing condition due to infinite resistivity of the oxide (ideal assumption).

These ideal MOS diode theory helps in understanding the practical MOS devices.

The band diagram at thermal equilibrium for ideal MOS after contact is shown in Fig. 7.3. At thermal equilibrium, the Fermi levels in the metal and semiconductor must be equal and the vacuum level must be continuous. These two requirements are essential to draw the energy band diagram for ideal as well as real MOS capacitors.

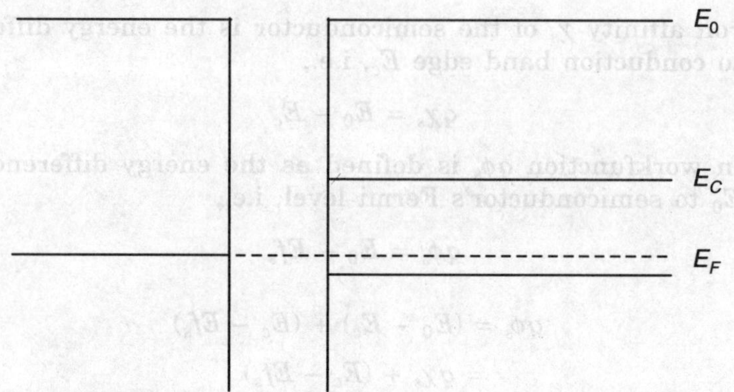

FIGURE 7.3 No biasing (flat-band condition).

Accumulation, depletion and inversion. The ideal MOS diode is biased with either positive or negative voltages. Depending on the polarity and the magnitude of the applied gate voltage V_G, three different operating regions, namely accumulation, depletion and inversion for the MOS system can be achieved. If negative voltage V_G is applied on the gate electrode, it produces a negative electric field. The holes from the p-substrate will get attracted towards the Si-SiO$_2$ interface, and the concentration of majority carriers (holes) near the surface becomes larger than its equilibrium concentration in the substrate. This condition is called **accumulation** and the resulting band diagram is shown in Fig. 7.4(a). The negative surface potential causes the bands to bend upwards near the surface. This is due to fact that number of holes are increasing near the interface or number of electrons are decreasing. Accumulation of holes near the interface causes the conduction band to move away from Fermi level, i.e., bend upwards as shown in Fig. 7.4(a).

Since, the applied negative potential on the metal will increase the energy of the electrons and hence Fermi-level on the metal side is above the semiconductor side by qV_G as seen from Fig. 7.4(a). The charge neutrality still holds in the bulk region, therefore, bands are flat. The corresponding charge distribution near the interface on semiconductor side and on metal side is shown on the right hand side of Fig. 7.4(a); where Q_s is the positive charge per unit area in the semiconductor side, and Q_m is negative charge per unit area in the metal side. By considering the MOS capacitor behaviour, one has

$$|Q_m| = Q_s$$

The potential at the silicon surface is called **surface potential**.

Consider another case in which small positive gate voltage is applied to the gate electrode of an ideal MOS capacitor. Since, the substrate bias is zero, the oxide electric field will be directed towards the substrate. Due to this positive electric field, the majority carrier holes will be repelled back into the substrate and leaves negatively charged immobile acceptor ions behind near the interface. This means the majority carrier holes are depleted from the surface. This is called **depletion** of carriers and a depletion region is created near the junction. Due to decrease in the majority

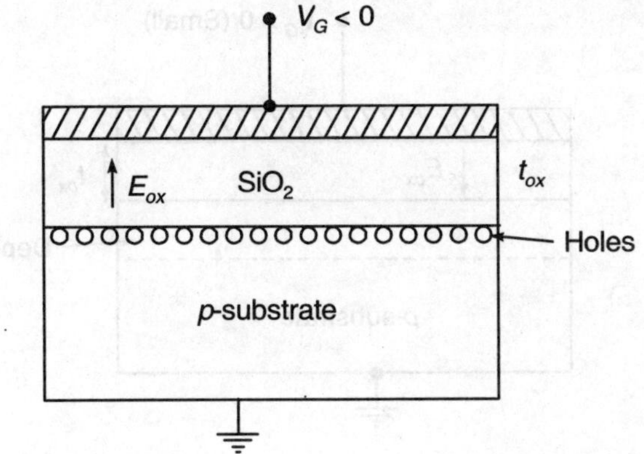

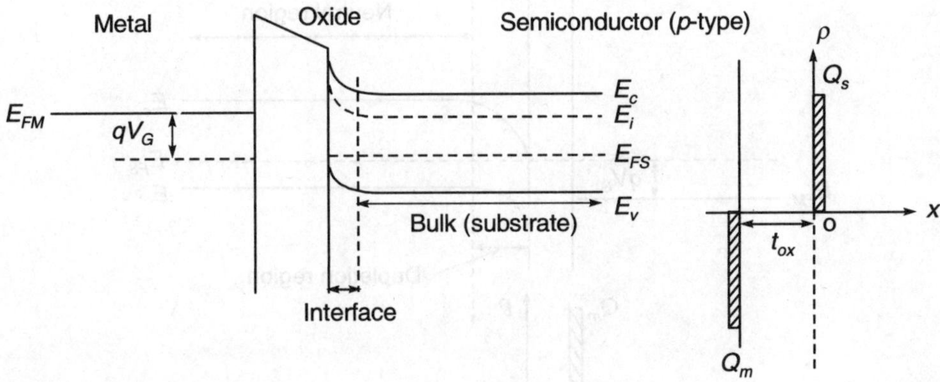

FIGURE 7.4(a) Band diagram and charge distribution in nMOSFET under accumulation.

carrier concentration holes near the interface, the conduction band moves near to Fermi-level and hence bands bend downwards as shown in Fig. 7.4(b). In the bulk region, the bands remain flat which is again due to charge neutrality. The charge distribution for ideal MOS capacitor is also shown in Fig. 7.4(b). The space charge per unit area Q_d in the semiconductor is given by

$$Q_d = -qN_aW_d \tag{7.5}$$

where N_a is substrate doping concentration and W_d is the depletion width. The applied positive gate voltage reduces the energy of the electron and hence Fermi-level in the metal side moves below the Fermi-level of semiconductor by qV_G (where $V_G > 0$). An equal amount of positive charge (Q_m) appears on the metal side of the MOS capacitor.

As the positive voltage on gate electrode increases, the band bending also increases resulting in a wider depletion region and more depletion charge. This goes on until the bands bend downward so much that the intrinsic potential at the surface becomes lower than the Fermi potential or in other words, the intrinsic level E_i at the surface crosses over the Fermi-level E_F as shown in Fig. 7.4(c). When this happens, all the

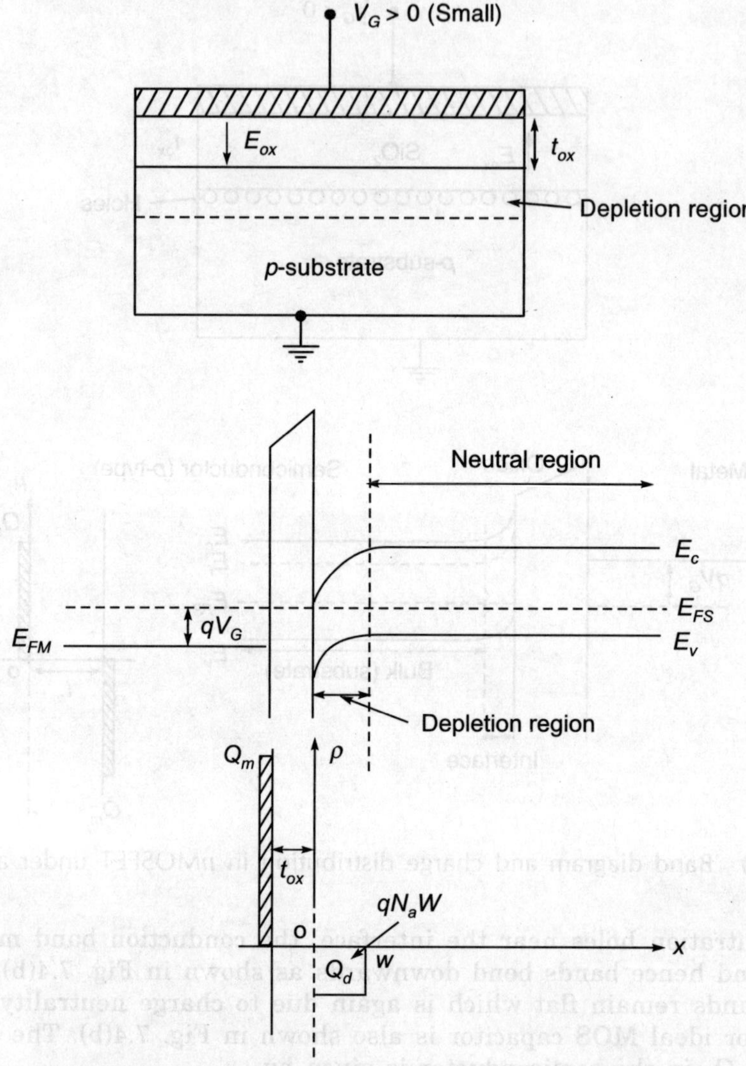

FIGURE 7.4(b)   Band diagram and charge distribution in an ideal *n*MOSFET under depletion condition.

holes are depleted from the surface and surface potential is such that it is energetically favourable for electrons to populate the conduction band near the surface. This can be explained as: the concentration of minority carrier electrons in *p*-type semiconductor is given by

$$n_p = n_i \, e^{(E_F - E_i)/kT} \tag{7.6a}$$

The majority carrier hole concentration is:

$$p_p = n_i \, e^{(E_i - E_F)/kT} \tag{7.6b}$$

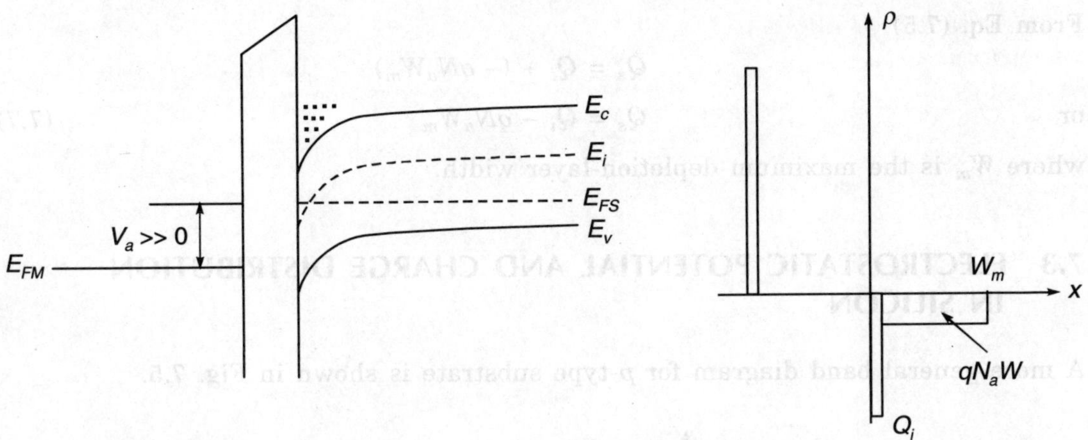

FIGURE 7.4(c) Band diagram and charge distribution in an ideal *n*MOSFET under inversion case.

Dividing Eq. (7.6a) and Eq. (7.6b), we have

$$\frac{n_p}{p_p} = e^{2(E_F - E_i)/kT} \tag{7.6c}$$

In the case, $(E_F - E_i) > 0$, means $n_p > p_p$, i.e., the concentration of electrons near the surface is larger than the concentration of majority carrier holes. This shows that the surface is inverted because earlier surface was *p*-type and now it becomes *n*-type. Important point here to note is that this *n*-type surface is formed not by diffusion of donor impurities, but instead by inverting the original *p*-substrate material with an applied electric field. This condition is called **inversion**.

The surface is inverted as soon as E_i crosses E_F and if $E_F > E_i$, i.e., Fermi-level is slightly above the intrinsic level at the surface then surface electron concentration is slightly larger than surface hole concentration and this is called **weak inversion**.

If the positive gate voltage is increased further so that $E_F \gg E_i$ then the concentration of electrons at the surface will exceed the hole concentration in the substrate. In other words, total surface is occupied by minority carrier concentration electrons and this condition is called **strong inversion**. For strong inversion,

$$n_s \gg p_s$$

The region near the Si-SiO$_2$ interface now behaves as a typical *n*-type material with electron concentration is given by

$$n_o = n_i\, e^{(E_F - E_i)/kT}$$

The inverted layer is separated from the underlying *p*-type material by a depletion region.

At strong inversion, depletion-layer width acquires its maximum value and the charge per unit area Q_s is given by sum of the inversion charge Q_i and depletion charge Q_d, i.e.,

$$Q_s = Q_i + Q_d$$

From Eq. (7.5)

$$Q_s = Q_i + (- qN_aW_m)$$

or

$$Q_s = Q_i - qN_aW_m \qquad (7.7)$$

where W_m is the maximum depletion-layer width.

7.3 ELECTROSTATIC POTENTIAL AND CHARGE DISTRIBUTION IN SILICON

A more general band diagram for p-type substrate is shown in Fig. 7.5.

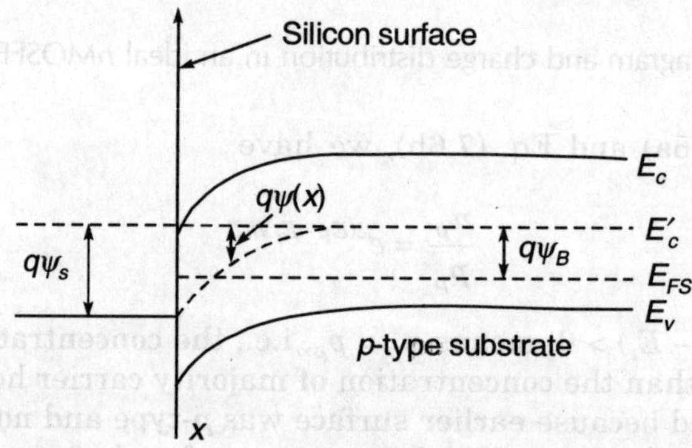

FIGURE 7.5 Band diagram of a p-type substrate.

Writing the Poisson's equation,

$$\frac{d^2\psi(x)}{dx^2} = - \frac{q}{\varepsilon_{Si}} [p(x) - n(x) + N_d^+(x) - N_a^-(x)] \qquad (7.8)$$

where $\psi(x)$ is electrostatic potential along the channel; and $\psi(x)$ [$= \psi_i(x) - \psi_i(x = \infty)$] is defined as the amount of band bending at x. Here $x = 0$ represents the surface or interface. $\psi_i(x = \infty)$ gives the intrinsic potential in the bulk.

Other parameters in Eq. (7.8) have their own significance.

In terms of electrostatic potential ψ, the $p(x)$ and $n(x)$ are expressed as:

$$p(x) = n_i \, e^{q(\psi_B - \psi)/kT} = N_a \, e^{-q\psi/kT} \qquad (7.9a)$$

where $\exp(q\psi_B/kT) = \dfrac{N_a}{n_i}$ and ψ_B is bulk potential $= |(\psi_f - \psi_i)|$ at $x = \infty$

and

$$n(x) = n_i \, e^{q(\psi - \psi_B)/kT} = \frac{n_i^2}{N_a} e^{q\psi/kT} \qquad (7.9b)$$

Assuming complete ionization,

$$N_d^+ = N_d(x) = \frac{n_i^2}{N_a} \tag{7.10a}$$

and

$$N_a^-(x) = N_a(x) \tag{7.10b}$$

Using Eq. 7.9(a, b) and Eq. 7.10(a, b) in Eq. (7.8), we have

$$\frac{d^2\psi(x)}{dx^2} = -\frac{q}{\varepsilon_{Si}}\left[N_a(e^{-q\psi/kT} - 1) + \frac{n_i^2}{N_a}(1 - e^{q\psi/kT})\right] \tag{7.11}$$

Multiplying both sides of Eq. (7.11) by $2(d\psi/dx)$ the resulting left-hand side can be recognized as $(d/dx)(d\psi/dx)^2$ and now integrating from bulk ($\psi = 0$) to the surface (ψ), one obtains (while replacing $-d\psi/dx = E(x)$).

$$E^2(x) = \frac{2kT\,N_a}{\varepsilon_{Si}}\left[\left(e^{-q\psi/kT} + \frac{d\psi}{dx} - 1\right)\right] + \frac{n_i^2}{N_a^2}\left(e^{q\psi/kT} - \frac{q\psi}{kT} - 1\right) \tag{7.12}$$

In the bulk $\psi(x = \infty) = 0$; $d\psi/dx = 0$ and at the surface $\psi(x = 0) = \psi_s$, (called **surface potential**). These are also known as Boundary conditions (BCs).

Substituting $x = 0$, and using $E(x = 0) = E_s$, the electric field at the surface can be obtained from Eq. (7.12). The total charge per unit area induced in the silicon is obtained by Gauss's law,

$$Q_s = -\varepsilon_{Si}\,E_s \tag{7.13}$$

and hence,

$$Q_s = \pm\sqrt{2\varepsilon_{Si}\,kT\,N_a}\left[\left(e^{-q\psi_s/kT} + \frac{q\psi_s}{kT} - 1\right) + \frac{n_i^2}{N_a^2}\left(e^{q\psi_s/kT} - \frac{q\psi_s}{kT} - 1\right)\right]^{1/2} \tag{7.14}$$

Figure 7.6 shows the variation of surface charge per unit area (Q_s) with the surface potential (ψ_s). At the flat band condition, $\psi_s = 0$ and hence, from Eq. 7.14, $Q_s = 0$. In accumulation, since band bends upwards, hence $\psi_s < 0$, and only first term of the square bracket will dominate. The surface charge per unit area in the accumulation is:

$$Q_{\text{sace}} = \pm\sqrt{2\varepsilon_{Si}\,kT\,N_a}\;e^{-q\psi_s/kT}$$

i.e.,

$$Q_{\text{sace}} \propto e^{-q\psi_s/kT}$$

In the depletion region, the band bends downwards means $\psi_s > 0$ and $q\psi_s/kT > 1$, but $e^{q\psi_s/kT}$ term in the square bracket is not large enough to make (n_i^2/N_a) term appreciable. Therefore, the term $(q\psi_s/kT)$ of second square bracket of Eq. 7.14 dominates. This shows that negative depletion charge density is proportional to $\sqrt{\psi_s}$.

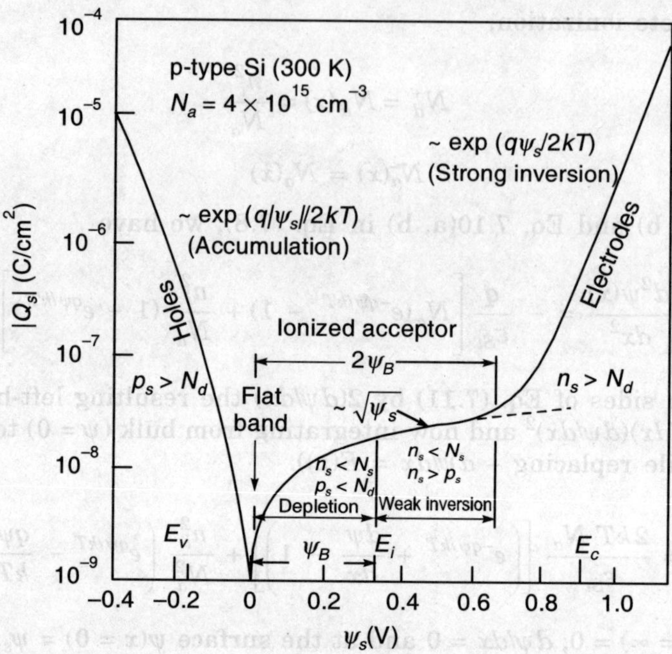

FIGURE 7.6 Variation of total charge density in silicon as a function of surface potential for a *p*-type MOS device.

When ψ_s increases further, the first term $e^{q\psi_s/kT}$ of square bracket becomes larger and dominates. This condition will result in larger value of $n_i^2/N_a = n_o$ at the surface and is called **inversion**.

For inversion,

$$e^{q\psi_s/kT} >> \left(1 + \frac{q\psi_s}{kT}\right)$$

Equation (7.14) becomes

$$Q_{\text{sinv}} = \pm \sqrt{\frac{2\varepsilon_{Si}\, kT n_i^2}{N_a}}\, e^{q\psi_s/2kT}$$

or

$$Q_{\text{sinv}} \propto e^{q\psi_s/2kT} \tag{7.15}$$

A popular criteria for the onset of the strong inversion is that the surface potential to reach a value such that

$$\frac{n_i^2}{N_a^2}\, e^{q\psi_s/kT} = 1$$

or

$$e^{q\psi_s/kT} = \frac{N_a^2}{n_i^2}$$

or
$$q\psi_s/kT = 2\ln\left(\frac{N_a}{n_i}\right)$$

or
$$\psi_s(\text{inversion}) = 2\,\frac{kT}{q}\,\ln\left(\frac{N_a}{n_i}\right) = 2\psi_B \tag{7.16}$$

Equation (7.16) gives the required condition for inversion to take place. Under this condition, the electron concentration at the surface, given by Eq. (7.9b), becomes equal to depletion charge density N_a. After inversion, even a slight increase in the surface potential results in a large electron density at the surface. The inversion layer effectively shields the silicon from further penetration of the gate field.

Equation (7.16) shows that a surface potential of ψ_B is required to bend the band down to the intrinsic condition at the surface ($E_i = E_F$), and E_i must be depressed another $q\psi_B$ at the surface to obtain the condition, called **strong inversion**.

7.3.1 Depletion Approximation and Maximum Depletion Width

In the depletion region, where $2\psi_B > \psi > kT/q$, and neglecting the mobile charges, Eq. (7.12) reduces to

$$E^2(x) = \left(\frac{d\psi(x)}{dx}\right)^2 \cong \frac{2kT}{\varepsilon_{Si}}\,N_a\,\frac{q\psi(x)}{kT}$$

or
$$\frac{d\psi(x)}{dx} \cong -\sqrt{\frac{2N_a}{\varepsilon_{Si}}\,q\psi(x)} \tag{7.17}$$

or
$$\frac{d\psi(x)}{\sqrt{\psi(x)}} = -\sqrt{\frac{2N_a}{\varepsilon_{Si}}\,q}\,\,dx$$

Integrating both sides from $x = 0$, $(\psi(x = 0) = \psi_s)$ to x, $(\psi(x) = \psi)$,

$$\int_{\psi_s}^{\psi}\frac{d\psi(x)}{\sqrt{\psi(x)}} = -\sqrt{\frac{2q}{\varepsilon_{Si}}\,N_a}\int_{0}^{x}dx \tag{7.18}$$

After solving,

$$\psi = \psi_s\left[1 - \frac{qN_a}{2\varepsilon_{Si}\psi_s}\,x\right]^2$$

or
$$\psi = \psi_s\left[1 - \frac{x}{W_d}\right]^2 \tag{7.19}$$

where W_d is depletion-layer width, and is given by

$$W_d = \sqrt{\frac{2\varepsilon_{Si}\psi_s}{qN_a}} \qquad (7.20)$$

Depletion-layer width is defined as the distance to which the band bending extends. From Eq. (7.20),

$$\psi_s = \frac{qN_a}{2\varepsilon_{Si}} W_d^2 \qquad (7.21)$$

Equation (7.21) shows that potential distribution is identical to one-sided n^+-p junction. The total depletion charge density in silicon-semiconductor is given as:

$$Q_d = -qN_aW_d$$

From Eq. (7.20),

$$Q_d = -\sqrt{2\varepsilon_{Si}qN_a\psi_s} \qquad (7.22)$$

In MOS case, the surface depletion-layer width W_d reaches to a maximum value W_{dm} when the surface is strongly inverted, i.e., $\psi_s = 2\psi_B$. Hence, from Eq. (7.20),

$$W_{dm} = \sqrt{\frac{2\varepsilon_{Si}(2\psi_B)}{qN_a}}$$

or

$$W_{dm} = 2\sqrt{\frac{\varepsilon_{Si}\psi_B}{qN_a}} \qquad (7.23)$$

Substituting Eq. (7.16) into Eq. (7.23), we have

$$W_{dm} = 2\sqrt{\frac{\varepsilon_{Si}}{qN_a}\frac{kT}{q}\ln\left(\frac{N_a}{n_i}\right)} \qquad (7.24)$$

Equation (7.24) shows that by controlling the dopant concentration of the substrate one can control the maximum depletion-layer width of MOS system. The electrons are distributed extremely close to the interface of Si-SiO$_2$ with an inversion-layer width less than 50 Å. A higher surface potential or field tends to confine the electrons even closer to the interface. Hence, MOS action is a surface phenomenon. In general, the inversion electrons are treated quantum-mechanically as a 2-dimensional gas. According to quantum mechanics, inversion-layer electrons occupy the discrete energy levels and have a peak distribution within few angstrom (Å) away from the surface.

7.3.2 Ideal MOS Curves

The bands are flat due to equal workfunctions of metal and semiconductor (ideal case) in absence of any applied gate voltage. Figure 7.7(a) shows the energy band diagram of an ideal MOS diode with band bending due to applied gate voltage V_G. A part of the applied gate voltage V_G appears as a potential drop V_{ox} across the silicon dioxide and rest of it appears as a band bending ψ_s in the silicon semiconductor, i.e.,

$$V_G = V_{ox} + \psi_s \tag{7.25}$$

The voltage across the insulator is related to the charge on the either side divided by the capacitance,

$$V_{ox} = -\frac{Q_s}{C_{ox}} \tag{7.26}$$

where $C_{ox} = \dfrac{\varepsilon_{ox}}{t_{ox}}$ is called **gate oxide capacitance** per unit area and t_{ox} is the thickness of gate oxide.

The negative sign in Eq. 7.26 shows that charge on the metal gate is always equal, but opposite to the charge in the silicon.

The charge distribution in an ideal MOS diode (capacitor) is shown in Fig. 7.7(b) where total charge per unit area induced in the silicon (Q_s) includes both depletion as well as inversion charges. For making discussion simpler, oxide and interface trapped charges are ignored. The electric-field distribution and potential distribution are shown in Figs. 7.7(c) and 7.7(d) respectively.

Small signal capacitances

In most cases, MOS capacitance is defined as the small-signal capacitances and can be easily measured by applying a small ac signal over the dc bias across the structure. The total MOS capacitance is defined as:

$$C = \frac{d(-Q_s)}{dV_G} \tag{7.27a}$$

Substituting Eq. (7.26) into Eq. (7.25), we have

$$V_G = -\frac{Q_s}{C_{ox}} + \psi_s \tag{7.27b}$$

Differentiating Eq. (7.27b) w.r.t. $(-Q_s)$, we have

$$\frac{dV_G}{d(-Q_s)} = \frac{1}{C_{ox}} + \frac{d\psi_s}{d(-Q_s)} \tag{7.27c}$$

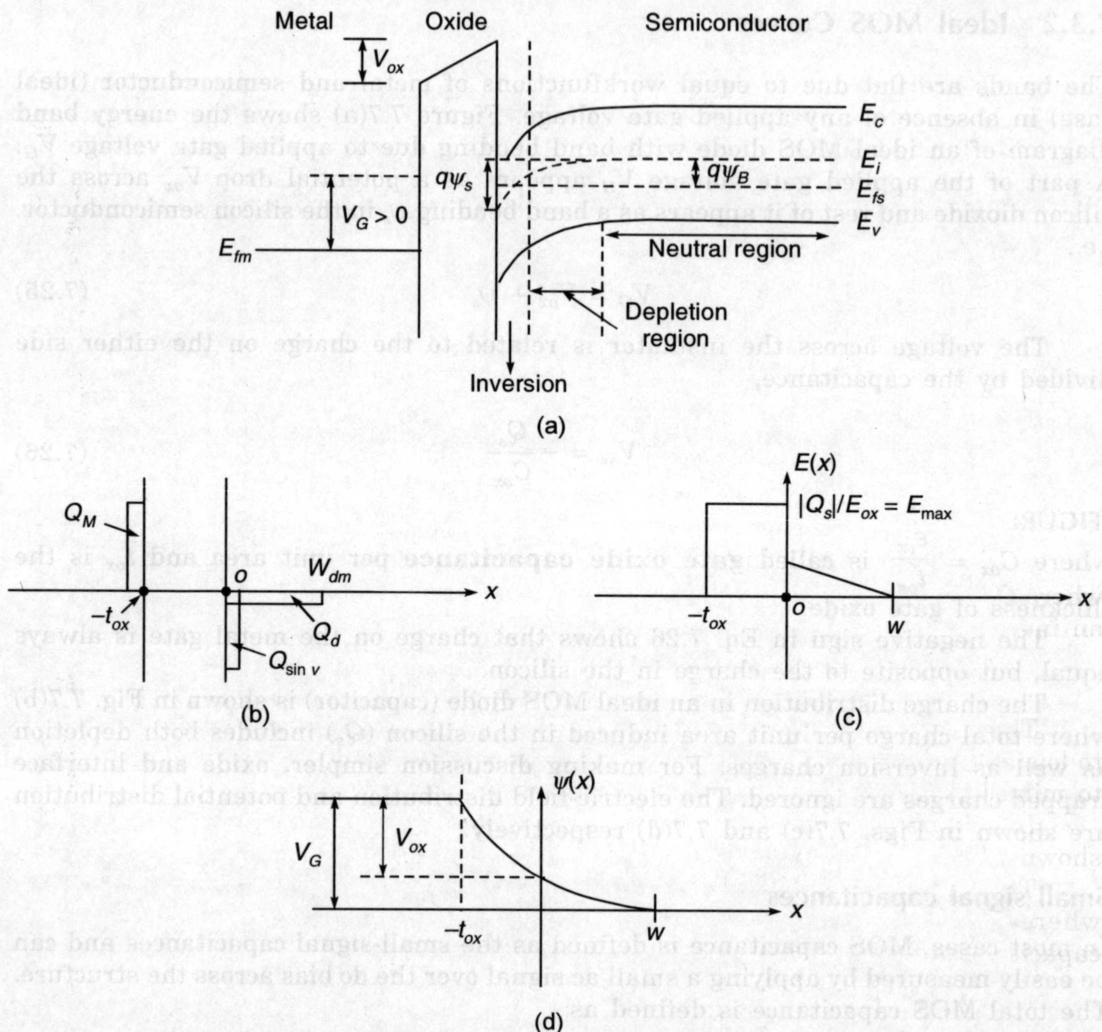

FIGURE 7.7 Energy band diagram of an ideal MOS device with band bending due to gate voltage, (b) Charge distribution in an ideal MOS diode, (c) Electric field distribution in an ideal MOS device, (d) Potential distribution in an ideal MOS device.

Defining $C_{Si} = \dfrac{d(-Q_s)}{d\psi_s}$ as a voltage-dependent silicon capacitance, Eq. (7.27c) can be further written as:

$$\frac{1}{C} = \frac{1}{C_{ox}} + \frac{1}{C_{Si}} \tag{7.27d}$$

Equation (7.27d) shows that the total capacitance of the MOS diode is given by the series combination of gate-oxide capacitance per unit area (C_{ox}) and silicon capacitance (C_{Si}) as shown in Fig. 7.8(a).

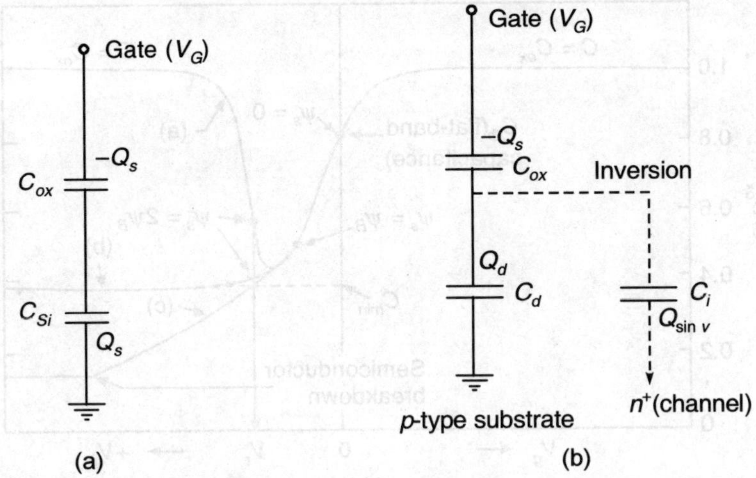

FIGURE 7.8 Combination of different types of small signal capacitances in MOS device.

where C_{Si} is taken as a parallel combination of depletion charge capacitance C_d and an inversion-layer capacitance C_i, i.e.,

$$C_{Si} = C_d + C_i \tag{7.27e}$$

The depletion charge capacitance C_d arises due to majority carriers and responds to low as well as high frequency signal. The inversion-layer capacitance C_i arises due to minority carriers and respond only to low-frequency signals.

In absence of inversion, $C_{Si} = C_d$, whereas due to inversion $C_{Si} = C_d + C_i$ as shown in Fig. 7.8(b).

The gate oxide capacitance per unit area is independent of applied gate voltage, whereas the silicon capacitance C_{Si} is voltage-dependent. Therefore, the total capacitance, as clear from Eq. (7.27d), is a voltage-dependent capacitor and given as:

$$C = \frac{C_{ox}C_{Si}}{C_{ox} + C_{Si}} \tag{7.28}$$

The variation of total capacitance C with the applied gate voltage is shown in Fig. 7.9.

As it is clear from Fig. 7.6, that in the accumulation regions, Q_s is very large at any given ψ_s means $d(-Q_s)/d\psi_s$ is very-very steep, i.e., the accumulation charge changes a lot with the surface potential, and gives a large value of semiconductor capacitance C_{Si}, i.e., $C_{Si} \gg C_{ox}$. Therefore, from Eq. (7.28), $C \simeq \frac{C_{ox}C_{Si}}{C_{Si}} \simeq C_{ox}$ as shown in Fig. 7.9.

When the gate bias is zero, the MOS is near the flat-band condition and $(q\psi_s)/kT \ll 1$. Under this approximation second term of Eq. 7.14 is neglected. Hence

$$Q_s = \pm\sqrt{2\varepsilon_{Si}kTN_a}\left[e^{-q\psi_s/kT} + \frac{q\psi_s}{kT} - 1\right]^{1/2}$$

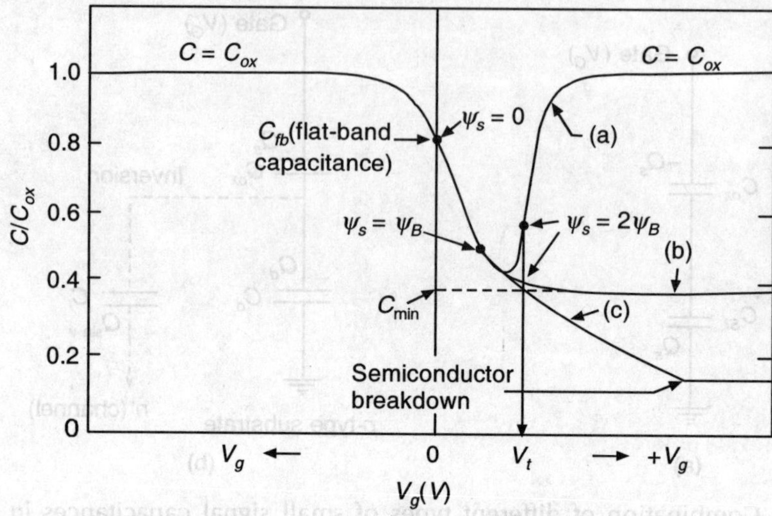

FIGURE 7.9 MOS capacitance-voltage curves: (a) Low frequency, (b) high frequency and (c) deep depletion.

Expanding the exponential term and taking only first three terms of the series, one has

$$Q_s = -\left(\frac{\varepsilon_{Si}q^2N_a}{kT}\right)^{1/2}\psi_s \qquad (7.29)$$

Using Eq. (7.29) and Eq. (7.27d), the flat-band capacitor is given as:

$$\frac{1}{C_{fb}} = \frac{1}{C_{ox}} + \sqrt{\frac{kT}{\varepsilon_{Si}q^2N_a}} \qquad (7.30)$$

From Eq. (7.30), it is clear that flat-band capacitor C_{fb} is somewhat less than C_{ox} as shown in Fig. 7.9. When the gate voltage is slightly positive for a p-type MOS capacitor, the surface starts to be depleted of holes (majority carriers). Thus, a depletion-layer capacitance C_d is added in series with C_{ox}. The depletion-layer capacitance C_d is given as:

$$C_d = \frac{\varepsilon_{Si}}{W_d} \qquad (7.31a)$$

Using Eq. (7.20) in Eq. (7.31a), we obtain

$$C_d = \sqrt{\frac{\varepsilon_{Si}qN_a}{2\psi_s}} \qquad (7.31b)$$

Since

$$V_G = V_{ox} + \psi_s = -\frac{Q_s}{C_{ox}} + \psi_s$$

and using

$$Q_s = -\, qN_aW_d$$

We have

$$V_G = \frac{qN_aW_d}{C_{ox}} + \psi_s \qquad\qquad (7.31c)$$

Using Eq. (7.27d), replacing C_{Si} by C_d eliminating ψ_s, using Eq. (7.31c) one has,

$$C = \frac{C_{ox}}{\sqrt{1 + \left(\dfrac{2C_{ox}^2 V_G}{\varepsilon_{Si} qN_a}\right)}} \qquad\qquad (7.31d)$$

Equation (7.31d) shows that MOS capacitance decreases with increasing V_G under depletion condition. This is shown in Fig. 7.9. As the gate voltage increases further, the capacitance stops decreasing when $\psi_s = 2\psi_B$ where inversion takes place. After inversion, the inversion charge dominates the total silicon change. After inversion the small-signal capacitance depends on whether the measurement are made at high frequency (~1 MHz) or at low frequency (~1–100 Hz).

If the gate bias is changed slowly, i.e., low frequency signal, there is enough time for minority carriers to be generated in the bulk, move across the depletion region to the inversion-layer or go back to the bulk and recombine. This means the minority carriers at the surface are able to follow the applied ac signal. This is true only if the frequency of the applied signal is slower than the reciprocal of the minority carrier response time. Since, the semiconductor capacitance C_{Si} is defined as the variation of the inversion charge with respect to ψ_s, therefore, C_{Si} becomes larger than the depletion capacitance at low frequency. From Eq. (7.28), it is clear that the low frequency MOS series capacitance in the strong inversion is C_{ox} as shown in Fig. 7.9.

For the frequency of the applied signal higher than 100 Hz, the inversion charge cannot respond to applied signal, and only the depletion charge will respond to signal variation. This shows that silicon capacitance C_{Si} is given by C_d with W_d equal to its maximum value W_{dm}. Thus, the high-frequency capacitance approaches a constant minimum value C_{min} at inversion, which is given by

$$\frac{1}{C_{min}} = \frac{1}{C_{ox}} + \sqrt{\frac{4\,kT\,\ln\,(N_a/n_i)}{\varepsilon_{Si}q^2 N_a}} \qquad\qquad (7.32)$$

This is already shown in the C-V curve of the MOS diode. The C-V curves for the MOS diode are traced by applying a slow-varying ramp voltage to the gate with a small ac signal superimposed on it. However, if the ramp signal is such that the ramping time is shorter than the minority-carrier response time then there is insufficient time for inversion-layer to form, and MOS capacitor is said to be biased into deep depletion as shown by curve (c) in Fig. 7.9. In this case, depletion width will further exceed to its maximum value and gives much lower value of depletion capacitance.

Therefore, the MOS capacitance decreases further below to $C_{\min}$ until impact ionization takes place.

Note: Deep-depletion is not a steady state condition because if MOS capacitor is held under such bias conditions, its capacitance will start increasing and reaches back to its value $C_{\min}$ due to thermally generated minority carriers in the inversion layer. This generation is continuous until an equilibrium state is further established. The time taken by MOS capacitor to recover from the deep depletion and return to equilibrium, is called **retention time**.

Ideal MOS diode threshold voltage V_t

The threshold voltage V_t is defined as the minimum required gate voltage to create the depletion charge plus the surface potential at the strong inversion and workfunction difference, i.e.,

$$V_t = -\frac{Q_d}{C_{ox}} + \psi_s(\text{inversion}) + V_{FB} \tag{7.33}$$

Since, for ideal MOS diode $V_{FB} = 0$ and using $\psi_s(\text{inversion}) = 2\psi_B$, Eq. (7.33) becomes

$$V_t = -\frac{Q_d}{C_{ox}} + 2\psi_B \tag{7.34}$$

For n MOS diode, where substrate is p-type semiconductor, the threshold voltage is given by V_{tn}, whereas for p-MOS diode the threshold voltage is represented by V_{tp}.

The threshold voltage is also defined as the minimum required gate voltage to achieve strong inversion.

EXAMPLE 7.1

For an ideal Si-MOS diode having $N_a = 10^{17}/\text{cm}^3$, calculate:

    (a) Maximum depletion width at room temperature
    (b) Surface potential required for strong inversion
    (c) Minimum gate voltage required for strong inversion take $t_{ox} = 15\text{Å}$

Solution: (a) Using Eq. (7.24),

$$W_{dm} = 2\sqrt{\frac{\varepsilon_{Si}}{q}\,\frac{kT}{q}\,\frac{1}{N_a}\ln\left(\frac{N_a}{n_i}\right)}$$

where at room temperature $kT/q = 26$ mV

$$n_i = 1.5 \times 10^{10}/\text{cm}^3$$

$$W_{dm} = 2 \sqrt{\frac{119 \times 8.854 \times 10^{-14}}{1.6 \times 10^{-19} \times 10^{17}} \times 26 \times 10^{-3} \ln\left(\frac{10^{17}}{1.5 \times 10^{10}}\right)}$$

$$= 2 \sqrt{\frac{2739.43 \times 10^{-17}}{1.6 \times 10^{-2}} \ln(0.67 \times 10^{7})}$$

$$= 2 \sqrt{1712.144 \times 10^{-15} \, (15.718)}$$

$$= 10.375 \times 10^{-6} \text{ cm} = 0.104 \text{ μm}$$

(b) Using $\psi_s(\text{in.}) = 2\psi_B = \dfrac{2kT}{q} \ln\left(\dfrac{N_a}{n_i}\right)$

$$\psi_s(\text{in.}) = 2 \times 26 \times 10^{-3} \ln\left(\frac{10^{17}}{1.5 \times 10^{10}}\right)$$

$$= \frac{52 \times 10^{-3}}{15.713 \text{ V}}$$

$$= 0.817 \text{ V}$$

(c) $\qquad V_t = -\dfrac{Q_d}{C_{ox}} + 2\psi_B$

$$= -\frac{t_{ox}}{\varepsilon_{ox}} [- qN_a W_{dm}] + 2\psi_B$$

$$= \frac{15 \times 10^{-8}}{3.9 \times 8.854 \times 10^{-14}} [1.6 \times 10^{-19} \times 10^{17} \times 0.104 \times 10^{-4}] + 0.817$$

$$= 0.4344 \times 106 \, [0.1664 \times 10^{-6}] + 0.817$$

$$= 0.889 \text{ V}$$

EXAMPLE 7.2

For an ideal MOS diode having $N_a = 10^{17}/\text{cm}^3$ and $t_{ox} = 5$ nm, calculate:

(a) Minimum capacitance of the C-V curve
(b) Flat-band capacitance value

Solution: (a) Using $C_{ox} = \dfrac{\varepsilon_{ox}}{t_{ox}} = \dfrac{3.9 \times 8.854 \times 10^{-14}}{5 \times 10^{-7}}$

$$= 6.91 \times 10^{-7} \text{ F}$$

Using Eq. (7.32),

$$\frac{1}{C_{min}} = \frac{1}{6.91 \times 10^{-7}} + 2\sqrt{\frac{26 \times 10^{-3} \ln\left(\dfrac{10^{17}}{1.5 \times 10^{10}}\right)}{11.9 \times 8.854 \times 10^{-14} \times 1.6 \times 10^{-19} \times 10^{17}}}$$

$$= \frac{1}{6.91 \times 10^{-7}} + 2\sqrt{\frac{408.53 \times 10^{-3}}{168.580 \times 10^{-16}}}$$

$$= \frac{1}{6.91 \times 10^{-7}} + 2\sqrt{2.423 \times 10^{13}}$$

$$= \frac{1}{6.91 \times 10^{-7}} + 9.845 \times 10^{6}$$

$$= 11.292 \times 10^{6}$$

$$C_{min} = 8.856 \times 10^{-8} \ \text{F/cm}^2$$

(b) Using Eq. (7.30),

$$\frac{1}{C_{fb}} = \frac{1}{6.91 \times 10^{-7}} + \sqrt{\frac{26 \times 10^{-3}}{11.9 \times 8.854 \times 10^{-14} \times 1.6 \times 10^{-19} \times 10^{17}}}$$

$$= \frac{10^{7}}{6.91} + 1.242 \times 10^{6}$$

$$= 1.45 \times 10^{6} + 1.242 \times 10^{6}$$

$$= 2.689 \times 10^{6}$$

$$C_{fb} = 37.186 \times 10^{-8} \ \text{F/cm}^2$$

7.3.3 The Real MOS Capacitor

For a real MOS diodes the flat-band voltage is generally not zero because workfunction of metal and semiconductor, in practice, is not equal. Also, there are various charges inside the oxide or at the interface. These non-ideal situation must be taken into consideration during the derivation of threshold voltage.

The workfunction difference

From Eq. (7.3) it is clear that the workfunction of a semiconductor material varies with the doping concentration because Fermi-level of semiconductor varies with dopant concentration. Therefore, we expect that the workfunction difference between the metal and semiconductor $q\phi_{ms}$ will also depend upon the doping of semiconductor material. The workfunction difference ϕ_{ms} is negative for the n^+-polysilicon on Si semiconductor.

The isolated band diagram of a real MOS diode is shown in Fig. 7.10(a). From band diagram, it is clear that $q\phi_{ms}$ = –ve because Fermi-level of metal is above the Fermi level of semiconductor. In the isolated situation, all bands are flat. As we know, that at thermal equilibrium, the Fermi-level must be spatially constant and the vacuum level must be continuous. To accommodate workfunction difference between metal and semiconductor the bands should move downwards at the interface of Si-SiO$_2$ as shown in Fig. 7.10(b). Thus, the metal will be positively charged and semiconductor surface will negatively charged at thermal equilibrium. As a result, the bands bend down near the semiconductor interface.

To achieve the ideal MOS capacitor behaviour at thermal equilibrium in absence of biasing, a voltage equal to workfunction difference $q\phi_{ms}$ must be applied. This

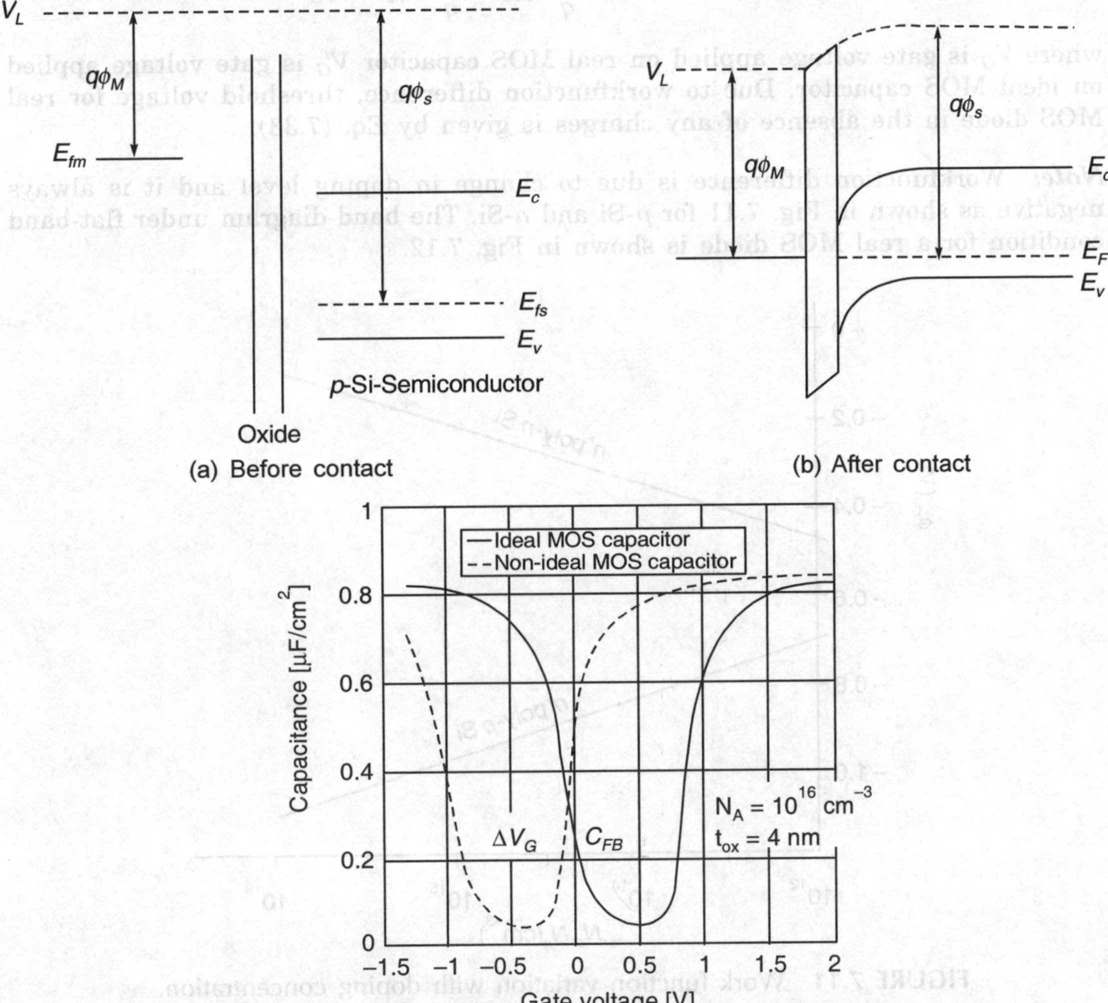

(c) Variation of capacitance with gate voltage for real and ideal MOS devices.

FIGURE 7.10 Energy band diagram of a real MOS device.

voltage is called **flat-band voltage** V_{FB}. In other words, flat-band voltage V_{FB} equals the required gate voltage to achieve the flat-band conditions.

Mathematically,

$$V_{FB} = \phi_{MS} = \phi_M - \phi_S \tag{7.35}$$

Equation (7.35) holds good only if there is no charge present in the oxide or at the oxide-semiconductor interface. The workfunction difference modifies the relationship between the surface potential and the applied gate bias. This causes a shift in the threshold voltage between the ideal and real C-V curves as shown in Fig. 7.10(c). Mathematically,

$$\Delta V_G = V_G - V_G' = \frac{1}{q}\,\phi_{MS} = \frac{1}{q}\,(\phi_M - \phi_S) \tag{7.36}$$

where V_G is gate voltage applied on real MOS capacitor V_G' is gate voltage applied on ideal MOS capacitor. Due to workfunction difference, threshold voltage for real MOS diode in the absence of any charges is given by Eq. (7.33).

Note: Workfunction difference is due to change in doping level and it is always negative as shown in Fig. 7.11 for p-Si and n-Si. The band-diagram under flat-band condition for a real MOS diode is shown in Fig. 7.12.

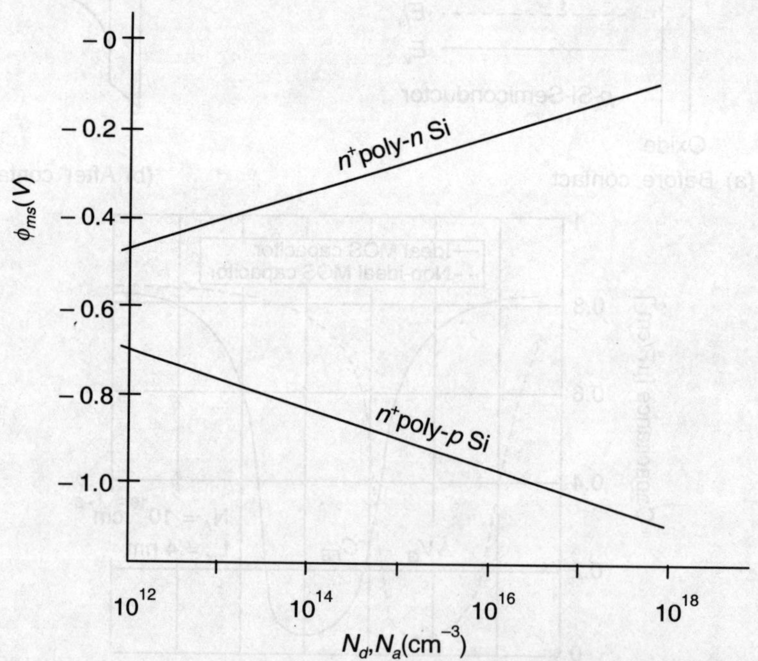

FIGURE 7.11 Work function variation with doping concentration.

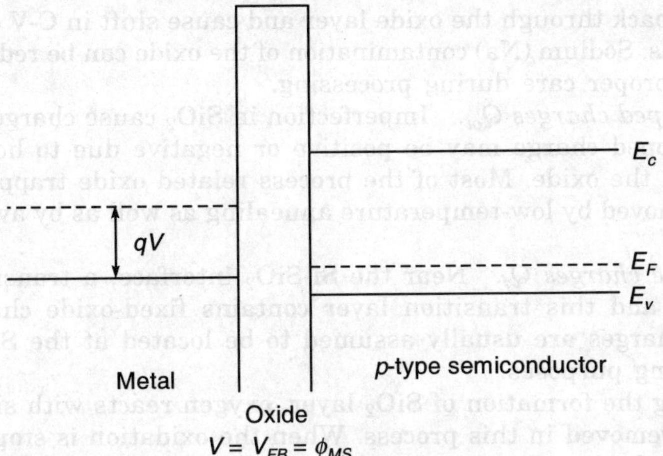

FIGURE 7.12 Energy band diagram of a real MOS capacitor biased at flat-band condition.

Interface charge

In addition to the workfunction difference (ϕ_{MS}), ideal MOS diode is also affected by presence of charges in the oxide layer as well as at the Se-SiO$_2$ interface. These are mobile ionic charge Q_m, Oxide trapped charge Q_{ot}, fixed oxide charge Q_f and interface trapped charge Q_{it} as shown in Fig. 7.13.

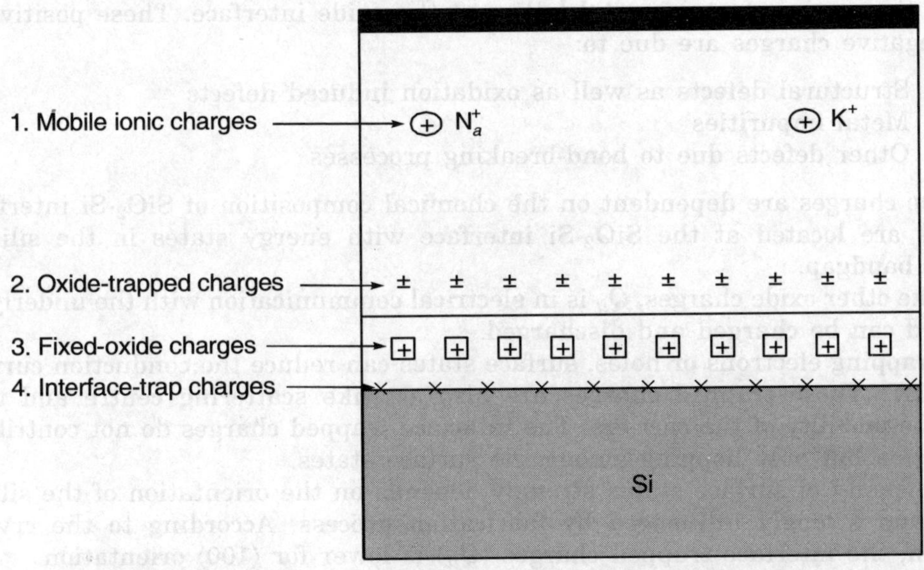

FIGURE 7.13 Charges and their location in thermally oxidized silicon.

(a) *Mobile ionic charges Q_m.* Mobile ionic charges in SiO$_2$ are due to sodium and potassium (K) contaminations which are introduced during fabrication process. Under high temperature and high electric field these mobile charges move forth and back through the oxide layer and cause shift in C-V curves along the voltage axis. Sodium (Na) contamination of the oxide can be reduced to tolerable levels by proper care during processing.

(b) *Oxide trapped charges Q_{ot}.* Imperfection in SiO_2 cause charges to be trapped. Oxide trapped charge may be positive or negative due to holes or electrons trapped in the oxide. Most of the process related oxide trapped charges (Q_{ot}) can be removed by low-temperature annealing as well as by avoiding radiation exposure.

(c) *Fixed-oxide charges Q_f.* Near the Si-SiO_2 interface, a transition layer SiOx is formed and this transition layer contains fixed-oxide charges Q_f. These positive charges are usually assumed to be located at the Si–SiO_2 interface for modeling purposes.

During the formation of SiO_2 layer, oxygen reacts with silicon and hence silicon is removed in this process. When the oxidation is stopped, some ionic silicon is left near the interface. These ions, along with uncompleted silicon bands (e.g. Si-Si or Si-O bonds) at the surface, may result in a positive interface charge Q_f. These charges are fixed and cannot be charged or discharged over a wide variation of surface potential ψ_s. Generally, Q_f is positive and depends on oxidation and annealing conditions and also on silicon orientation. Q_f can be regarded as the charge sheet located at the SiO_2-Si interface. These charges affect the potential in the silicon and reduces the mobility of carriers in the inverted channel.

(d) *Interface-trapped charges Q_{it}.* These charges result from the sudden termination of the semiconductor crystal lattice at the oxide interface. These positive or negative charges are due to:

- Structural defects as well as oxidation induced defects
- Metal impurities
- Other defects due to bond-breaking processes

These charges are dependent on the chemical composition of SiO_2-Si interface. The traps are located at the SiO_2-Si interface with energy states in the silicon-forbidden bandgap.

Unlike other oxide charges, Q_{it} is in electrical communication with the underlying silicon and can be charged and discharged.

By trapping electrons or holes, surface states can reduce the conduction current in MOSFET. These trapped charges are also act like scattering centre and thus reduces the mobility of the carriers. The interface trapped charges do not contribute in conduction but only hopping among the surface states.

The density of surface states strongly depends on the orientation of the silicon material and strongly influenced by fabrication process. According to the crystal orientation, the interface trapped charges (Q_{it}) is lower for $\langle 100 \rangle$ orientation.

i.e., $$\langle 100 \rangle < \langle 110 \rangle < \langle 111 \rangle$$

The interface trapped charges can be reduced by post metallization anneal in hydrogen at $T = 400°C$.

EXAMPLE 7.3

Why MOS devices are generally made on [100] Si?

Solution: The value of Q_{it} for [100]–oriented silicon can be as low as $10^{10}/cm^2$ which amounts to about one interface-trapped charge per 10^{15} surface atoms. For [111] oriented silicon Q_{it} is about $10^{11}/cm^2$.

The typical fixed oxide charge densities for a carefully treated SiO_2-Si interface system are about $10^{10}/cm^2$ for [100] surface and $5 \times 10^{10}/cm^2$ for a [111] surface. Because of the lower value of Q_{it} and Q_f, the [100] oriented Si is preferred for MOS devices.

Calculation of flat-band voltage in the presence of various charges

In a real MOS diode, to achieve the flat-band condition, one should apply a negative potential on the metal. As the negative potential on the metal increases, more negative charges are introduced on the metal and thereby, the electric-field shift downward until the electric field at the interface is zero. Under this condition, the area within the electric-field distribution corresponds to flat-band voltage V_{FB} as shown in Fig. 7.14.

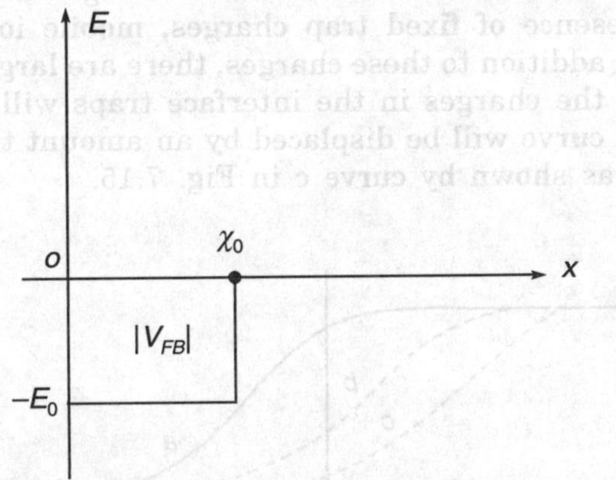

FIGURE 7.14 Schematic illustrating the effect of a flat band condition.

Therefore,

$$V_{FB} = -\,E_0 x_0 = -\,\frac{Q_0}{\varepsilon_{ox}}\,x_0 = -\,\frac{Q_0}{C_{ox}}\cdot\frac{x_0}{t_{ox}} \qquad (7.37)$$

where

Q_0 is a positive sheet charge per unit area within the oxide, C_{ox} is gate oxide capacitance per unit area and t_{ox} is oxide thickness.

Equation (7.37) clearly shows that the flat-band voltage depends on the density of the sheet charge and its location x_0 within the oxide.

Now, consider two cases:

(i) If $x_0 = 0$ means no induce charges in the silicon, i.e., $Q_0 = 0$ and from Eq. (7.37),

$$V_{FB} = 0$$

(ii) If $x_0 = t_{ox}$ means Q_0 is located very close to semiconductor then

$$V_{FB} = -\frac{Q_0}{C_{ox}} \cdot \frac{t_{ox}}{t_{ox}} = -\frac{Q_0}{C_{ox}} \qquad (7.38)$$

For simplicity, it is assumed that the oxide as well as interface charge are the effective net charges per unit area and given by Q_i (C/cm^2). The effect of this charge is to induce an equal amount of negative charge in the semiconductor. In presence of effective positive charge Q_i as well as non-zero workfunction difference, the experimental C-V curve will be shifted from the ideal theoretical curve by an amount,

$$V_{FB} = \phi_{MS} - \frac{Q_i}{C_{ox}} \qquad (7.39a)$$

$$= \phi_{MS} - \frac{(Q_f + Q_m + Q_{ot})}{C_{ox}} \qquad (7.39b)$$

Here, it is assumed that the interface-trap charge is negligible. Curve in Figure 7.15 shows the ideal MOS C-V curve, whereas curve b in Fig. 7.15 shows the shift in the C-V curve due to presence of fixed trap charges, mobile ionic charges and oxide trapped charges. If, in addition to these charges, there are large amounts of interface-trapped charges then the charges in the interface traps will vary with the surface potential and the C-V curve will be displaced by an amount that itself changes with the surface potential as shown by curve c in Fig. 7.15.

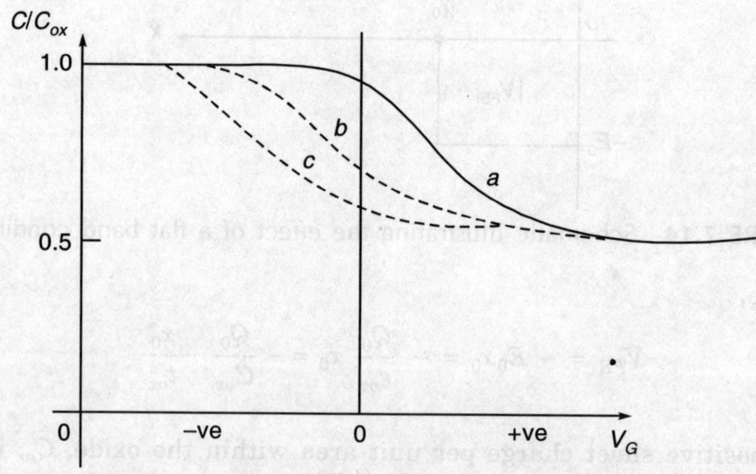

FIGURE 7.15 Capacitance versus gate voltage for ideal and real MOS capacitor.

Non-ideal threshold voltage

The flat-band voltage V_{FB} must be added to the ideal MOS threshold voltage as given by Eq. (7.34) to get the threshold voltage for the real MOS diode.

Therefore,

$$V_t = -\frac{Q_d}{C_{ox}} + 2\psi_B + V_{FB}$$

Using Eq. (7.39a), we get

$$V_t = -\frac{Q_d}{C_{ox}} + 2\psi_B + \phi_{MS} - \frac{Q_i}{C_{ox}} \tag{7.40}$$

Equation (7.40) tells that to create the strong inversion, the gate voltage must be large enough to first achieve the flat-band condition then accommodate the charges in the depletion region $(-Q_d/C_{ox})$ and finally to induce an inverted region $(2\psi_B)$. For n-substrate,

$$\phi_{MS} \Rightarrow \text{Negative}$$

$$\frac{Q_i}{C_{ox}} \Rightarrow \text{Positive}$$

$$\frac{Q_d}{C_{ox}} \Rightarrow \text{Positive}$$

$$2\psi_B \Rightarrow \text{Negative}$$

Therefore, the threshold voltage for a p-channel MOS will be negative and given as:

$$V_{tp} = V_{FB} - 2|\psi_B| - \frac{Q_d}{C_{ox}} \tag{7.41a}$$

whereas for p-type substrate, which will be used for fabrication of n-MOS device, the sign of various parameter used in Eq. (7.40) are:

$$\phi_{MS} \Rightarrow \text{Negative}$$

$$\frac{Q_i}{C_{ox}} \Rightarrow \text{Positive}$$

$$\frac{Q_d}{C_{ox}} \Rightarrow \text{Negative}$$

$$2\psi_B \Rightarrow \text{Positive}$$

and the threshold voltage for an n-MOS transistor is positive quantity

$$V_{tn} = \frac{Q_d}{C_{ox}} + 2\psi_B + V_{FB} \tag{7.41b}$$

Equations (7.41a) and (7.41b) are true for a special MOS transistor called enhancement-mode MOS transistor which will be discussed later.

where
$$\psi_B = \frac{kT}{q} \ln\left(\frac{N_a}{n_i}\right) \quad n\text{-MOS}$$

$$\psi_B = \frac{kT}{q} \ln\left(\frac{N_d}{n_i}\right) \quad p\text{-MOS}$$

with

$$\frac{kT}{q} \cong 26 \text{ mV at } T = 300 \text{ K}$$

The variation of threshold voltage with substrate (dopant) concentration for n-channel and p-channel MOS is shown in Fig. 7.16.

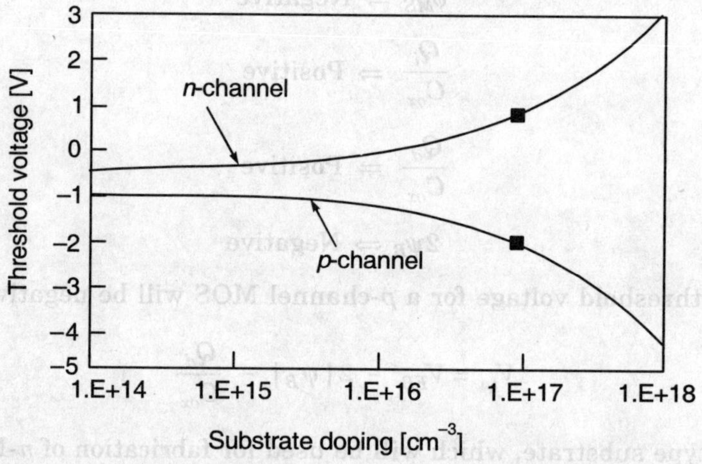

FIGURE 7.16 Threshold voltage variation with substrate doping for n-channel and p-channel MOS devices.

Control of threshold voltage

(a) *Ion Implantation:* Boron ions can be implanted in a two-dimensional sheet just below the oxide layer. These ions are negatively charged and change the value of Q_d, which in turn, change the threshold voltage, i.e.,

$$V_{t(\text{new})} = V_{t(\text{old})} + \frac{qF_B}{C_{ox}} \tag{7.42}$$

where F_B is boron ion dose (B^+) which is measured in ions/cm^2.

$$V_{t(\text{new})} = V_{t(\text{old})} - \left(\frac{qF_P}{C_{ox}}\right)$$

where F_P is phosphorous ion dose.

The negatively charged boron acceptors reduce the effect of the positive depletion charges in a p-MOS and causes threshold voltage to be negative.

A shallow boron implant into a p-type substrate (for a n-channel MOS) can shift threshold voltage from negative to positive.

The implantation energy required for shallow V_t adjustment implants is low and of order of (50–100 keV), and relatively low doses are required.

Typical required dose is for 10 seconds.

(b) *Control of C_{ox}* For a higher current drive of the MOS device, low threshold voltage is required. From Eq. (7.40), it is clear that lower threshold voltage can be achieved by taking larger value of C_{ox}. The thin gate oxide layer will result in larger C_{ox}. From Eqs. (7.41a) and (7.41b), one can see that increasing C_{ox} makes threshold voltage less positive for n-channel device and less negative for p-channel device. Now in modern technology, the gate thickness for sub-micron region is typically varies from 10 Å–100 Å.

The gate oxide capacitor C_{ox} can also be controlled by varying the ε_{ox}. Higher value of ε_{ox} means higher value of C_{ox}. A SiO_2 layer incorporated with some nitrogen (N) will result in a larger value of ε_{ox} than SiO_2. The resulting material is Silicon oxynitrides. The other materials such as Ta_2O_5, ZrO_2 and ferroelectrics can also be used in place of SiO_2 because of their higher value of ε_{ox}.

(c) *Control of Q_i* Threshold voltage of a MOS device can also be controlled by controlling the effective net charge per unit area Q_i. For that it is required to grow the SiO_2 layer on [100] oriented wafers which results in

- Less dangling bonds
- Slower growth rate which leads higher quality layer

(d) *Silicon Gate Technology* Since V_t depends on ϕ_{MS}, the choice of the gate electrode material has a impact on the threshold voltage. To reduce the threshold voltage it is required to reduce the ϕ_{MS} which can be achieved by using poly-silicon as the gate material. Poly-silicon is a heavily doped silicon. Using poly- silicon material as a gate electrode, the workfunction difference between poly- and silicon material is very close to zero. Polysilicon is more process-friendly and withstands upto higher temperatures than Al.

EXAMPLE 7.4

(a) Calculate the flat-band voltage of a silicon n-MOS capacitor with a substrate doping of $10^{17}/cm^3$ and an aluminium gate ($\phi_M = 4.1$ V). Assume that there is no fixed charge in the oxide or at the Si-SiO_2 interface.

(b) Calculate the threshold voltage for Part (a). Take $t_{ox} = 20$ nm.

Solution: (a) Since $\qquad\qquad \phi_{MS} = \phi_M - \phi_S$

From Eq. (7.3),

$$\phi_{MS} = \phi_M - x_s - \frac{E_g}{q} - V_{Th} \ln \frac{N_a}{n_i}$$

Substituting all values, taking $x_s = 4.05$ V

$$\phi_{MS} = -0.93 \text{ V}$$

Since $Q_i = 0$. Therefore, using (7.39a),

$$V_{FB} = -0.93 \text{ V}$$

(b) From Eq. (7.40),

$$V_t = V_{FB} + \frac{Q_d}{C_{ox}} + 2\psi_B$$

$$C_{ox} = \frac{\varepsilon_{ox}}{t_{ox}} = \frac{3.9 \times 8.854 \times 10^{-14}}{20 \times 10^{-7}} = 1.726 \times 10^{-7} \text{ F}$$

$$Q_d = 2(\varepsilon_{Si} q N_a \psi_B)^{1/2}$$

$$= 2\sqrt{11.9 \times 8.854 \times 10^{-14} \times 1.6 \times 10^{-19} \times 10^{17} \times 0.41} = 16.63 \times 10^{-7}$$

$$\psi_B = \frac{kT}{q} \ln\left(\frac{N_a}{n_i}\right)$$

$$= 26 \times 10^{-3} \ln\left(\frac{10^{17}}{1.5 \times 10^{10}}\right) = 0.41 \text{ V}$$

Substituting all values,

$$V_t = -0.93 + \frac{16.63 \times 10^{-8}}{1.726 \times 10^{-7}} + 0.82 = 0.854 \text{ V}$$

7.4 NON-IDEAL OXIDE CAPACITANCE-VOLTAGE (C-V) BEHAVIOUR

As discussed earlier, the fixed oxide charge $Q_f = q N_f$ is very close to Si-SiO$_2$ interface and causes shift in flat-band voltage, i.e.,

$$V_{FB} = V_{FB}^o - \left(\frac{q N_f}{C_{ox}}\right) \tag{7.43}$$

where V_{FB} is flat-band voltage in presence of fixed-oxide charges and V_{FB}^o is flat-band voltage of ideal MOS capacitor. Here N_f is taken as a positive charge.

The gate voltage V_G required for depletion and inversion is:

$$V_G = V_{FB} + \frac{|Q_{Si}'|}{C_{ox}} + \psi_s \tag{7.44}$$

where Q'_{Si} is charge per unit area in the semiconductor. Using Eq. (7.43) in Eq. (7.44), we have

$$V_G = \left[V^o_{FB} + \frac{|Q'_{Si}|}{C_{ox}} + \psi_s \right] - \frac{qN_f}{C_{ox}}$$

or

$$V_G = V^o_G - \frac{qN_f}{C_{ox}}$$

or

$$V_G - V^o_G = - \frac{qN_f}{C_{ox}} = \Delta V_G \tag{7.45}$$

where V^o_G is applied gate voltage on the ideal MOS capacitor for inversion and depletion.

Equation (7.45) simply tells that in the presence of fixed-oxide charges, the gate voltage is shifted from its ideal value by $(-qN_f/C_{ox})$. A similar expression can be obtained for the accumulation. This whole discussion is carried out for p-substrate. Therefore, the calculated C-V curve for real MOS is shifted by the change in flat-band voltage due to fixed oxide charges. Figures 7.17(a) and 7.17(b) show the C-V curve for a real MOS capacitor in presence of fixed oxide charges for two cases:

Case 1:   When Q_f is positive
Case 2:   When Q_f is negative

This curve is mainly observed at high frequency. A similar but opposite C-V curve shift at high frequency will be observed for n-type semiconductor as a substrate.

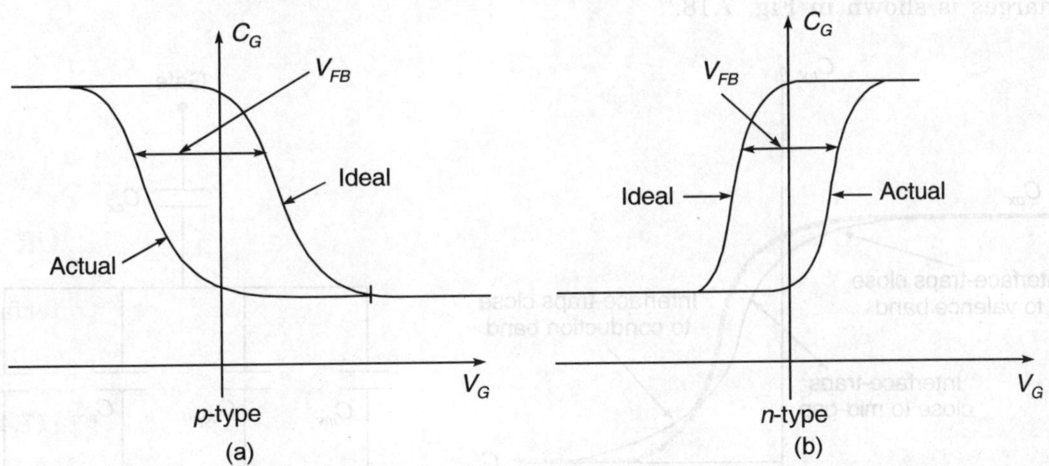

FIGURE 7.17   C-V curve for real MOS for (a) fixed positive charge, (b) fixed negative charge.

EXAMPLE 7.5

In a n-channel MOSFET, where do the electrons that form the inversion layer come from?

Solution: Due to vertical field nature electrons get attracted from substrate towards the interface and holes are pushed into the substrate. However, electrons could also come from a *n*-doped regions which act as source and drain respectively.

EXAMPLE 7.6

Calculate the boron dose F_B required to reduce the threshold voltage of *n*MOS from 1.1 V to 0.7 V. Assume that the implanted acceptors form a sheet of negative charge just below the Si surface. Take t_{ox} = 0.01 μm.

Solution: Using Eq. (7.42),

$$0.7 = 1.1 + \frac{qF_B}{C_{ox}}$$

or

$$F_B = \frac{(0.7 - 1.1)\,C_{ox}}{q} = -\frac{0.4 \times 10^{19}}{1.6} \times \frac{3.9 \times 8.854 \times 10^{-14}}{10^{-6}}$$

$$= -\frac{1}{4} \times 3.9 \times 8.854 \times 10^{11}$$

$$= 8.383 \times 10^{11}/\text{cm}^2$$

$$\sim 0.84 \times 10^{12}/\text{cm}^2$$

Note: 1. The modification of high-frequency (H_F) C-V curve due to interface trapped charges is shown in Fig. 7.18.

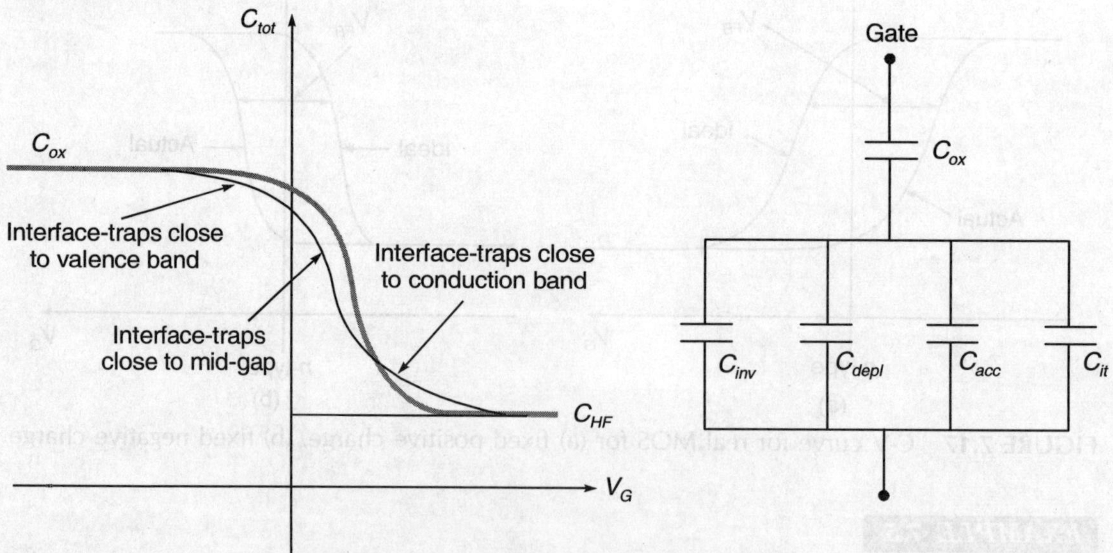

FIGURE 7.18 Modification of high-frequency C-V curve due to interface trapped charges.

2. In real MOS capacitor, the gate is usually made of heavily-doped poly-silicon. Although, poly-silicon is heavily doped, but there is always some finite depletion region which gives rise to poly-gate capacitance (C poly) that degrades the total lumped capacitor of MOS, as shown in Fig. 7.19.

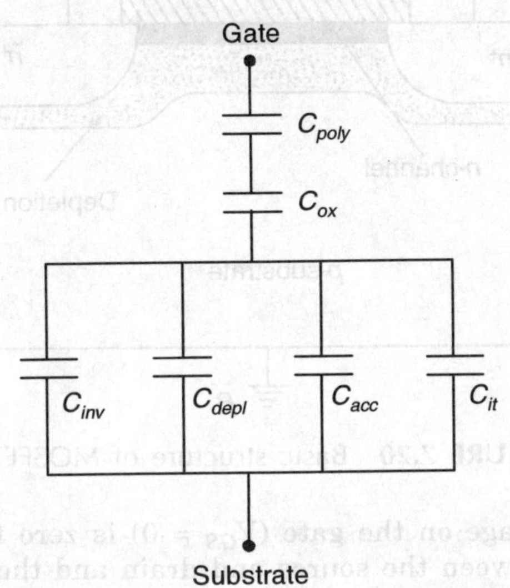

FIGURE 7.19 Total lumped capacitor of real MOS device.

7.5 LONG CHANNEL MOSFET

MOSFET stands for metal-oxide semiconductor field effect transistor which is the building block of VLSI circuits in microprocessors and dynamic memories. Because current in a MOSFET is mainly dominated by carriers of one polarity only, either electrons or holes, the MOSFET is usually referred to as a unipolar or majority carrier device. If current flows in the MOSFET is due to electrons, the MOSFET is termed as *n-MOS* or *n-channel MOSFET*. Similarly, if current is due to holes in the MOSFET, it is called *p-channel MOSFET* or *p-MOS*. In modern days, mainly *n*-channel MOSFETs are widely used because of high mobility of electrons. The basic structure of a MOSFET is shown in Fig. 7.20. MOSFET is a four-terminal device with the terminals designated as gate G, source S, drain D and body, bulk or substrate B.

For an *n*-channel MOSFET, the starting semiconductor material, which is called **substrate, body** or **bulk,** is *p*-type Si semiconductor. Two-highly doped *n*-regions (n^+) are formed either due to ion implantation or diffusion. These two regions serve as source S and drain D. The gate electrode is usually made of metal or heavily doped poly-silicon and is separated from the substrate by a thin silicon-dioxide film, called **gate oxide**. The gate oxide is usually formed by thermal oxidation of silicon. The surface region under the gate oxide between the source and drain region is called **channel**, and its formation is necessary for the current flow from the drain to source. This channel is due to inversion of substrate carriers near the interface of the Si-SiO$_2$

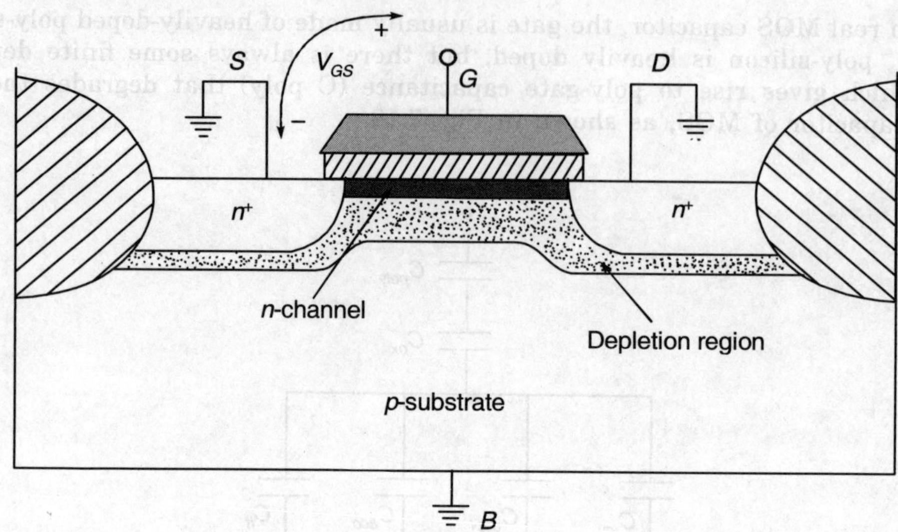

FIGURE 7.20   Basic structure of MOSFET.

layer. If the applied voltage on the gate ($V_{GS} = 0$) is zero for a n-MOS transistor, no channel is formed between the source and drain and the drain current I_D (from drain-to-source) is zero and MOS transistor is OFF.

For an applied negative bias on the gate, MOS transistor is either in accumulation or depletion and again results no current from source to drain.

Let us apply positive potential on the gate of n-MOS transistor, a vertical positive field E_{ox} will attract the electrons from the two heavily doped n-regions and from substrate to near the Si-SiO$_2$ interface and repel the existing holes at the interface into the bulk. If gate positive bias is continuously increasing then the population of electrons at the surface becomes more than the majority carrier holes. At some particular gate voltage, the total area near the interface will be filled with electrons. This is called **inversion** and a channel of electrons will form between source and drain. Since, total region from source to drain is occupied by the electrons near the surface and hence the resulting device is called **n-channel MOSFET**. MOSFET is a surface phenomenon. The minimum positive gate voltage required to achieve inversion is called **threshold voltage V_t**. Although the area near the interface from source to drain is occupied by the free electrons, but still no current will flow from drain to source due to zero bias between source and drain. A drain current will flow in the device, when a voltage is applied between source and drain V_{DS}. The drain to source voltage V_{DS} along the channel will exert a lateral field E_x on the electron. Therefore, electron will move from source to drain. This resultant lateral field causes electrons to move from source to drain and results in drain current due to the resultant field of longitudinal and lateral. Structurally, drain and source are interchangeably due to symmetric nature but electrially drain should be at higher positive potential than source. When current will flow in MOS device, the MOSFET is said to be ON. A MOSFET device, therefore, ideally operates like a switch. Since, the gate electrode is electrically insulated from channel, therefore, no current flows from channel to gate or vice versa, means $I_G = 0$ and provides zero leakage current. Therefore, for an ideal MOS device, the input impedance is infinite. It is of order of megaohm for real MOSFET. The channel is capacitively coupled to the gate via electric field in the oxide.

The conductance in the channel can be modulated by varying the transverse electric field, i.e., by changing applied potential at the gate electrode.

For a n-channel MOS, the biasing conditions $V_{GS} > V_t$ and $V_{DS} > 0$ results source/substrate and drain/substrate junctions in reverse biased.

A long channel MOSFET is assumed to have sufficiently long channel regions where *edge effects* can be neglected. The channel length is much larger than the depletion widths of the source or the drain. Also, the channel width W is assumed to be much larger than the depletion depth under the gate.

7.5.1 Gradual-Channel Approximation (GCA)

The long channel MOSFET behaviour can be analyzed by 1-dimensional model. One of the key assumptions in 1-D MOSFET model is the gradual channel approximation (GCA), which assumes that the variation of lateral electric field (i.e., along the channel) is much less than the variation in the transverse field (i.e. perpendicular to the channel). In other words, $dE_x/dx << dE_y/dy$. This allows one to reduce the Poisson's equation to the 1-D form by neglecting the transverse component of the field. The GCA is valid for most of the channel region except at the pinch-off point.

It is also assumed that both the hole current as well as generation and recombination current are negligible in the channel and current continuity is maintained along the channel. In other words, we can say that the drain-to-source current I_{DS} is same at any point along the channel.

Current-voltage characteristics of MOSFET

In this section, we will derive the I_{DS}-V_{DS} characteristics as a function of gate-source voltage V_{GS} for a n-channel MOSFET. Let us consider a drain voltage that is small enough so that it does not affect the inversion mobile charge Q_i in the inversion layer. The geometry of the n-MOS transistor is shown in Fig. 7.21 with the y direction perpendicular to the channel and x-direction along the channel. The coordinate $x = 0$ is taken at the source, whereas $x = L$ gives the coordinate of drain. The origin in the y-direction for $y = 0$ is at the Si-SiO$_2$ interface, while the interface between the inversion layer and the space-charge region in the substrate is $y = x_j$. Let us consider a small part of the channel of length dx as shown in Fig. 7.21. The incremental voltage $dV(x)$ across the incremental channel length dx is:

$$dV(x) = I_{DS}\, dR(x) \tag{7.46}$$

where I_{DS} is drain-to-source current and assumed to be constant as discussed in section 7.5.1, and $dR(x)$ is the incremental channel resistance which is given as:

$$dR(x) = \frac{\rho}{Wdy} \cdot dx \tag{7.47}$$

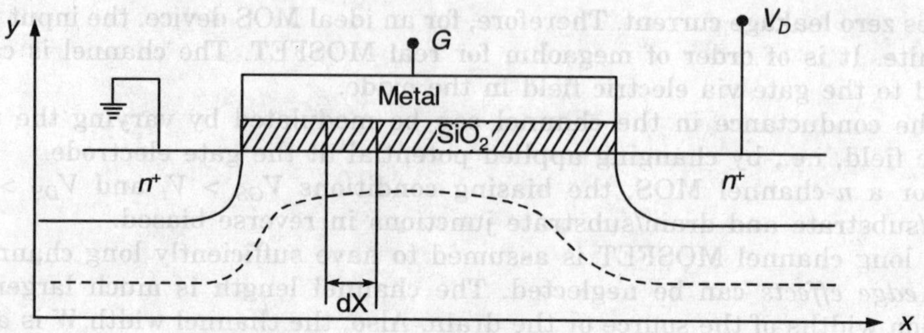

FIGURE 7.21 Geometry of *n*MOSFET.

where ρ is the resistivity of the semiconductor material and Wdy gives the cross-sectional area of the incremental volume.

Using $\sigma = 1/\rho = q\,\mu_{neff}\,n$ in Eq. (7.47), we get

$$dR(x) = \frac{dx}{(Wq\mu_{neff}n)\,dy} \qquad (7.48)$$

where μ_{neff} is effective mobility of electron at the surface and it is lower than the mobility in the bulk, due to additional surface scattering mechanisms and n is the electron concentration at the interface.

Assuming carrier mobility in the inversion layer is constant, Eq. (7.48) may be written as:

$$\frac{1}{dR(x)} = \frac{W}{dx}\,\mu_{neff} \int_0^{x_1} qn(x)\,dy \qquad (7.49)$$

where the integral represents the charge in the inversion layer per unit area and expressed as Q_i for MOSFET analysis. To avoid the sign complication, Eq. (7.49) can be reduced as:

$$dR(x) = \frac{dx}{W\mu_{neff}|Q_i|} \qquad (7.50)$$

Substituting Eq. (7.50) into Eq. (7.46) and rearranging the terms, we have

$$I_{DS} = W\mu_{neff}|Q_i|\,\frac{dV(x)}{dx}$$

or

$$I_{DS}\,dx = W\mu_{neff}|Q_i|\,dV(x) \qquad (7.51)$$

Using the boundary conditions,

$$V(x = 0) = 0$$

$$V(x = L) = V_{DS}$$

Neglecting built-in-potential of the *p-n* junction, the LHS of Eq. (7.51) is integrated from $x = 0$ to $x = L$ and RHS to its corresponding voltage, one has

$$I_{DS} \int_0^L dx = W \mu_{eff} \int_0^{V_{DS}} |Q_i| \, dV(x)$$

or

$$I_{DS} = \left(\frac{W}{L}\right) \mu_{eff} \int_0^{V_{DS}} |Q_i| \, dV(x) \tag{7.52}$$

Equation (7.52) is the basic equation for I_{DS}-V_{DS} curves for various MOSFET models. To solve the I_D equation, we have to assume charge-sheet approximation in which the inversion-layer thickness is treated as zero. According to this model, all the inversion charges are located at the silicon surface like a sheet of charge, and there is no potential drop or band bending across the inversion layer. After the onset of inversion, the surface potential is pinned at $\psi_s = 2\psi_B + V(x)$ and hence, bulk depletion charge density is:

$$Q_d = -qN_a W d_m = -\sqrt{2\varepsilon_{Si} q N_a [2\psi_B + V(x)]} \tag{7.53a}$$

Assuming real MOS transistor, the total charge density in the silicon is:

$$Q_S = -C_{ox}(V_{GS} - V_{FB} - \psi_s)$$

or

$$Q_S = -C_{ox}[V_{GS} - V_{FB} - 2\psi_B - V(x)] \tag{7.53b}$$

Hence, the inversion charge density $Q_i = Q_S - Q_d$ is given by using Eqs. (7.53a) and (7.53b), i.e.,

$$Q_i = -C_{ox}\left[V_{GS} - V_{FB} - 2\psi_B - V(x) + \frac{1}{C_{ox}}\sqrt{2\varepsilon_{Si} q N_a [2\psi_B + V(x)]}\right] \tag{7.54}$$

Neglecting the variation of Q_d with bias $V(x)$ and replacing,

$$V_{FB} + 2\psi_B - \frac{1}{C_{ox}}\sqrt{2\varepsilon_{Si} q N_a \psi_B}$$

by V_{to}, where V_{to} is threshold voltage when body and source are at the same potential. Equation (7.54) can be re-written as:

$$Q_i = -C_{ox}[V_{GS} - V_{to} - V(x)] \tag{7.55}$$

This equation describes the mobile charge in the channel at the point *x*.

Substituting Eq. (7.55) into Eq. (7.52), we obtain

$$I_{DS} = \left(\frac{W}{L}\right) \mu_{neff} \int_0^{V_{DS}} C_{ox} \left[V_{GS} - V_{to} - V(x)\right] dV(x)$$

or

$$I_{DS} = \left(\frac{W}{L}\right) \mu_{neff} \, C_{ox} \int_0^{V_{DS}} \left[V_{GS} - V_{to} - V(x)\right] dV(x)$$

or

$$I_{DS} = \left(\frac{W}{L}\right) \mu_{neff} \, C_{ox} \left[(V_{GS} - V_{to})V_{DS} - \frac{V_{DS}^2}{2}\right] \tag{7.56}$$

Substituting $k_N = (W/L) \, \mu_{neff} \, C_{ox}$, defined as the transconductance parameter of the n-channel MOSFET, we get;

$$I_{DS} = k_N \left[(V_{GS} - V_{to})V_{DS} - \frac{V_{DS}^2}{2}\right] \tag{7.57}$$

In this analysis, we have assumed that depletion charge Q_d does not vary with drain voltage although $Q_d(x)$ varies considerably when V_{DS} is applied, to reflect the variation in $V(x)$. Let us consider some special cases:

Case I: If $V_{GS} = 0$ or $V_{GS} < V_{to}$

No inversion occurs at the surface, the drain current closes to zero and the MOSFET is in OFF state. This region of operation is called **cut-off region**.

Case II: $V_{GS} \geq V_{to}$

The inversion occurs at the surface and conduction channel is formed.

(a) Assuming $V_{DS} = 0$ V, it is observed from Eq. (7.57), that no current will flow from drain-to-source. The source and drain are connected by a channel through two reversed biased p-n junctions. The induced channel is surrounded by a depletion region.

(b) Small positive V_{DS}, i.e., $V_{DS} < (V_{GS} - V_{to})$ is applied, then from Eq. (7.57),

$$I_{DS} = k_N(V_{GS} - V_{to}) \, V_{DS} \tag{7.58a}$$

For a constant V_{GS} and threshold voltage V_{to}, the drain-source current is linearly related to the drain voltage, i.e.,

$$V_{DS} = \left[\frac{1}{k_n(V_{GS} - V_{to})}\right] I_{DS}$$

or

$$V_{DS} = R_{ON}I_{DS} \tag{7.58b}$$

This is just a Ohm's law and hence, for $V_{GS} \geq V_{to}$ and $V_{DS} < (V_{GS} - V_{to})$, the n-MOS transistor operates in the linear region or ohmic region or triode region. So, the channel behaves like a simple resistor. This relation shows that MOSFET behaves as a resistor when it is biased in the linear region for low V_{DS}.

In this case, the drain end in the channel becomes less inverted than the source end as shown in Fig. 7.22(b). Therefore the induced layer carriers correspondingly decreases towards the drain and depletion region surrounding the channel and the drain region become wider than the source end.

(c) For $V_{DS} = V_{Dsat} = (V_{GS} - V_t)$

From Eq. (7.57),

$$I_{DS} = k_N \left[(V_{GS} - V_t)(V_{GS} - V_{to}) - \frac{(V_{GS} - V_{to})^2}{2} \right]$$

$$= \frac{k_N}{2} \left[(V_{GS} - V_{to}) \right]^2 = I_{Dsat} \tag{7.59}$$

Equation (7.59) shows that the drain-source current is independent of the V_{DS}, and this current is called **drain saturation current**. This region of operation $V_{GS} > V_{to}$ and $V_{DS} \geq (V_{GS} - V_{to})$ is called **saturation region**. The voltage $V_{DS} = V_{Dsat}$, called **saturation drain voltage** and at this voltage, the inverted surface near the drain side vanishes as shown in Fig. 7.22(c). This is called **pinch-off condition**.

(d) Large positive V_{DS}, i.e., $V_{DS} > V_{Dsat}$. As the V_{DS} increases beyond V_{Dsat}, the effective channel length decreases, as shown in Fig. 7.22(d). This is called **channel length modulation (CLM)** and current I_{DS} increases slightly.

The excess voltage $(V_{DS} - V_{Dsat})$ is dropped across the depletion region and compensates for the increased resistance.

From the current-voltage relations, it is clear that (W/L) ratio is a primary design variable and by controlling the value of (W/L) one can control the drain current at the desired level. Therefore, in VLSI design, designer has two degrees of freedom (W, L). The channel length modulation, affects the drain saturation current more aggresively for short channel devices, whereas for long channel device their effect is ignored.

Since, the inversion charge density is given as:

$$Q_i = - C_{ox}[V_{GS} - V_t - V(x)]$$

At the saturation point, $V(x) = V_{DS} = V_{Dsat} = V_{GS} - V_{to}$ and hence $Q_i = 0$. Therefore, one can conclude that the surface channel vanishes at the drain end of the channel when saturation occurs. This is called **pinch-off**. Beyond the pinch-off, current remains constant because for $V_{DS} > V_{Dsat}$, the voltage at the pinch-off point remain at V_{Dsat}.

Channel Length Modulation (CLM):
From Fig. 7.22(d), it is clear that

$$L_{eff} = L - \Delta L$$

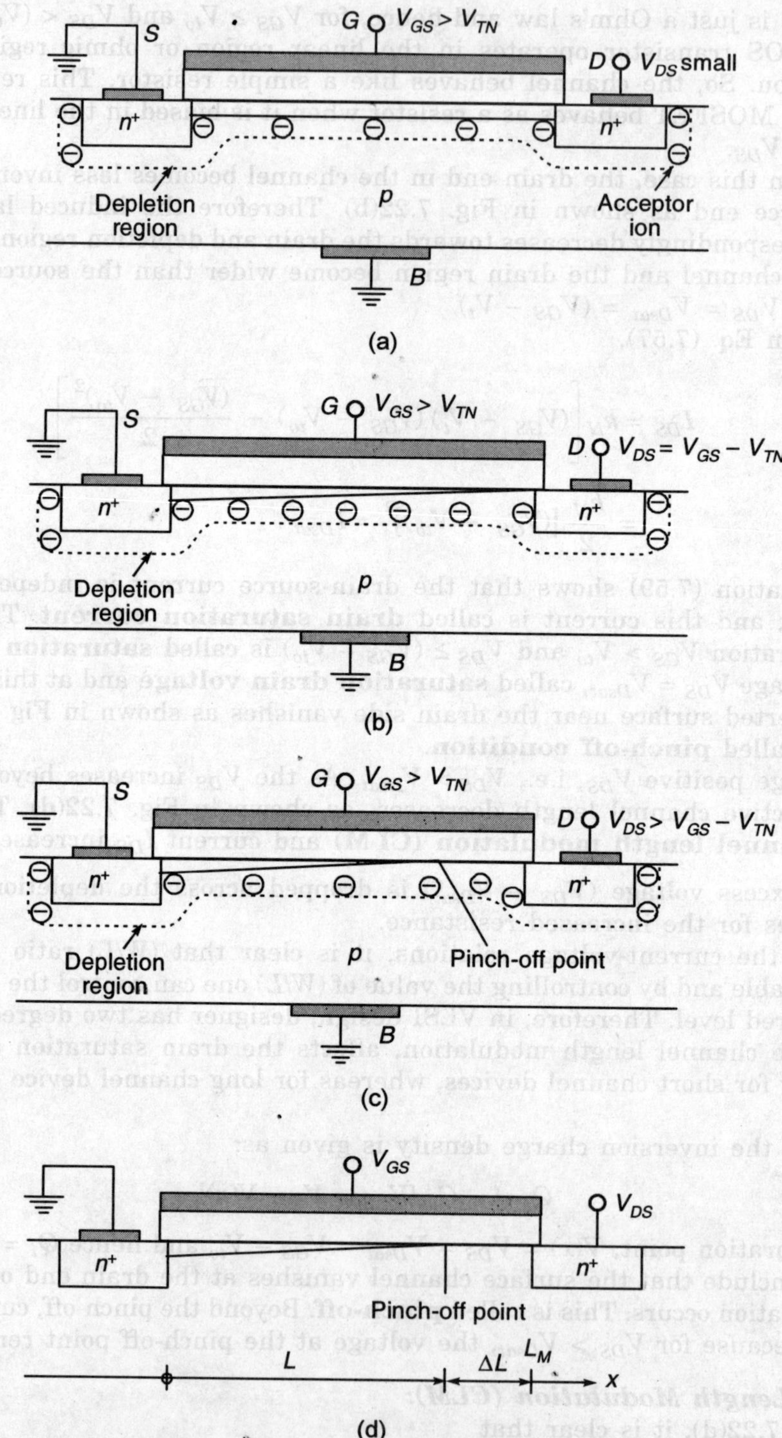

FIGURE 7.22 MOSFET under different bias voltages.

where L_{eff} is effective channel length available before saturation and ΔL is the channel length available after pinch-off.

After pinch-off, drain-saturation current,

$$I_{Dsat1} = \frac{\mu_n C_{ox}}{2} \left(\frac{W}{L_{eff}} \right) (V_{GS} - V_t)$$

$$I_{Dsat1} = \frac{\mu_n C_{ox}}{2} \left(\frac{W}{L - \Delta L} \right) (V_{GS} - V_t)$$

or

$$I_{Dsat1} = \frac{\mu_n C_{ox}}{2} \left(\frac{W}{L} \right) \left(1 - \frac{\Delta L}{L} \right)^{-1} (V_{GS} - V_t)$$

$$= \frac{I_{Dsat}}{\left[1 - \dfrac{\Delta L}{L} \right]}$$

For $\Delta L \ll L$,

$$I_{Dsat1} = \left(1 + \frac{\Delta L}{L} \right) I_{Dsat} \tag{7.60}$$

where I_{Dsat} is the saturation current without channel length modulation.

Introducing an empirical relation,

$$1 + \frac{\Delta L}{L} = 1 + \lambda V_{DS} \tag{7.61}$$

and using Eq. (7.61) in Eq. (7.60), one has

$$I_{Dsat1} = I_{Dsat} (1 + \lambda V_{DS}) \tag{7.62}$$

where λ is defined as the channel length modulation parameter and representing a small influence of drain voltage on drain current. This is true for a real MOSFET shown in Fig. 7.24.

The parameter λ can be determined by extrapolating the $I_{DS} - V_{DS}$ curves backward and the point at which all curves meet is represented by voltage V_A which is called **early voltage** and given as:

$$V_A = \frac{1}{\lambda} \tag{7.63}$$

The ideal current-voltage characteristics for n-MOS enhancement transistor is shown in Fig. 7.23(a) where we have ignored the channel length modulation due to infinite input impedance of the transistor

By plotting I_{DS} versus V_{GS} (for a given small V_{DS}), (as shown in Fig. 7.23(b), known as Input characteristics), the threshold voltage V_t can be deduced from the linearly extrapolated value at the V_{GS} axis. In reality, this extrapolated threshold voltage is slightly larger than the actual threshold voltage.

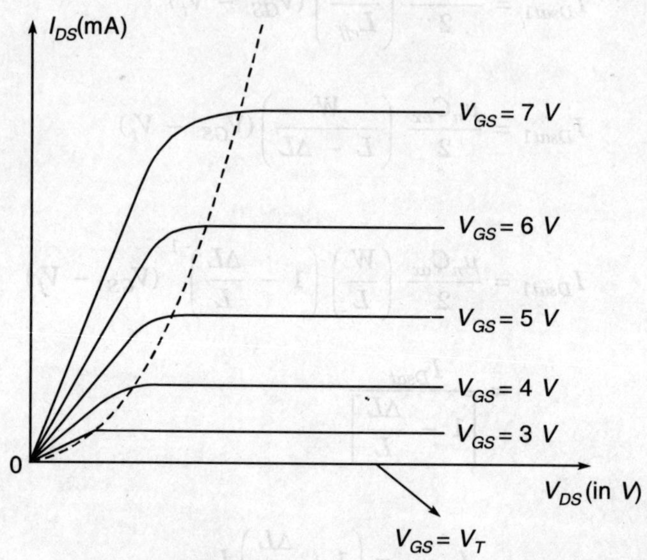

FIGURE 7.23(a)   Voltage-current curve for ideal *n*MOS.

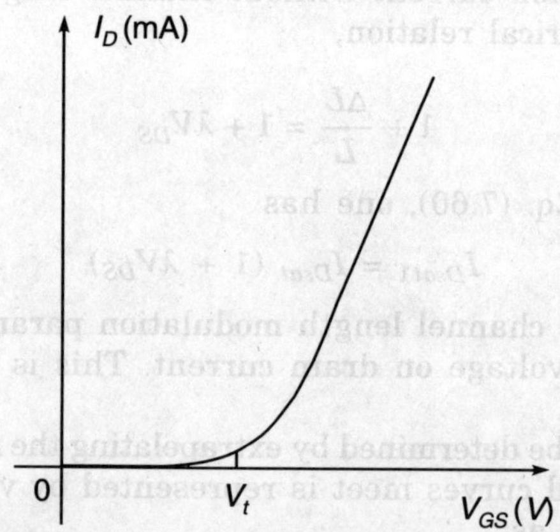

FIGURE 7.23(b)   Input characteristics of *n*MOSFET.

In linear region, using Eq. (7.58a), the channel conductance g_d and transconductance g_m are given as:

$$g_d = \left. \frac{\partial I_{DS}}{\partial V_{DS}} \right|_{V_{GS} = \text{constant}} \cong k_N(V_{GS} - V_{to})$$

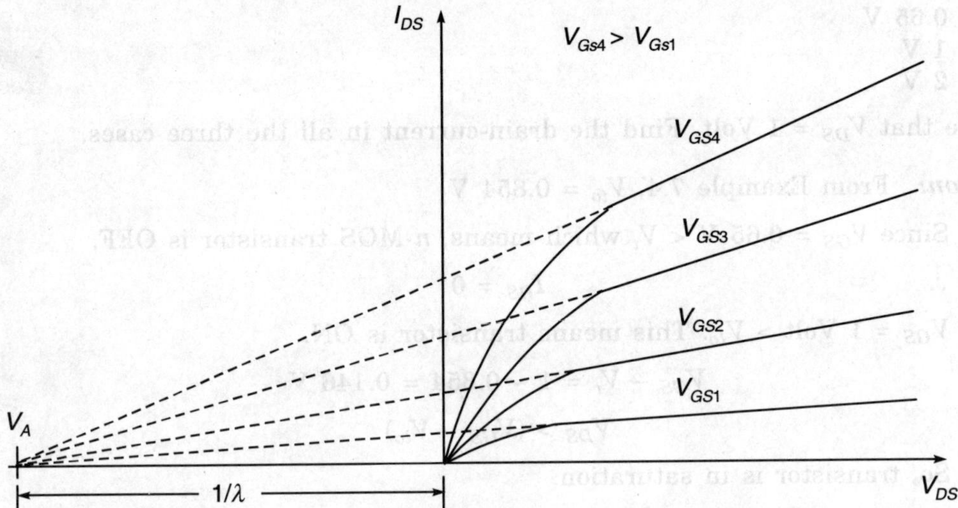

FIGURE 7.24   Voltage-current characteristics for real MOSFET.

and

$$g_m = \left.\frac{\partial I_{DS}}{\partial V_{GS}}\right|_{V_{DS}\,=\,\text{constant}} \cong k_N V_{DS} \tag{7.64}$$

Similarly, the transconductance g_m in saturation region can be obtained by using Eq. (7.59), i.e.,

$$(g_{msat}) = \left.\frac{\partial I_{DS}}{\partial V_{GS}}\right|_{V_{DS}\,=\,\text{constant}} = k_N (V_{GS} - V_{to}) \tag{7.65}$$

The output resistance is given as:

$$r_0 = \left(\frac{\partial I_D}{\partial V_{DS}}\right)^{-1} = (g_d)^{-1}$$

$$= \frac{|V_A|}{I_{Dsat}} = \frac{1}{\lambda I_{Dsat}}$$

EXAMPLE 7.7

Consider a n-MOS transistor formed with a substrate doping of 10^{17}/cm^3 and aluminium gate ($\phi_M = 4.1$ V). The gate oxide thickness is 20 nm and it is assumed that there is no fixed charge either in the oxide or at the Si-SiO$_2$ interface. The MOS device biased with the gate voltage w.r.t. source is 0.65. Take = μn_{eff} = 1000 cm^2/V-sec; W = 100 μm, L = 5 μm.

   (i) 0.65 V

  (ii) 1 V

 (iii) 2 V

Assume that $V_{DS} = 1$ Volt. Find the drain-current in all the three cases.

Solution: From Example 7.4, $V_{to} = 0.854$ V

  (i) Since $V_{GS} = 0.65$ V $< V_t$ which means, n-MOS transistor is OFF.

$$\therefore \qquad\qquad\qquad\qquad I_{DS} = 0$$

 (ii) $V_{GS} = 1$ Volt $> V_{to}$. This means transistor is ON.

$$V_{GS} - V_t = 1 - 0.854 = 0.146 \text{ V}$$

$$V_{DS} > (V_{GS} - V_{to})$$

So, transistor is in saturation.

$$I_{DS} = \frac{\mu_{neff} C_{ox}}{2}\left(\frac{W}{L}\right)(V_{GS} - V_{to})^2$$

$$= \frac{1000 \times 3.9 \times 8.854 \times 10^{-14}}{2 \times 20 \times 10^{-7}}\left(\frac{100}{5}\right)(1 - 0.854)^2$$

$$= \frac{3.9 \times 8.854 \times 10^{-4}}{40}(20)(0.146)^2$$

$$= 0.37 \times 10^{-4} \text{ A}$$

$$= 37 \text{ }\mu\text{A}$$

 (iii) $V_{GS} = 2$ V > 0.854 V

      Transistor is ON.

$$V_{GS} - V_{to} = 2 - 0.854 = 1.146 \text{ V}$$

$$(V_{GS} - V_{to}) > V_{DS}$$

Transistor is in linear region.

$$I_{DS} = \frac{1000 \times 3.9 \times 8.854 \times 10^{-14}}{20 \times 10^{-7}}\left(\frac{100}{5}\right)\left[(1.146) \times 1 - \frac{1}{2}\right]$$

$$= 34.53 \times 10^{-4}\,[0.646]$$

$$= 2.231 \text{ mA}$$

Note: At some position x along the source-drain channel.

Current in channel at x = Mobility $\times$ Electric charges at x $\times$ Electric field at x

Classification of MOSFET

MOSFETs are classified mainly into two categories:

Enhancement-mode (E-MOSFET).   It is most widely used MOSFET where initially at zero gate bias (for n-MOS) there is no channel exists between drain and source. Therefore, E-MOSFET are normally OFF. Channel is formed after inversion only. The threshold voltage for n-MOS transistor is always positive, whereas the threshold voltage of p-MOS transistor is negative because to create a channel between source and drain, i.e., for inversion, one should apply positive gate voltage in n-MOS transistor similar argument is true for p-MOS enhancement-mode. The operation of E nMOS transistor is same as discussed earlier. The symbol for n-MOS enhancement and p-MOS enhancement transistors are shown in Fig. 7.25.

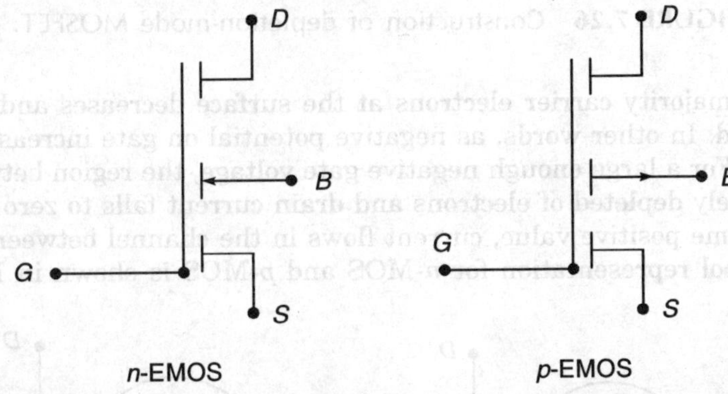

FIGURE 7.25 Symbol representation of enhancement MOSFETs.

The broken line indicates that channel is initially not formed. The gap between gate terminal and channel shows that the channel is isolated from gate by an insulating material. The I_{DS}–V_{DS} curve for E nMOS is shown in Fig. 7.23.

Depletion-mode MOSFET (D-MOSFET).   The construction of an n-channel depletion MOSFET (D-MOSFET) is shown in Fig. 7.26. As clear from figure, channel (n-channel) is already exist in the transistor, i.e., the region near the interface is occupied by electrons from source to drain at the time of fabrication. Therefore, D-MOSFETs are generally ON and channel doping is opposite to the substrate doping, i.e., n-MOS depletion mode transistor, surface is doped with donor impurities and substrate is doped with acceptor impurities (p-type) and reverse is true for p-MOS depletion mode transistor.

A negative voltage is applied on the gate of n-MOS depletion mode transistor. The negative field across the oxide layer repels the electrons from the surface into the bulk, i.e., electrons are depleted from the surface and hence the transistor is called *Depletion-mode transistor*. Similar argument is true for depletion-mode p-MOS. Therefore, the threshold voltage of n-MOS depletion mode is negative and p-MOS depletion mode is positive. Due to applied negative voltage on the gate of n-MOS, the

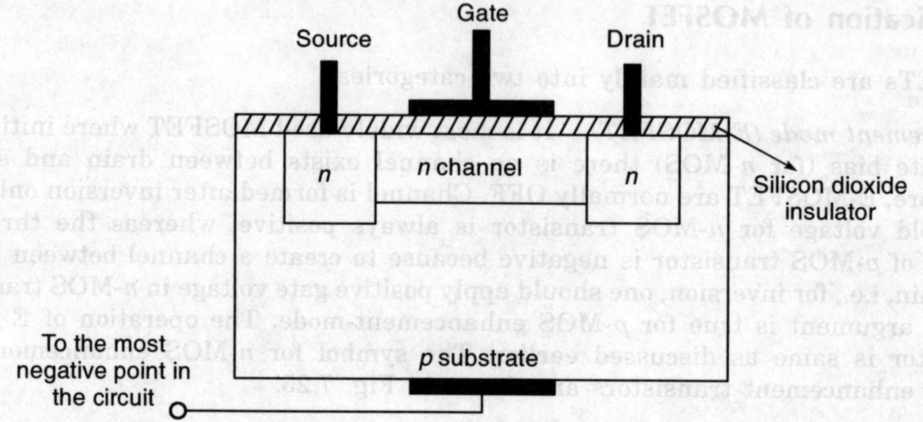

FIGURE 7.26 Construction of depletion-mode MOSFET.

concentration of majority carrier electrons at the surface decreases and hence channel current is reduced. In other words, as negative potential on gate increases, the channel starts depleting. For a large enough negative gate voltage, the region between drain and source is completely depleted of electrons and drain current falls to zero. When $V_{GS} = 0$ V and V_{DS} has some positive value, current flows in the channel between the drain and source. The symbol representation for n-MOS and p-MOS is shown in Fig. 7.27.

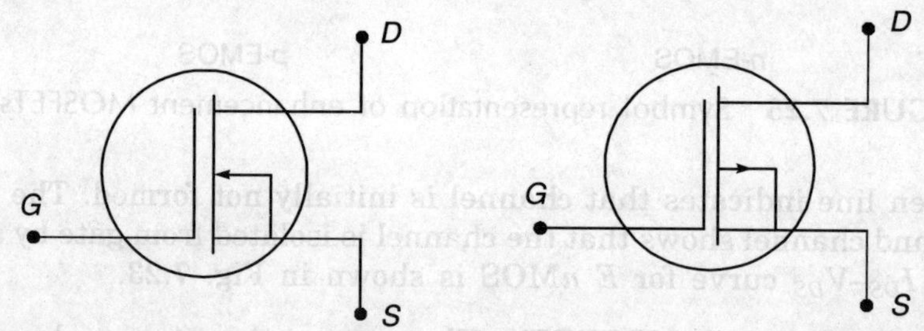

FIGURE 7.27 Symbol representation of depletion mode MOSFETs.

In order to avoid the body-effect, the bulk is generally connected to the source, i.e., $V_{BS} = 0$. The E-MOSFET can be operated only in the enhancement-mode, whereas the D-MOSFET can be operated either in enhancement or depletion-mode. For this reason, the D-MOSFET is also called **DE-MOSFET** (Depletion Enhancement MOSFET).

A JFET can be visualized as a depletion-mode transistor. The expressions derived earlier can be applied to the enhancement-mode transistor, whereas the current expression of JFET can be used for depletion-mode transistor.

For the proper operation of an enhancement-mode transistor,

- Source should be at most negative potential and drain should be at most positive potential for n-MOS, i.e., drain should be at higher potential than the source.

- Source should be at most positive potential and drain should be at most negative potential for *p*-MOS, i.e., source is at higher potential than the drain.

Normally, the EMOS is represented as three-terminal device by the symbol (assuming bulk is connected to source) as shown in Fig. 7.28.

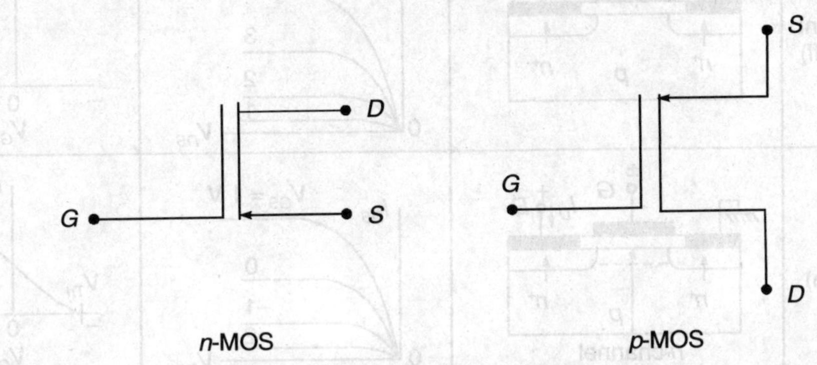

FIGURE 7.28 Three terminals symbol of enhancement mode MOSFETs.

The output characteristic, input or transfer characteristic and cross-section for enhancement-mode as well as depletion-mode for *n*-MOS and *p*-MOS are shown in Fig. 7.29.

7.6 SUBTHRESHOLD CHARACTERISTICS

When the gate voltage on the MOSFET is slightly lower than the threshold voltage so that semiconductor surface is weakly inverted then the MOSFET is said to be biased in the subthreshold region and corresponding drain current is called **subthreshold current**. The various operating regions of a MOSFET are shown in Fig. 7.30. Subthreshold region reflects how fast the MOSFET can switch.

As shown in Fig. 7.31, the drain current on a linear scale appears to approach zero immediately below the threshold voltage. On the other hand, on a logarithmic scale, the descending drain current remains at non-negligible levels for several tenths of a volt below V_t. This is due to fact that the inversion charge density does not drop to zero abruptly. Rather, it follows an exponential dependence on the surface potential or gate voltage.

The subthreshold current is dominated by diffusion current component. Therefore,

$$I_{DS(x)} = WD_n \frac{dQ_i}{dx} \tag{7.66}$$

where D_n is diffusion coefficient $= \mu_{eff}\, kT/q$.

Equation (7.66) further simplified as:

$$I_{DS(x)} = W\mu_{eff} \frac{kT}{q} \frac{dQ_i}{dx} \tag{7.67}$$

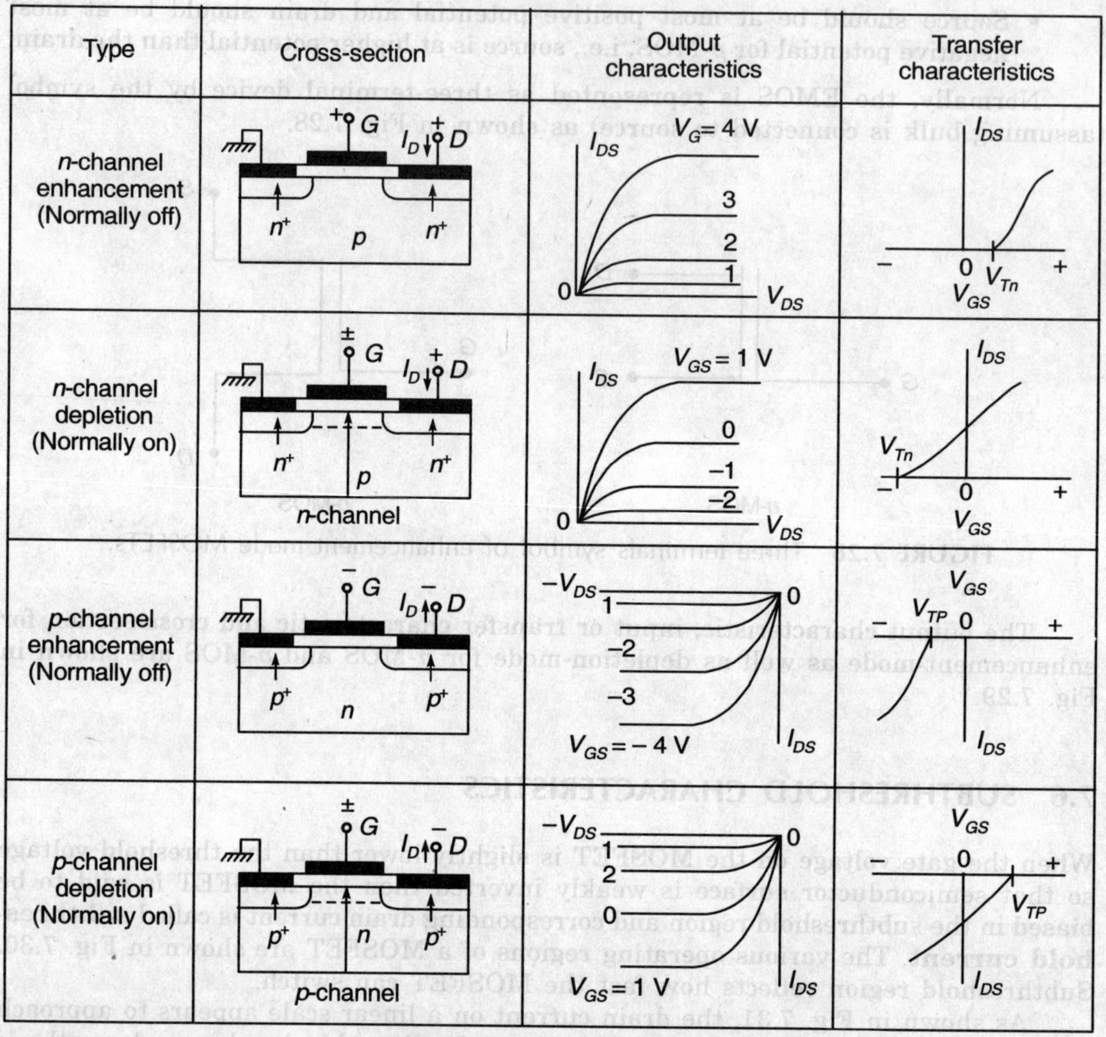

Type	Cross-section	Output characteristics	Transfer characteristics
n-channel enhancement (Normally off)	G, I_D, D; n^+ p n^+	I_{DS} vs V_{DS}; $V_G = 4$ V, 3, 2, 1	I_{DS} vs V_{GS}; 0, V_{Tn}
n-channel depletion (Normally on)	G, I_D, D; n^+ p n^+; *n*-channel	I_{DS} vs V_{DS}; $V_{GS} = 1$ V, 0, -1, -2	I_{DS} vs V_{GS}; V_{Tn}, 0
p-channel enhancement (Normally off)	G, I_D, D; p^+ n p^+	$-V_{DS}$ vs I_{DS}; 1, -2, -3, $V_{GS} = -4$ V	V_{GS}; V_{TP}, 0; I_{DS}
p-channel depletion (Normally on)	G, I_D, D; p^+ n p^+; *p*-channel	$-V_{DS}$ vs I_{DS}; 1, 2, $V_{GS} = 1$ V	V_{GS}; 0, V_{TP}; I_{DS}

FIGURE 7.29 Cross-section, output characteristic and input characteristic of enhancement and depletion mode *n*MOSFET and *p*MOSFET.

Integrating Eq. (7.67) from $x = 0$ to $x = L$,

$$I_{DS} = \mu_{eff}\, \frac{W}{L}\, \frac{kT}{q} \int_{Q_i(x=0)}^{Q_i(x=L)} dQ_i$$

or

$$I_{DS} = \mu_{eff}\left(\frac{W}{L}\right)\frac{kT}{q}\,[Q_i(x=L) - Q_i(x=0)] \tag{7.68}$$

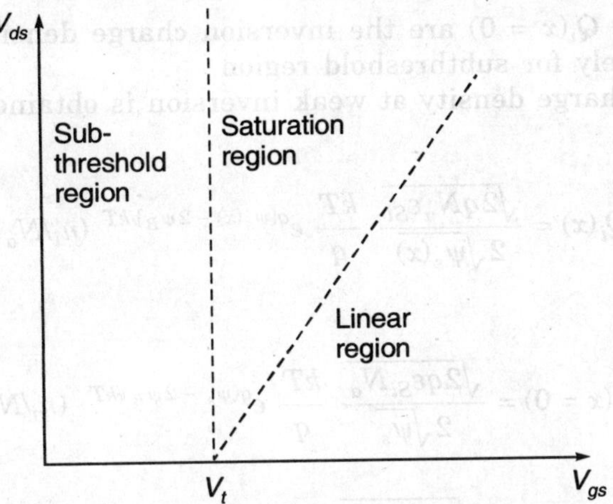

FIGURE 7.30 Various operating regions of MOSFET.

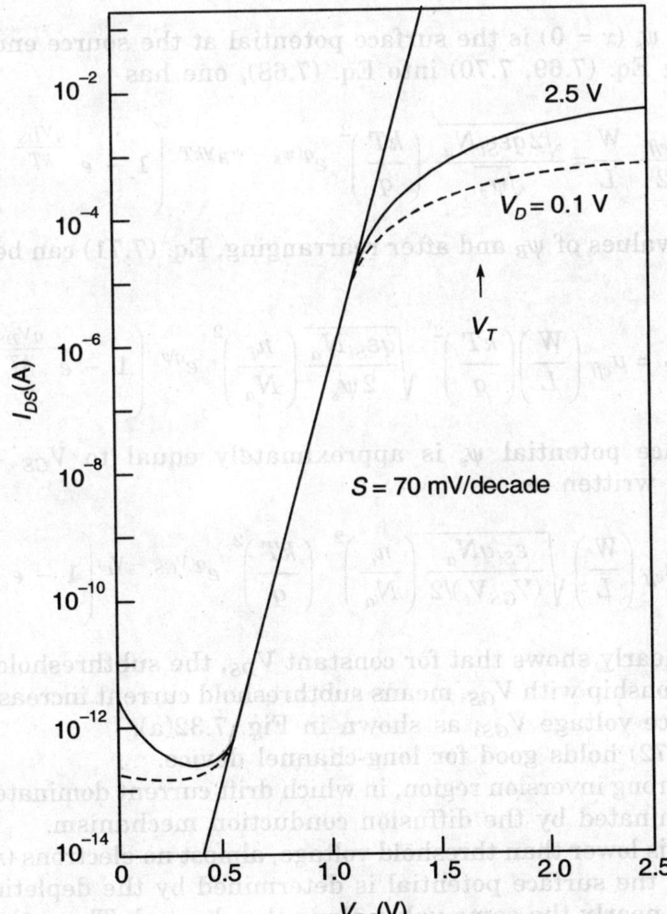

FIGURE 7.31 I_D-V_{GS} characteristic of MOSFET on log scale.

where $Q_i(x = L)$ and $Q_i(x = 0)$ are the inversion charge densities at the drain and source end respectively for subthreshold region.

The inversion charge density at weak inversion is obtained by MOS C-V curve and is given as:

$$Q_i(x) = \frac{\sqrt{2qN_a\varepsilon_{Si}}}{2\sqrt{\psi_s(x)}} \; \frac{kT}{q} \; e^{q[\psi_s(x) - 2\psi_B]/kT} \; (n_i/N_a)^2$$

Hence

$$Q_i(x = 0) = \frac{\sqrt{2q\varepsilon_{Si}N_a}}{2\sqrt{\psi_s}} \; \frac{kT}{q} \; e^{q(\psi_s - 2\psi_B)/kT} \; (n_i/N_a)^2 \tag{7.69}$$

$$Q_i(x = L) = \frac{\sqrt{2q\varepsilon_{Si}N_a}}{2\sqrt{\psi_s}} \; \frac{kT}{q} \; e^{q(\psi_s - 2\psi_B - V_{DS})/kT} \; (n_i/N_a)^2 \tag{7.70}$$

where $\psi_s = \psi_{so} = \psi_s(x = 0)$ is the surface potential at the source end of the channel. Substituting Eq. (7.69, 7.70) into Eq. (7.68), one has

$$I_{DS} = \frac{\mu_{eff}}{2} \; \frac{W}{L} = \frac{\sqrt{2q\varepsilon_{Si}N_a}}{\sqrt{\psi_s}} \left(\frac{kT}{q}\right)^2 e^{q(\psi_s - \psi_B)/kT} \left(1 - e^{-\frac{qV_{DS}}{kT}}\right)\left(\frac{n_i}{N_a}\right)^2 \tag{7.71}$$

Substituting the values of ψ_B and after rearranging, Eq. (7.71) can be further written as:

$$I_{DS} = \mu_{eff}\left(\frac{W}{L}\right)\left(\frac{kT}{q}\right)^2 \sqrt{\frac{q\varepsilon_{Si}N_a}{2\psi_s}}\left(\frac{n_i}{N_a}\right)^2 e^{q\psi_s}\left(1 - e^{-\frac{qV_{DS}}{kT}}\right) \tag{7.72}$$

Since, surface potential ψ_s is approximately equal to $V_{GS} - V_t$ and hence Eq. (7.72) can be written as:

$$I_{DS} = \mu_{eff}\left(\frac{W}{L}\right)\sqrt{\frac{\varepsilon_{Si}qN_a}{(V_{GS}V_t)/2}}\left(\frac{n_i}{N_a}\right)^2 \left(\frac{kT}{q}\right)^2 e^{q(V_{GS} - V_t)}\left(1 - e^{-\frac{qV_{DS}}{kT}}\right) \tag{7.73}$$

Equation (7.73) clearly shows that for constant V_{DS}, the subthreshold current has an exponential relationship with V_{GS}, means subthreshold current increases exponentially with gate-to-source voltage V_{GS}; as shown in Fig. 7.32(a).

Equation (7.72) holds good for long-channel device.

Unlike the strong inversion region, in which drift current dominates, sub-threshold conduction is dominated by the diffusion conduction mechanism.

Because V_{GS} is lower than threshold voltage, almost no electrons (n-MOS) inverted at the surface, so the surface potential is determined by the depletion region under the gate, and has nearly the same value along the channel. Thus, electric field along

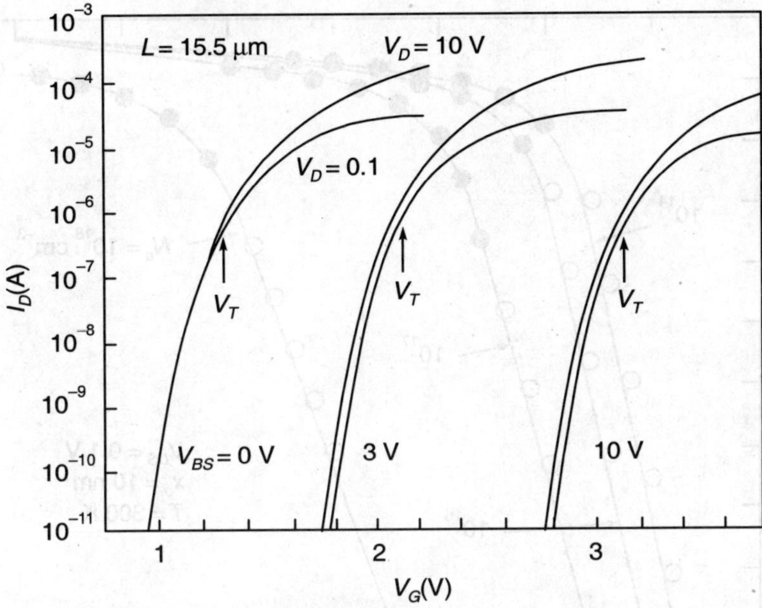

FIGURE 7.32(a) Variation of subthreshold current with gate voltage at different bias conditions.

the channel direction approaching zero, which makes almost no drift current in subthreshold region.

7.6.1 Subthreshold Swing (S)

Subthreshold swing is an important device characteristics. It is defined as the change in gate voltage V_{GS} required to reduce subthreshold current I_{DS} by one decade, i.e.,

$$S = \frac{dV_{GS}}{d(\log I_{DS})} \tag{7.74}$$

After detailed calculation, subthreshold swing S is given as:

$$S \cong 2.3 \frac{kT}{q} \eta \tag{7.75}$$

where $\eta = (1 + C_{dm}/C_{ox})$ with C_{dm} is the bulk depletion capacitance at $\psi_s = 2\psi_B$ and C_{ox} is gate-oxide capacitor per unit area. Subthreshold slope S is typically 60–100 mV/decade. Smaller the value of S, better turn-ON performance of the device. For a VLSI circuits, subthreshold slope is desirable for the case of switching the transistor current off. However, except for a slight dependence on bulk doping concentration through C_{dm}, the subthreshold slope is rather insensitive to device parameters. Lower substrate doping can have a thick depletion layer, which results in lower depletion capacitance C_{dm} and hence smaller value of S. This lower value of S reflects that it is easier for the gate electrode to control the lower doping substrate as shown in Fig. 7.32(b).

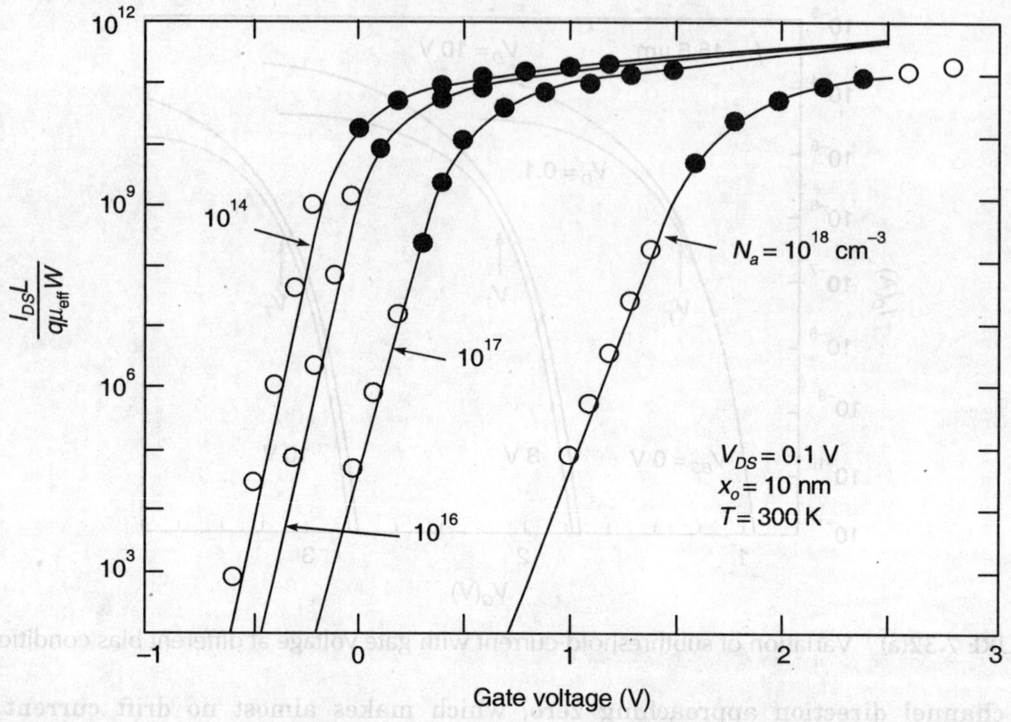

FIGURE 7.32(b) Variation of subthreshold current with gate voltage at different doping concentration.

If there is significant trap density then

$$\eta = 1 + \frac{C_{dm}}{C_{ox}} + \frac{C_{it}}{C_{ox}} \tag{7.76}$$

as shown in Fig. 7.33.

From Eq. (7.76), it is clear that

(i) If t_{ox} reduces, C_{ox} will increase. The increase in C_{ox} gives lower value of η and we have sharper subthreshold swing S. In other words,

$$t_{ox} \downarrow \; \rightarrow \; C_{ox} \uparrow \; \eta \downarrow \; \rightarrow \; S \downarrow \; \rightarrow \; \text{better turn-0}$$

(ii) $N_a \uparrow \; \rightarrow \; C_{dm} \uparrow \; \rightarrow \; \eta \uparrow \; \rightarrow \; S \uparrow \; \rightarrow \;$ softer subthreshold slope

(iii) $|V_{SB}|$, (i.e., substrate bias) $\uparrow \; \rightarrow \; C_{dm} \downarrow \; \rightarrow \; \eta \downarrow \; \rightarrow \; S \downarrow \; \rightarrow \;$ sharper subthreshold as shown in Fig. 7.34

(iv) T(temperature) $\uparrow \; \rightarrow \; S \uparrow \; \rightarrow \;$ softer subthreshold slope

Note: At room temperature (300 K), the ideal limit of S is 60 mV/decade.

Normally, most of the devices always operate in a higher temperature ambient due to heat dissipation. Higher temperature means larger value of S than the room

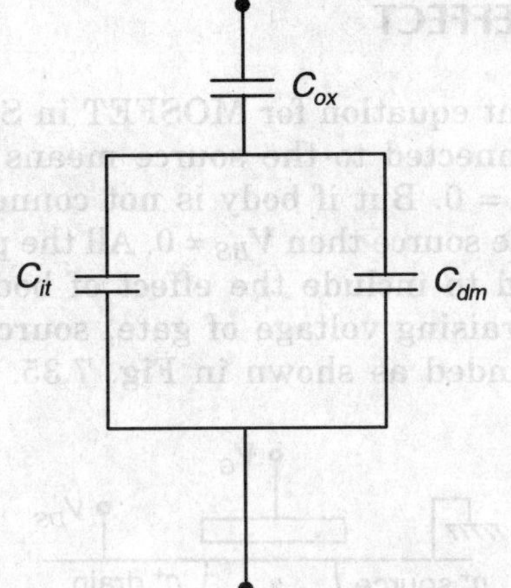

FIGURE 7.33   Different parasitic capacitances in MOSFET.

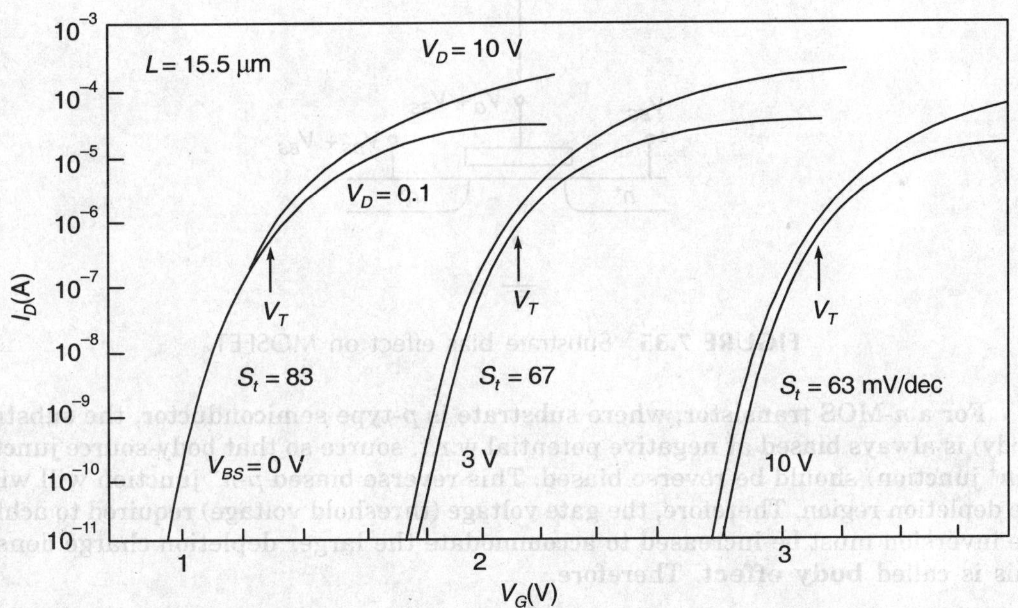

FIGURE 7.34   Subthreshold current variation with gate voltage for different subthreshold slope.

temperature value. For improving the switching action of MOS, it is required to reduce the value of S which can be achieved by lowering the temperature.

The subthreshold region is particularly important when the MOSFET is used as a low voltage, low-power device, such as device in a digital logic and memory applications because in the subthreshold region only a small current flows.

7.7 SUBSTRATE BIAS EFFECT

In deriving the drain current equation for MOSFET in Section 7.5, we have assumed that body (substrate) is connected to the source means body and source are at the same potential so that $V_{BS} = 0$. But if body is not connected to source or body is at some other potential than the source then $V_{BS} \neq 0$. All the previously derived MOSFET's equations must be modified to include the effect of body because applying $-V_{BS}$ to substrate is equivalent to raising voltage of gate, source and drain by $+ V_{BS}$ while keeping the substrate grounded as shown in Fig. 7.35.

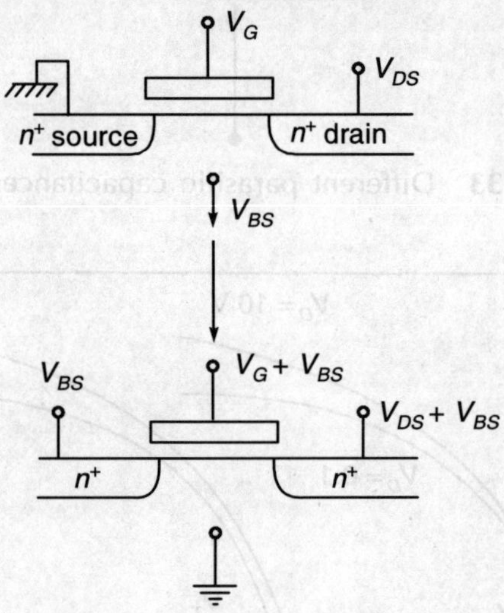

FIGURE 7.35 Substrate bias effect on MOSFET.

For a n-MOS transistor, where substrate is p-type semiconductor, the substrate (body) is always biased at negative potential w.r.t. source so that body-source junction (p-n^+ junction) should be reverse biased. This reverse biased p-n^+ junction will widen the depletion region. Therefore, the gate voltage (threshold voltage) required to achieve the inversion must be increased to accommodate the larger depletion charge density. This is called **body effect**. Therefore,

$$Q'_d = - [2\varepsilon_{Si}qN_a(2\psi_B - V_{BS})]^{1/2} \tag{7.77}$$

The change in threshold voltage due to substrate bias is:

$$\Delta V_t = \frac{\sqrt{2\varepsilon_{Si}qN_a}}{C_{ox}} [(2\psi_B - V_{BS})^{1/2} - (2\psi_B)^{1/2}] \tag{7.78}$$

The influence of substrate bias is like a second gate. From Eq. (7.78), it is clear that threshold voltage due to body effect is:

$$V_t = V_{to} + \gamma \left[\sqrt{(2\psi_B + V_{BS})^{1/2}} - \sqrt{(2\psi_B)} \right] \qquad (7.79)$$

where V_{to} is threshold voltage for $V_{BS} = 0$ and γ is called **body effect** coefficient, and is given as:

$$\gamma = \sqrt{\frac{2\varepsilon_{Si}qN_a}{C_{ox}}}$$

This is process dependent parameter.

If the substrate bias V_B is much larger than $2\psi_B$ (typically ~ 0.6 V) then

$$\Delta V_t = \frac{\sqrt{2\varepsilon_{Si}qN_a}}{C_{ox}} \, [-V_B]^{1/2} \qquad (n\text{-MOS}) \qquad (7.80)$$

and

$$\Delta V_t = \frac{\sqrt{2\varepsilon_{Si}qN_d}}{C_{ox}} \, [V_B]^{1/2} \qquad (p\text{-MOS}) \qquad (7.81)$$

Figure 7.36 shows the variation of threshold voltage as a function of V_{BS}. It is clear from figure that substrate sensitivity is higher for higher bulk doping concentration and substrate sensitivity decreases as the substrate bias voltage increases. The body effect is used to raise the threshold voltage of a marginally enhancement device to somewhat larger and more manageable value.

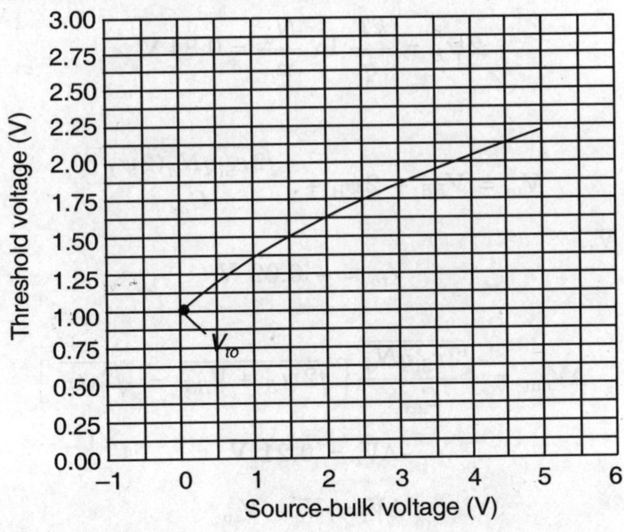

FIGURE 7.36 Threshold voltage variation with substrate bias voltage.

The variation of threshold voltage of a MOS device due to body effect is utilized mainly in VLSI design where millions of transistors are placed on the chip. Since, at the same time many transistors are not required to ON because these transistors are not doing any function at that time. To avoid them to become ON, one can apply body effect to enhance their threshold voltage which is very difficult during fabrication process. These transistors are said to be in sleep mode. Due to that power consumption of the circuit is reduced tremendously.

The threshold voltage variation with temperature has a typical slope of order of $dV_t/dT \sim 1$ mV/K. Temperature coefficient decreases slightly as N_a increases. At an elevated temperature of 100°C, the threshold voltage is 55–75 mV lower than at the room temperature. Since, digital VLSI circuits often operate at very high temperature due to heat generation of the circuit. This effect along with the degradation in the subthreshold slope at high temperature, causes leakage current in the MOS devices at $V_{GS} = 0$. This leakage current increases considerably to its room temperature value. Typically, the offset leakage current of a MOSFET at 100°C is 30–35 times larger than the leakage current at 25°C.

EXAMPLE 7.8

For a n-channel n^+-polysilicon-SiO$_2$-Si MOSFET with $N_a = 10^{17}/\text{cm}^3$ and $Q_f/C_{ox} = 5 \times 10^{11}/\text{cm}^2$, having the gate oxide thickness 5 nm, calculate

    (a) The increase in threshold voltage if the substrate bias is increased from zero to 2 V.

    (b) The threshold voltage due to body effect.

Solution: (a) $C_{ox} = \dfrac{t_{ox}}{t_{ox}} = \dfrac{3.9 \times 8.854 \times 10^{-14}}{5 \times 10^{-7}} = 6.9 \times 10^{-7}$ F/cm^2

$$2\psi_B = \frac{kT}{q} \ln \frac{N_a}{n_i} = 0.84 \text{ V}$$

For $V_{BS} = 0$ V,

$$V_{to} = V_{FB} + 2\psi_B + \frac{\sqrt{2\varepsilon_{Si}qN_a(2\psi_B)}}{C_{ox}}$$

Using $V_{FB} = -1.1$ V,

$$V_{to} = -0.02 \text{ V}$$

Using expression

$$\Delta V_{out} = \frac{\sqrt{2\varepsilon_{Si}qN_a}}{C_{ox}}\left[\sqrt{2\psi_B + V_{BS}} - \sqrt{2\psi_B}\right]$$

$$\Delta V_t = 0.21 \text{ V}$$

(b)

$$\Delta V_t = V_t - V_{to}$$

or

$$V_t = \Delta V_t + V_{to} = 0.21 \text{ V} + 0.02 \text{ V}$$

$$= 0.23 \text{ V}$$

7.8 MOSFET CHANNEL MOBILITY

The surface mobility of electron in a MOSFET is considerably lower than that in bulk due to additional scattering mechanisms. The mobility is primarily a function of effective gate field rather than of the carrier density. At low effective field, the mobility is limited by the ionized-impurity scattering (both donors and interface charges). At medium E_{eff}, the mobility is affected by phonon (acoustic, optical and intervelley) scattering. At high effective field and low temperature, mobility is limited by surface roughness scattering. Channel mobility is also affected by the processing conditions that alter the Si-SiO$_2$ interface properties, e.g. oxide charge.

In deriving the current-voltage relation of a MOSFET, we have assumed that mobility of the carriers in the channel is constant and defined as the effective mobility.

The electron mobility is usually reported as a function of the transverse effective field E_{eff}, i.e., the average electric field pushing the carriers against the Si/SiO$_2$ interface and is given as:

$$E_{eff} = \frac{Q_{\text{dep}} + \eta Q_{\text{inv}}}{\varepsilon_{Si}} \tag{7.82}$$

where $\eta = 1/2$ for (100) electrons and $\eta = 1/3$ for holes and (111) and (110) electrons, Q_{inv} is the inversion carrier density and Q_{dep} is the bulk depleted charge density. The variation of effective mobility with E_{eff} for three different concentration profile at a given temperature is shown in Fig. 7.37. In reality, the dependence of the mobility on the applied gate-to-source voltage E_{eff} can be given by the following conventional mobility model:

$$\mu_G = \frac{\mu_0}{1 + \theta(V_{GS} - V_t)} \tag{7.83}$$

where μ_G is the mobility as a function of V_{GS}, V_t is the threshold voltage, μ_0 is the low-field mobility and θ is the mobility degradation coefficient which is given as:

$$\theta = \frac{\beta_\theta}{t_{ox}} \tag{7.84}$$

where the parameter β_θ typically varies between 0.001 and 0.004 µm/V and t_{ox} is gate oxide thickness. For $V_{GS} = V_t$,

$$\mu_G = \mu_0$$

Physically, it was also found that mobility would increase with larger applied V_{DS} and smaller channel length. The effect of substrate doping on carrier's mobility can be ignored for substrate doping concentration below $10^{17}/cm^3$, and hence the mobility is primarily determined by the effective field* because the coulomb scattering is mainly due to interface charges rather than the localized impurity charges. The

---

*Electron mobility in inversion and accumulation layers on thermally oxidized silicon surfaces. S.C. Sun and James D. Plummer, *IEEE Trans. On Electron Devices*, vol. ED-27, No. 8, August 1980, *pp.* 1947.

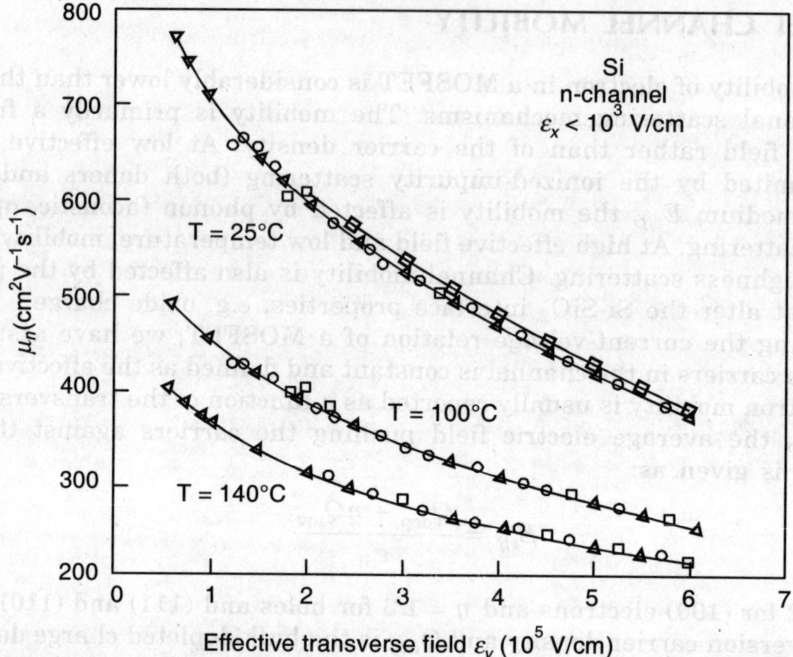

FIGURE 7.37 Variation of mobility of carrier with effective transverse field.

dependence of the effective mobility on the effective field can be described by the empirical relation,

$$\mu_{eff} = \mu_{max} \left(\frac{E_c}{E_{eff}} \right)^{C_1} \tag{7.85}$$

where μ_{max} is maximum value of mobility for a given doping level and Q_f value, and given as;

$$\mu_{max} = \frac{\mu_0 N_a}{1 + \alpha N_a Q_f} \tag{7.86}$$

where,

$$\mu_0 = 3490 - 164 \log N_a$$
$$\alpha = 1.04 \times 10^{-1} + 1.93 \times 10^{-2} \log N_a$$

C_1 is an empirical constant and found to be independent of Q_f. And weakly depending on the substrate doping as,

$$C_1 = 0.341 - 5.44 \times 10^{-3} \log N_a \quad \text{(Wet oxidation)}$$
$$C_1 = 0.313 - 6.05 \times 10^{-3} \log N_a \quad \text{(Dry oxidation)}$$

Typically C_1 is on the order of 0.26 for steam oxide and 0.22 for a dry oxidation.

The constant E_C depends on both N_a and Q_f and empirically given as:

$$E_C = E_{Co}(Na) \times Ae^{BQ_f} \qquad (7.87)$$

where $E_{Co}(Na) = 2.054 \times 10^{-4} \, Na^{0.25}$
$\qquad$ A = 2.79×10^4 V/cm for wet oxidation
$\qquad$ B = 8.96×10^{-2} for wet oxidation
$\qquad$ A = 2.61×10^4 V/cm for dry oxidation
$\qquad$ B = 0.13 for dry oxidation

These constants are found for $N_a = 4 \times 10^{14}/cm^3$ and $Q_f = 10^{11}/cm^2$.

Beyond $E_{eff} = 5 \times 10^5$ V/cm, μ_{eff} decreases much more rapidly with increasing E_{eff} because of increased surface roughness scattering of carriers and given as:

$$\mu_{SR} \; \propto \; \frac{E_{eff}^{-2}}{(\Delta L')^2} \qquad (7.88)$$

where Δ and L' are the mean asperity heights and the correlation length of the surface roughness.

For each doping concentration, there exists an effective field below which the mobility falls off the universal curve due to coulomb (or impurity) scattering which becomes more important when doping concentration is high and the gate voltage or normal effective field is low at high temperature. The coulomb scattering limited mobility is expressed as:

$$\mu_{coul.} = \frac{\mu_0}{[N_{it}\alpha + N_a \, L_{Th}]} \qquad (7.89)$$

where N_{it} is interface charged states per unit area and $L_{Th} = \left(\dfrac{\hbar^2}{2m^*kT}\right)^{1/2}$ is thermal length from Si/SiO$_2$ interface. α typically assumes value of order of 10^{-2} for N_a larger than $10^{11}/cm^2$.

There is less effect of coulomb scattering on mobility when the inversion charge density is high because of charge screening effects.

The phonon scattering limited mobility is:

$$\mu_{Ph} = \left(\frac{AT}{E_{eff}} + \frac{B}{E_{eff}^{1/3}}\right)\frac{1}{T} \qquad (7.90)$$

where A and B are fitting parameters.

Using Mathiessan-rule, the effective mobility in the presence of all scattering mechanism is:

$$\frac{1}{\mu_{eff}} = \frac{1}{\mu_{Ph}} + \frac{1}{\mu_{coul.}} + \frac{1}{\mu_{SR}} \qquad (7.91)$$

- Phonon scattering is dominating at high temperature and medium effective field E_{eff}.
- Coulomb scattering dominates at low temperature and low E_{eff}.
- Since, the Si/SiO$_2$ interface is not ideally flat (smooth), but shows irregularities (roughness) with a typical amplitude of one or two atomic layers. Scattering by this potential fluctuation, called **surface roughness scattering** degrades the carrier mobility at high effective field and low temperature.

7.9 SMALL SIGNAL MODEL OF MOSFET

AC response of a MOSFET is expressed in terms of small signal equivalent circuits. This circuit can be derived from the two-port network as shown in Fig 7.38.

FIGURE 7.38 Two port representation of MOSFET.

The input appears as an open circuit (because of high input impedance) except for the presence of the gate capacitor.

At output, drain current is obtained which is a function of gate and drain voltages w.r.t. source, i.e.,

$$I_D = f(V_{GS}, V_{DS}) \tag{7.92}$$

Any ac signal in output voltage V_{GS} or output voltage V_{DS} will result in corresponding ac variation in drain current I_D, i.e.,

$$\Delta I_D = \left.\frac{\partial I_D}{\partial V_{GS}}\right|_{V_{DS}} \Delta V_{GS} + \left.\frac{\partial I_D}{\partial V_{DS}}\right|_{V_{GS}} \Delta V_{DS}$$

or

$$I_d = g_m V_{gs} + g_d V_{dS} \tag{7.93}$$

where

$$g_m = \left.\frac{\partial I_D}{\partial V_{GS}}\right|_{V_{DS}} \text{ is called } \textbf{transconductance, and}$$

$$g_d = \left.\frac{\partial I_D}{\partial V_{DS}}\right|_{V_{GS}} \text{ is called } \textbf{drain or channel conductance}.$$

and I_d, V_{gs} and V_d are small-signal ac current and voltages. Small signal means that the ac voltages and currents are small as compared to dc values.

The equivalent ac model at low-frequency, after neglecting gate capacitance, is shown in Fig. 7.39.

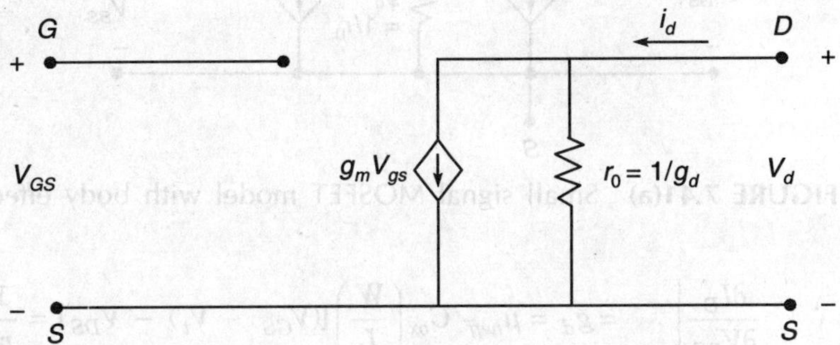

FIGURE 7.39 Small-signal model of MOSFET.

The high-frequency ac equivalent model after including capacitive effects is shown in Fig. 7.40. Here C_{gs} and C_{gd} are called **stray capacitors**.

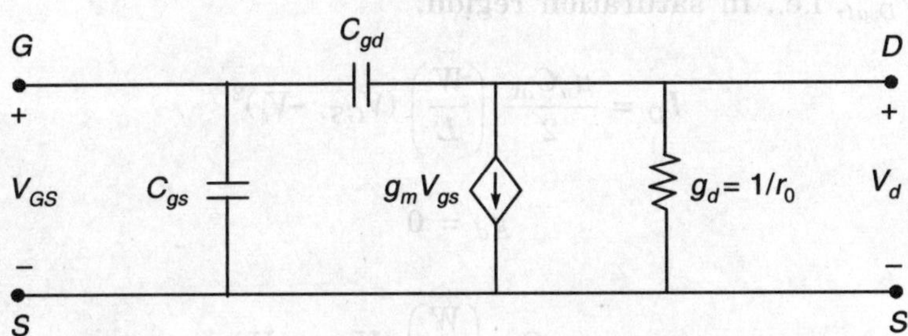

FIGURE 7.40 High frequency ac model of MOSFET.

Small-signal equivalent pi-circuit, as shown in Fig. 7.39, must be modified to include the body-effect, because source and gate are connected at different ac voltages. The modified model is shown in Fig. 7.41(a).

Here

$$g_{mb} = \frac{\partial I_D}{\partial V_{BS}}\bigg|_{\substack{V_{GS} \\ V_{DS}}} = \left(\frac{\partial I_d}{\partial V_t}\right)\left(\frac{\partial V_t}{\partial V_{BS}}\right)$$

and

$$\frac{\partial V_t}{\partial V_{BS}} = \frac{\gamma}{2\sqrt{2\psi_B + V_{BS}}}$$

When $V_{DS} < V_{Dsat}$, i.e., below pinch-off region or linear region,

$$I_D = \mu_{neff}\, C_{ox}\left(\frac{W}{L}\right)\left[(V_{GS} - V_t)\,V_{DS} - \frac{V_{DS}^2}{2}\right]$$

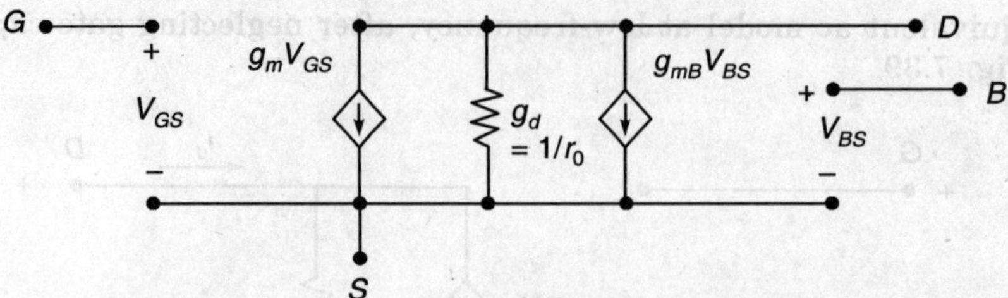

FIGURE 7.41(a) Small signal MOSFET model with body effect.

Therefore,
$$\left.\frac{\partial I_D}{\partial V_{DS}}\right|_{V_{GS}} = g_d = \mu_{neff}\, C_{ox}\left(\frac{W}{L}\right)[(V_{GS} - V_t) - V_{DS}] = \frac{1}{r_0} \qquad (7.94a)$$

$$\left.\frac{\partial I_D}{\partial V_{GS}}\right|_{V_{DS}} = g_m = \mu_{neff}\, C_{ox}\left(\frac{W}{L}\right)[V_{DS}] \qquad (7.94b)$$

When $V_{DS} > V_{Dsat}$, i.e., in saturation region,

$$I_D = \frac{\mu_n C_{ox}}{2}\left(\frac{W}{L}\right)(V_{GS} - V_t)^2$$

means
$$g_d = 0 \qquad (7.94c)$$

and
$$g_m = \mu_n\, C_{ox}\left(\frac{W}{L}\right)(V_{GS} - V_t) \qquad (7.94d)$$

Equation (7.94) shows that the conductance parameters depend on the dc bias V_{GS} and V_{DS}.

Cut-off frequency f_T is defined as the frequency at which current gain is unity.

Since
$$\text{Input current} = \omega C_{gs} V_{gs}$$
$$\text{Output current} = g_m V_{gs}$$

where V_{gs} is the input ac signal and ω is angular frequency.

Since,
$$C_{GS} \approx (WL)\, C_{ox}$$

where W is width of the channel.

and
$$\text{Current gain} = \frac{\text{Output current}}{\text{Input current}}$$

$$= \frac{g_m\, V_{GS}}{C_{GS} V_{GS}\omega}$$

Neglecting C_{Gd} in saturation region of long channel because the channel is very thin near to drain and V_{DS} has little effect on the channel or gate charge, otherwise input current = $W(C_{GS} + C_{GD})V_{GS}$

According to condition,

$$\frac{g_m \, V_{GS}}{\omega C_{GS} V_{GS}} = 1$$

or

$$\omega C_{GS} = g_m$$

where, $\omega = 2\pi f$
Therefore,

$$2\pi f_T = \frac{g_m}{C_{GS}}$$

or

$$f_T = \frac{g_m}{2\pi \, C_{GS}} \tag{7.95}$$

where, f_T is the common figure of merit to evaluate the ability of a device to operate at high frequencies.

Equation (7.95) shows that for high-speed operation of the devices, a larger value of g_m and lower value of C_{GS} are required. Since, $C_{GS} \cong C_{ox}$ W.L., therefore, higher speed implies shorter device length L.

MOSFET Related Capacitances

The switching behaviour of a MOSFET is decided by the capacitances associated with the MOS device. The value of the capacitance decides how fast the gate can be charged or discharged to a certain voltage level which turns the MOSFET either ON or OFF. The MOSFET capacitance is broadly divided into two categories: intrinsic capacitance or oxide related capacitances and other is parasitic capacitances. The presence of the inversion and depletion charges in the channel gives rise to intrinsic capacitances. The intrinsic capacitances are C_{gs} (gate-to-source capacitance), C_{gd} (gate-to-drain capacitance) and C_{gb} (gate-to-bulk capacitance). The parasitic capacitances are mainly junction capacitances or diffusion capacitances and overlap capacitances. All these capacitances are shown in Fig. 7.41(b). Since, total gate capacitance is determined by parallel-plate capacitor between polysilicon gate and underlying structure, hence magnitude of oxide related capacitance is strongly affected by gate oxide thickness and dimensions of device (i.e., length and width of the channel). Reducing oxide thickness will increase the capacitance, whereas reducing either L or W or both will reduce the capacitance value. The total oxide related capacitance is given as:

$$C_g = C_{gs} + C_{gd} + C_{gb}$$

These all capacitances change with bias conditions and region of operation as shown in Fig. 7.41(b). The values of these capacitances in the different region of operations are given as follows:

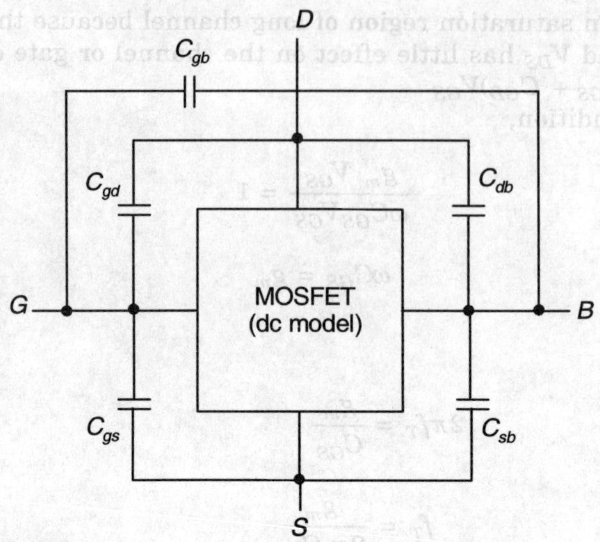

FIGURE 7.41(b)   Representation of various capacitors associated with MOSFET.

Cut-off region: In cut-off region, channel is not inverted and no inversion charge exists at source or drain end, therefore, $C_{gd} = C_{gs} = 0$.

and
$$C_{gb} \approx C_{ox}WL$$

Linear region: In linear region, inversion charge is uniformly distributed along the channel. This means, gate-to-channel capacitances are equally shared between source and drain;

i.e.,
$$C_{gd} = C_{gd} = 1/2\ C_{ox}WL$$

No capacity coupling occurs between the gate and the body due to the screening by the inversion charges. This results $C_{gb} = 0$.

Strong inversion: In strong inversion, MOSFET operates in saturation region where no channel exists at the drain end, i.e., inversion charge density at drain end is zero. The total inversion charge exists at the source side. Thus, these two arguments results;

$$C_{gd} = 0$$

$$C_{gs} \approx 2/3\ C_{ox}WL \text{ and again due to screening effect; } C_{gb} \approx 0.$$

Another two capacitors C_{db} and C_{sb} arise due to depletion charge between source/drain and substrate. These capacitors are known as junction capacitors or diffusion capacitors. These capacitors are voltage dependent because as the voltage changes the depletion charge also changes. The source-substrate and drain-substrate junctions are reverse biased under normal opperating conditions. The junction capacitance associated with depletion region is defined as:

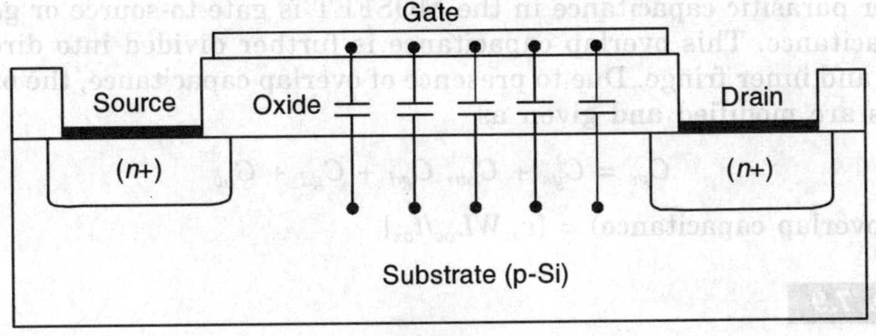

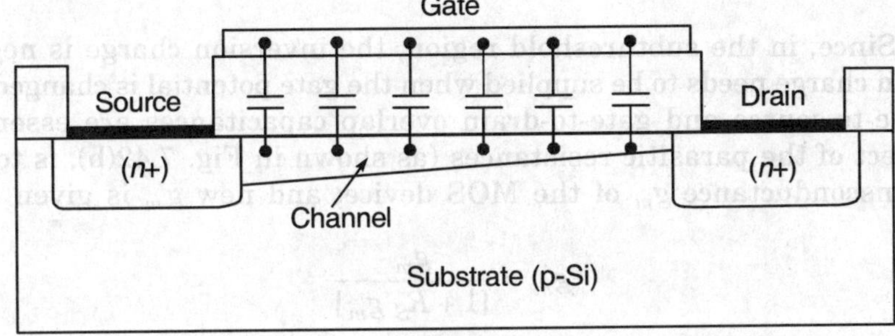

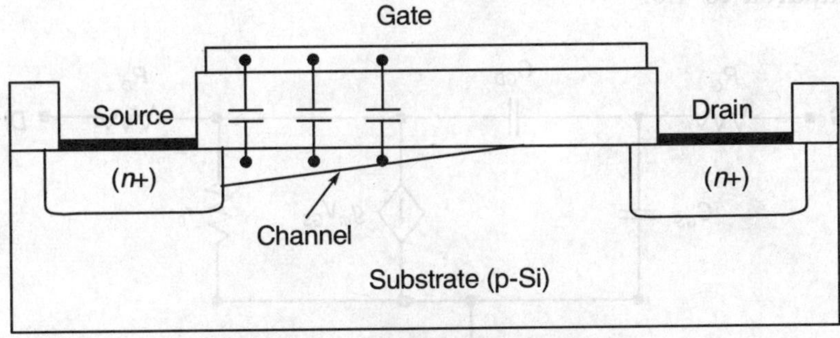

FIGURE 7.42(a) Oxide related capacitances in MOSFET at different bias conditions.

$$C_j = \left(\frac{dQ}{dV}\right) = \left[\frac{\varepsilon_{si} q N_d}{2V_{bi} + V_R}\right]^{1/2} = \frac{\varepsilon_{si}}{W_{dj}}$$

where W_{dj} is depletion layer width, V_{bi} is built-in-potential and V_R is applied reverse bias voltage. The total diffusion-to-substrate capacitance is:

$$C_j = W.d.C_j$$

where W is device width and d is diffusion width.

Another parasitic capacitance in the MOSFET is gate-to-source or gate-to-drain overlap capacitance. This overlap capacitance is further divided into direct overlap, outer fringe and inner fringe. Due to presence of overlap capacitance, the oxide related capacitances are modified and given as

$$C_{gs} = C_{gs} + C_{ov}, \; C_{gd} + C_{gd} + C_{ov}$$

where C_{ov} (overlap capacitance) $= [\varepsilon_{ox} W L_{ov}/t_{ox}]$

EXAMPLE 7.9

Why the parasitic capacitances C_{gs} and C_{gd} are zero in subthreshold region or weak inversion?

Solution: Since, in the subthreshold region, the inversion charge is negligible and only depletion charge needs to be supplied when the gate potential is changed. Therefore, intrinsic gate-to-source and gate-to-drain overlap capacitances are essentially zero.

The effect of the parasitic resistances (as shown in Fig. 7.42(b), is to reduce the intrinsic transconductance g_m of the MOS device; and new g'_m is given as:

$$g'_m = \frac{g_m}{[1 + R_S \, g_m]} \tag{7.95b}$$

This shows the intrinsic transconductance is degraded as $R_S \, g_m$ becomes significant value as compared to 1.0.

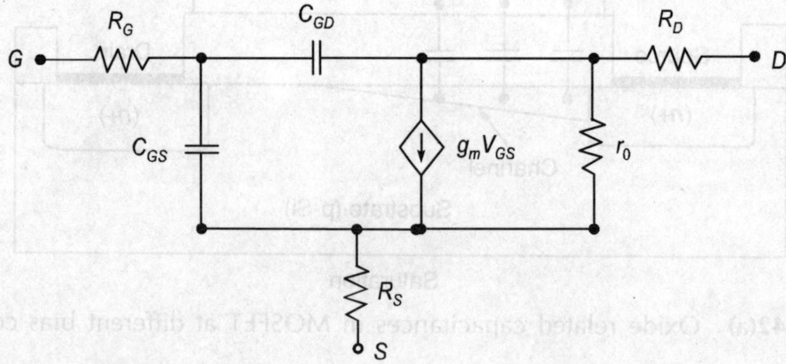

FIGURE 7.42(b) Neglecting overlap capacitance.

7.10 SHORT CHANNEL MOS DEVICE

A MOS device is said to be short channel device when the channel length L is comparable to the junction depth of the source and drain diffusion.

Also, short channel MOSFET is defined as the device where the length and width of the device is short enough to neglect the edge effects. In other definition, we can say that a short channel MOSFET is that device in which gate length L is comparable to depletion widths associated with the drain and source. The short channel MOSFET is shown in Fig. 7.43.

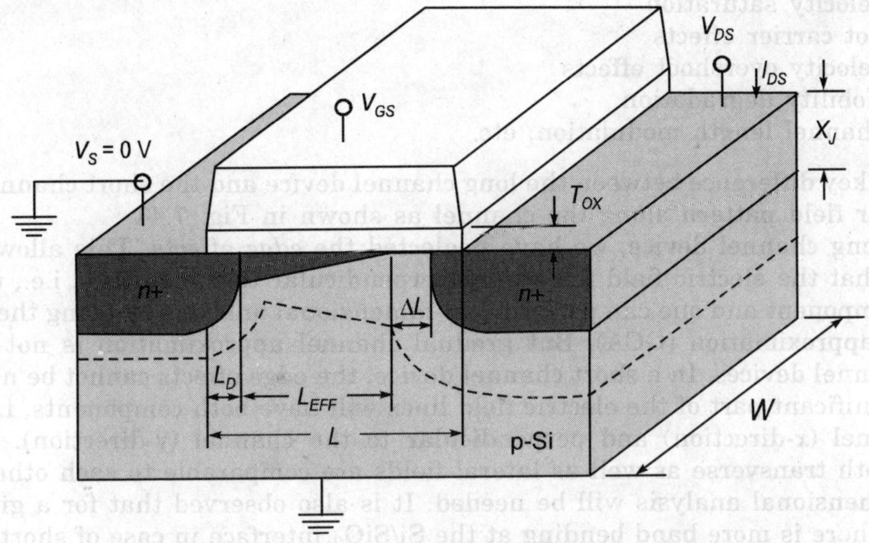

FIGURE 7.43 Structure of short-channel MOSFET.

In short channel device, the source-drain potential has a strong effect on the control of the current in the device. It is clear from current-voltage relation of a long channel MOSFET that as the length of the channel reduces, the drain current of the device increases. The intrinsic capacitance of a short channel MOSFET is also lower which makes it faster. MOSFETs are continually downscaled for higher package density on the chip, higher device speed (because lowering of L means μ_{eff} increases) and lower power consumption. Reducing all the physical parameters by same factor is called **scaling** (which we will discuss later). However, for a given process, the channel length can not be arbitrarily reduced even if allowed by lithography due to various limitations (we will discuss in detail later).

In this section we will cover the basic features of a short-channel devices that are important for device design consideration.

7.10.1 Short Channel Effects

The short channel effects (SCEs) are attributed to two physical phenomena:

(i) The limitation imposed on electron drift characteristics in the channel.

(ii) The modification of the threshold voltage due to the shortening of the channel length.

The different types of SCEs are:

(a) Threshold voltage roll-off
(b) Drain induced brarrier lowering
(c) Velocity saturation
(d) Hot carrier effects
(e) Velocity overshoot effects
(f) Mobility degradation
(g) Channel length modulation, etc.

The key difference between the long channel device and the short channel device is in their field pattern along the channel as shown in Fig. 7.44.

In long channel device, we have neglected the *edge* effects. This allows one to assume that the electric field lines were perpendicular to the surface, i.e., they had only y-component and one can perform one-dimensional analysis by using the gradual channel approximation (GCA). But gradual channel approximation is not true for short channel devices. In a short channel device, the edge effects cannot be neglected, and a significant part of the electric field lines will have both components, i.e., along the channel (x-direction) and perpendicular to the channel (y-direction). In other words, both transverse as well as lateral fields are comparable to each other. Thus, a two-dimensional analysis will be needed. It is also observed that for a given gate voltage, there is more band bending at the Si/SiO$_2$ interface in case of short channel device.

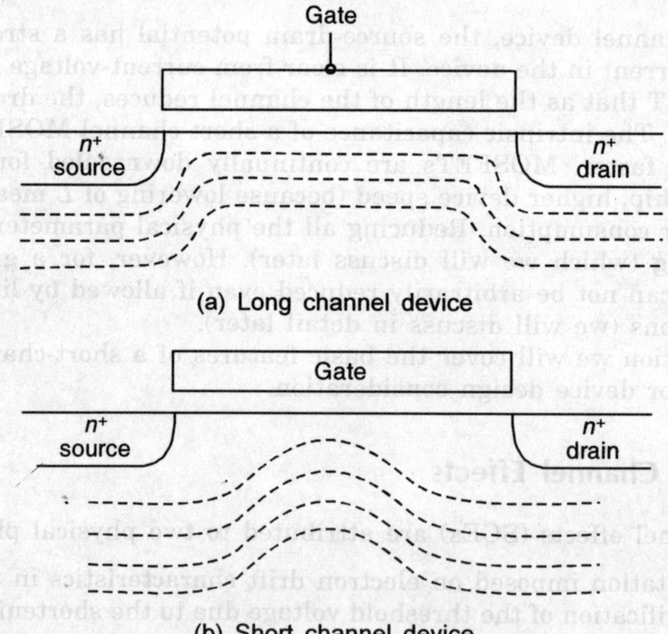

FIGURE 7.44 Field patterns in (a) long channel, (b) short channel MOSFETs.

The two-dimensional field pattern in a short-channel MOSFET is due to proximity of the source and drain regions. There are also depletion regions surrounding the source and drain junctions apart from depletion region under the gate. Since, in a long channel device, source and drain are very far away from each other so that their depletion regions have no effect on the potential or field pattern. Since, in short channel device, the channel length is comparable to depletion width in the vertical direction, therefore, source-drain potential has a strong effect on the band bending over a significant portion of the device.

Let us consider Fig. 7.45 where we can see the net charge (either ionized donor or acceptor impurities) in the depletion region of the device. The field lines which are originating either from the source or the drain will terminate on these fixed charges.

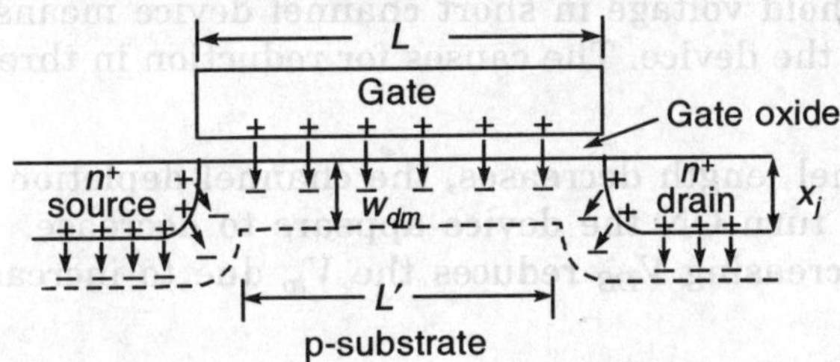

FIGURE 7.45 Representation of charge sharing in short channel MOSFET.

This is called the charge-sharing model, first discussed by You in 1974. At low drain voltage, only the field lines terminating on the depletion charges within the trapezoidal region are assumed to originate from the gate and rest of the fields are either originate from source or drain. Therefore, the total charge within the trapezoidal is given as,

$$Q'_B \propto W_{dm} \times \frac{(L + L')}{2} \tag{7.96}$$

where W_{dm} is maximum depletion width, L is the gate length and L' is electrical gate length or channel length.

If there is no overlapping then

$$L' = L$$

This total depletion charge Q'_B in case of short channel device is less than the total depletion charge Q_B of the long channel device, where

$$Q_B \propto W_{dm} \times L \tag{7.97}$$

As a result, short channel device takes lower gate voltage to reach the condition of inversion, i.e.,

$$V_{t\,\text{short}} = V_{FB} + 2\psi_B + \frac{Q'_B}{WL\,C_{ox}}$$

In other words, the threshold voltage in the case of short channel device becomes lower than the long channel device.

$$\Delta V_t = V_t \text{ long channel} - V_t \text{ short channel}$$

This effect is called **threshold voltage roll-off** and given by,

$$\Delta V_t = \frac{1}{C_{ox}} \sqrt{2\varepsilon_{Si} q N_a |2\psi_B|} \cdot \frac{X_j}{2L} \left[\left\{ \sqrt{1 + \frac{2X_d}{X_j}} - 1 \right\} + \left\{ \sqrt{1 + \frac{2X_{DS}}{X_j}} - 1 \right\} \right] \quad (7.98)$$

where, X_j is junction depth, X_d junction depletion region at the drain side and X_{DS} junction depletion region width at the source side.

In long channel device, depletion region under gate appears as rectangular in shape, whereas for short channel device it appears as the trapezoidal in shape.

Reducing threshold voltage in short channel device means the less charges are required to turn ON the device. The causes for reduction in threshold voltage of short channel device are:

- As the channel length decreases, the channel depletion becomes smaller and V_t needed to turn ON the device appears to decrease.
- Similarly, increasing V_{DS} reduces the V_{to} due to increase in the drain space charge layer.

7.10.2 Drain-Induced Barrier Lowering (DIBL)

The electrons (for n-MOS device) in the channel face a potential barrier at the surface between the source and drain due to p-n^+ junction formation. This barrier blocks their movement. Under off condition (i.e., cut-off region) this potential barrier prevents any movement of carriers from source to drain. In long channel device, the surface potential is mainly controlled by the gate voltage because the potential barrier is flat over most part of the device as shown in Fig. 7.44(a), and source/drain fields affect can be neglected. As the channel length is shortened, the source and drain fields penetrate deeply into the middle of the channel and potential barrier is controlled by a two-dimensional electric field vector, i.e., by both V_{GS} and V_{DS}. If the drain voltage is increased, the potential barrier in the channel decreases. Since, the potential barrier height between the source and channel is lowered by the drain voltage so this effect is called **drain-induced barrier lowering (DIBL)**. Due to this effect, gate loss its control over the channel. The DIBL effect causes a substantial increase of the subthreshold current. In other words, under DIBL effect, electrons can flow between the source and drain even if $V_{GS} < V_t$.

The threshold voltage becomes lower than the long channel value and the region of maximum potential barrier shrinks to a single point near the centre of the device. The threshold voltage, thus, decreases further with increasing drain bias in short channel device because DIBL effect leads to a substantial increase in electron injection from the source to the drain. DIBL effect explains the experimentally observed increase in subthreshold current with the increase in drain voltage in short channel device. The subthreshold characteristics of a long channel device and short channel MOSFET is shown in Fig. 7.46.

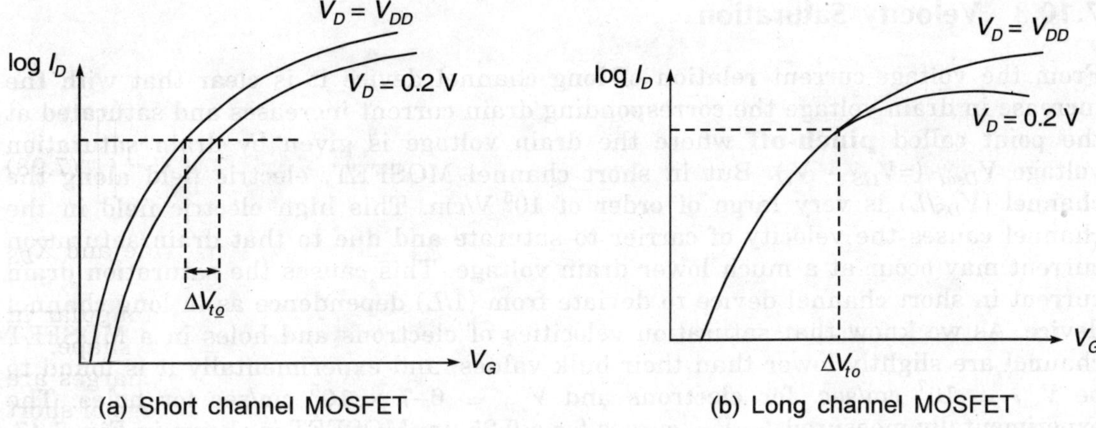

FIGURE 7.46 Subthreshold characteristics.

For long channel device, the subthreshold current is independent of the drain voltage ($\geq 2kT/q$), whereas for short channel MOSFET, there is parallel shift of the curve to a lower threshold voltage for high drain bias conditions.

Punch through

The DIBL phenomenon can be accompanied by the so called **punch through**. The punch through occurs when the depletion region surrounding the drain extends to the source when the drain voltage is increased. As a result leakage current may flow from the drain to source through the bulk. At punch through gate totally loses control of the channel and high drain current persists independent of the gate voltage. High leakage current limits the device operation. Punch through can be minimized with thinner oxide and larger substrate doping. This is achieved by performing anti-punch through implant in the channel. But these will give another constraints to the device. In case of long channel device, punch through is not a serious problem because source and drain are far away from each other.

How to avoid DIBL effect:

- DIBL is caused by the lowering of the source-junction potential barrier below the built-in potential V_{bi} for a p-n^+ junction. Hence, if one gets DIBL in a MOSFET for grounded substrate then to avoid this problem (because it will give leakage current), one should apply a reverse bias on the substrate so that the potential barrier height at the source can be raised.
- To prevent the DIBL effect (because the gate being unable to shut off the device), the source/drain junctions must be made sufficiently shallow.
- Sometimes it is required to perform localized implant only near the source/drain instead of throughout the channel. This type of implantation is called **halo** or **pocket implants**. The higher doping of source/drain reduces the depletion width of the respective junctions and prevents their interaction.

7.10.3 Velocity Saturation

From the voltage-current relation of long channel device it is clear that with the increase in drain voltage the corresponding drain current increases and saturated at the point called **pinch-off** where the drain voltage is given by drain saturation voltage V_{Dsat} ($=V_{GS} - V_t$). But in short channel MOSFET, electric field along the channel (V_{DS}/L) is very large of order of 10^5 V/cm. This high electric field in the channel causes the velocity of carrier to saturate and due to that drain saturation current may occur at a much lower drain voltage. This causes the saturation drain current in short channel device to deviate from ($1/L$) dependence as in long channel device. As we know that saturation velocities of electrons and holes in a MOSFET channel are slightly lower than their bulk values, and experimentally it is found to be $V_{sat} \simeq 10^7$ cm/sec. for electrons and $V_{sat} \simeq 6\text{--}7 \times 10^6$ cm/sec for holes. The experimentally measured I_D–V_{DS} curves for a 0.25 μm MOSFET is shown in Fig. 7.47. It is clear from the curves that the drain current saturates at a drain voltage much lower than the saturation voltage. This severely limit the saturation current of a short channel MOSFET.

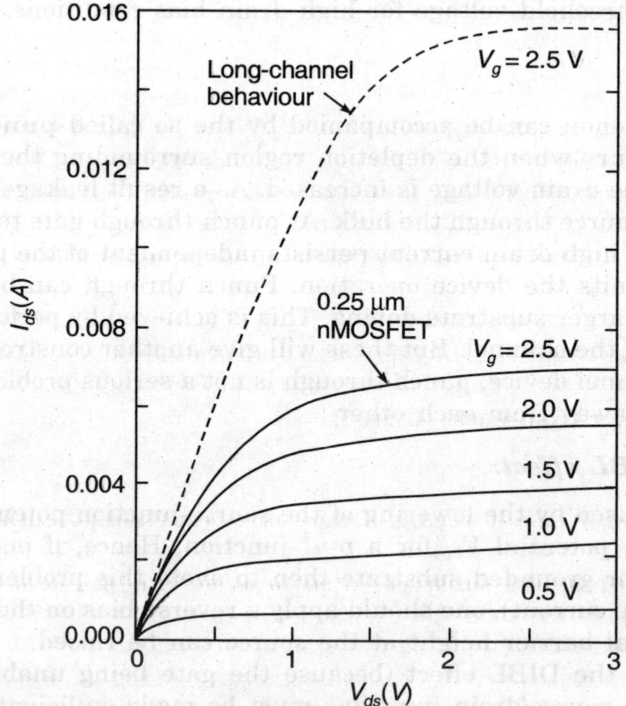

FIGURE 7.47 Experimentally measured I_D-V_{DS} curves for 0.25 μm MOSFET.

The empirical relation between the velocity and field for carriers (electrons/holes) was first developed by Caughey and Thomas in 1967, which matched with the experimental measurements. This empirical relation is given as:

$$v = \frac{\mu_{eff}\, E_x}{[1 + (E_x/E_c)^n]^{1/n}} \qquad E_x \leq E_c \qquad (7.99)$$

$$= V_{sat} \qquad E_x \geq E_c$$

where $|E_x| = dV(x)/dx$ is electric field along the channel and E_c is called **critical field**, given by

$$E_c = \frac{v_{sat}}{\mu_{eff}}$$

for $E_x \to \infty$, $V \to V_{sat}$

$n = 1$ for hole and 2 for electron and n measures how rapidly the carriers approach to saturation. In all analytic MOS transistor models for IC simulators, Eq. (7.99) has been employed with $n = 1$. This is due to fact that integration in the dc current expressions cannot be carried out in closed form for $n > 1$.

The current in short channel MOS device is given as:

$$I_D = -v\, Q_i(V)\, W$$

where v is carrier's velocity. Using Eq. (7.99) for $n = 1$, the drain current is expressed as:

$$I_D = -\, Q_i(V)\, W \frac{\mu_{eff} E_x}{[1 + E_x/E_c]} \qquad (7.100)$$

Using $E_x = -\, dV/dx$ and $E_c = v_{sat}/\mu_{eff}$, Eq. (7.100) becomes

$$I_D = -W\mu_{eff} Q_i(V) \frac{dV/dx}{\left[1 + \dfrac{\mu_{eff}}{v_{sat}}\dfrac{dV}{dx}\right]} \qquad (7.101)$$

where V is the quasi-Fermi potential at a point x in the channel and $Q_i(V)$ is inversion charge density at that point.

Rearranging Eq. (7.101) and integrating from $x = 0$ to $x = L$ and $V = 0$ to $V = V_{DS}$, we have

$$I_D = \frac{-\,\mu_{eff}\,(W/L) \displaystyle\int_0^{V_{DS}} Q_i(V)dV}{\left[1 + \mu_{eff}\dfrac{V_{DS}}{V_{sat}\,L}\right]} \qquad (7.102)$$

Numerator of Eq. (7.102) simply gives the drain current for long channel device without velocity saturation, i.e.,

$$I_{D\,\text{short}} = \frac{I_{D\,\text{long}}}{\left[1 + \mu_{eff}\,\dfrac{V_{DS}}{V_{sat}\,L}\right]} \tag{7.103}$$

If the average field (V_{DS}/L) along the channel is much less than the critical field E_c, the drain current is hardly affected by velocity saturation. When $V_{DS}/L \geq E_c$, the drain current is significantly reduced due to velocity saturation. Using,

$$Q_i(V) = -\,C_{ox}(V_{GS} - V_t - V)$$

in Eq. (7.102) and after solving, one has

$$I_{D\,\text{short}} = \kappa(V_{DS})\,\mu_{eff}\,C_{ox}\left(\frac{W}{L}\right)\left[(V_{GS} - V_t)\,V_{DS} - \frac{V_{DS}^2}{2}\right] \tag{7.104}$$

where $\kappa(V_{DS}) = \dfrac{1}{\left[1 + \dfrac{V_{DS}}{L}\,\dfrac{\mu_{eff}}{v_{sat}}\right]}$ and measure the degree of velocity saturation.

For large values of L or small V_{DS}, κ approaches 1 and Eq. (7.104) reduces to long channel MOSFET's current. For short channel device, $\kappa < 1$, means $I_{Dsat} < I_{D\,\text{long}}$.

If drain-source voltage increases continuously then electric field increases to E_c and the carriers at the drain become velocity saturated. Assuming the drain-voltage at the point of saturation is V_{Dsat}; the drain-current is:

$$I_{Dsat} = V_{sat}\,W\,C_{ox}(V_{GS} - V_t - V_{Dsat}) \tag{7.105}$$

Putting $V_{DS} = V_{Dsat}$ in Eq. (7.104), we have

$$I_{Dsat} = \kappa(V_{Dsat})\,\mu_{eff}\,C_{ox}\left(\frac{W}{L}\right)\left[(V_{GS} - V_t)\,V_{Dsat} - \frac{V_{Dsat}^2}{2}\right] \tag{7.106}$$

Equating Eqs. (7.105) and (7.106), we have

$$V_{Dsat} = \kappa'(V_{GS} - V_t) \tag{7.107}$$

where κ' is a function of $(V_{GS} - V_t)$. For short channel device and large value of $(V_{GS} - V_t)$, $\kappa' < 1$ means

$$I_{Dsat}\big|_{\text{short}} < I_{Dsat}\big|_{\text{long}}$$

Hence the device enters into the saturation before V_{DS} reaches to $(V_{GS} - V_t)$ as shown in Fig. 7.48.

Note: In the limit of $L \to 0$, the drain-current will become velocity-saturation-limited current and given by Eq. (7.105). Equation (7.105) is independent of L and varies

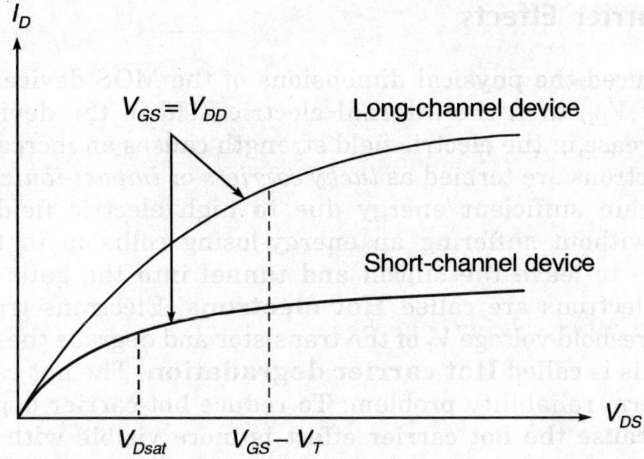

FIGURE 7.48   Velocity saturation effect in short channel MOSFET.

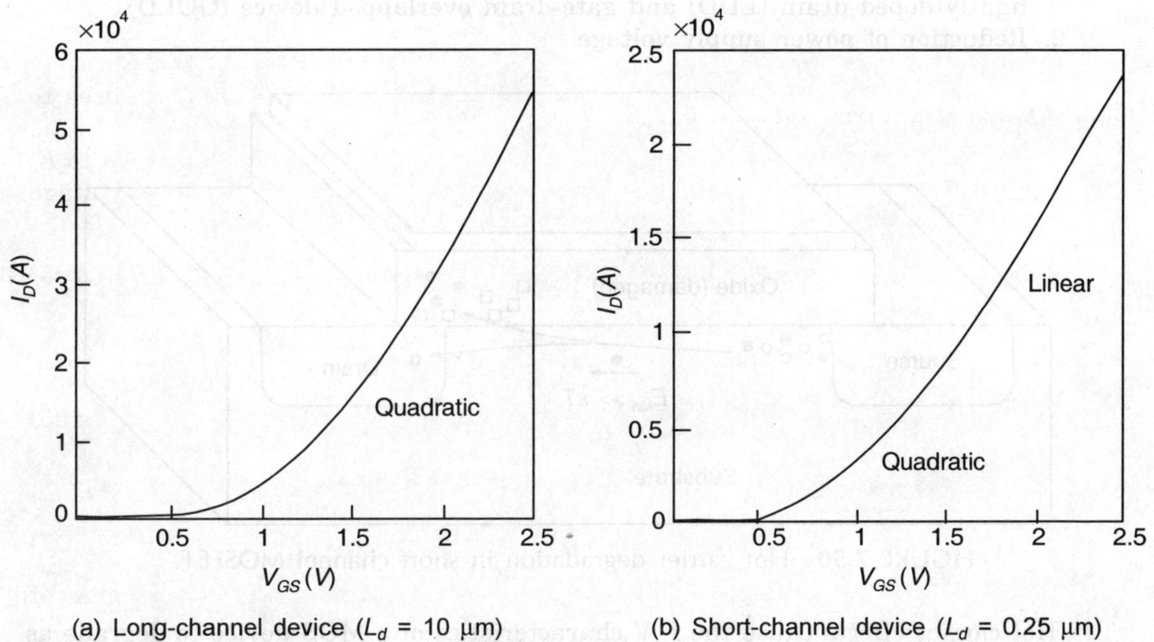

(a) Long-channel device (L_d = 10 μm)   (b) Short-channel device (L_d = 0.25 μm)

FIGURE 7.49   Transfer characteristics of (a) Long channel MOS devices; (b) Short channel MOS device.

linearly with $(V_{GS} - V_t)$ instead of quadratically as in long channel as shown in Fig. 7.49 device.

- A key difference between pinch-off in long channel device and short channel device is that in short channel device, the inversion charge density Q_i does not vanish at the drain.

7.10.4 Hot Carrier Effects

If one simply reduced the physical dimensions of the MOS devices without reducing the power supply V_{DD} then the internal electric field in the device would increase. The resulting increase in the electric field strength causes an increase in the electron's energy. These electrons are termed as *lucky carriers* or *impact-ionized carriers* because these electrons gain sufficient energy due to high electric field to surmount the Si–SiO$_2$ barrier without suffering an energy-losing collision in their paths. These electrons are able to leave the silicon and tunnel into the gate oxide as shown in Fig. 7.50. Such electrons are called **Hot electrons**. Electrons trapped in the oxide will change the threshold voltage V_t of the transistor and degrade the C–V characteristics of a MOSFET. This is called **Hot carrier degradation**. The hot carriers degradation leads to a long term reliability problem. To reduce hot-carrier degradation in scaled MOS devices (because the hot carrier effect is more visible with a channel lengths below 1 μm), two approaches have been suggested:

1. Hot-carrier resistant device structures such as double diffused drain (DDD), lightly doped drain (LDD) and gate-drain overlapped device (GOLD)
2. Reduction of power supply voltage

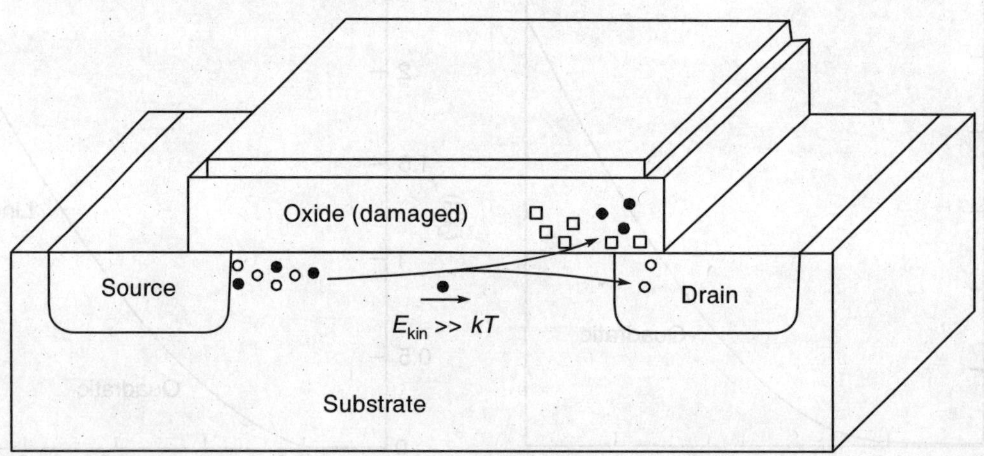

FIGURE 7.50 Hot carrier degradation in short channel MOSFET.

Hot carrier effects cause the *I–V* characteristics of a MOS device to degrade as shown in Fig. 7.51.

Hot carrier effects are less problematic for holes in a *p*-channel MOSFETs than the electrons in *n*-channel due to two reasons

(i) Lower mobility of holes
(ii) The barrier for hole injection in the valence band between Si and SiO$_2$ is higher than electrons in the conduction band

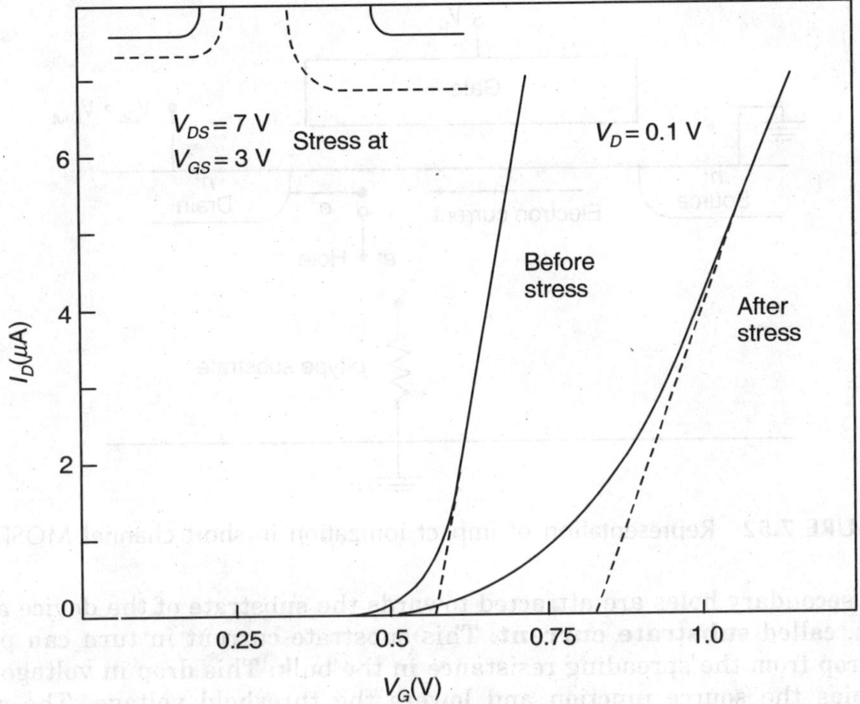

FIGURE 7.51 Transfer characteristics before and after stress.

EXAMPLE 7.10

Why L_{DD} is used for reducing the hot carrier effect?

Solution: Lightly doped drain L_{DD} means the concentration of drain is reduced. By reducing the concentration in the drain one can increase the depletion width at the reverse-biased drain-channel junction. This increase in depletion width reduces the electric field at the drain side. Reduced electric field cause the velocity of carriers to lower and avoid hot carrier effect. Hot carriers effect is more pronounced near the drain end.

Substrate current

When the field in the MOS device exceeds 10^5 V/cm, impact ionization takes place at the drain leading to an abrupt increase of drain current because the presence of high longitudinal field can accelerate the electrons towards the drain region. These electrons are termed as hot electrons. These hot electrons may be able to ionize the silicon atoms by impact ionization. Due to impact ionization secondary electron-hole pairs are generated as shown in Fig. 7.52. These secondary electrons are attracted towards the drain. Due to secondary electrons, drain current increases with bias voltage which results in decrease in output impedance of the MOSFET.

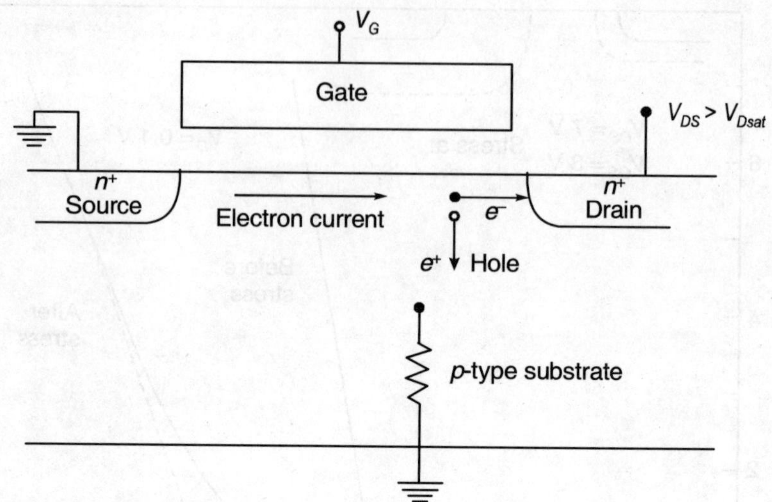

FIGURE 7.52 Representation of impact ionization in short channel MOSFET.

The secondary holes are attracted towards the substrate of the device and gives a current, called **substrate current**. This substrate current in turn can produce a voltage drop from the spreading resistance in the bulk. This drop in voltage tends to forward-bias the source junction and lowers the threshold voltage. The substrate current can create circuit problems such as noise or latch up in CMOS circuits. Substrate current is usually a good indicator of hot carriers generation by low-level impact ionization before runaway breakdown occurs.

If more electron-holes pairs are generated due to impact ionization then drain current increases sharply at a constant voltage in saturation region, called **breakdown**. The breakdown voltage of n MOSFET is usually lower than that of p MOSFET because electrons have a higher rate of impact ionization and because n^+ source and drain junctions are more abrupt than p^+-junctions. Shorter device has lower breakdown voltage. Breakdown often results in a permanent damage to the MOSFET because a large amounts of hot carriers are injected into the oxide in the gate-to-drain overlap region. MOSFET breakdown is a problem in VLSI design. Breakdown problem can be reduced to some extent by lightly doped drain L_{DD} structure.

The MOSFET operating in breakdown mode can be used as an effective ON-chip protection devices against electrostatic discharge.

Note: 1. Let us consider $L = 0.25$ μm (known as **short channel**), for a MOS device which is operated at a power supply of 5 V. Average electric field along the channel for such device is:

$$E_x = \frac{V_{DS}}{L} = \frac{5 \times 10^4}{0.25}$$

$$= \frac{5}{25} \times 10^6 = \frac{1}{5} \times 10^6 = 2 \times 10^5 \text{ V/cm}$$

This field is a very high electric field in absence of scaling of power supply V_{DD}. So, why all the secondary effects are very important in the short channel device.

2. Now, let us consider a long channel device of length 2.5 µm operated at $V_{DD} = 5$ V. Then electric field along the channel for such device is:

$$E_x = \frac{5\,V}{2.5\,\mu m} = \frac{50}{25} \times 10^4 = 2 \times 10^4 \text{ V/cm}$$

and this field is not very high and hence the secondary effects can be ignored.

7.10.5 Velocity Overshoot Effect

Non-local effects are becoming more and more prominent as the MOSFET dimensions shrink to ultra short regime (submicron regime). In this regime drift-diffusion model breaks down where high field or rapid spatial variation of potential is present. In such cases, the scattering events are no longer localized, but some fraction of carriers may acquire sufficient thermal energy near to drain region (called **hot carriers**). Hot carriers are not in thermal equilibrium with the silicon lattice. It is possible that the velocity of hot electrons can exceed the saturation velocity. This phenomenon is called **velocity overshoot**. Velocity overshoot is one of the most important effects from the practical point of view because it is directly related with the increase of current drive. The velocity overshoot was first predicted by Ruch using Monte-Carlo simulations. It was observed that experimentally measured transconductance in submicron MOS devices was found to be higher than the theoretically predicted value. This result has been shown for channel lengths under 0.15 µm which means the electron velocity along the channel is higher than the saturation velocity. Therefore, by controlling the velocity overshoot, the performance of the submicron devices can be improved with respect to the performance of the long devices.

It has been observed that an electric field step for shorter period than the relaxation time causes the electron velocity to overshoot. Relaxation time is defined as the time required by electron to once again reach equilibrium with the lattice. Therefore, as the longitudinal electric field increases, the electron gas starts to be in unequilibrium with the lattice, and there is insufficient number of phonon-scattering events experienced by electron during its motion towards drain. As a result of that electron loses less energy and can be accelerated to velocity higher than the saturation velocity; approaching ballistic transport conditions. This effect is due to the nonequivalence of momentum and energy-relaxation times and can be observed for a period shorter than the energy relaxation time. Overshoot is a nonequilibrium effect and cannot be explained by simple drift-diffusion model. Once the device approaches the ballistic limit, the current depends only on the injection velocity from the source, independent of field and scattering parameters. In other words, there is an upper limit on the MOSFET current which is set by thermal injection from source, and velocity overshoot in the channel and the current should not exceed than this limit.

7.10.6 Mobility Degradation

Carrier mobility in deep submicron MOSFETs is limited by the morphology of the Si/SiO$_2$ interface. The Si/SiO$_2$ interface is not ideally flat, but shows irregularities with a typical amplitude of one or two atomic layers. Scattering of carriers by this potential fluctuation degrades the carrier mobility at high electric field. Increasing efforts are, therefore, being devoted to improve the surface quality and find a correlation between the features of the interface roughness and mobility values. It is most desirable to suppress the surface scattering of carriers in order to further integrate the devices on the chip and operate them at higher speed.

7.10.7 Channel Length Modulation Effect

Channel length modulation (CLM) is caused by the increase of the depletion layer width near the drain region as the drain voltage is increased. This reduces the electrical channel length compared to the physical length which results in an increase in drain current. The CLM effect is more pronounced in low-doped substrate devices. The CLM effect can be decreased using the scaling rules such as increasing the doping concentration as the gate length is reduced.

7.11 NARROW WIDTH EFFECTS

In this effect the threshold voltage V_t goes up as the channel width W is reduced for very narrow devices. This is due to fact that fringing field causes channel depletion region to extend beyond the gate in the width direction. Thus, additional gate charge is required which causes an increase in threshold voltage. This effect is not important for very wide devices, but as the channel becomes smaller it is more important (i.e., as the widths are reduced below 1 μm).

Note: Unlike short channel effect, where the effective depletion charge is reduced due to charge sharing, in narrow width effect, depletion charges belonging to gate is increased.

7.12 GATE INDUCED DRAIN LEAKAGE (GIDL)

As the gate bias voltage is further reduced below threshold voltage and becomes more negative, an excess drain current starts to flow. This current is called **gate induced drain leakage (GIDL) current**. In other words, the off-state leakage current actually goes up as one tries to turn OFF the MOSFET for high V_{DS}.

The GIDL current dominates at negative bias of V_{GS} and positive bias of V_{DS}. The larger the difference between V_{DS} and V_{GS}, higher the GIDL. Since, GIDL current generates the excessive heat dissipation, therefore, it needs to be maintained below some specified value, e.g., 10 pA/μm.

Consider a MOS structure as shown in Fig. 7.53(a). The band diagram is shown in Fig. 7.53(b). In this structure gate overlaps the drain junction. As gate becomes more

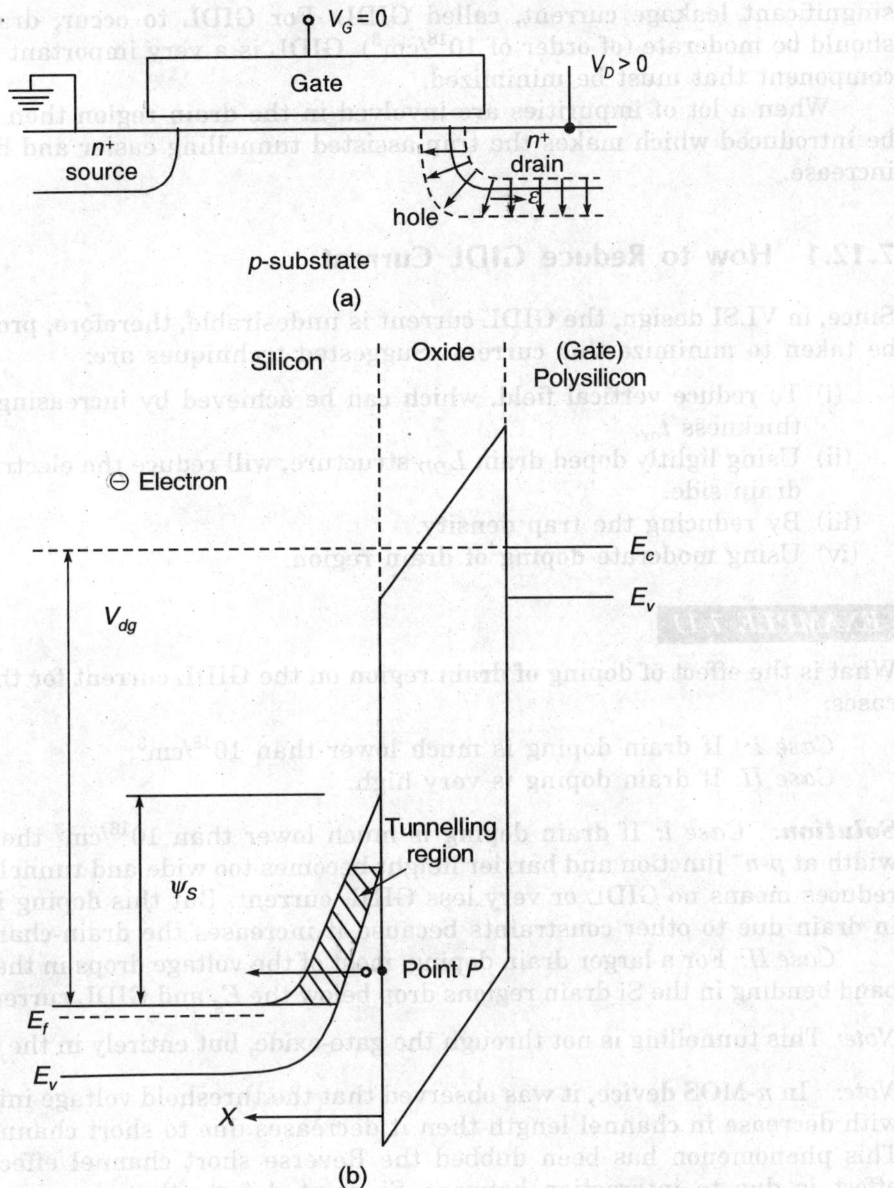

FIGURE 7.53 (a) Short channel MOSFET's structure, (b) Band diagram.

negative or we can say drain becomes more positive at a fixed gate voltage then p-n^+ junction will become reverse biased. Due to reverse biased p-n^+ junction a depletion region forms near the drain junctions. Since, drain is heavily doped so the depletion width tends to be narrow oxd electric field approaches to 10^6 V/cm, and hence band-to-band tunnelling will start in this region, i.e. electrons tunnel from the valence band of p-region into conduction band of n^+ region. This tunnelling will result in a singnificant leakage current, called GIDL. For GIDL to occur, drain doping level should be moderate (of order of $10^{18}/cm^3$). GIDL is a very important leakage-current component that must be minimized.

When a lot of impurities are involved in the drain region then more traps will be introduced which makes the trap-assisted tunnelling easier and hence GIDL will increase.

7.12.1 How to Reduce GIDL Current

Since, in VLSI design, the GIDL current is undesirable, therefore, proper care should be taken to minimize this current. Suggested techniques are:

(i) To reduce vertical field, which can be achieved by increasing the gate oxide thickness t_{ox}.
(ii) Using lightly doped drain L_{DD} structure, will reduce the electric field near the drain side.
(iii) By reducing the trap density.
(iv) Using moderate doping of drain region.

EXAMPLE 7.11

What is the effect of doping of drain region on the GIDL current for the following two cases:

Case I: If drain doping is much lower than $10^{18}/cm^3$.
Case II: If drain doping is very high.

Solution: *Case I:* If drain doping is much lower than $10^{18}/cm^3$ then the depletion width at p-n^+ junction and barrier height becomes too wide and tunnelling probability reduces means no GIDL or very less GIDL current. But this doping is not preferred in drain due to other constraints because it increases the drain-channel resistance.

Case II: For a larger drain doping, most of the voltage drops in the gate oxide and band bending in the Si drain regions drop below the E_g and GIDL current will increase.

Note: This tunnelling is not through the gate-oxide, but entirely in the Si drain region.

Note: In n-MOS device, it was observed that the threshold voltage initially increases with decrease in channel length then it decreases due to short channel effect (SEC). This phenomenon has been dubbed the Reverse short channel effect (RSCF). This effect is due to interaction between Si point defects that are created during the source/drain implant and the boron doping in the channel. This causes boron to pile up near the source and drain, causes the increase in V_t.

7.13 SOURCE-DRAIN SERIES RESISTANCES

In deriving the relations for current and voltage of a MOSFET, it was assumed that source and drain regions were perfectly conducting, i.e., their resistances are neglected. But in real practice, there is small voltage drop in the source and drain regions due to the finite silicon resistance and metal contact resistance as the current flows from the channel to the terminal contact as shown in Fig. 7.54. In long channel device, the source/drain parasitic resistance is negligible compared to channel resistance because channel length L is much larger than the source/drain area. But in short channel length device, the source-drain series resistance can be an appreciable fraction of the channel resistance and, therefore, cause a significant current degradation.

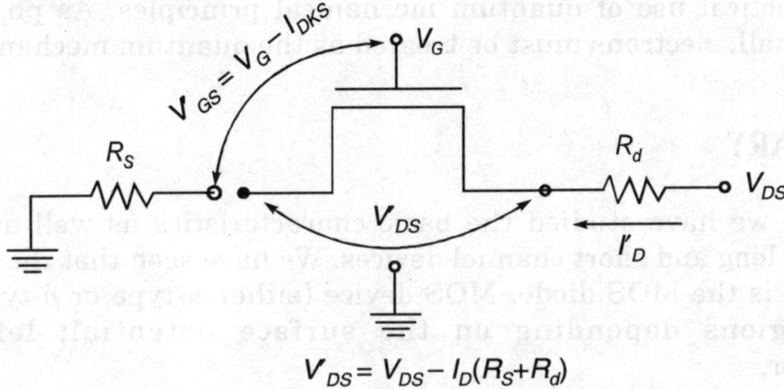

$$V_{DS} = V_{DS} - I_D(R_S + R_d)$$

FIGURE 7.54 Parasitic resistances in MOSFET.

The most significant current degradation due to source drain series parasitic resistance occurs in the linear region (low V_{DS}) when the gate voltage is high. The MOSFET current in the saturation region is least affected by the source-drain resistance because I_D is essentially independent of V_{DS}. The saturation current is only affected through gate-voltage degradation which is due to voltage drop between the source contact and source end of the channel.

7.14 QUANTUM MECHANICAL EFFECT IN MOS DEVICES

Carriers trapped within semiconductor structures with dimensions between 1 and 50 nm exhibit strong quantum mechanical effects. In classical physics, an electron must have sufficient energy to surmount a barrier. Otherwise, it recoils from the barrier. Quantum mechanics gives the probability of finding the electrons on the other side of the barrier even their energy is lower than the barrier energy (called Quantum tunnelling). Semiconductor statistics includes both classical statistics (Maxwell–Boltzmann statistics) and quantum statistics (Fermi–Dirac statistics). If the difference between the quasi-Fermi energy and the band energy is small, (i.e., the lightly doped case), one can use Maxwell–Boltzmann statistics to achieve the simpler and better approximation for the carrier concentrations. However, the validity of Boltzmann statistics fails for heavily doped devices. The devices making use of quantum mechanical principles include: resonant tunneling diodes, quantum tunnelling transistors, metal insulator metal diodes, and quantum dot transistors.

When the dielectric thickness of the MOSFET device is reduced below 4 nm, quantum mechanical (QM) effects near the silicon/silicon-oxide interface become significant. QM tunnelling (responsible for leakage current) and carrier confinement affects the profile of the inversion charge in the direction perpendicular to the interface. According to quantum mechanics, the electron density must vanish at the interface, but the classical theory based drift–diffusion model gives maximum charge density at the interface. This is the main discrepancy between the classical and quantum models.

Quantum mechanical effects are a hindrance to ever smaller conventional MOS transistors. The path to ever smaller geometry devices involves unique active devices which make practical use of quantum mechanical principles. As physical geometry becomes very small, electrons must be treated as the quantum mechanical equivalent.

7.15   SUMMARY

In this chapter, we have studied the basic characteristics as well as the operation principle of both long and short channel devices. We have seen that the core component of the MOSFET is the MOS diode. MOS device (either n-type or p-type) can operate in various regions depending on the surface potential; let us consider n-MOS capacitor,

$\psi_s < 0$	Hole accumulation	$p_s > N_a$
$\psi_s = 0$	Flat band	$p_s = N_a$
$\psi_s > 0$	Depletion	$n_i < p_s < N_a$
$\psi_s = -\psi_B$	Intrinsic	$n_i = p_s = n_s$
$\psi_s = -2\psi_B,\ \psi_s > 0$	Strong inversion	$n_s > N_a$

For weak inversion $n_i < n_s < N_a$,
Similarly for a p-MOS capacitor where n-Si semiconductor acts as a substrate,

$\psi_s > 0$	Electron accumulation	$n_s > N_d$
$\psi_s < 0$	Flat band	$n_s = N_d$
$\psi_s < 0$	Depletion	$n_i < n_s < N_d$
$\psi_s = -\psi_B$	Intrinsic	$n_i = n_s = p_s$
$\psi_s < 0$ $-\psi_s = 2\psi_B$	Strong inversion	$p_s > N_d$

These regions can be controlled by applying the voltage on the gate. The minimum voltage required to turn ON the MOS devices, i.e., to make the surface inverted, is called **threshold voltage** and represented by V_{to} for without body effect and V_t with body effect. In MOS device, drain and source regions are heavily doped and can be physically interchangeable. The MOS devices are mainly used as switch in digital design either operate in cut-off (OFF) or saturation region (ON). Adjustment of threshold voltage can be done by chosing suitable substrate doping, oxide thickness, substrate bias and gate materials.

The flat-band voltage is the gate voltage that must be applied to achieve the flat-band condition in which the conduction and valence bands in the semiconductor donot bend, and there is no space charge region in the semiconductor. The flat-band voltage is the difference between metal and semiconductor workfunctions. For ideal

MOS curve, flat-band voltage is zero because of equal workfunction of metal and semiconductor. In real MOS it is not possible.

From C-V curve of MOSFET one can extract the flat-band voltage, threshold voltage, semiconductor doping concentration and oxide thickness. The current in MOS device is due to flow of carriers from source to drain. The current is only due to majority carrier flows (either electron or hole). For a proper device operation, electrically drain should be at higher potential than source for n-MOS enhancement and reverse is true for p-MOS.

MOSFET (either p-MOS or n-MOS) further classified as enhancement-mode and depletion-mode. In enhancement-mode, channel is formed after proper application of gate voltage, whereas in depletion-mode MOS channel is already introduced during the fabrication. The threshold voltage for n-MOS enhancement-mode transistor is positive and for depletion-mode it is negative. Reverse is true for p-MOS. n-MOS enhancement-mode is widely used due to large mobility of electrons. The surface mobility of carrier is lower than the bulk mobility because of different scattering processes. MOS devices are mainly used in VLSI design because of low fabrication cost, small size and low power consumption. Their applications are mainly in microprocessor, memories and power devices.

It was found that the cut-off frequency, which is a figure of merit for the frequency response of the device, is inversely proportional to channel length and a reduced channel length will give an increased frequency capability of the MOSFET.

Any MOS device whose gate length is $\geq 1\ \mu m$ is considered as long channel device. In long channel device, the transverse field is larger in comparison to lateral field and hence lateral field is ignored. This shows the applicability of gradual channel approximation in long channel device. Any MOS device of gate length $< 1\ \mu m$ is taken as short channel device. In short channel device, both lateral and longitudinal fields are comparable to each other and hence gradual channel approximation breaks down. The surface potential analysis can be carried out by the two-dimensional analysis. Various short channel effects were considered and observed that the threshold voltage of short channel devices are lower than the long channel devices and this effect is called **threshold-voltage roll off**. In long channel device, current saturates due to pinch-off and pinch-off voltage moves towards the source side as drain voltage increases. The current after saturation voltage varies slightly. In short channel device, current mainly saturate at lower saturation drain voltage due to velocity saturation. The short-channel saturation current is lower than the long channel saturation current.

In short channel device, channel length modulation is more pronounced, but in long channel device its effect can be ignored. DIBL effect explains the experimentally observed increase in subthreshold current with the increase in drain voltage in short channel device. In short channel device mobility is mainly degraded due to surface roughness scattering. Due to high electric field at the drain of the short channel device, electrons get sufficient energy to surmount the $Si\text{-}SiO_2$ barrier without suffering any collisions. These electrons are called **hot electrons**. These hot electrons reduce the drain current, and the effect is known as **hot electron effect**. Hot carrier effect is less problematic for holes in a p-channel MOS than the electrons in n-channel MOS.

It is always advisable to operate the MOS device below the breakdown voltage in saturation region because breakdown causes a permanent damage to device. The basic properties of MOS are:

(a) Unipolar device
(b) Very high input impedance
(c) Capable of power gain
(d) Four-terminal device: Source (S), Drain (D), Gate (G), Body (B)
(e) Either enhancement or depletion
(f) Depending upon the majority carrier, either p-MOS or n-MOS

Useful relations: long channel devices.

n-MOS

$$V_{GS} < V_t \qquad \text{Cut-off transistor OFF } I_D = 0$$
$$V_{GS} \geq V_t \qquad \text{Transistor ON } I_D \neq 0$$

If $V_{GS} - V_t > V_{DS}$, Transistor in linear region

$$I_D = \mu_{neff}\, C_{ox}\, \frac{W}{L}\left[(V_{GS} - V_t')\,V_{DS} - \frac{V_{DS}^2}{2}\right]$$

If $(V_{GS} - V_t) < V_{DS}$, Transistor in saturation

$$I_D = \frac{\mu_{neff}\, C_{ox}\,(W/L)}{2}(V_{GS} - V_{to})^2$$

p-MOS

$$V_{GS} > |V_{tp}'| \qquad \text{Transistor OFF, } I_D = 0$$
$$V_{GS} \leq |V_{tp}'| \qquad \text{Transistor ON, } I_D \neq 0$$

- $(V_{GS} - |V_{tp}'|) < V_{DS}$ Transistor in linear

$$I_D = \mu_p C_{ox}\left(\frac{W}{L}\right)\left[(V_{GS} - |V_{tp}|)\,\frac{V_{DS} - V_{DS}^2}{2}\right]$$

- $(V_{GS} - |V_{tp}|) > V_{DS}$ Transistor in saturation

$$I_D = \frac{\mu_p C_{ox}(W/L)}{2}\,\frac{[V_{GS} - V_{tp}']}{2}$$

$$V_{Dsat} = V_{GS} - V_t$$

where V_t' is either V_{to} for $V_{SB} = 0$
$$V_t \text{ for } V_{SB} \neq 0$$
V_{tp}' is either V_{top} for $V_{SB} = 0$
$$V_{tp} \text{ for } V_{SB} \neq 0$$

Channel length modulation: effects only in saturation region

$$I_D = \frac{\mu_{neff}\, C_{ox}\,(W/L)}{2}\,[(V_{GS} - V_t')]^2\,(1 + \lambda V_{DS})$$

where λ is channel length modulation parameter.

$$\frac{\Delta L}{L} = \lambda$$

Effect of transverse field on mobility,

$$\mu_{eff} = \frac{\mu_0}{1 + \theta(V_{GS} - V_t')}$$

where μ_0 is low field mobility.

Short channel with velocity saturation,

$$I_D = \frac{I_D\ \text{(Neglecting velocity saturation) long}}{\left(1 + \dfrac{\mu_{eff} V_{DS}}{L v_{sat}}\right)} \qquad V_{DS} \le V_{Dsat}$$

$$I_{Dsat} \simeq \frac{I_{Dsat}\ \text{(long channel)}}{\left(1 + \dfrac{\mu_{eff} V_{Dsat}}{L v_{sat}}\right)} \qquad V_{DS} \ge V_{Dsat}$$

For short channel,

$$I_D = W\, C_{ox}\, v_{sat}\, [V_{GS} - V_t' - V(x)]$$

where $V(x)$ is potential along channel.

$$I_{Dsat} = W\, C_{ox}\, v_{sat}\, (V_{GS} - V_t' - V_{Dsat})$$

where

$$V_{Dsat} = \frac{v_{sat}}{\mu_{eff}}\, L\left[\left(1 + \frac{2\mu_{eff}\,(V_{GS} - V_t')}{v_{sat}\, L}\right)^{1/2} - 1\right]$$

Effect of R_S and R_D (short channel),

$$I_D = \frac{W\, C_{ox}\, \mu_{eff}\, [V_{GS} - V_t' - V_{DS}/2]\,(V_{DS} - 2 I_D R_S)}{L\left[1 + \dfrac{\mu_{eff}\,(V_{DS} - 2 I_D R_S)}{L v_{sat}}\right]} \qquad (V_{DS} \le V_{Dsat})$$

$$I_D = \frac{W\, C_{ox}\, \mu_{eff}\, [V_{GS} - V_t' - V_{Dsat}/2]\,(V_{Dsat} - 2 I_{Dsat} R_S)}{L\left[1 + \dfrac{\mu_{eff}\,(V_{Dsat} - 2 I_{Dsat} R_S)}{L v_{sat}}\right]} \qquad (V_{DS} > V_{Dsat})$$

High frequency performance of MOSFETs,

$$t_d \ (n\text{-MOS discharge}) = \frac{C_L V_{DD}}{1/4 \ (W/L)_n \ \mu_{neff} \ C_{ox} (V_{DD} - V_{tn})}$$

$$t_d \ (p\text{-MOS charge}) = \frac{C_L V_{DD}}{1/4 \ (W/L)_p \ \mu_{neff} \ C_{ox} (-V_{DD} + |V_{tp}|)^2}$$

where C_L is lumped capacitor.

Threshold voltage can be determined by plotting I_D, V_S, V_{GS} curve for low V_{DS}.

REVIEW QUESTIONS

1. What are the basic differences between a JFET and MOSFET?

2. Why a depletion mode MOSFET is called JFET?

3. Describe the working of a p-MOS.

4. What do you mean by threshold voltage of a MOSFET and how it is differ from pinch-off voltage?

5. What will happen if a negative potential is applied on the gate of n-MOS transistor?

6. Draw the band diagram for a p-MOS capacitor, when

     (i) $V_G = 0$ (ii) $V_G = +ve$ (iii) $V_G = -ve$

7. Plot the $I_D - V_{DS}$ curve for a p-MOS enhancement-mode transistor.

8. Show the input characteristics for a n-MOS enhancement-mode and p-MOS enhancement-mode transistors.

9. What is a basic difference between MOS capacitor and parallel plate capacitor?

10. Why polysilicon is preferred as the gate material?

11. Define flat-band voltage and calculate V_{FB} for ideal MOS capacitor.

12. The requirement for the strong inversion is $\psi_s = 2\psi_B$. Explain this.

13. Why an enhancement-mode transistor is called OFF transistor and a depletion mode transistor is ON?

14. Why MOSFET is called majority carrier transistor?

15. Why the MOSFET substrate is lightly doped and source/drain are heavily doped?

16. For a proper operation of a n-MOS enhancement mode, drain should be at higher potential than the source end. Explain this statement.

17. Ideal low-frequency capacitance versus gate voltage of an MOS capacitor with p-type substrate is shown in Fig. 7.55.

     (a) Find the threshold voltage and flat-band voltage.
     (b) Whether it is p-type or n-type? Why?

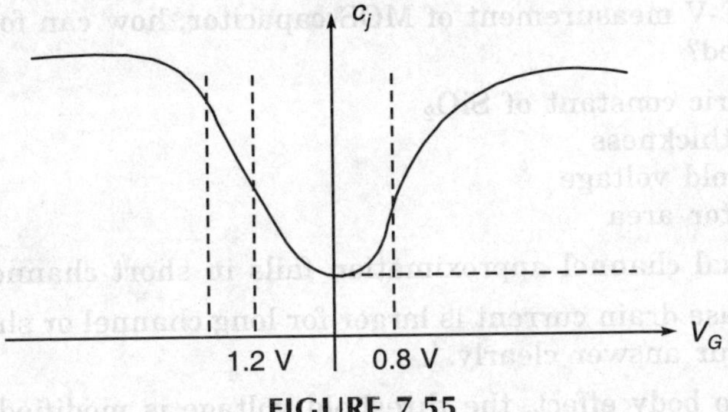

FIGURE 7.55

18. Draw the variation of magnitude of charge density in the Si (semiconductor as a function of surface potential for p-type si) with $N_a = 4 \times 10^{15}/ \text{cm}^3$ at room temperature. Clearly show all the regions of operation.

19. Write application of MOS devices.

20. Explain why the inversion charge can be neglected in the determination of the threshold voltage.

21. The variation of electric field along the channel at the thermal equilibrium is shown in Fig. 7.56. Find

    (a) Type of MOSFET and sketch the charge density
    (b) Max depletion width
    (c) Find the

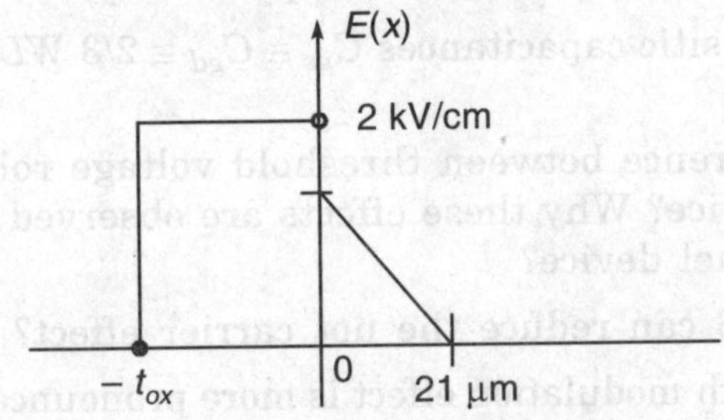

FIGURE 7.56

22. Why for current flow in the channel, inversion is required?

23. What are the sources for oxide trapped charges in MOS capacitor and how can be removed?

24. Explain how and why the current saturates in long channel as well as short channel MOS devices.

25. Explain why the threshold voltage of n-MOS increases with doping concentration (N_d) and decreases with N_a for p-MOS, as shown in Fig. 7.16.

26. With the C-V measurement of MOS capacitor, how can following parameters be extracted?

    (a) Dielectric constant of SiO_2
    (b) Oxide thickness
    (c) Threshold voltage
    (d) Capacitor area

27. Why gradual channel approximation fails in short channel device?

28. In which case drain current is larger for long channel or short channel device? Explain your answer clearly.

29. Why due to body effect, the threshold voltage is modified?

30. Why pinch-off mainly occurs at the drain side?

31. If the built-in potential at the drain and source junctions is not neglected then how the current-voltage relations for long channel device will be modified?

32. Why subthreshold current is not desirable?

33. Why is the mobility of the channel carriers not a constant? How does this variable mobility change the I-V characteristics?

34. Discuss the charge sharing effect on the threshold voltage of a short channel devices.

35. What are the advantages of *LDD* transistor?

36. In VLSI design, a steep subthreshold slope is required. Why?

37. Subthreshold current is treated as a leakage current still this region is required particularly for low power devices. Explain why.

38. Explain why parasitic capacitances $C_{gs} \cong C_{gd} \cong 2/3\ WL\ C_{ox}$ in saturation region and $C_{db} = 0$.

39. What is the difference between threshold voltage roll-off and DIBL effect in short channel device? Why these effects are observed in short channel device, not in long channel device?

40. How a *LDD* MOS can reduce the hot carrier effect?

41. Why channel length modulation effect is more pronounced in low-doped substrate devices?

42. Why the cut-off frequency is called figure of merit of MOSFET?

43. Why MOS is main building block for VLSI design?

44. Why the electrons are not moving from channel to gate or vice-versa?

45. What is the effect of stray MOS capacitors on the performance of the MOS devices?

46. As a VLSI designer how can you control the current in the MOS devices?

47. Why generally enhancement-mode MOS is preferred?

48. What are the advantages of using short channel MOS devices?

49. How the device is degraded due to hot carrier effects?

50. Explain in your own words, why the pinch-off point shifts towards the source side as drain voltage increases w.r.t. source.

51. Discuss the breakdown process in MOS devices.

NUMERICAL PROBLEMS

1. Consider an isolated MOS structure as shown in Fig. 7.57 having metal (Al), substrate of p-type semiconductor and SiO_2. The equilibrium fermi-potential of the doped Si substrate is given as 0.2 eV. Calculate the built-in potential difference across the MOS system. Assume that MOS system contains no other charges in the oxide or on the Si-SiO$_2$ interface.

Also find that after contact whether MOS is in accumulation, inversion or depletion and why? Find shift in threshold voltage between the real and ideal C-V curves.

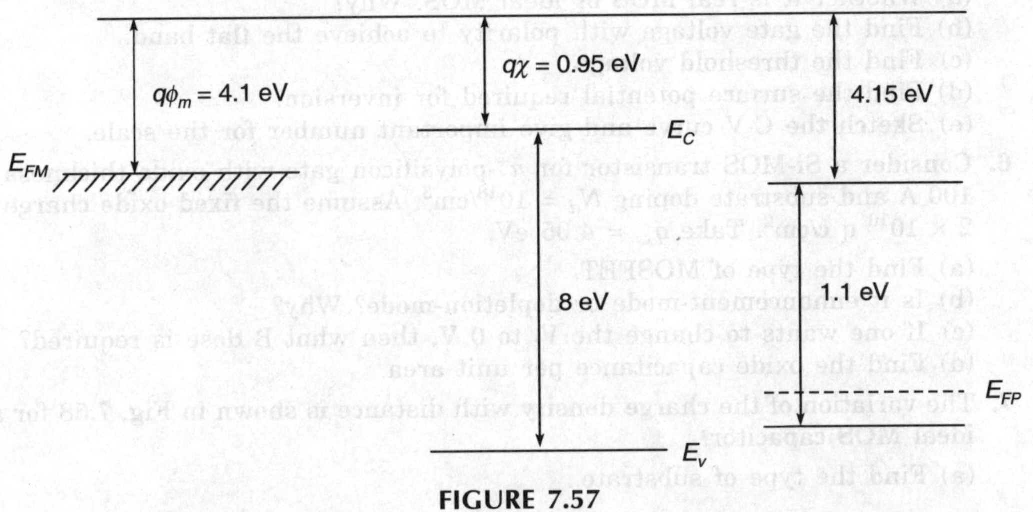

FIGURE 7.57

2. A MOS capacitor at room temperature has an Al (Aluminium) gate and a silicon substrate of acceptor doping concentration $3 \times 10^{16}/cm^3$. The oxide thickness is 550 Å. Assume there is no charge at the interface as well in the oxide. Find

(a) Whether it is n-MOS or p-MOS

(b) Find the flat-band voltage

(c) Fermi-potential

(d) Maximum depletion width

What will be the voltage applied on the gate to achieve flat-band? Also find the polarity.

3. In Problem 2, find

(a) Flat-band capacitor
(b) Capacitor in accumulation
(c) Minimum capacitance value
(d) Draw the C-V curve for the MOS system also.

4. An ideal MOS capacitor has an Al gate, an oxide thickness of 1000 Å and substrate of acceptor doping of $4 \times 10^{15}/cm^3$.

(a) Find the surface potential at the onset of strong inversion.
(b) Find the flat-band voltage.
(c) Find the threshold voltage.
(d) Find the maximum depletion width.
(e) Discuss qualitatively, on the threshold voltage if some oxide charges are there.

5. An n^+-polysilicon-gate n-channel MOS transistor is made of a p-type Si-substrate with $N_a = 5 \times 10^{15}/cm^3$. The SiO_2 thickness is 200 Å in the gate region and the effective interface charge is 4×10^{10} c/m^2.

(a) Whether it is real MOS or ideal MOS. Why?
(b) Find the gate voltage with polarity to achieve the flat band.
(c) Find the threshold voltage.
(d) Find the surface potential required for inversion.
(e) Sketch the C-V curve and give important number for the scale.

6. Consider a Si-MOS transistor for n^+-polysilicon gate with oxide thickness of 100 Å and substrate doping $N_d = 10^{18}/cm^3$. Assume the fixed oxide charge of 2×10^{10} q c/cm^2. Take $q_{xs} = 4.05$ eV.

(a) Find the type of MOSFET.
(b) Is it enhancement-mode or depletion-mode? Why?
(c) If one wants to change the V_t to 0 V, then what B dose is required?
(d) Find the oxide capacitance per unit area.

7. The variation of the charge density with distance is shown in Fig. 7.58 for an ideal MOS capacitor.

(a) Find the type of substrate.

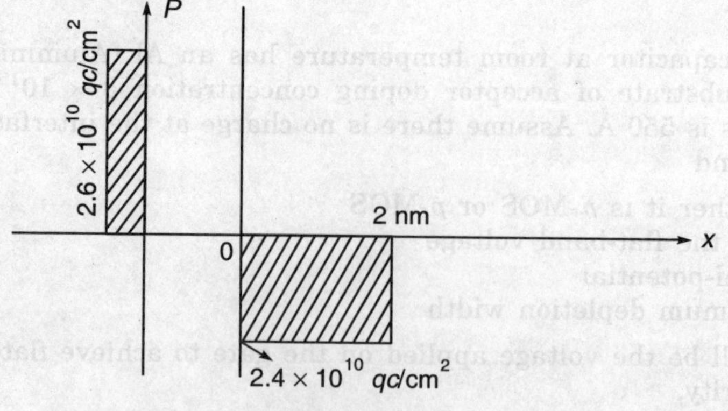

FIGURE 7.58

(b) Whether MOS system is in accumulation, depletion or inversion.

(c) Draw the energy band diagram.

(d) Find the substrate doping concentration.

8. For an ideal MOS capacitor at 300 K with $N_a = 8 \times 10^{15}/cm^3$ and $t_{ox} = 350$ Å. Find the voltage across the oxide and electric field in the oxide at threshold.

9. Show that the maximum depletion width (W_{dm}) in a p-type semiconductor is given by $2L_D \sqrt{\ln (N_a/n_i)}$ where L_D is the debye length.

10. The variation of the electric field with the distance at thermal equilibrium is shown in Fig. 7.59 for an ideal MOS with p-Si substrate.

   (a) Sketch the charge density V_S x.

   (b) Sketch the potential V_S x.

   (c) Draw the energy band diagram corresponding to the electric field shown in Fig. 7.59.

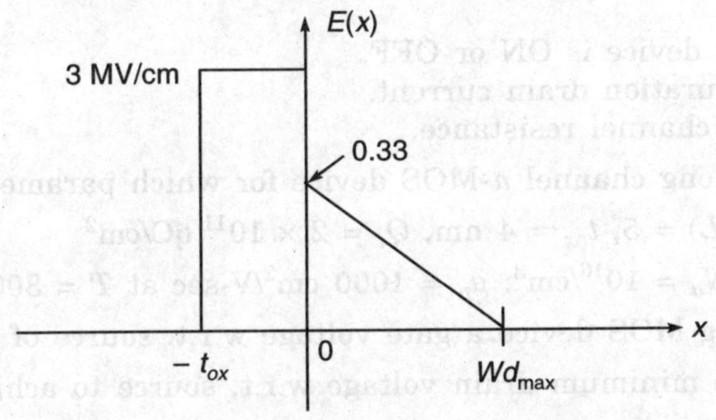

FIGURE 7.59

11. If the thickness of the gate oxide is 100 Å and applied gate potential is 5 V. in Fig. 7.59.

   (a) Find the surface potential.

   (b) Find the surface charge density.

   (c) To get the value of maximum depletion width, what other information do you want and why?

12. The electric field distribution to flat-band voltage is shown in Fig. 7.60.

   (a) Find the flat-band voltage at $x = 0$.

   (b) Find the electric field E_0 at $x_0 = 2$ nm for getting flat-band voltage of 60 V.

   (c) Discuss the significance of the threshold voltage.

13. If the shift in the gate voltage bias from its ideal case is 1 V for a n-MOS capacitor, find the fixed-oxide charge.

14. Consider a long channel n-MOS device for which gate oxide thickness is 500 Å and $N_a = 4 \times 10^{15}/cm^2$. Assume there is no interface as well as oxide charges. A gate voltage of -2 V is applied in the MOS. Find

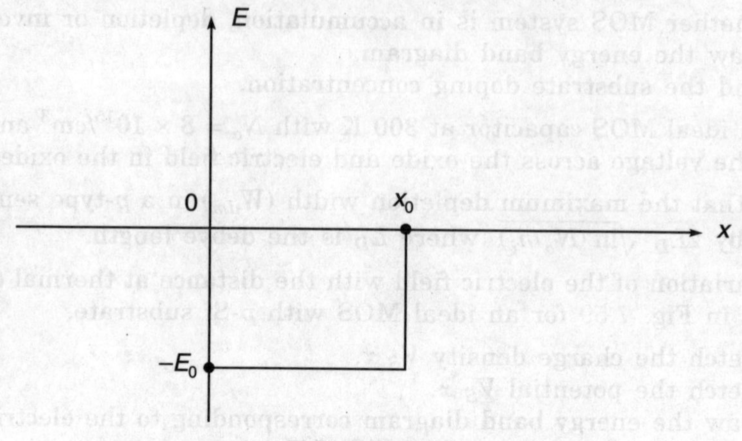

FIGURE 7.60

(a) Whether device is ON or OFF.

(b) Find saturation drain current.

(c) Find its channel resistance.

15. Consider a long channel n-MOS device for which parameters are given as:

$$(W/L) = 5, \ t_{ox} = 4 \text{ nm}, \ Q_i = 2 \times 10^{11} \text{ qC/cm}^2$$

$$N_a = 10^{16}/\text{cm}^3; \ \mu_n = 1000 \text{ cm}^2/\text{V-sec at } T = 300 \text{ K}$$

On such long MOS device a gate voltage w.r.t. source of 2 V is applied.

(a) Find the minimum drain voltage w.r.t. source to achieve the saturation condition.

(b) Find the drain current in the channel when device operates in the saturation region.

(c) Repeat the calculation of parts (a) and (b) for p-MOS transistor.

16. Consider an n^+-polysilicon gate and p-type silicon substrate doped to $N_a = 5 \times 10^{16}/\text{cm}^3$. Assume $Q_i = 10^{11}/\text{cm}^2$. Determine the thickness of gate oxide such that $V_{to} = 0.40$ V. Take $\phi_{ms} = -1.15$ V.

17. Calculate the electric field and voltage drop across the oxide of $t_{ox} = 150$ Å.

18. Find the minimum gate voltage w.r.t. source for long channel required to turn the n-MOS transistor ON for which parameters are:

$$N_a = 10^{16}/\text{cm}^3, \ Q_i = 10^{11}/\text{cm}^2, \ V_{to}n = 0.6 \text{ V}$$

Calculate the required minimum drain voltage w.r.t. source bring the transistor in linear region?

If $\mu_n = 1000$ cm^2/V-sec at $T = 300$ K, find the drain current in the linear region. Find also the average electric field in the channel. What is the channel resistance? Take $W = 10$ μm and $L = 2$ μm.

19. Find the voltage drops at the source and drain junction for $V_{GS} = 1$ V and drain-to-source voltage is 5 V. Neglect the built-in potential.

20. How the current-voltage equations for a long channel device is modified when one includes the built-in potential?

21. How the drain current in linear is modified for the Problem 7.18 due to channel length modulation for long channel device? Assume $\lambda = 2$ V.

22. Calculate:
   (a) Drain current
   (b) Output resistance
   (c) Transconductance
   (d) Saturation current

for a n-channel Si MOSFET at $T = 300$ K with an oxide thickness of 200 Å and $\mu_{neff} = 1000$ cm^2/V-sec. Take $W = 100$ μm and $L = 5$ μm, $N_a = 10^{16}$/cm^3. The bias conditions are:

$$V_G = 2 \text{ V}, \quad V_S = 0 \text{ V}$$
$$V_D = 5 \text{ V}, \quad V_S = 0 \text{ V}$$

Assume $V_{to} = 0.67$ V.

Determine the change in threshold voltage if bulk is connected to –2 V.

23. A MOSFET has a threshold voltage of 0.5 V (without body effect) and a subthreshold swing of 100 mV/decade and drain current of 0.1 μA at V_{to}. Find the subthreshold leakage current at $V_G = 0$. Also find the reverse substrate-source voltage required to reduce the leakage current by one order of magnitude. Take $N_a = 5 \times 10^{17}$/cm^2 and $t_{ox} = 5$ nm.

24. Consider a n-channel MOS device of length $L = 2$ μm which is biased for $V_{GS} = 5$ V. Take $\mu_{neff} = 1000$ cm^2/V-sec at $T = 300$ K, $C_{ox} = 6.9 \times 10^{-8}$ F/cm^2 and $V_{ton} = 0.65$ V. Design the MOS channel so that $I_{Dsat} = 4$ mA.

25. Consider the following p-channel MOSFET process substrate doping = 10^{15}/cm^3 polysilicon gate doping density = 10^{20}/cm^3, $t_{ox} = 650$ Å, oxide interface charge density = 2×10^{10}/cm^2. Use $\varepsilon_{si} = 11.7\ \varepsilon_0$, $\varepsilon_{ox} = 3.97\ \varepsilon_0$ where ε_0 is the free space permittivity.

(a) Calculate the threshold voltage.
(b) Determine the type and amount of channel ion implantation which are necessary to achieve the threshold voltage of $V_{to} = -2.0$ V.

26. Consider an n-channel MOSFET with an Al metal gate on p-Si. The hole concentration is $N_a = 4 \times 10^{15}$/cm^3 and $t_{ox} = 500$ Å. The flat-band voltage is found to be -0.842 V, the gate capacitance per unit area is 6.903×10^{-8} F/cm^2, and the threshold voltage is 0.230 V. The debye length is 6.485×10^{-6} cm and $\mu_{neff} = 600$ cm^2/V-sec. Take $W = 40$ μm and $L = 10$ μm.

(a) Find V_{Dsat} and I_{Dsat} for $V_{GS} = 1, 2, 3, 4$ V.
(b) Find g_m values for V_{GS} in Part (a).
(c) For g_m at $V_{GS} = 4$ V, find the value of source resistance R_s that reduces g_m to $0.85\ g_m$.
(d) As g_m gate larger, what influence does R_s have?
(e) Find the approximate value of f_T for V_{GS} in Part (a).

(f) If the gate length is reduced from 10 μm to 2 μm, how much does f_T increase or decrease?

(g) Draw the small-signal model.

27. A n-MOS has a threshold voltage of 0.8 V and subthreshold swing of 100 mV/decade. The drain current at $V_{GS} = 0$ is 0.1 μA. Find the subthreshold leakage current for $V_{GS} = V_{to}$ V. Explain how the increase in substrate concentration will result in softer subthreshold slope.

28. The high-frequency C-V curve of a MOS capacitor is shown in Fig. 7.61. The area of the MOS device is 2×10^{-3} cm^2. The workfunction difference between metal and semiconductor ϕ_{ms} is -0.45 V. The substrate doping concentration is 10^{16}/cm^3.

(a) Is the substrate p-type or n-type?
(b) Find the oxide thickness.
(c) What is flat-band capacitance and trapped oxide charge density?

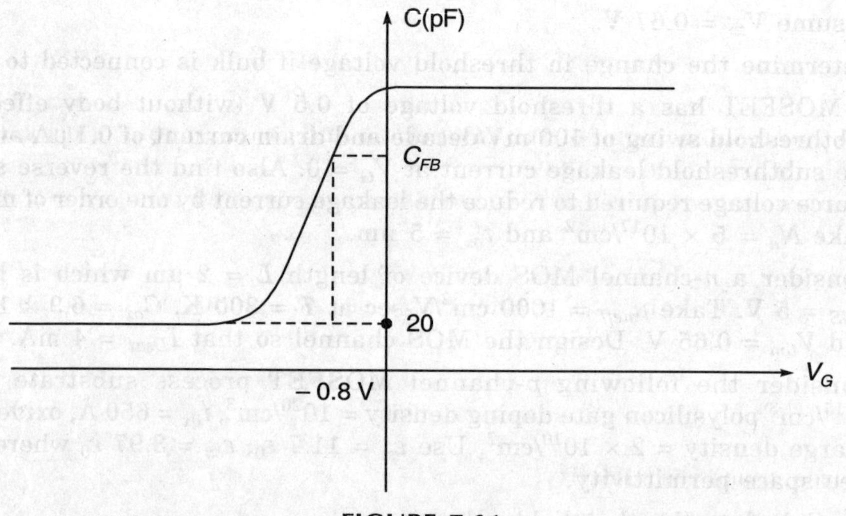

FIGURE 7.61

29. Consider an n-channel Si MOSFET with the following parameters at room temperature:

μ_{neff} = 1000 cm^2/V-sec, t_{ox} = 100 Å, W = 40 μm, L = 5 μm, L_D = 6.845 × 10^{-6} cm, Q_i = 2 × 10^{10}/cm^2, V_{FB} = − 0.842 V, $|\psi_B|$ = 0.335 V.

Assume V_{DS} be greater than $3\,kT/q$. Calculate and plot the subthreshold current for V_{GS} = −0.2 V to threshold voltage value.

30. Consider an n-channel MOSFET with t_{ox} = 400 Å, W = 50 μm, L = 2 μm, μ_{neff} = 1000 cm^2/V-sec at room temperature. Assume the substrate doping is 10^{16}/cm^3. Take V_{FB} = −0.846 V and fixed oxide charge of 5 × 10^{10} q c/cm^2.

(a) Calculate the drain saturation current for V_{GS} = 0 V and V_{SB} = 0 V.

(b) If the gate-voltage has raised to 3 V, w.r.t. source and bulk is still tied to source, find the new saturation drain current and discuss the effect of gate.

(c) Find the minimum drain voltage w.r.t. source so that n-MOS should operate in saturation region.

(d) Find the bulk voltage w.r.t. source so that the threshold voltage to be lowered by 2 V.

31. For an ideal n-channel MOSFET, the ratio of width-to-channel is 10 and the electron effective mobility is 400 cm^2/V-sec. Take $t_{ox} = 500$ Å and $V_{to} = 0.80$ V.

Determine the maximum value of source resistance so that the saturation transconductance g_m is reduced by no more than 20% from its ideal value when $V_{GS} = 5$ V.

Calculate the cut-off frequency. Discuss the effect of overlapping of source/drain region with gate on the cut-off frequency.

32. Consider an n-channel MOSFET with the following parameters: $V_{to} = 0.42$ V, $\mu_{neff} = 600$ cm^2/V-sec, $t_{ox} = 500$ Å, $(W/L) = 5$, $L_D = 6.485 \times 10^{-6}$ cm. Assume the source of the MOSFET is connected to ground and take $V_{SB} = 0$, let $V_G = 1$ V, $V_D = 3$ V.

Draw the low-frequency small signal model and show the value of each parameter.

If the drain voltage is reduced to 900 mV then draw the small signal low frequency model with value of each parameter.

33. Consider an n-channel Si-MOSFET of length $L = 2$ μm, with aluminium gate ($q\phi_M = 4.1$ eV). Assume that there is no fixed charges either in the oxide or interface. Take substrate doping as 10^{17}/cm^3 and $x_s = 4.05$ V. If the channel length of the MOS device is reduced to 0.25 μm, so that the threshold voltage is reduced to 0.54 V. Find the change in threshold voltage and also find percentage change in threshold voltage. Explain the main cause of threshold voltage reduction.

34. Find the channel length L of Si-n channel MOSFET for which velocity saturation effect reduces the submicron current by a factor of 2 for $V_{DS} = 2$ V. Assume $\mu_{eff} = 500$ cm^2/V-sec, and $V_{sat} = 1 \times 10^7$ cm/sec. If the channel length reduces to 0.25 μm then find the ratio of drain current of long channel to short channel.

35. Plot I_D–V_{DS} characteristics for a long channel Si n-MOS transistor for which $W = 20$ μm, $L = 2$ μm, $t_{ox} = 100$ Å, $V_{to} = 1$ V and the channel length modulation parameter is 0.04/V. Take V_{GS} from 1 V to 5 V.

36. Consider a p-channel MOSFET with $t_{ox} = 600$ Å and $N_d = 5 \times 10^{15}$/cm^3. Determine the body-to-source voltage such that shift in threshold voltage from $V_{BS} = 0$ is −1.5 V.

37. Consider an n-channel MOS process with the following parameters: substrate doping is 10^{16}/cm^3, polysilicon gate doping is 2×10^{20}/cm^3, $t_{ox} = 50$ nm, interface oxide fixed charge density $= 4 \times 10^{10}$/cm^2 and source/drain diffusion density

$= 10^{17}/\text{cm}^3$. The channel region is implanted with p-type impurities of $10^{11}/\text{cm}^2$ to adjust the threshold voltage. The junction depth of the source and drain diffusion regions is 1.0 μm. Find V_{to} for $L = 0.7$ μm, $V_{DS} = $ Si $V_{SB} = 0$.

38. Consider an enhancement n-MOS transistor with following parameters:

$$V_{to} = 0.8 \text{ V}, \ \gamma = 0.2\sqrt{V}, \ \lambda = 0.05 \text{ V}, \ |2\psi_B| = 0.58 \text{ V and } \kappa'_n = 20 \text{ μA/V}^2$$

(a) When the transistor is biased with $V_G = 3$ V, $V_D = 5$ V, $V_S = 1$ V, $V_B = 0$, a drain current of 0.24 mA flows. Find (W/L).

(b) Calculate I_D for $V_G = 3$ V, $V_D = 1.5$ V, $V_S = 1$ V and $V_B = 1$ V.

(c) Find the output resistance of the MOS for $I_{Dsat} = 0.30$ V.

39. For an n-channel MOSFET, the given parameter are: $V_{FB} = -0.84$ V, $V_B = 0$ V. $Q_i = 10^{12}q$ C/cm^2, $N_a = 10^{16}/\text{cm}^3$, $L = 2$ μm, $W = 20$ μm and $\mu_{neff} = 1000$ cm^2/V-sec. If $V_G \geq 2$ V, $V_D = 4$ V, $V_S = 0$ V, find the drain current for two cases:

(a) When $t_{ox} = 50$ Å
(b) When $t_{ox} = 100$ Å

Discuss the effect of oxide thickness on the current qualitatively with suitable reason.

40. Consider the cross-section of the n-MOS device that includes source and drain resistances R_S and R_D respectively. The transistor parameters are $V_{to} = 0.8$ volt, $\mu_{neff} = 1000$ cm^2/V-sec, $t_{ox} = 100$ Å and $(W/L) = 5$.

(a) Find drain current I_D for $R_S = 0 = R_D$ when

    (a) $V_{GS} = 4$ V, $V_{DS} = 3$ V
    (b) $V_{GS} = 2$ V, $V_{DS} = 5$ V
    (c) $V_{GS} = .6$ V, $V_{DS} = 5$ V

(b) Find drain current I_D when $R_S = R_D = 50$ Ω for

    (a) $V_{GS} = 4$ V, $V_{DS} = 3$ V
    (b) $V_{GS} = 2$ V, $V_{DS} = 5$ V
    (c) $V_{GS} = 0.6$ V, $V_{DS} = 5$ V

Compare these two results and discuss the effect of R_S and R_D on drain current.

41. For a p-MOS enhancement-mode transistor following parameters are given: $|V_{top}| = 1$ volt, $\mu_{neff} = 1000$ cm^2/V-sec, $t_{ox} = 400$ Å, $V_S = 0$ V, $V_{DS} = 5$ V, $I_{Dsat} = 1.5$ mA.

(a) Find the (W/L) ratio to achieve the minimum saturation drain current.
(b) Find transconductance.
(c) What will happen to drain saturation current if the (W/L) ratio is increased to 5 times of the value of (a)? Discuss the result qualitatively.

42. Discuss the effect of substrate doping concentration on the drain current qualitatively. Consider an n-channel MOSFET with the following parameters:

$V_{FB} = -0.846$ V, $t_{ox} = 100$ Å, $\mu_{neff} = 800$ cm^2/V-sec,

$(W/L) = 4$, $Q_i = 10^{10}$ qc/cm^2,

$V_{GS} = 2$ V, $V_{DS} = 5$ V

Find the drain current for two-substrate doping considerations:
$N_a = 10^{16}/cm^3$, $N_a = 10^{15}/cm^3$.

43. Design an ideal n-channel silicon enhancement-mode transistor with a polysilicon gate of threshold voltage $V_{to} = 0.8V$. Assume an oxide thickness of 400 Å, a channel length of 1.5 µm and $Q_i = 1.5 \times 10^{11}/cm^2$. It is desired to have a drain current of 50 µA at $V_{GS} = 2.5$ V and $V_{DS} = 100$ mV. Determine the substrate doping concentration and channel width. Take $\mu_{neff} = 800$ cm^2/V-sec.

44. For an n-channel n^+-polysilicon-SiO$_2$-Si MOSFET with gate oxide = 10 nm, $N_a = 10^{16}/cm^3$, $V_G = 3$ V and $V_S = 0$. Calculate V_{Dsat}.

45. For an n-channel field effect transistor with $N_a = 10^{16}/cm^3$, $Q_f/q = 5 \times 10^{11}/cm^2$, $t_{ox} = 500$ nm and $V_B = 0$. Calculate the charge in threshold voltage if substrate bias is increased from 0 V to 2 V. Why change in threshold voltage occurs due to body effect? Discuss it qualitatively.

46. For an ideal Si-SiO$_2$ MOS capacitor with $t_{ox} = 100\,$Å and $N_a = 10^{17}/cm^3$, find the applied voltage required inversion.

47. Consider a Si-SiO$_2$ n-MOS capacitor at $T = 300$ K with $V_{FB} = 0.846$, $N_a = 10^{16}/cm^3$ and $Q_i/q = 5 \times 10^{11}/cm^2$. Find the minimum gate voltage required to turn the transistor ON.

48. Assume that the oxide trapped charge in an oxide layer has a triangular distribution like

$$\rho_{ot}(y) = q \times (5 \times 10^{23}\ y)\ cm^{-3}$$

The oxide thickness is 10 nm. Find the change in the flat-band voltage due to Q_{ot}.

49. One model used to describe the variation in electron mobility in the channel is given by

$$\mu_{eff} = \frac{\mu_0}{1 + \theta(V_{GS} - V_{tn})}$$

where θ is mobility degradation factor.

Compare the mobility of the carrier for $\theta = 0$ and $\theta = 0.5/V$. Assume $V_{GS} - V_{tn} \geq 1$ and $\mu_0 = 500$ cm^2/V-sec. Discuss the result qualitatively.

50. The threshold voltage of an n-MOS transistor with $t_{ox} = 400$ Å, needs to be shifted in the negative direction by 1.4 V. Determine the type and dose of an ion implant required.

CHAPTER
8

CMOS: Scaling and Its Limitations

8.1 INTRODUCTION

The phenomenal growth in the electronics industry over last decades are mainly due to the rapid advances in the integration technologies and very large-scale system design. The use of integrated circuits in high performance areas like computing, telecommunication, etc. has been growing rapidly. The requirement of high speed computation and information processing power of these applications is driving force for the fast grow of this field. The development of various leading-edge-technologies is summarized in Table 8.1. As seen from Table 8.1 until 1960s logic circuits were constructed with bulky components such as transistors and resistors that came as individual parts. These circuits were called **discrete circuits**, but the advent of the integrated circuits made it possible to place a number of the components, (thus an entire circuit) on a single chip. In the beginning, the number of circuits on the chip were few, but as the technology advanced a large number of transistors as well as circuits are placed on the chip. This was possible due to the invention of CMOS technology. CMOS stands for complimentary metal oxide semiconductor and widely used in modern technology. A CMOS device is formed by the combination of a p-MOS and an n-MOS transistor. CMOS is an advanced IC manufacturing process technology characterized by high integration, low cost, low power consumption and high performance. In other words, CMOS is an integrated circuit that uses both p-MOS and n-MOS devices on the same substrate, resulting in extremely low power dissipation. But main drawback of CMOS technology is that it is slower than transistor transistor logic (TTL) and more susceptible to electrical static discharge circuits are mainly useful in battery-powered devices such as portable electronic devices.

Table 8.1 Development of Various Leading-edge Technologies

Integration level	*No. of transistors*
Zero scale integration (ZSI)	1
Small scale integration (SSI)	2–30
Medium scale integration (MSI)	$30–10^3$
Large scale integration (LSI)	$10^3–10^5$
Very large scale integration (VLSI)	$10^5–10^7$
Ultra large scale integration (ULSI)	$10^7–10^9$
Giga-scale integration (GSI)	$10^9–10^{11}$
Tera-scale integration (TSI)	$10^{11}–10^{13}$

8.2 MOORE'S LAW

As seen from Fig. 8.1(a), the number of transistors per chip has continued to increase at an exponential rate over the last three decades. This confirms the Gorden Moore's prediction on the growth rate of chip complexity. According to Moore's prediction in 1960s, which is also called **Moore's law**, the number of transistors on a chip (die) doubles every 18 to 24 months. Digital CMOS integrated circuits have been the

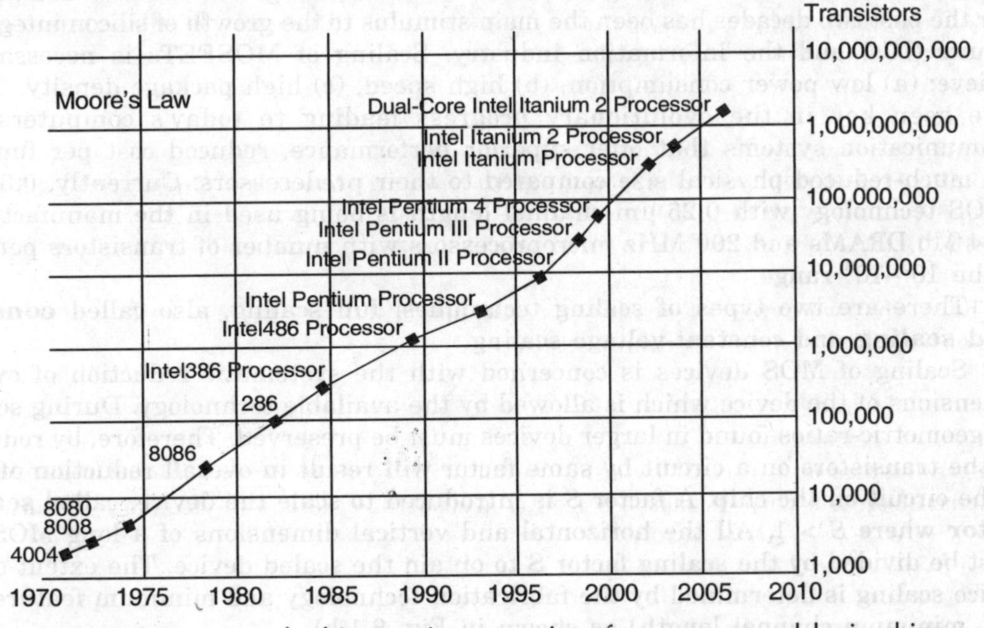

FIGURE 8.1(a) Level of integration over time for memory and logic chips.

driving force behind the development of VLSI (very large scale integration) circuits. These circuits are used for high-performance computing and other scientific and engineering applications. The demand for digital CMOS ICs will be continuously increasing due to salient features of CMOS such as low power consumption, better performance, high speed and better processing technology.

Historically, one of the advantages of CMOS technology over competing technologies like transistor-transistor logic (TTL) and emitter coupled logic (ECL) has been its negligible standby power.

EXAMPLE 8.1

Why in VLSI, main circuit component is CMOS?

Solution: The prevailing VLSI technology comprises CMOS devices because of their unique characteristic of negligible power consumption during switching action as well as standby power, which allows the integration of tens of millions of transistors on the chip.

8.3 MOSFET SCALING

The design of high-density chips on MOS VLSI technology requires that package density (i.e., number of transistors on the chip) should be as large as possible. This is only possible if the size (lateral as well as vertical) of the MOS transistors are as small as possible. The reduction of the size, i.e., the dimensions of the MOSFETs is commonly referred to as scaling. The steady downscaling of transistor dimensions over the past two decades has been the main stimulus to the growth of silicon integrated circuits (ICs) and the information industry. Scaling of MOSFETs is necessary to achieve: (a) low power consumption, (b) high speed, (c) high package density. These have been key in the evolutionary progress leading to today's computers and communication systems that offer superior performance, reduced cost per function and much-reduced physical size compared to their predecessors. Currently, 0.35 μm CMOS technology with 0.25 μm channel length is being used in the manufacturing of 64 Mb DRAMs and 200 MHz microprocessors with number of transistors per chip in the 10^7–10^8 range.

There are two types of scaling techniques: full scaling, also called **constant field scaling** and constant-voltage scaling.

Scaling of MOS devices is concerned with the systematic reduction of overall dimensions of the device which is allowed by the available technology. During scaling the geometric-ratios found in larger devices must be preserved. Therefore, by reducing all the transistors on a circuit by same factor will result in overall reduction of area of the circuit on the chip. A factor S is introduced to scale the device, called **scaling factor** where $S > 1$. All the horizontal and vertical dimensions of a long MOSFET must be divided by the scaling factor S to obtain the scaled device. The extent of the device scaling is determined by the fabrication technology and minimum feature size (i.e., minimum channel length) as shown in Fig. 8.1(b).

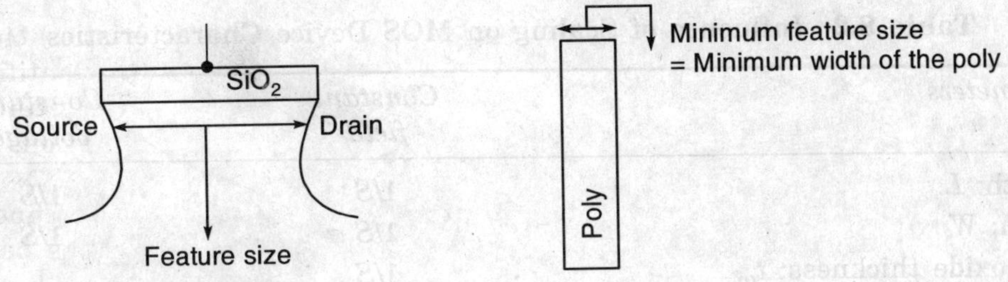

FIGURE 8.1(b)   Minimum feature size.

8.3.1 Full Scaling

In this scaling mode, the power supply and the device dimensions (both lateral and vertical) are scaled by the same factor S such that the internal electric fields remain unchanged. This results in scaling of the drawn current (or current drive). At the same time gate capacitance is also scaled due to reduced device size. This provides a reduction on gate charging time which directly results into higher speed. Power-dissipation per transistor is also reduced. Constant-field scaling provides a good framework for CMOS scaling without degrading the reliability of the device.

There are many parameters like thermal voltage (kT/q) and energy band gap that cannot be scaled. Since, the built-in-potential ϕ_{bi} and surface potential ψ_s are determined by energy gap and hence they cannot scale either. Consequently, depletion widths do not scale as much as other parameters which results in worsened short-channel effects. Another parameter which cannot be easily scaled is the threshold voltage. This sets a lower limit to the power supply. Off-state current and subthreshold slope do not also scale well.

Although, constant field scaling provides a reasonable guideline but scaling power supply by same factor as the dimension is not always practical. This is due to our inability to scale the subthreshold slope properly.

8.3.2 Constant Voltage Scaling

In constant voltage scaling, all dimensions of the MOSFET are reduced by a scaling factor S whereas the power supply voltage and the terminal voltages remain unchanged. The doping densities must be increased by a factor of S^2 to preserve the charge-field relations. One problem with constant voltage scaling is that the oxide-field increases because t_{ox} (oxide thickness) is scaled. To avoid this problem one should scale t_{ox} slower than other parameters. Modifications of the constant field scaling and constant voltage scaling have also been tried to avoid the high-field effects. For example, on third type of scaling, called **quasi-constant voltage scaling,** physical dimensions and voltages are scaled by different factors. Table 8.2 gives the scaling effects on various parameters for constant voltage and constant field scaling.

Table 8.2 Influence of Scaling on MOS Device Characteristics

Parameters	Constant field	Constant voltage
Length: L	$1/S$	$1/S$
Width: W	$1/S$	$1/S$
Gate oxide thickness: t_{ox}	$1/S$	1
Supply voltage: V_{DD}	$1/S$	1
Threshold voltage	$1/S$	1
Substrate doping: N_a, N_d	S	S^2

EXERCISE 8.2

Why constant voltage scaling is introduced?

Solution: While the full scaling method dictates that the power supply voltage and all terminal voltages be scaled down properly with the device dimensions, but in many cases the scaling of power supply is not practical, e.g. the peripheral and interface circuitary require certain voltage levels for all the input and output voltages. For these reasons, constant voltage scaling is introduced.

EXERCISE 8.3

Find the scaled linear drain current using constant field scaling.

Solution: Linear drain current before scaling,

$$I_D = \frac{\mu_n C_{ox}}{2} \left(\frac{W}{L} \right) [2(V_{GS} - V_{tn}) V_{DS} - V_{DS}^2]$$

After scaling,

$$I_D' = \frac{\mu_n C_{ox}}{2} \left(\frac{W'}{L'} \right) [2(V_{GS}' - V_{tn}') V_{DS}' - V_{DS}'^2]$$

Since $\mu_n' = \mu_n$,

$$C_{ox}' \Rightarrow SC_{ox}$$

$$\frac{W'}{L'} \Rightarrow \left(\frac{W}{S} \right) \left(\frac{S}{L} \right) \Rightarrow \left(\frac{W}{L} \right)$$

$$V_{DD}' \Rightarrow \frac{V_{DD}}{S} \qquad (V_{GS}' - V_{tn}') \Rightarrow \frac{V_{GS} - V_{tn}}{S}$$

Substituting all these values, we have

$$I'_n = \frac{s\mu_n \, C_{ox}}{2}\left(\frac{W}{L}\right)\frac{1}{S}\,[2(V_{GS} - V_{tn})\,V_{DS} - V_{DS}^2]$$

$$= \frac{1}{S}\left[\frac{\mu_n \, C_{ox}}{2}\left(\frac{W}{L}\right)\{2(V_{GS} - V_{tn})\,V_{DS} - V_{DS}^2\}\right]$$

$$= \frac{I_D}{S} \Rightarrow \frac{\text{Before scaling}}{\text{Scaling factor}} = \text{Current after scaling}$$

This clearly shows that the linear current of a MOS device is reduced by a factor S after scaling.

EXERCISE 8.4

Find the scaled linear drain current with constant voltage scaling.

Solution: Without scaling,

$$I_D = \frac{\mu_n \, C_{ox}}{2}\left(\frac{W}{L}\right)[2(V_{GS} - V_{tn})\,V_{DS} - V_{DS}^2]$$

After scaling,

$$I'_D = \frac{\mu_n C'_{ox}}{2}\left(\frac{W'}{L'}\right)[2(V'_{GS} - V'_{tn})\,V'_{DS} - V_{DS}'^2]$$

$$\mu'_n \to \mu_n$$
$$C'_{ox} \to SC_{ox}$$
$$W'/L' \to (W/L)$$
$$(V'_{gs} - V'_{tn}) \to (V_{gs} - V_{tn})$$
$$V'_{DS} \to V_{DS}$$

Hence

$$I'_D = \frac{S\mu_n C_{ox}}{2}\left(\frac{W}{L}\right)[(V_{GS} - V_{tn})\,(V_{DS} - V_{DS}^2)]$$

$$= SI_D \text{ linear (without scale)}$$

Therefore, drain current in linear region increases after scaling by constant voltage.

8.4 SCALING LIMIT OF CMOS

Silicon CMOS has emerged over the last 25 years as the predominant technology of the microelectronics industry. The concept of scaling, as proposed by Dennard, et al., has been consistently applied over many technological generations, resulting in consistent improvement in device density (package density), speed and power consumption (low power). Device dimensions are well below the micrometer scale and enter into the nanometer regime (called sub-micron devices). The industry roadmap for CMOS technology development suggests that CMOS technology is nearing some fundamental physical limits which is not too distant future.

As the CMOS dimensions (particularly channel length) is scaled to the nanometer regime (< 100 nm), the electrical barriers on the device begin to lose their insulating properties because of the thermal injection and quantum-mechanical tunnelling of the carriers. This will result in the leakage current which causes a rapid rise of the stand-by power of the chip. This places a limit on the integration level as well as on the switching speed.

The key challenges on further scaling of bulk CMOS technology into nanometer (sub. 0.1 μm) regime are: lithography, power supply, threshold voltage, short channel effects, gate oxide thickness, high field effects, dopant number fluctuations and interconnection delays.

8.4.1 Lithography

This is the ability of lithography patterning to continually reduce the lateral dimension of the device. This has directly led us into the ultra-large scale integration (ULSI) era. Therefore, improvement on the lithography technologies is needed in order to bring CMOS into the Nano-scale regime.

Optical lithography has exceeded previously predicted resolution limits many times over the combination of improved lenses with larger numerical aperture and use of shorter wavelength for illumination. With the exception of near-field techniques, which may be impractical for device fabrication applications, there are no current expectations that optical lithography techniques will extend into the sub-100 nm regime.

For the fabrication of such ultrasmall devices, mainly X-ray lithography is a prime candidate, but the challenges in implementing an X-ray lithography technology lie primarily in mask fabrication. X-ray mask consists of thin (~ 2–5 μm) membrance of Si or a Si compound (such as Si_3N_4 or SiC) patterned with an X-ray absorbing material. Main absorber is electroplated gold. Precise control of mechanical stress on the absorber-covered membrance must be maintained. Also more stringent control of defects is required as compared to system using image reducing optics.

Another X-ray lithography, called **extreme ultraviolet (EUV) lithography**, is used for fabrication of sub 100 nm devices.

This technique uses reflective optics at the 13-nm wavelength with LIX reduction scheme. Key challenges to this technology lie in the radiation source, in the multilayer thin-flim mirror optics and in mask fabrication.

Electron beam lithography is also used for the nanostructure patterning within the research environment, but main challenge for e-beam lithography lies on throughput. An alternative techniques which has received recent interest is electron beam projection lithography.

Sublithographic feature size may be obtained by etching techniques or side wall image-transfer techniques. However, these are largely experimental techniques and never been used in manufacturing environment.

The development of a reliable, manufacturable, cost effective lithographic technique is essential to continued progress in the CMOS technology.

8.4.2 Power Supply and Threshold Voltage

It is clear from Fig. 8.2(a) that as the channel length is scaled down, power supply V_{DD} is reduced as well to keep the electric field (reliability) and active power within a reasonable limit. The active power of a CMOS chip is given by

$$P_{ac} \sim C_{SW} \left(\frac{V_{DD}^2}{2} \right) f \tag{8.1}$$

where C_{SW} is total node capacitance being switched at a clock cycle of frequency f.

As CMOS technology advances, clock frequency f goes up and the switching capacitance C_{SW} also increases as more and more circuits integrated into the chip. This means active power increases. The active power of today's microprocessor is already 10–20 watt range.

Therefore, a power management system is required which can be achieved by architectural innovation. From Eq. (8.1), it is clear that the active power depends on the square of the power supply and hence one of the effective methods to curb the growth of the active power is to reduce the power supply. But as the power supply voltage is reduced, the gate delay will increase. This gives a tradeoff between power supply and gate delay. This tradeoff limits the scaling of power supply. As clear from Fig. 8.2(b) in each successive year power supply is reduced. As seen in Fig. 8.2(a) in contrast to power supply, the threshold voltage has not been scaled nearly as much. This is because of the standby power requirement. For room temperature CMOS devices, a minimum threshold voltage of order of 0.3 V–0.4 V is required; below which the standby power due to offstate leakage current becomes very high. The increasing (V_t/V_{DD}) ratio signifies a loss of gate overdrive, which degrades the CMOS circuit performance. Since V_t cannot be scaled below 0.3 V and CMOS delay increases rapidly when $(V_t/V_{DD}) > 0.3$, there is very little performance gain by scaling the power supply significantly below 1 V.

Another problem due to the reduction of power supply is increased sensitivity to soft errors due to ionization radiation.

8.4.3 Short Channel Effects

Short channel effect is to reduce the threshold voltage due to two-dimensional electrostatic charge sharing between the gate and the source-drain regions in short channel devices. Short channel effect plays an important role in deciding the lowest acceptable threshold voltage. To scale down the effective channel length of the MOS devices without short

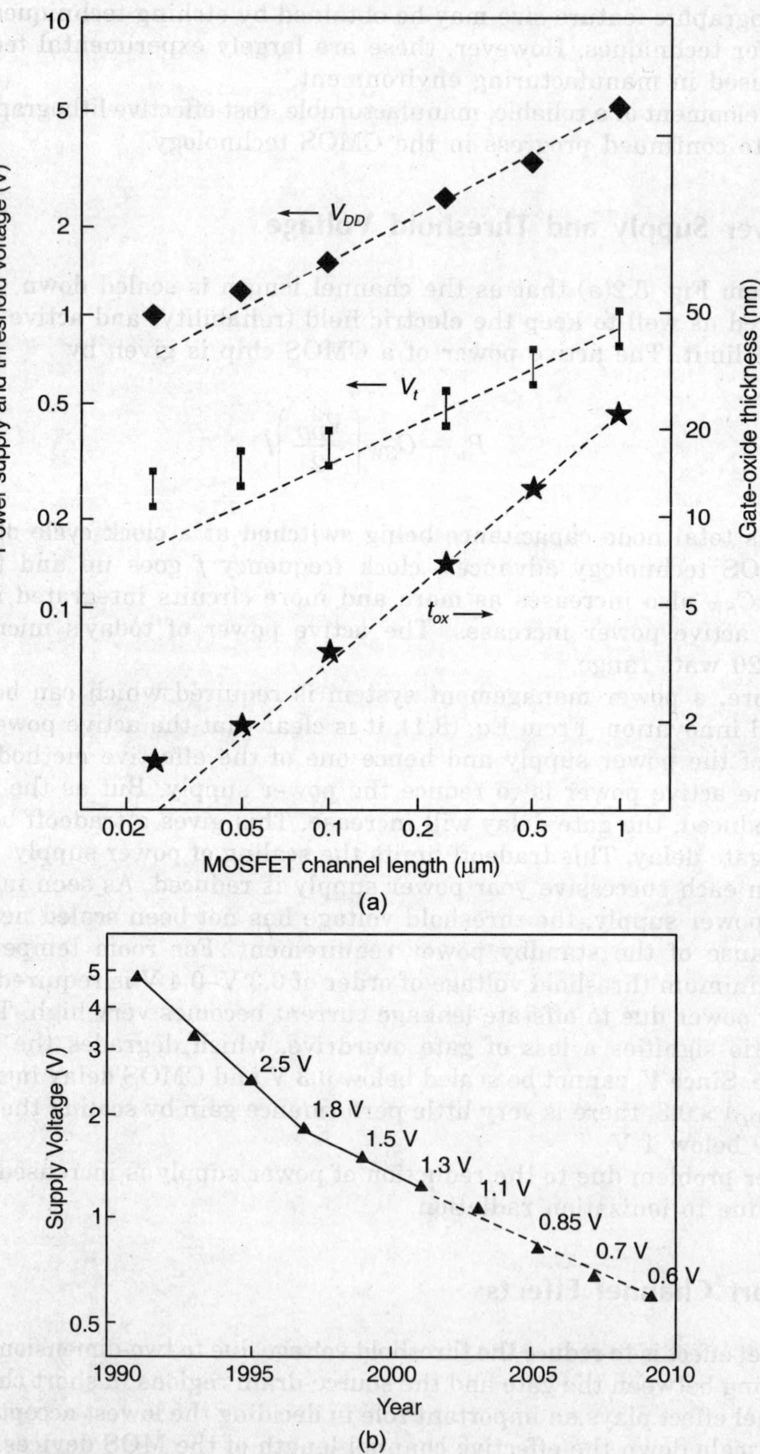

FIGURE 8.2 (a) Effect of scaling on power supply and threshold voltage, (b) Reduction of power supply over the time period.

channel effect, it is required that the oxide thickness t_{ox} and gate-controlled depletion width W_{dm} must be also scaled by same factor as the channel length L. To reduce W_{dm} one should increase the charge doping concentration, for a uniformly doped channel. But the increase in doping concentration leads to higher depletion charge and electric field at the silicon surface. These in turn causes the potential across the oxide and hence threshold voltage rises. To reduce the gate-controlled depletion width (while keeping the reduction on threshold voltage in same trend as shown in Fig. 8.2) a retrograde (low-high) channel doping is needed below 0.2 μm channel length.

For a given t_{ox}, reduction in W_{dm} improves short channel effect, but one has to pay the price of substrate sensitivity and subthreshold slope. Halo doping or non uniform channel profile in the lateral direction gives the freedom for reduction of V_t. Fundamental limitations on the maximum halo doping come from band-to-band tunnelling.

8.4.4   Gate Oxide Thickness

In order to control the short channel effect and to maintain a good subthreshold turn-off slope, one should scale the gate oxide thickness in same amount as the channel length: For example, for a 0.1 μm CMOS devices operating at 1.5 V, an oxide thickness of about 30 Å is needed. Such a thin oxide layer causes quantum mechanical tunnelling. This tunnelling leads to a gate leakage current that increases exponentially as the oxide thickness is scaled down. It is also observed that below 20 Å, oxide tunnelling current quickly becomes unmanageable. Unless a new gate dielectric material is developed, this sets a channel length scaling limit of 25–50 nm for bulk CMOS. Dynamic memory devices have a more stringent leakage requirement and hence a thicker oxide limit.

8.4.5   High Field Effects

Since the power supply voltage has not been scaled down in proportion to the channel length, therefore, electric field strength is increasing as the CMOS channel lengths are scaled down.

For 0.1 μm channel length devices, the oxide field has reached to its maximum value of 5 MV/cm while the field in the silicon has exceeded 1 MV/cm. The fields are expected to increase further when the channel length is scaled into deep submission region. At such high electric fields several undesirable effects occur. These effects either increase the leakage current (so power consumption) or degrade the device performance. At high fields main concerns are hot-carrier effect, quantum effect on threshold voltage, band-to-band tunnelling and mobility degradation.

8.4.6   Depletion Depths and Junction Depths

Depletion depths and junction depths have been controlled by ion-implantation of appropriate dopants into selected regions and limiting their movement during subsequent heat cycles. Future generation of technology requires ever steeper doping profiles. Present annealing procedures are unable to produce such steep profiles. Therefore, it is required to find out some alternative method if such devices are ever to be manufactured.

Another important concern is that very thin depletion depths (needed in future CMOS) require very high substrate doping concentration (of order of 10^{18}/cm^3). At these high doping levels direct body-to-drain tunnelling results and imposes a serious problem.

8.4.7 Random Fluctuation of Device Properties

Random fluctuation of device properties may ultimately limit the number of devices which can be integrated on the chip. Fluctuations of device properties result in variations of transistor current drive capabilities and propagation delays. These effects will lead to malfunctioning of the circuit. The random fluctuation of the threshold voltage of a MOSFET due to random fluctuation of the placement and number of dopant atoms are most studied effect. As MOSFET scaling approaches to sub-100 nm regime, the number of dopants is of order of hundreds in the depletion region for minimum geometry devices. As a result, the detailed microscopic dopant distribution in the MOSFET channel will have non-negligible effects on device threshold voltages. Threshold voltage matching is important for certain circuits like SRAM and sense amplifiers.

The effects of discrete random dopants become more important as the channel length and width are scaled down because high drain I-V of a narrow width devices were asymmetric upon interchanging the source and drain terminals. The asymmetry of threshold voltage is about 20–40 mV. The discrete and random nature of dopant atom results in an unhomogeneous channel potential.

8.4.8 Interconnection Delays

If the performance of processors is to keep pace with the speed improvement of devices, one should give special attention on the interconnection wires.

The delay associated with a wire, known as **RC delay**, is given as:

$$\tau = RL_W \left(C_L + \frac{1}{2} CL_W \right) \tag{8.2}$$

Here, R and C are resistance and capacitance of unit wire length, L_W and C_L are the wire length and load at the end of wire. In long channel device, the wire delay is rarely noticed, but as the channel length shrinks the delay associated with wire starts dominating. Therefore, one must include the delay associated with the interconnection when calculating the delay of overall circuit.

RC delay due to the wire alone does not decrease spite of scaling to smaller dimensions. The factor that improves the $RCL^2\omega/2$ term by reducing L_W is negated by the increase in R due to wire-cross sectional shrinkage because $R = PL_W/A$. High performance processors need two types of wires: first one is the *short wires* so that *RC* delay can be reduced. Second there is a need for *long wires*, where density is secondary to delay consideration. Long wires mainly run between distant parts of the chip.

One consequence of having low *RC* wires is that one will observe transmission line characteristics not only on the package but also on the chips themselves.

Any future technology (not only CMOS) has to deal with the non-scaling nature of wire capacitance. Fast devices are useless unless they can drive capacitances that are comparable to wire capacitance.

8.4.9 Need for Low Power VLSI Design

One of the most serious concerns in future VLSI design is the exponentially increased standby power caused by the subthreshold (OFF) current in the MOSFETs. Power consumption has become an important issue not only for battery operated devices, where the need for low power implementations is directly related to the operation time of the device, but also for high speed integrated circuits because there is a well known trade-off between high speed and low power. In order to suppress the active power consumption, the supply voltage V_{DD} must be reduced which reduces the speed of the circuit.

The dramatic decrease on device size with the corresponding increase in the number of devices in the chip and increase in the operating frequency put power dissipation into the pool of critical parameters of many circuit design. Therefore, power consumption is rapidly becoming one of the key limiting factors for the number of components that can be placed on a single chip. Power estimation and low power design styles are vital not only for explicit low power devices but also for all kinds of VLSI circuits because many other key parameters are benefitted from a low power design style. The reliability drops by 50% and performance decreases by approximately 3% for every 10°C rise in substrate temperature. The main sources for power dissipation are:

(i) Capacitive power dissipation due to charging and discharging of the load capacitor because

$$P_{av} = C_{\text{load}} \times V_{DD}^2 \times f$$

(ii) Short circuit currents due to existence of conducting path between the power supply and ground for brief period during which a logic gate makes a transition.

(iii) Leakage current which consists of
   (a) Reverse-bias diode current: Due to stored charge between the drain and bulk of active transistors.
   (b) Subthreshold current: Due to carrier diffusion between the source and drain of the OFF transistors.

The short circuit power dissipation can be reduced to 10% of the total power dissipation by designing a circuit which has equal rise and fall times. Redefining the problem, architecture, algorithms and/or the protocol can often save several orders of magnitude in power dissipation. A new functional CMOS device called **variable**

threshold voltage MOSFET (VTCMOS) has promised to be among the next generation of ultra low power devices operating at low voltage. The operating principle of the VTCMOS is that its threshold voltage is controlled by the substrate bias. In standby mode threshold voltage V_t is increased to suppress the OFF current by applying the substrate bias.

8.5 PERCEIVED LIMITS OF CMOS TECHNOLOGY

The ultimate limit of CMOS technologies has been the subject of discussion for a long time. Although all the technological predictions in the past have proven to be too pessimistic. Now, it is believed that Si CMOS technologies are approaching towards their physical fundamental limits.

From the device design point of view, the depletion depth for bulk CMOS is limited to about 10 nm due to body-to-drain tunnelling current limitations on maximum body doping and to difficulty on controlling the extremely abrupt doping profiles necessary. For most advanced structures quantum mechanical effects would render control of threshold voltage below 5 nm channel thickness.

The international technology roadmap for semiconductor (ITRS) predicts that the source/drain extension junction depth below sub-50 nm CMOS will be scaled further down around 10 nm to maintain acceptable short channel performance, but this may lead to a high series resistance problem threatening the ultimate device performance.

8.5.1 Advantages of CMOS over Bipolar Junction Transistor

    (i) Lower static power dissipation due to high input impedance
    (ii) Higher noise margins
    (iii) Higher package density due to scaling of CMOS
    (iv) Simpler fabrication process and lower manufacturing cost per device
    (v) Higher yield with large integrated complex functions

8.5.2 Other CMOS Advantages

    (i) Low drive current due to high input-impedance (MΩ)
    (ii) Scalable threshold voltage
    (iii) High delay sensitivity to load
    (iv) Low output drive current
    (v) Bi-directional capability, i.e., source and drain are interchangeable
    (vi) A near ideal switching device

Major disadvantage of CMOS is that its g_m value is smaller than BJT and CMOS circuit is slower than BJT.

Other disadvantages of CMOS over MOS is that it occupies larger area on the chip than MOS devices and its processing is lengthy than MOS devices.

8.6 CMOS INVERTER

The most basic element of digital static CMOS circuits is a CMOS inverter. A CMOS inverter consists of an enhancement-type n-MOS transistor and an enhancement-type p-MOS transistor as shown in Fig. 8.3. These two transistors operate in the complimentary manner and hence this configuration is called **complimentary MOS**. As seen in Fig. 8.3, the source of p-MOS is connected to power supply and source of n-MOS is connected to ground. The gates of two are tie together and used as the input node. The two drains are tied together and used as the output node. The circuit has two stable states either logic low or logic high.

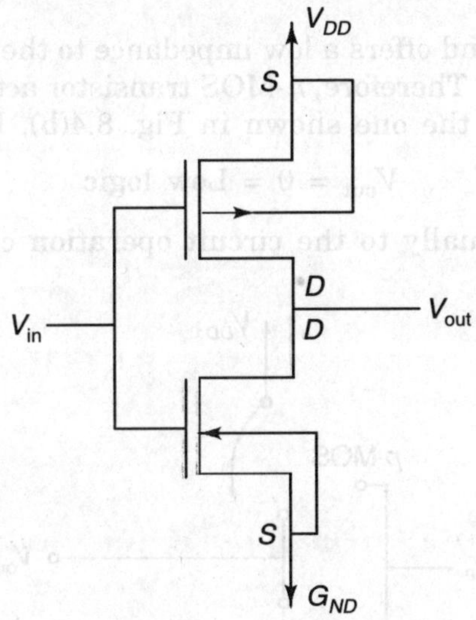

FIGURE 8.3 CMOS inverter.

Consider two cases:

Case I: V_{in} = Low

For low input voltage [i.e., $V_{in} < (V_{tn})$], the n-MOS transistor is OFF and p-MOS transistor is ON means n-MOS offers a high impedance to the current whereas p-MOS offers low input impedance. Therefore, n-MOS acts as load and p-MOS acts as a driver. Also, p-MOS pull up the output node to the high voltage, and is called **pull-up transistor**. Figure 8.3 reduces to circuit as shown in Fig. 8.4(a).

Let current through p-MOS transistor is I_{DP}, then according to current continuity,

$$I_{DP} = I_{DN}$$

Since, n-MOS is offering a very high input impedance, means

$$V_{out} = \text{high}$$

Case II: V_{in} = high $(\geq V_{tn}) = V_{DD}$

p-MOS transistor turns OFF and offers a high impedance to current flow because it acts as open circuit. Therefore, p-MOS in this case serves the purpose of load,

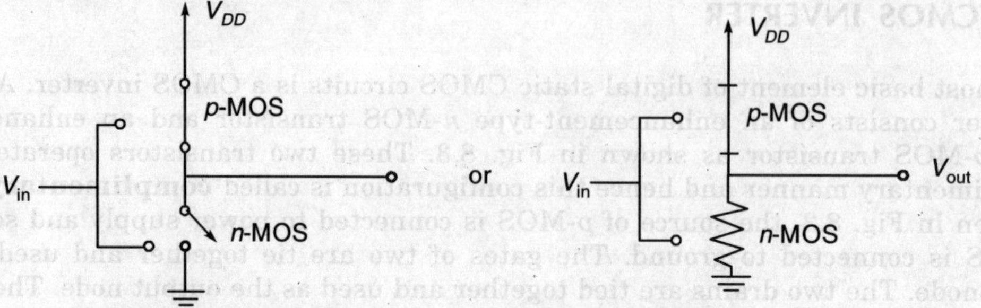

FIGURE 8.4(a) CMOS inverter for low input voltage.

whereas n-MOS turns ON and offers a low impedance to the current flow and pull down the output node at low logic. Therefore, n-MOS transistor acts as a pull-down transistor. Figure 8.3 reduces to the one shown in Fig. 8.4(b). Hence

$$V_{\text{out}} = 0 = \text{Low logic}$$

Both devices contribute equally to the circuit operation characteristics.

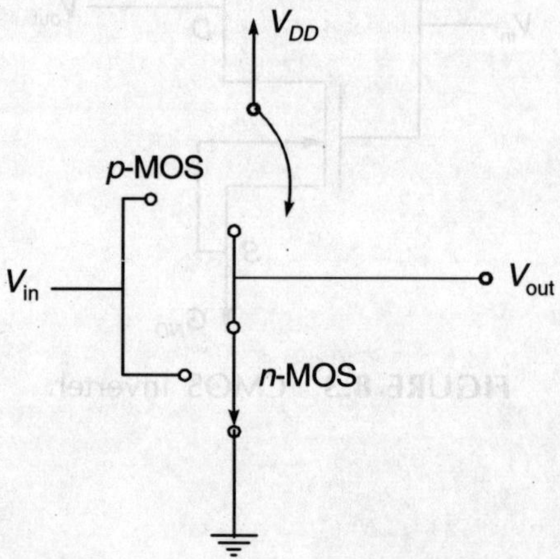

FIGURE 8.4(b) CMOS inverter for high input voltage.

Note: As we know n-MOS and p-MOS transistors are mainly used as a switch in digital electronics. Their switching actions are

(i) For high V_{GS}
n-MOS turn ON → Means switch is closed.
p-MOS turn OFF → Means switch is open.

(ii) For low V_{GS}, i.e., input
n-MOS turn OFF → Acts as open switch R_{in} = high
p-MOS turn ON → Acts as closed switch means short circuit and hence R_{in} = 0 = low

To make transistor ON to OFF, we have to move the transistor from saturation to cut-off region. The CMOS inverters have two basic advantages over other inverters. First one is the steady state power dissipation of the CMOS inverter circuit is virtually negligible. Only a small power dissipates due to switching. The second advantage is sharp voltage transfer characteristic, i.e., it exhibits a full voltage swing between 0 V and V_{DD}; as shown in Fig. 8.5.

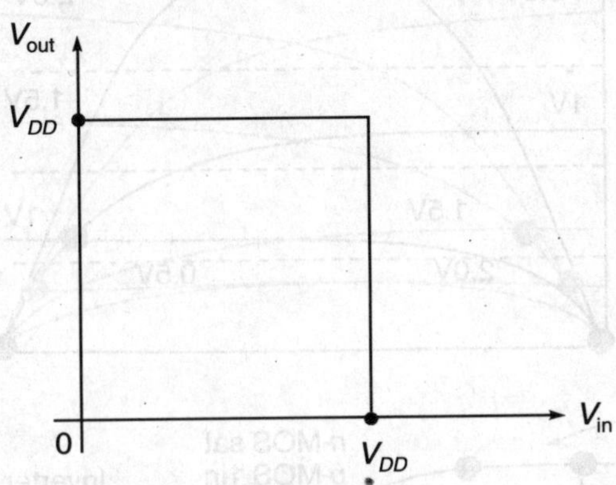

FIGURE 8.5 Transfer characteristic of ideal CMOS inverter.

The VTC of CMOS inverter resembles that of an ideal inverter.

Note: 1. For $V_{in} = V_{DD}/2 = V_{TH}$ = Inversion threshold voltage, both the transistors turn ON (i.e., both enters into saturation regions) and there is a direct path from power supply V_{DD} to GND. Therefore, a large current will flow in the circuit and power dissipation becomes very high. Therefore, one should avoid this situation as shown in Fig. 8.6. In CMOS inverter, one transistor should be ON at one time.

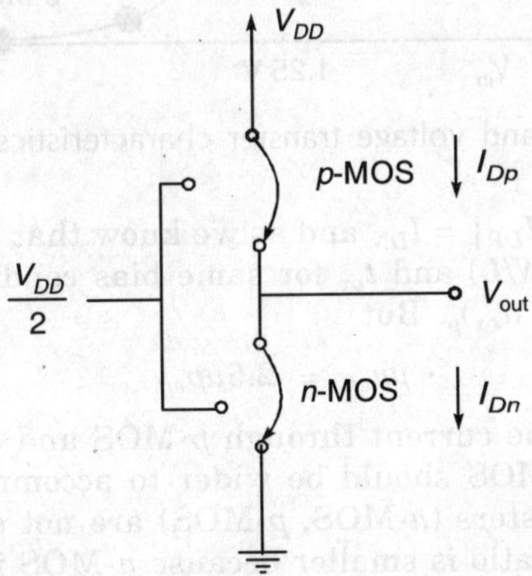

FIGURE 8.6 CMOS inverter when $V_{in} = V_{TH}$.

The characteristics (I_D–V_{OUT}) and VTC (voltage transfer characteristics) of a real CMOS inverter is shown in Fig 8.7(a).

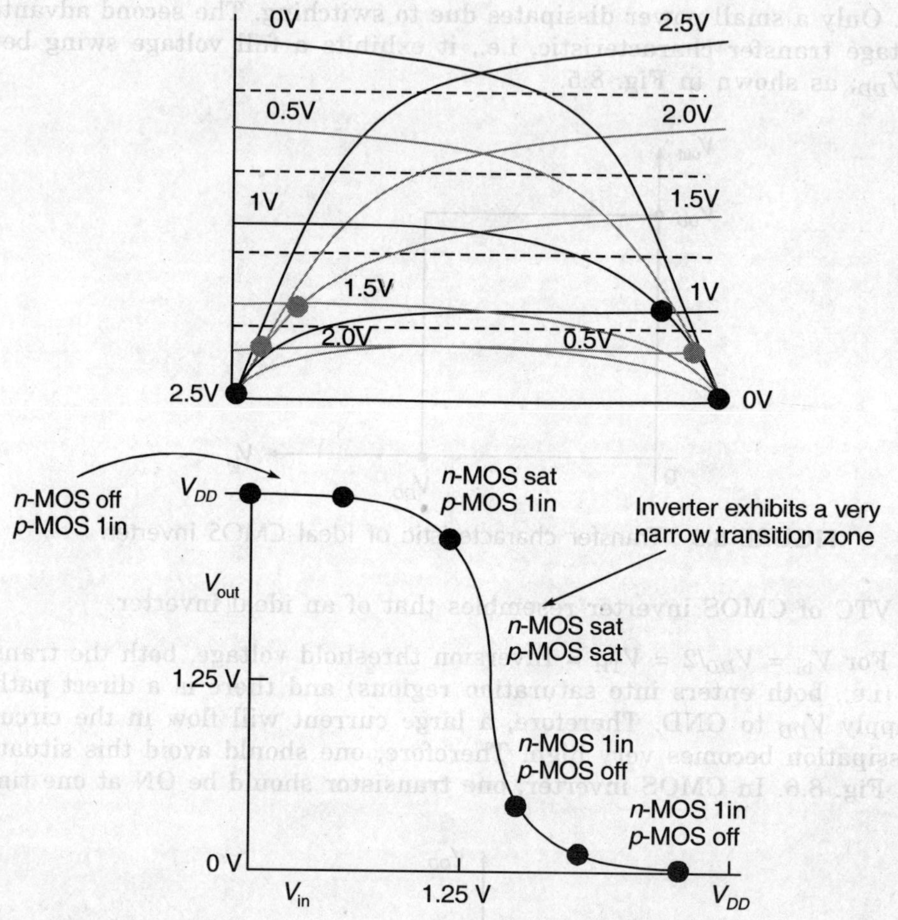

FIGURE 8.7(a) (I_D–V_{OUT}) and voltage transfer characteristics of a real CMOS inverter.

2. In CMOS inverter $|I_{DP}| = I_{DN}$ and as we know that the drain current depends on the effective mobility, (W/L) and t_{ox} for same bias condition. Since, for the given technology $L_n = L_p$, $(t_{ox})_n = (t_{ox})_p$. But

$$\mu n_{eff} \simeq 2.5 \mu p_{eff}$$

Therefore, to equate the current through p-MOS and n-MOS one should choose (W_p/W_n) ≃ 2.5, i.e., the p-MOS should be wider to accommodate the same current. Therefore, those two transistors (n-MOS, p-MOS) are not symmetrical. However for short channel devices, the ratio is smaller because n-MOS is more velocity saturated than p-MOS.

The inverter threshold voltage V_{TH} is identified as one of the most important design parameter. The CMOS inverter can provide a full swing output voltage swing between 0 V and V_{DD}, therefore, the noise margin which is defined as:

$$N_{\text{ML}} = V_{\text{IL}} - V_{\text{OL}}$$

$$N_{\text{MH}} = V_{\text{OH}} - V_{\text{IH}}$$

must be very high. Here N_{ML} is low level noise margin, N_{MH} is high level noise margin, V_{IL} is input voltage at low logic, where n-MOS is in saturation and p-MOS is in linear region, V_{OL} low logic output, for V_{in} = high, V_{IH} is high input logic where n-MOS is in linear and p-MOS is in saturation regions and $V_{\text{OH}} \simeq V_{DD}$ for CMOS inverter.

Figure 8.7(b) shows the noise margin of a CMOS inverter.

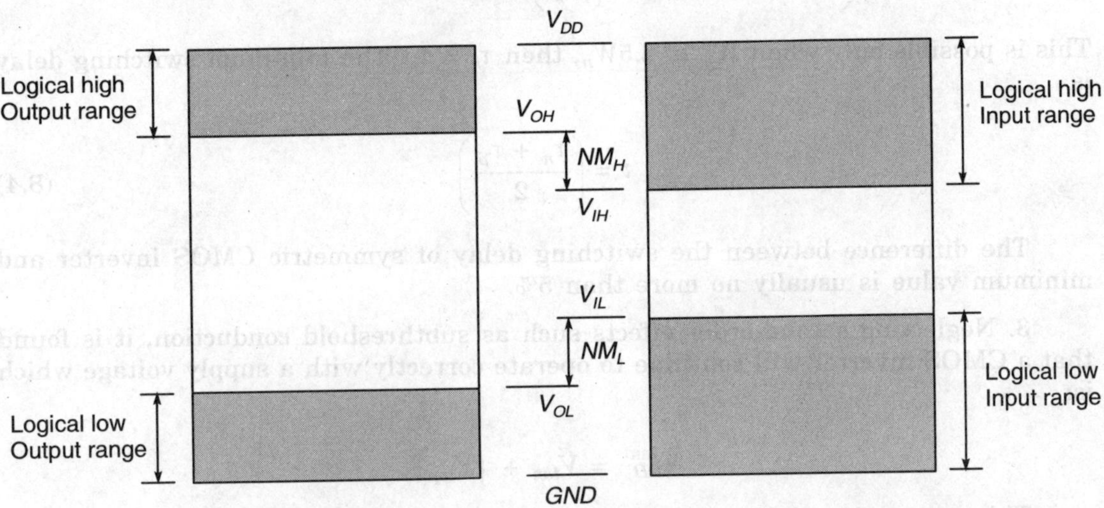

FIGURE 8.7(b)   Schematic representation of noise margin in CMOS inverter.

This means any voltage below or equal the N_{ML} value is treated as logic low and in same way any voltage either equal or greater than N_{MH} is treated as logic high. If any voltage is lying in between these two ranges is known as **forbidden voltage**.

Since, only one transistor is ON at one time, i.e., in the steady state, therefore, there is almost negligible static current or static power dissipation. Power dissipation occurs only during switching transients when a charging and discharging current is flowing through the circuit.

It is conventional to define an nMOSFET pull-down delay τ_n as the time taken for the output voltage to reach $V_{DD}/2$, i.e.,

$$\tau_n = \frac{C_L V_{DD}}{2 W_n I_{Dn\,\text{sat}}} \tag{8.3a}$$

Similarly, for *p*MOSFET pull-up delay is:

$$\tau_p = \frac{C_L V_{DD}}{2 W_p I_{Dp\,sat}}$$

(8.3b)

where I_{Dnsat} and I_{Dpsat} are saturation current from *n*-MOS and *p*-MOS.

or

$$\frac{\tau_n}{\tau_p} = \left(\frac{W_p}{W_n}\right)\left(\frac{I_{Dp\,sat}}{I_{Dn\,sat}}\right)$$

If

$$\frac{I_{Dp\,sat}}{I_{Dn\,sat}} = \left(\frac{W_n}{W_p}\right)$$

This is possible only when $W_p \simeq 2.5 W_n$, then $\tau_n = \tau_p$. The minimum switching delay is

$$\tau = \left(\frac{\tau_n + \tau_p}{2}\right)$$

(8.4)

The difference between the switching delay of symmetric CMOS inverter and minimum value is usually no more than 5%.

3. Neglecting second-order effects such as subthreshold conduction, it is found that a CMOS inverter will continue to operate correctly with a supply voltage which is:

$$V_{DD}^{min} = V_{ton} + |V_{top}|$$

This means that the correct inverter operation will be sustained if at least one of the transistors remains in conduction for any given input.

EXAMPLE 8.5

Why CMOS inverter is preferred over other types of inverters like resistive load inverter or MOS inverter?

Solution: Since the CMOS inverter does not draw any significant current from the power supply V_{DD} on both of its steady states, i.e., $V_{out} = V_{ON}$ and $V_{out} = V_{OL}$ because at one time only one transistor is ON. Therefore, the DC power dissipation of the circuit is almost negligible. The drain current that flows through the *n*-MOS and *p*-MOS transistors in both cases is essentially limited to reverse leakage current of the source and drain *p-n* junctions and in short channel device, the relatively small subthreshold current. Therefore, in any applications where low power consumption is required, this CMOS inverter is preferred over others.

EXAMPLE 8.6

Consider the circuit as shown in Fig. 8.8. Given, $V_{ton} = 0.8 = |V_{top}|$

$$\mu_n C_{ox} = \mu_p C_{ox} = 45.0 \ \mu\text{m/V}^2 \quad \text{and} \quad L_n = L_p = 0.25 \ \mu\text{m}$$

Find $\gamma = 0$, $\lambda = 0$
  (a) $(W_p/W_n) = ?$
  (b) V_{OH} and V_{OL}
  (c) If $V_{\text{in}} = V_{\text{OH}}$ find V_{out} and $I_D = I_{DP} = I_{DN}$

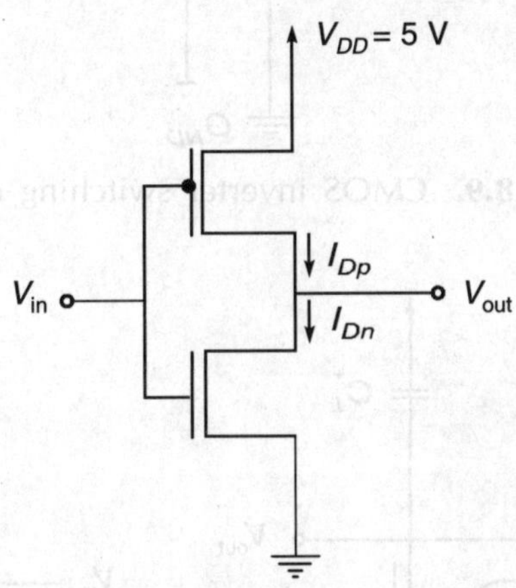

FIGURE 8.8   Circuit for Example 8.6.

Solution: (a) $(W_p/W_n) = 2.5$
  (b) Assuming ideal CMOS inverter $V_{\text{OH}} = V_{\text{DD}}$ and $V_{\text{OL}} = 0$.
  (c) For $V_{\text{in}} = V_{\text{OH}} = V_{\text{DD}}$, pMOS – OFF, nMOS ON and hence, $V_{\text{out}} = V_{\text{OL}} = 0$ Volt.

8.6.1   Switching Energy and Power Dissipation in CMOS Inverter

When a CMOS inverter switches from one logic state to other (true also for other logic gates) then only it consumes power from power supply. Let us consider Fig. 8.9, where C_L is lumped (or load) capacitor and assumed to be fully discharged in the beginning.

If input is low then p-MOS turns ON and n-MOS turn OFF. This means the current I_{DP} will charge the capacitor from zero to $C_L^- V_{DD}$ as shown in Fig. 8.10(a). Therefore, in other words, there is an energy of $C_L^- V_{DD}$ flowing out of the power supply in the pull-up transition. Half of this energy ($1/2 \ V_{DD}^2 \ C_L^-$) is dissipated by charging current in the p-MOS resistance. And other half is stored energy in the capacitor which is $(1/2) \ C_L^- V_{DD}^2$. This energy stays in the capacitor until the next pull down takes place.

Now, if $V_{\text{in}} = V_{DD} = $ high, n-MOS turns ON and p-MOS turns OFF. This means new capacitor C_L^- starts discharging through n-MOS resistance as shown in Fig 8.10(b) and finally capacitor voltage C_L^- reaches to zero. Likewise, for the capacitor C_L^+ between

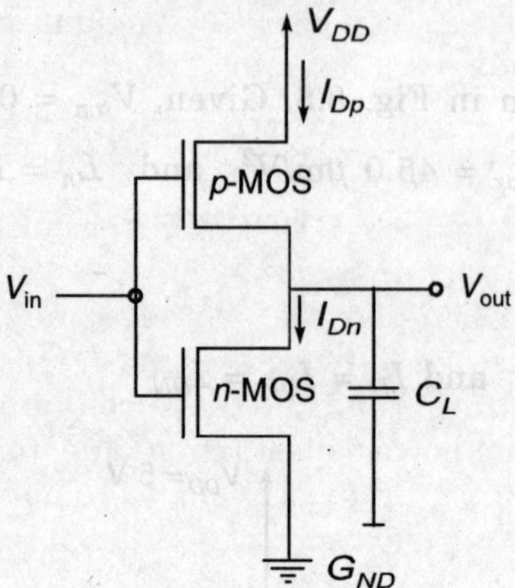

FIGURE 8.9 CMOS inverter switching circuit.

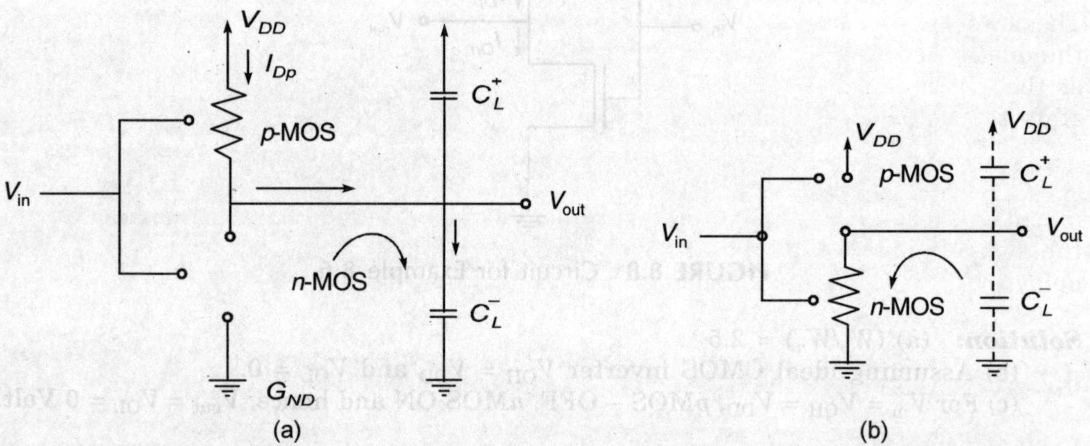

FIGURE 8.10 Schematic representation of switching behaviour of CMOS inverter.

output and V_{DD}, and amount of energy $C_L^+ V_{DD}^2$ is supplied by the power source during pull-down transition. Again half of this energy is dissipated through n-MOS resistance and other half ($C_L^+ V_{DD}^2/2$) is stored in the capacitor C_L^+. The stored energy is later dissipated through the p-MOS resistance during next pull-up transition. Therefore, it is clear that any capacitor C_L^+ or C_L^- to be charged or discharged, an energy of ($C_L V_{DD}^2/2$) is dissipated. It is always desirable to take one complete cycle consisting of a pair of transitions either up-down ($0 \rightarrow V_{DD} \rightarrow 0$) or down-up ($V_{DD} \rightarrow 0 \rightarrow V_{DD}$). In that case, dissipation energy is given per cycle which is equal to $V_{DD}^2 C_L$.

Since, the dc power dissipation is negligible in CMOS circuits, the only power consumption comes from switching. In a CMOS processor, the switching of logic gates is controlled by a clock generator of frequency f. If on average a total equivalent capacitance C_L is charged or discharged within a clock cycle of frequency f then the average power dissipation is:

$$P = C_L \, V_{DD}^2 \, f \tag{8.5}$$

For a high speed operation, frequency increases means power dissipation per cycle increases. Therefore, low power advantage of CMOS circuits become less prominent. One important observation of Eq. (8.5) is that the average power dissipation of the inverter circuit is independent of transistor size and transistor characteristics.

Note: Equation (8.5) will hold for any CMOS logic circuits, when the leakage power is neglected and short circuit current is zero for real situation, the signal is not a ideal signal, i.e., input has nonzero rise and fall times, then n-MOS and p-MOS conduct simultaneously a certain amount of current during the switching event. This is called **short circuit current** and in this case a conducting path will form between the V_{DD} and the ground. The power dissipation due to this short circuit current cannot be calculated by using Eq. (8.5) because the short circuit cannot be used to discharge or charge the internal capacitors.

8.6.2 Power-delay Product

The power-delay product (PDP) is a fundamental parameter which is used for measuring the quality and performance of a CMOS gate design. Power-delay product is defined as the average energy required for a gate to switch its output voltage from one logic state to other logic state, and is given as:

$$PDP = C_L \, V_{DD}^2 \tag{8.6}$$

The energy is dissipated in form of heat when n-MOS and p-MOS conduct during the switching action. As a designer, one is always interested to reduce the Power-delay product, i.e., keep C_L as well as V_{DD} as small as possible.

8.7 LATCH-UP PROBLEM IN CMOS

Latch-up is a failure mechanism of CMOS integrated circuits characterized by excessive drain current coupled with functional failure, parametric failure and/or device destruction. Latch-up is a controllable phenomenon.

The cause of the latch-up exists on all junction-isolated or bulk CMOS processes: parasitic p-n-p-n paths. Figure 8.11 shows a basic n-substrate CMOS cross-section in which there are two parasitic bipolar transistors: n-p-n and p-n-p. These two transistors are basically responsible for latch-up. The p^+-regions (acting as source and drain) of p-MOS devices act as a emitters (or sometimes collector) of lateral p-n-p devices where n-substrate acts as a base of transistor and collector of a vertical n-p-n device. The p-well acts as a collector of the p-n-p and base of the n-p-n transistors. Finally, the n^+-sources and drains of the n-channel MOS devices serve as the emitter of the n-p-n. The substrate is normally connected to V_{DD} (the most positive circuit voltage) via an n^+ diffusion tap while the p-well is terminated at the ground (the most negative circuit voltage) through a p^+ diffusion.

As long as MOS source and drain junctions remain reverse biased, CMOS is well behaved. But in the presence of intense ionizing radiation, thermal or over voltage-stress, current can be injected into the *p-n-p* emitter-base (BE) junction, making it forward biased and hence current *I* begins to flow. This provides base current I_{BN} to the *n-p-n* transistor because this current will flow through the substrate and into *p*-well. This base current I_{BN} turns device ON. Collector current I_{CN} ($= I_{BP}$) from the *n-p-n* transistor force *p-n-p* transistor to conduct more current. The transistors form a feedback path with each transistor configured so that its collector drives the base of the other. This is a positive feedback, thus sustains current flow and sinks current directly from V_{DD} to ground. This type of action is called **latch-up**. In order for *n-p-n/p-n-p* pair of transistors to latch the product of their β values should exceed unity.

The use of CMOS at or beyond its rated maximum voltage range and the presence of inductive transients can trigger latch-up. Environment including thermal stress, poorly regulated or noisy supplies and radiation incidence can also contribute or cause latch-up. Latch-up is generally resulted from the regenerative switching (positive feedback) along a *p-n-p-n* path.

One should remember that there are three conditions responsible for latch-up to occur.

(i) Both parasitic transistors must be used in active region.
(ii) The product of their common emitter current gain β must be larger than unity.
(iii) The terminal network must be capable of supplying a current greater than holding current required by the *p-n-p-n* path.

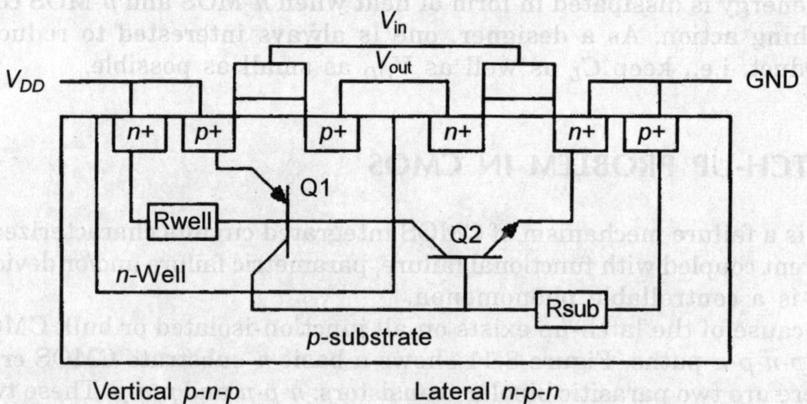

FIGURE 8.11 Latch-up problem in CMOS.

8.7.1 Effects Which Can Trigger Latch up

1. Hot plug-in of un-powered circuit board or module
2. Electrostatic discharge (ESD)
3. Sudden transients on ground or power buses
4. Leakage currents across well junction
5. Radiator

8.7.2   Prevention and Suppression of Latch up

Use of layout and process adjustment such as following:

(i) Reduce β parameter of both parasitic transistor so that their product should be lower than unity. This can be achieved by placing heavily doped guard rings around MOSFETs as shown in Fig. 8.12.

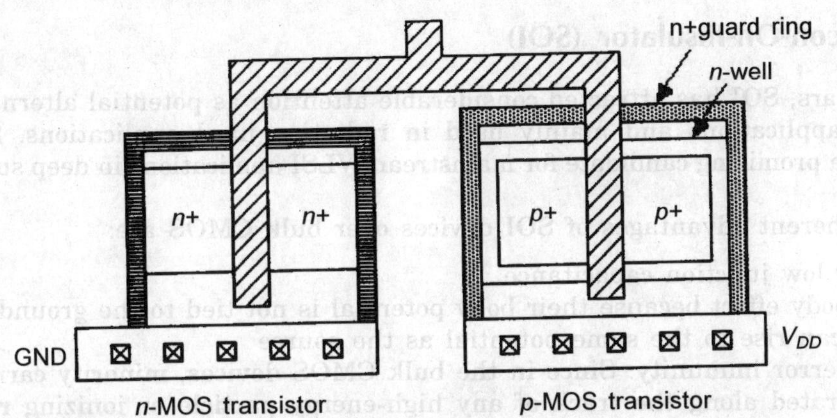

FIGURE 8.12   Schematic for preventation of latch-up in CMOS circuit.

(ii) Avoid the biasing of the parasitic transistor into active regions, i.e., make V_{BE} (Active) below than 0.7 V.

(iii) Reduce R_S (series resistance) and R_W (wire resistance) to ensure that they will bypass base current to supply rails. This can be achieved by adding more tub-ties.

(iv) Choose maximum distance between tub-ties (contacts to the tub/well region) to ensure that chip will not latch up during normal operation.

(v) R_W can be lowered by using low sheet resistivity, e.g. use deep implant to achieve a retrograde well doping profile lower the sheet resistivity of the well. This can also be achieved by using buried layer.

(vi) Oxide trench deep enough to reach the low resistivity substrate is used to isolate the well from the substrate region. Trench isolation automatically preclude the possibility of latch up by blocking the current flow path. No guard rings are now required.

(vii) Silicon On Sapphire (SOS) is a CMOS technology which uses epitaxial silicon layers grown on sapphire (insulator) substrate. This technique is used to avoid latch up. This also reduces device capacitance levels tremendously allowing higher speed circuits. But price we are paying for that is its higher cost.

Also through innovative and careful layout one can eliminate the latch-up problem.

8.8 NOVEL DEVICES

In this section many unconventional device structures that may offer further improvement in performance and power over bulk CMOS and take us to the outermost limit of silicon scaling. These devices are silicon-on-insulator (SOI), double-gate MOSFET (DG), SiGe MOSFET, carbon-nanotube MOSFET.

8.8.1 Silicon-On-Insulator (SOI)

In recent years, SOI has attracted considerable attention as potential alternative for low power applications and mainly used in radiation hard applications. SOI has emerged as a promising candidate for mainstream VLSI applications in deep submicron era.

The inherent advantages of SOI devices over bulk CMOS are:

(i) Very low junction capacitance

(ii) No body effect because their body potential is not tied to the ground or V_{DD}, but can rise to the same potential as the source

(iii) Soft-error immunity: Since in the bulk CMOS devices, minority carriers are generated along the track of any high-energy particle or ionizing radiation that strikes through the silicon. If the collected charge of the junction node exceeds a certain voltage then it may cause damage the stored logic which is commonly known as **soft error**. SOI devices offer an improvement in the soft error rate because of the presence of the buried oxide. The volume susceptible to ionizing radiation greatly reduces.

Many SOI devices have been proposed, including silicon-on-sapphire (SOS), silicon-on-spinel, silicon-on-nitride and silicon-on-oxide.

SOI uses the same substrate, same material set and the same fabrication processes as the bulk CMOS, but still it gives extra degree of freedom, i.e., almost perfect isolation which can be used to gain higher performance, lower power and other desirable attributes.

The CMOS devices (or MOS devices) which we have covered in Chapter 7 are called **bulk-CMOS devices** because they are fabricated directly on the substrate by using well method.

EXAMPLE 8.7

Why SOI CMOS is preferred over bulk CMOS?

Solution: As the scaling of MOS transistors approaches the submicron regime, various parasitic effects become very serious. These secondary effects include short channel effects such as source/drain punch through, hot carrier effects and mobility degradation. One of the methods to circumvent these problems is to fabricate MOS transistors on silicon-on-insulator (SOI) materials.

SOI wafers

SOI starting material is usually made by one of the two processes: SIMOX (separation by implanted oxygen) or bonded SOI. For SIMOX, oxygen is implanted at sufficiently high density that can react with silicon during a post-implant high temperature bake, to form a continuous silicon dioxide film buried under the thin silicon surface that effectively isolates the surface crystalline Si layer from that of the substrate. This is now the leading commercial process for making SOI. The burried oxide (BOX) is typically a few hundred nanometers thick and SOI is about 200 nm thick. It can be further thinned down by oxidation, e.g. for an *n*-channel MOSFET, oxygen is ion implanted beneath the surface of *p*-substrate. The wafer is then annealed at high temperature so a layer of SiO_2 is formed that effectively isolates the silicon surface from that of the substrate. The device is then formed in the surface of silicon by standard fabrication process. Finally Si surface is selectively etched to isolate the devices as shown in Fig. 8.13.

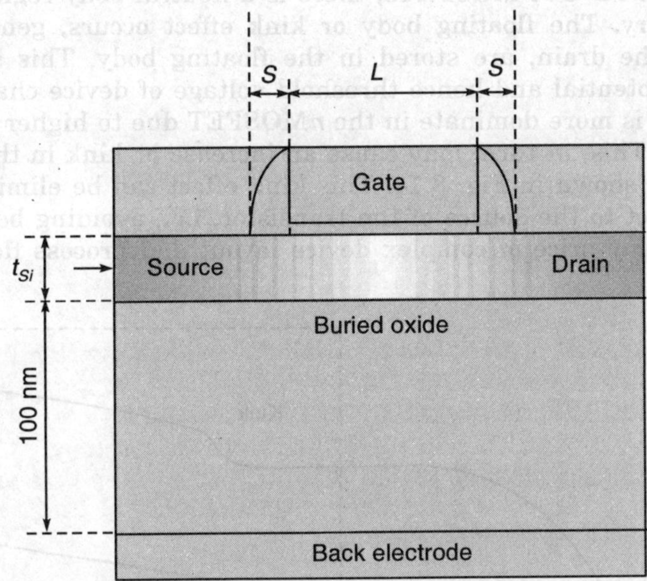

FIGURE 8.13 Schematic of SOI MOSFET.

Bonded SOI (BESOI) is achieved based on the fact that very smooth and clean silicon and oxide surfaces will bond together when forced into intimate contact when subsequently heated bond becomes permanent. One BESOI process involves bonding a wafer with an epitaxial silicon layer including a buried heavily boron doped etch stop. After bonding, the back of the wafer is grounded and then etched off down to the etch stop.

Another BESOI process is much easier to scale up to a large production volume. For this process hydrogen is implanted under the surface of one wafer. After bonding, wafer is heated and reacts with the implanted hydrogen and forms buried gas-filled cavities which join together, splitting off the back of the wafer and leaving the thin SOI layer still bonded.

Another method of making SOI is to selectively grow silicon epitaxially from a local area (the seed) and have it grown over an adjacent isolation oxide or even through predefined regions or tunnels. An essential adjunct to the bonding and epitaxial processes are chemical-mechanical polishing processes to resmooth the surface.

Depending on the thickness of the Si channel layer, SOI can be classified into partially deplected (PD) and fully deplected (FD) SOI.

Partially depleted (PD) SOI. SOI MOSFETs are called partially depleted (PD) when the silicon film is thicker than the maximum gate depletion width and devices exhibit a floating body effect. PD-SOI uses a thicker Si channel layer so that the depletion width of the channel does not exceed the thickness of the silicon layer. Device design and performance of a PD-SOI is similar to that of bulk CMOS. One major difference between PD-SOI and bulk CMOS is that floating substrate is used in SOI devices. In PD-SOI MOSFETs, there is a neutral body region below the gate depletion boundary. The floating body or kink effect occurs, generated by impact ionization near the drain, are stored in the floating body. This storage of charge alters the body potential and hence threshold voltage of device changes due to body effect. This effect is more dominate in the nMOSFET due to higher impact ionization rate of electrons. This, in turn, may cause an increase or kink in the current voltage characteristics as shown in Fig. 8.14. This kink effect can be eliminated by forming a substrate contact to the source of the transistor, i.e., avoiding body effect. But for that we have to pay price of complex device layout and process flow.

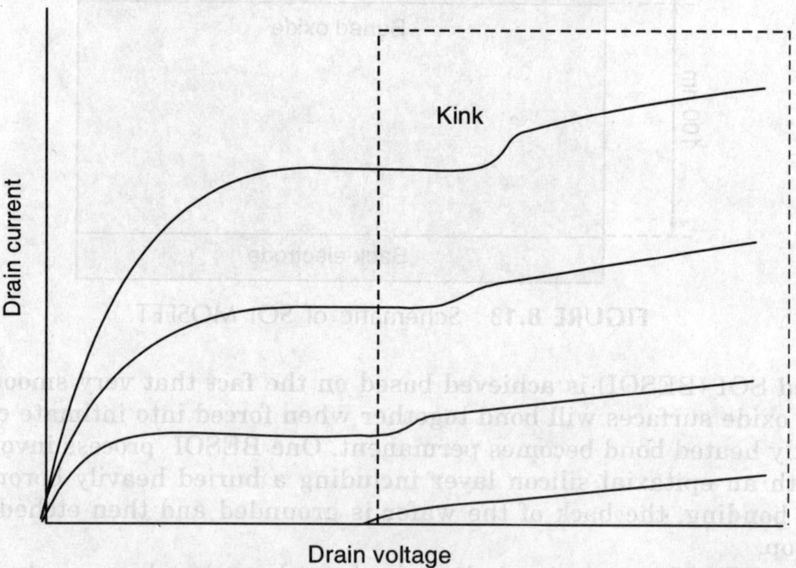

FIGURE 8.14 Kink effect in partially-doped SOI MOSFET.

Eventhough, floating body effects tend to enhance the speed in certain conditions, but the drain current overshoot (kink effect) is history dependent. The floating body potential depends on how recently and how often the device has been switched through its impact ionization conditions. Let us consider a case that the device has been in the ON state for a long time and is then turned OFF, but it is again turned ON before the body charge reaches to its equilibrium value. This process causes a great problem in the circuit design. The other undesirable consequence of the floating body effect in PD SOI is the higher OFF current due to forward-biased body-to-source junction when the drain voltage is high.

Note: By using body contact, one can restore the properties of bulk CMOS in the SOI MOSFETs, but one has to compromise with the body-effect advantage of SOI devices.

Fully depleted SOI. FD SOI uses a very thin Si layer so that the channel of the transistor is completely depleted before the threshold is reached. This allows the device to be operated at a lower field. The subthreshold slope of a long-channel FD SOI MOSFET can be near the ideal 60 mV/decade at room temperature. This is because the effective gate depletion width is very large in FD SOI and body-effect coefficient is equal to unity. The steeper subthreshold slope permits a lower threshold voltage for the same OFF current, which in turns allows the device to be used at low supply voltages and hence FD SOI is very attractive for low-power applications.

In short-channel FD SOI MOSFET, the thick buried oxide acts like a wide gate depletion region, which leaves them vulnerable to source-drain field penetration and results in poor short-channel effects. The high electric field in FD SOI devices eliminate the kink effect.

Note: Thin film fully depleted (FD) SOI devices are attracting a great attention as a high-performance, low power IC components due to their inherent properties such as sharp subthreshold slope, reduced I_D–V_{DS} kink effect and suppressed short-channel effects. However, the threshold voltage of FD devices is sensitive to intrinsic variation of the SOI thickness across a wafer and it becomes uncontrollable.

Note: 1. A partially depleted SOI circuit is easier to fabricate than the fully depleted one. It is believed to be the most suitable for practical use. Therefore, an SOI MOSFET can be operated as FD or PD depending on the film thickness and SOI doping.

2. SOI devices continue to have a non-uniform electric field distribution in the channel with the peak of the field profile occurring near the drain. Thus, the charge carriers move with a low velocity near the source, gradually accelerating towards the drain resulting in a lower mean carrier transport velocity. Another undesirable phenomenon at the shorter channel length is the hot carrier effect which is precipitated due to electric field peak near the drain.

Advantages of FD SOI over bulk MOS devices. Thin film fully depleted SOI offer several advantages over bulk MOS devices like reduced junction capacitance, increased channel mobility, excellent latch up immunity and reduced SCEs. As a consequence, deep submicrometer SOI circuit design and simulation are increasingly becoming important on VLSI technology research. In contrast to bulk CMOS devices, the front gate of the SOI device has better control over its active device region in the thin film and hence charge sharing effects from source/drain regions are subsequently reduced. However, the thin-film thickness has to reduce to the order of 10 nm to significantly improve the device performance which becomes prohibitively difficult to manufacture and causes large device external resistance due to shallow source/drain extensions (SDE) depths.

Note: Applications of dual-material gate (DMG) on bulk MOSFET lead to a simultaneous transconductance enhancement and suppression of SCEs compared to single material gate (SMG). This is due to introduction of step function in the channel potential. In DMG MOSFET, the workfunction of metal gate 1 (M1) is greater than the metal gate 2(M2), i.e., ϕM1 > ϕM2 for an n-channel MOSFET and vice versa for p-channel MOSFET.

EXAMPLE 8.8

Is floating body effect desirable?

Solution: The floating body can charge up, causing dc effects such as premature breakdown and enhanced subthreshold slope at high drain bias. It also causes transient effects on circuit performance (some good and some bad). On balance, floating body effect (FBE) are not desirable because they lead to irreproducible circuit performance. Body charge may be generated either by capacitive coupling to gate and drain or by impact ionization of the drain current and is drained away by recombination at the body/source depletion region.

Promising solution to the FBE involves increasing the recombination on the source by implanting to create recombination centres or lower the bandgap.

Note: FD SOI has a poorer scaling potential than bulk because of the lack of screening from the back of the channel. FD SOI has much stronger DIBL than PD SOI for all but the thinnest (< 40 nm) SOI layers.

EXAMPLE 8.9

Why a large parameter spread has been noted for FD SOI device during experiment?

Solution: Since, the threshold voltage in FD SOI CMOS devices depends on the SOI thickness and boundary conditions at the back SOI interface.

Note: 1. If any FD SOI circuit is built on a wafer with non-uniform Si thickness, its operation will be unstable.
2. In brief,
   - When the silicon film thickness (t_{Si}) > depletion width: Partially depleted SOI and device exhibit floating-body effect.
   - When the silicon film thickness (t_{Si}) < depletion width: Fully depleted SOI exhibits superior subthreshold characteristics, attracting attention for low-voltage operation.

There is no abrupt boundary between PD and FD. When the silicon film thickness is reduced, the barrier height of the body-to-source junction is lowered, and there is less change in both the body potential and threshold voltage. The floating body effect becomes rather complex in the ac conditions.

Another type of SOI device, the dynamic threshold voltage MOSFET (DTMOS), is obtained by tying the body to the gate so that they both switch in a transition. Such devices can have high V_t when V_G is low to suppress OFF current and low V_t when V_G is high to increase ON current. DTMOS is essentially a double-gate MOSFET whose subthreshold slope is close to ideal, value of 60 mV/decade. However, DTMOS can be operated below the *p-n* junction turn-on voltage to avoid significant forward bias currents at the body to source/drain junctions, hence it is more suitable for very low voltage operation.

8.8.2 SiGe MOSFET or Strained MOSFET

The transition from Si to a different material with better transport properties appears to be an attractive solution to enhance the mobility of the carriers. It is essential that the material system should be compatible with the state-of-art technology of fabrication. The epitaxial growth of strained SiGe compounds on Si substrates produces the strain in the Si and SiGe layers to be tuned. Theoretically, the hole mobility increases in silicon under either tensile or compressive strain, due to breaking of valence-band degeneracy and reduction of conductivity mass. The electron mobility is enhanced in Si under the tensile strain because of electron increased population in two lower energy valleys with reduce conductive mass. Tensile strain can be created by growing an epitaxial silicon layer on a relaxed SiGe film. The lattice constant of SiGe is slightly larger than the bulk silicon, i.e., mismatch of these two materials.

Due to confinement of the charge carriers in a quantum well there is no measurable roughness scattering at room temperature; also alloy scattering on SiGe has a negative effect on the mobilities of both electrons as well as holes.

Saturation velocities of electrons and holes in strained Si and SiGe are comparable to bulk Si, but are reached at much lower electric fields. This translates into lower bias requirement and hence lower power consumption and higher reliability.

The parasitic resistances and velocity saturation effect restricts the performance improvement and eventually limit the total gain attainable from high mobilities.

A possible design of Si/SiGe CMOS is shown in Fig. 8.15. A higher mobility material can be viewed as an addition/alternative to scaling. These materials improve the power and performance of CMOS technology at the same channel length. But there are many challenges in integrating such a strained film or films in an advanced CMOS process. In particular, one may be constrained to low temperature processing ($T < 800°C$) to avoid strain relaxation and Ge segregation problems, i.e., growing or depositing gate oxide layers at $T < 800°C$ and annealing doped polysilicon gates and source-drain implants at $T < 800°C$ while trying to achieve the high dopant activation and low junction leakage. Therefore, to exploit the performance advantage of SiGe devices one should keep the source-drain series resistance as small as the bulk silicon devices. The scaling of Si and SiGe channels layer thickness is limited to 30 Å due to quantum mechanical effects.

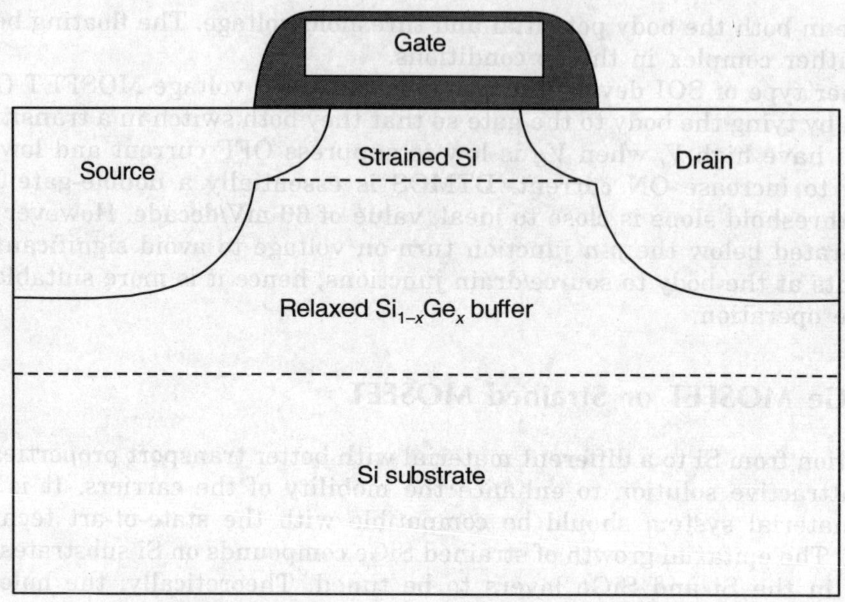

FIGURE 8.15   Schematic of SiGe MOSFET.

Note: The strain effects generally increases mobilities of the carriers. In SOI, compressive strain is generated when LOCOS is used as isolation between SiO$_2$ and silicon. It was found experimentally that there is 50% increase in hole's mobility for SOI island widths of 1 μm.

8.8.3   Double-Gate (DG) MOSFET

Double-gate MOSFET is becoming a subject of interest in VLSI design because in principle, DG MOSFET can be scaled to the shortest channel length possible for a given gate oxide thickness. The advantages of DG MOSFETs are:

    (i) Ideal 60 mV/decade subthreshold slope
    (ii) Scaling of silicon film thickness without high doping
    (iii) Setting of threshold voltage by gate work functions
    (iv) Good control of SCEs such as mobility enhancement
    (v) Provides inherent electrostatic coupling in the channel. This intimate coupling between the gates and the channel makes DG MOSFET technology the most scalable of all FET designs.
    (vi) The DG MOSFET is electrostatically superior to a single-gate MOSFET because two gates are used to control the channel from both sides as shown in Fig. 8.16(a), thus allowing for additional gate length scaling by nearly a factor of two.
    (vii) The two gates control roughly twice as much current as single gate, resulting in stronger switching signals.
    (viii) The ballistic transport of carriers can be achieved with DG MOSFETs due to the low doping in the channel and the DG features.

The key factors that limit how far a DG MOSFET can be scaled come from short-channel effects such as threshold voltage roll-off and DIBL effect. Another challenge is to obtain a suitable threshold voltages for high-speed logic devices, while controlling the extrinsic resistance.

Contrary to bulk MOSFETs, depletion charges in DG MOSFETs are negligible because the silicon film is undoped or lightly doped. Thus, only mobile charge term needs to be included in Poisson's equation.

As clear from Fig. 8.16(a) the dual gate MOSFET has two gates with work-functions of $\Delta\phi_1$ and $\Delta\phi_2$, where $\Delta\phi_1$ is the workfunction difference between top gate and intrinsic, silicon. Similarly, $\Delta\phi_2$ is workfunction difference between bottom gate and intrinisic silicon. If $\Delta\phi_1 = \Delta\phi_2$ then DG MOS is called **symmetric DG MOSFET**. And if $\Delta\phi_1 \neq \Delta\phi_2$, it is an **asymmetric DG MOSFET**.

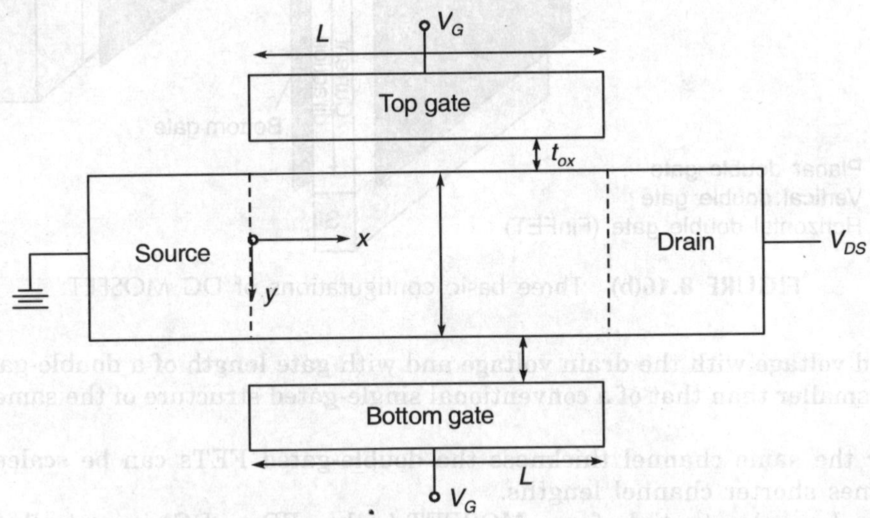

FIGURE 8.16(a) Schematic of Double-Gate (DG) MOSFET.

Actually DG MOSFET can be made in three basis configurations as shown in Fig. 8.16(b) labelled as Type I, II and III.

Type I has the advantage that the channel layer is in the plane of the silicon wafer surface so that channel thickness is controlled by the thickness of uniform planar layers roller than by lithography. In type II, which has the channel in the vertical direction, where one can get low leakage current. Type III has the highest package density for high-speed logic applications since the channel width is perpendicular to the plane of wafer. Increasing height increases gate area without affecting "real silicon area".

The two gates in DG MOSFET are electrically connected so that they both serve to modulate the channel. Short-channel effects are greatly suppressed because the two gates very effectively terminate the drain field lines preventing the drain potential from being felt at the source end of the channel. Consequently, the variation of the

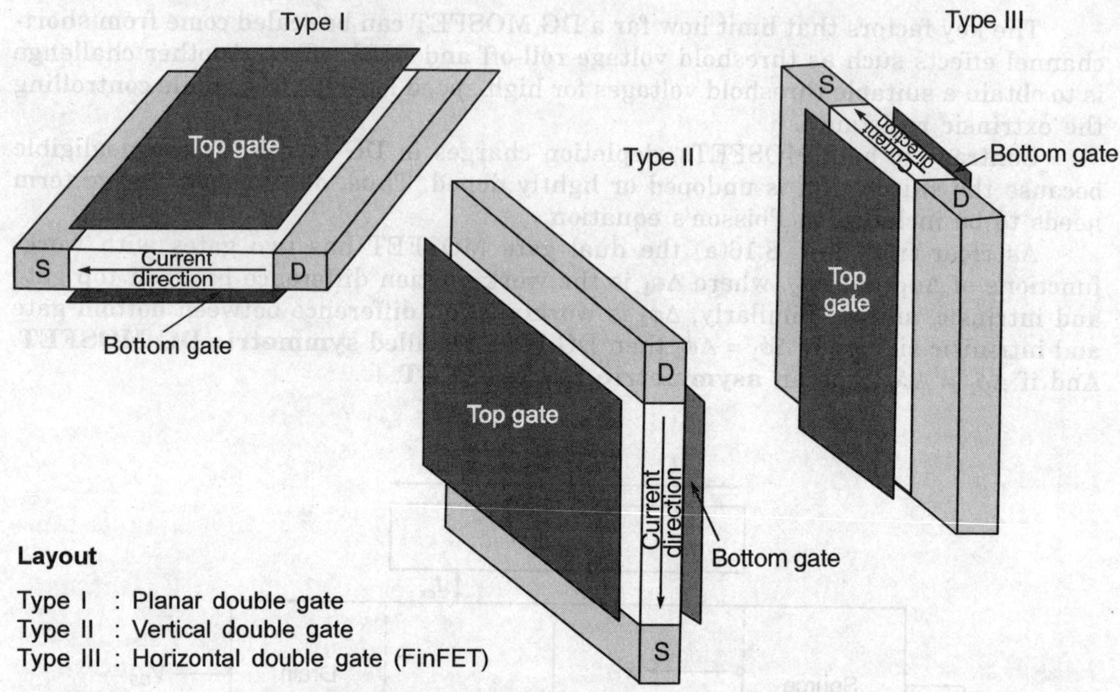

Layout

Type I   : Planar double gate
Type II : Vertical double gate
Type III : Horizontal double gate (FinFET)

FIGURE 8.16(b)   Three basic configurations of DG MOSFET.

threshold voltage with the drain voltage and with gate length of a double-gated MOS is much smaller than that of a conventional single-gated structure of the same channel length.

For the same channel thickness the double-gated FETs can be scaled to two-three times shorter channel lengths.

The channel potential of any MOSFET (either FD or DG) is controlled by both the front gate and back gate. For double-gate MOSFETs, the ideal subthreshold slope occurs due to control of channel by both gates, which are switched simultaneously. However, in FD SOI MOSFETs, nearly ideal subthreshold slope occurs because the back gate does not have much control over the channel. The lack of back gate control leaves a wide depletion or dielectric region vulnerable to source-drain field penetration, which results in poor SCE.

Ultimate double-gate limits

- Thermionic emission above the channel potential barrier i.e. short channel effects lower the potential barrier.
- Band to band tunnelling between body and drain *p-n* junction.
- Quantum mechanical tunnelling directly between source and drain.
- Other effects of quantum confinement in the thin body.

These limitations are shown in Fig. 8.17(a).

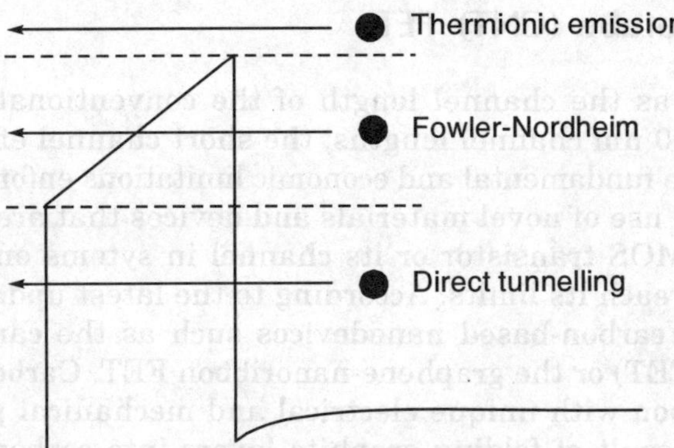

FIGURE 8.17(a) Ultimate double gate limits.

Challenges facing double-gate technology:

- Identically sized gates and connecting these two gates with a low resistance path
- Self-alignment of source and drain to both gates and alignment of both gates to each other

Leakage current problem in CMOS. Leakage current is a primary concern for low-power, high performance digital CMOS circuits for portable applications. Industry trends show that leakage will be roughly 50% of the total power in future technologies. New leakage mechanisms, such as tunnelling across thin gate oxides, leading to gate oxide leakage current come into play at the 90 nm technology and remain a daunting challenge for a number of technology nodes.

According to ITRS (The International Technological Roadmap for Semiconductors) the physical oxide thickness t_{ox} of 7–12 Å will be required for high performance CMOS circuits. Quantum effects that cause tunnelling of carriers will play a dominant role in such ultrathin oxide devices. Since, the probability of electron tunnelling depends on the barrier height and oxide thickness, therefore, any small change in t_{ox} can have a tremendous impact on I_{gate}. Moreover the other component to leakage, subthreshold leakage I_{sub}, forms a reducing fraction of the total leakage as t_{ox} is reduced. The most effective way to control I_{gate} is to use of high K-dielectric materials.

Leakage power can be broadly divided into (a) Standby leakage which corresponds to the situation when the circuit is in a non-operating mode or sleep mode. (b) Active leakage which relates to leakage during normal operation. Standby leakage can be reduced by state assignment technique, use of multiple threshold CMOS (MT CMOS) sleep transistors, body-biasing and dual t_{ox} combined with state assignment. Leakage power dissipation in the active mode can be reduced by the use of dual t_{ox} assignments.

EXAMPLE 8.10

Why should one take the care of power dissipation?

Solution: Because power consumption will generate heat which will affect the reliability and functionality of the circuit. Also it increases the cost of the device.

8.8.4 Carbon Nanotube (CNT) FET

It is well known that as the channel length of the conventional MOSFET is scaled down in sub-micron 100 nm channel lengths, the short channel effects (SCEs) become more pronounced. These fundamental and economic limitations enforce the semiconductor industry to explore the use of novel materials and devices that are able to complement or even replace the CMOS transistor or its channel in sytems on chip before silicon-based technology will reach its limits. According to the latest update of ITRS the most promising devices are carbon-based nanodevices such as the carbon nanotube field-effect transistor (CNTFET) or the graphene-nanoribbon FET. Carbon nanotubes (CNTs) are a new form of carbon with unique electrical and mechanical properties. They can be considered as the result of folding graphite layers into carbon cylinders and may be composed of a single shell, single wall nanotubes (SWNTs) or of several shells, multi-wall nanotubes (MWNTs). Depending on the folding angle and the diameter, nanotubes can be metallic or semiconducting. Simple theory also shows that the band gap of semiconducting NTs decreases with increasing diameter. These predictions have been verified by scanning tunnelling spectroscopy experiments.

The carbon nanotube field-effect transistor (CNTFET) is a promising candidate for future electron devices. Rapid progress in the field has recently made it possible to fabricate CNTFET based circuits, such as logic gates, static memory cells, and ring oscillators, The structure of a CNT-FET is similar to the structure of a typical MOSFET where the CNT forms the channel between two electrodes that work as the source and the drain of the transistor. The structure is build on top of an insulating layer and a substrate wafer that works as the back gate. The device uses zirconium oxide rather than silicon dioxide as the gate insulator. High-dielectric-constant materials are useful as gate insulators because they can provide efficient charge injection into transistor channels and reduce direct-tunnelling leakage currents. The basic structure of a CNT-FET based on a SWNT is shown in Fig. 8.17(b).

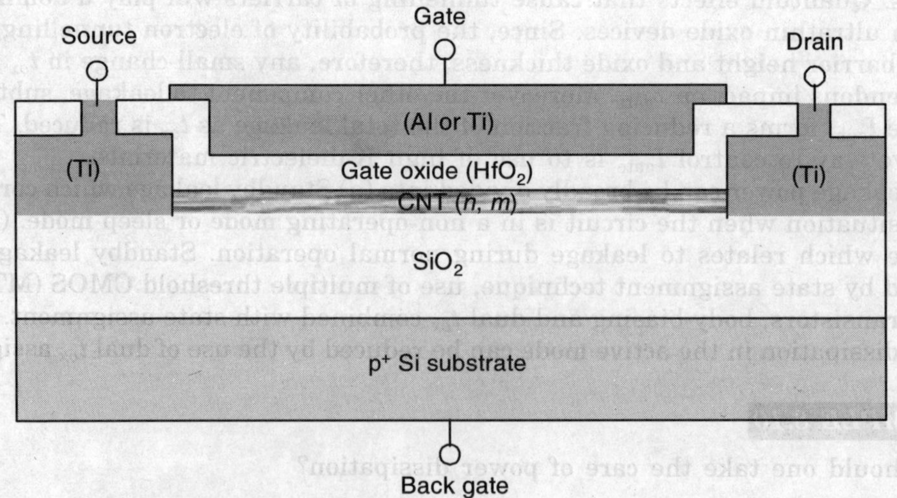

FIGURE 8.17(b) Structure of CNT MOSFET.

Primarily, the typical CNTFET is a Schottky-barrier device, i.e., one whose performance is determined by contact resistance rather than channel conductance, owing to the presence of tunnelling barriers at both or either of the source and drain contacts. These barriers occur due to Fermi-level alignment at the metal-semiconductor junction, and are further modulated by any band bending imposed by the gate electrostatics.

8.9 SUMMARY

Scaling is defined as the reduction of MOS device dimension (vertical and horizontal) along with various parameter by a predefined factor. The factor is called **scaling factor** and chosen by technology. Due to scaling, number of components on the chip has increased tremendously and also power consumption reduced and device becomes faster. There are mainly two scaling methods one is constant voltage scaling where power supply remains constant and second is constant field scaling. Thermal voltage (kT/q), built-in-potential, energy gap and threshold voltage cannot be scaled down.

A CMOS device is formed by the combination of a p-MOS and n-MOS whose standby power consumption is negligible (almost zero). There is only power consumption during the switching action. At one time only one transistor either n-MOS or p-MOS will operate. The scaling limit of a typical MOS device is shown in Fig. 8.18.

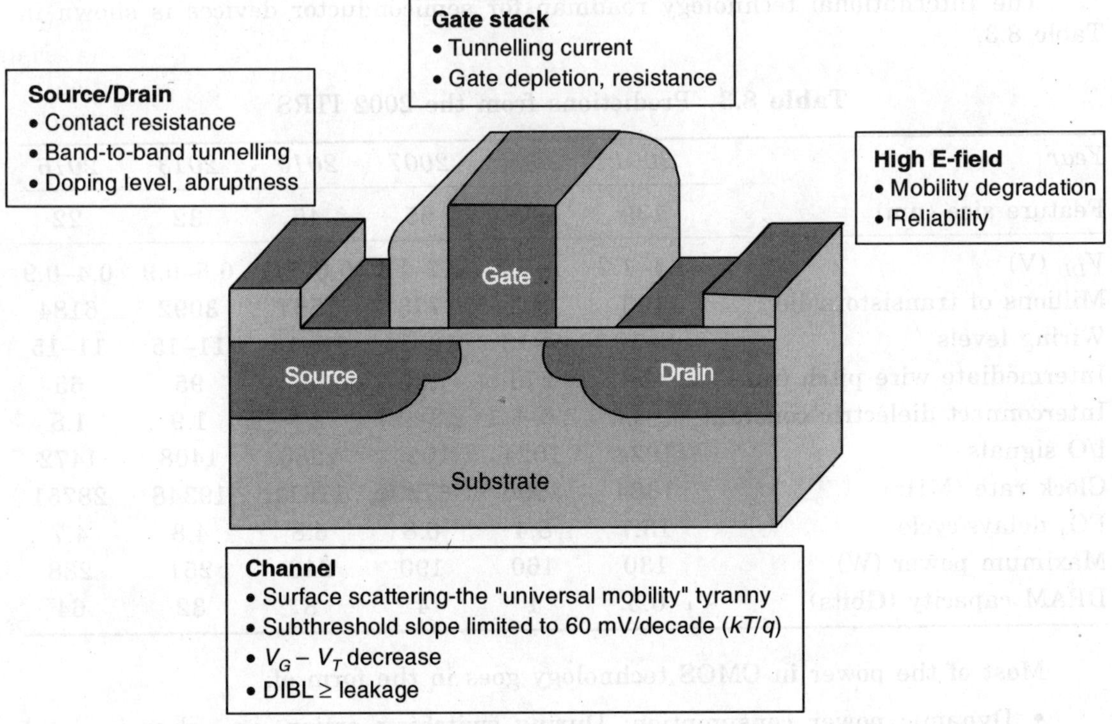

FIGURE 8.18 Scaling Limit of a typical MOS device.

One of the most serious concerns in future VLSI design is the exponential increased standby power caused by the subthreshold current in the device. Power consumption in the circuit can be reduced drastically by reducing the power supply V_{DD}, but that causes the slowdown of the device, i.e., there is a trade-off between power consumption and delay. So, as a designer one has to set his/her priority.

CMOS inverter is a basic digital static circuit. When input is low, p-MOS turns ON and n-MOS turns OFF, and circuit provides a high resistance path to the current from p-MOS and hence output is high. When input is high (= V_{DD}), n-MOS turns ON, p-MOS turns OFF and charge stored in load capacitor will discharge upto zero means V_{out} is low. If one applies $V_{in} = V_{DD}/2$ for an ideal CMOS inverter both transistors (p-MOS and n-MOS) turn ON (because both will enter into saturation) and hence a direct path will be available from V_{DD} to ground and a large current will flow in the circuit which causes a large power consumption. Therefore, one should avoid this input, called **inversion threshold** voltage. Noise margin of a CMOS inverter is very high because $V_{OH} = V_{DD}$.

and
$$V_{OL} = \text{Very low}$$

Latch-up is a failure mechanism in CMOS integrated circuits and can be avoided be guard ring or FOX (field oxide) methods.

There are many alternative devices available which can give more freedom of scaling than CMOS, i.e., single gate CMOS devices like silicon-on-insulator, double-gate MOSFET, SiGe MOSFET (Strained MOS).

The International technology roadmap for semiconductor devices is shown in Table 8.3.

Table 8.3 Predictions from the 2002 ITRS

Year	2001	2004	2007	2010	2013	2016
Feature size (nm)	130	90	65	45	32	22
V_{DD} (V)	1.1–1.2	1–1.2	0.7–1.1	0.6–1.0	0.5–0.9	0.4–0.9
Millions of transistors/die	193	385	773	1564	3092	6184
Wiring levels	8–10	9–13	10–14	10–14	11–15	11–15
Intermediate wire pitch (nm)	450	275	195	135	95	65
Interconnect dielectric constant	3–3.6	2.6–3.1	2.3–2.7	2.1	1.9	1.8
I/O signals	1024	1024	1024	1280	1408	1472
Clock rate (MHz)	1684	3990	6739	11511	19348	28751
FO$_4$ delays/cycle	13.7	8.4	6.8	5.8	4.8	4.7
Maximum power (W)	130	160	190	218	251	288
DRAM capacity (Gbits)	0.5	1	4	8	32	64

Most of the power in CMOS technology goes in the form of:

- Dynamic power consumption: During switching action, i.e., charging and discharging capacitors

- Short circuit currents: When $V_{in} = V_{DD}/2$
- Leakage current: due to leaking diodes and transistors

Variable threshold voltage MOSFET (VTCMOS) has promised to be among the next generation of ultra-low-power devices operating at low supply voltage. The operating principle of the VTCMOS is that the threshold voltage (V_t) is controlled by the substrate bias, leading to lower standby OFF current or higher active ON current. In the standby, mode V_t is increased to suppress the OFF current, while in the active mode V_t is lowered to enhance the ON current applying the substrate bias of an opposite sign. Therefore, in VTCMOS it is required to reduce the OFF current and keeping the active ON current high. Both are required for high speed low-power performance.

Challenges for analog CMOS. Power supply is steadily decreasing, which is a requirement for digital applications where power dissipation is a serious problem but analog circuits are adversely affected because the reduction in power supply directly translates into a lower dyanamic range.

The design style for analog circuit is determined by the low supply voltage. Passive devices such as capacitors and inductors, which is mainly used in analog application, cannot be scaled down and hence size and cost of the analog part increases in comparison to digital part. Special analog steps are sometimes introduced in the process such as thick-gate-oxide transistors, thick metal layers, thin oxides and so on which increases the overall cost of the system.

REVIEW QUESTIONS

1. What are the needs of scaling?
2. What is Moore's law?
3. Write the advantages of CMOS over BJT. And also list basic drawback of CMOS.
4. Why CMOS is preferred over MOS?
5. List down all the scaling limits of CMOS.
6. What are the advantages of full scaling on constant voltage scaling?
7. If power supply V_{DD} is reduced, what will be the effect on delay and power consumption of the circuit?
8. What will happen if one scale down the thickness of the gate oxide? Suggest some method to avoid these problems.
9. Why interconnection delay is not important for long channel devices, whereas for short channel devices it must be considered?
10. What is the need for low power VLSI design and how one can achieve a circuit for low power? Discuss its consequences.
11. Why a CMOS inverter is assumed basic building element in digital circuit?
12. Why noise margin of a CMOS inverter is very high?

13. Why $(W_p/W_n) = 2.5$ is taken for a design of a CMOS inverter?

14. Why should one avoid $V_{in} = V_{DD}/2$ on the CMOS inverter?

15. As a designer, how can one reduce the average power consumption of the circuit in VLSI design?

16. What is latch-up problem in the CMOS and how it degraded the performance of the device?

17. What is the significance of power delay product and why it is required to keep its value as small as possible?

18. How one can reduce the load capacitance of a device and what will be its effect on the circuit performance?

19. What are the advantages of SOI over bulk MOS devices?

20. Why in SOI MOS, one gets more freedom to scale the device which is not possible in bulk CMOS?

21. What are the advantages of FDSOI over bulk CMOS?

22. How in the DGMOSFET, the mobility of the carrier is increased in comparison to bulk MOS?

23. Why in the case of double-gate MOSFET, one gets ideal subthreshold slope?

24. What is the basic differences between the SOI MOS and DGMOSFET?

NUMERICAL PROBLEMS

1. Find the gate-oxide capacitance per unit area after constant field scaling and constant voltage scaling. Discuss the result in terms of speed of device.

2. Find the effect of scaling (both) on the power density and discuss the result qualitatively.

3. How the power dissipation affected by the scaling?

4. Consider a n-MOS device with the following parameters:

$$\mu_n = 1000 \text{ cm}^2/\text{V-sec}, \ t_{ox} = 50 \text{ Å}, \ (W/L) = 10$$

$$V_G = 1.2 \text{ V}, \ V_S = 0 \text{ V}, \ V_D = 5 \text{ V}, \ V_{t0} = 1.0 \text{ V}$$

Find

(a) The minimum gate voltage required to turn the transistor ON.
(b) Whether the transistor is in linear region or saturation region?
(c) The drain current before scaling.
(d) The drain current after scaling due to constant voltage scaling, if S = 2.

5. When a linear dimensions of a MOSFET are scaled down by a factor of 10 based on the constant field scaling, what is the scaling factor for the corresponding switching energy?

6. For an n-channel FD–SOI device having $N_a = 10^{17}/cm^3$ and $W = 4$ nm, calculate maximum allowable thickness for Si channel layer t_{si}?

7. A CMOS inverter drives a load that consists of the gate of another FET. The gate area is $.02 \times 5$ μm. Find the dynamic power discipation if the clock frequency is 350 MHz, $V_{DD} = 2.5$ V and $t_{ox} = 5$ nm.

8. Find the (W_p/W_n) ration needed to match the drain current for CMOS inverter, if

$$\mu_n = 1000 \ cm^2/V\text{-sec}$$
$$\mu_p = 400 \ cm^2/V\text{-sec}$$

Minimum feature size $= 0.5$ μm

and $V_{GS} - V_t = 2.5$ V. Assume $v_{sat} = 4 \times 10^6$.

9. Consider an ideal CMOS inverter with the following parameter:

$$V_{DD} = 3.3 \text{ V}, \ V_{ton} = 1 \text{ V}, \ V_{top} = -0.7 \text{ V}$$
$$K_n = 200 \ \mu A/V^2, \ K_p = 80 \ \mu A/V^2$$

Find the value of V_{OH} and V_{OL}.

10. Consider a CMOS inverter circuit with the following parameters

$$V_{DD} = 3.3 \text{ V}, \ V_{ton} = 0.8 \text{ V}, \ V_{top} = -0.7 \text{ V}, \ K_n = 200 \ \mu A/V^2, \ K_p = 80 \ \mu A/V^2$$

If an input voltage of 2.5 V is applied, find the output voltage.

CHAPTER 9

Bipolar Transistors

- Construction and operation of BJT
- Qualitative analysis of BJT
- Current-voltage characteristics of real BJT and ideal BJT
- Ebers-Moll and Gummel-Poon models
- Non-ideal behaviours of bipolar transistor
- Basic ac model
- Switching behaviour of BJT
- Design consideration of BJT
- Heterojunction bipolar transistor (HBT)
- The thyristor or silicon controlled rectifier (SCR)
- The unijunction transistor (UJT)
- The Schottky-clamped transistor
- BiCMOS technology

9.1 INTRODUCTION

The transistor, which stands for transfer resistor, is a multijunction semiconductor device. Normally, the transistor is integrated with other circuit elements on the chip for voltage gain, current gain or signal power gain in analog electronics and for switching action in digital electronics. There are basically two types of transistors namely field-effect transistor (FET) which is a unipolar device and other is bipolar junction transistor (BJT) where both majority carriers as well as minority carriers are responsible for current flow in the device. BJT is also a current controlled current device, i.e., by controlling input current one can control output current in contrast to FET which is voltage controlled current device. The bipolar junction transistor was invented by William Schockley at Bell Laboratory in 1947.

BJT is basically used in high speed circuits, analog circuits and power applications. The bipolar junction transistor is basically a three-terminal device namely, emitter (which emits carriers and should be heavily doped), base (which is lightly doped region) and collector (which mainly collects the carrier and it is lightly doped region). Unlike the FET, where source and drain are interchangeable physically, in BJT all the three terminals are fixed and can not be interchanged. On the basis of the majority carriers, a bipolar junction transistor can be classified as *n-p-n* transistor (where collector and emitter is n-type semiconductor and base is p-type semiconductor) and *p-n-p* transistor (where emitter and collector is p-type semiconductor and base is n-type semiconductor). In this chapter we will concentrate mainly on the *n-p-n* transistor. Since it is a widely used transistor because of high mobility of carriers.

Although most circuits in microelectronics are made up of the CMOS transistors, but bipolar junction transistors are also gaining importance in VLSI design due to inherently high speed for digital circuit application and superior characteristics for Analog circuit applications. Commonly used bipolar devices are either lateral transistors, where the three regions of transistor are arranged horizontally adjacent to one another and the active current flow laterally, or vertical transistors where the active three regions of transistor are arranged vertically one on top of another and current flows in the vertical direction. Practically all high speed bipolar transistors used in digital circuits are of the vertical types. The equations for a *n-p-n* transistor can be extended to a *p-n-p* transistor simply by reversing the voltage and dopant polarities and using the appropriate device parameters' values.

9.2 CONSTRUCTION AND FUNDAMENTALS OF BJT OPERATION

A prespective view of a discrete *n-p-n* transistor is shown in Fig. 9.1. The transistor is formed by starting with a n-type substrate. A p-type region is thermally diffused through an oxide window into the n-substrate. A heavily doped n^+ region is then diffused into the p-region. Metallic contacts are made to the n^+ and n-regions through a window opened in the oxide layer and to the n-region at the bottom. Actually, as seen from Fig. 9.1, a bipolar transistor consists of a pair of p-n junction diodes that are joined back-to-back, i.e., a sort of sandwich where one kind of semiconductor is placed in between two others. An idealized, one-dimensional structure of a *n-p-n* bipolar transistor is shown in Fig. 9.2(a). The three regions of transistor are doped separately. The heavily doped region (n^+) is called **emitter** because it supplies carriers. Lightly doped region (n) serves the purpose of collector whereas moderate doped region (p) (larger than collector but smaller to emitter) is called **base region**. The base region should be as narrow as possible, whereas collector region should be more wider to dissipate more heat. The emitter region is of moderate width, i.e., larger than base and smaller than collector. The symbol for *n-p-n* transistor is shown in Fig. 9.2(b), where arrow indicates the direction of current flow.

Figures 9.2(a) and (b) also shows the polarities of various voltages applied under normal condition, also called **active mode**.

Here, $V_{BE} = V_B - V_E$ is voltage across base to emitter.

$V_{BC} = V_B - V_C$ is voltage across base to collector.

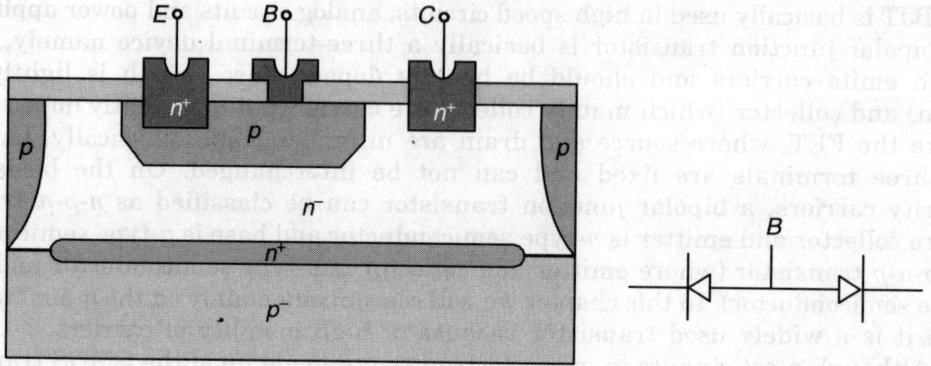

FIGURE 9.1 A perspective view of *n-p-n* transistor.

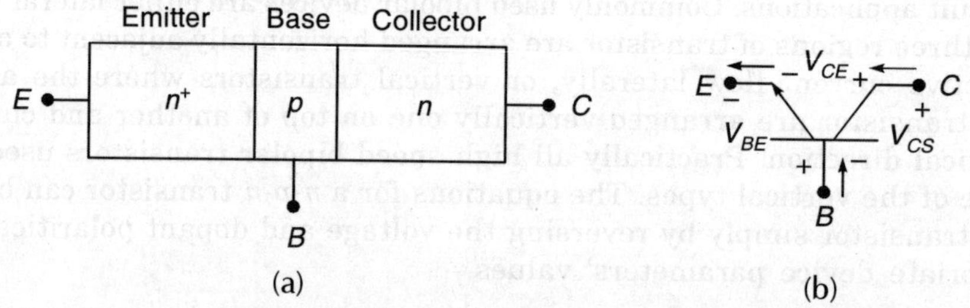

(a)

(b)

FIGURE 9.2 (a) One dimensional view of *n-p-n* transistor, (b) Symbol of *n-p-n* transistor.

Under forward active mode, the emitter-base junction should be forward biased and collector-base junction should be reverse biased. The various currents components I_B (base current), I_E (emitter current) and I_C (collector current) and their directions are shown in Fig. 9.2(b).

For forward bias of the emitter-base junction $V_{BE} \simeq 0.7$ V for silicon.

When both junctions of the transistor is in forward bias then transistor is said to be in saturation region and $V_{CE} = 0.2$ V for Si substance. If both junctions are reverse biased then transistor is said to operate in cut-off region and $I_E = 0$. A transistor operates in reverse active mode when emitter base junction is reverse biased and collector base junction is forward biased as shown in Fig. 9.3.

To use bipolar junctions transistor as an amplifier in analog electronics, it must be biased in active region.

In digital electronics, BJT is used as switch, i.e., BJT is OFF, when it is operating in cut-off region. BJT is ON when it is operating in saturation region.

For FET, saturation region and active region are same whereas in BJT, these two regions are separate.

According to Kirchhoff's law, total current is:

$$I_E = I_C + I_B \qquad (9.1)$$

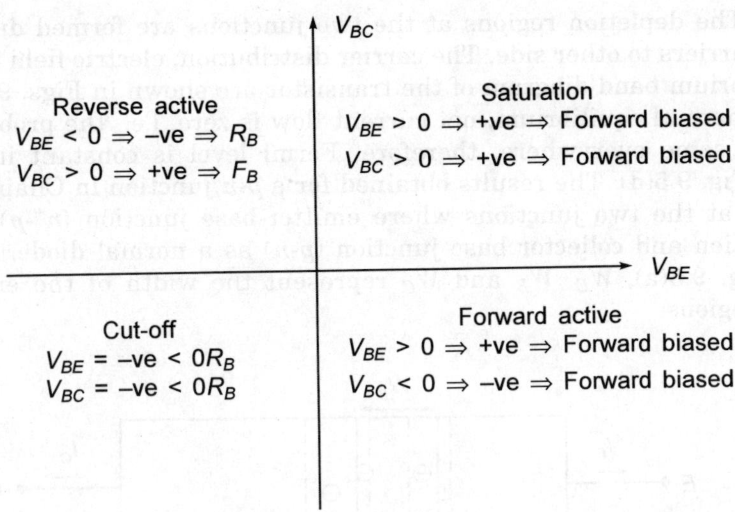

FIGURE 9.3 Bias conditions for operation in different regions.

The *p-n-p* transistor is complimentary structure of the *n-p-n* bipolar transistor. The structure and symbol, with voltage polarities and current direction for *p-n-p* transistor is shown in Fig. 9.4.

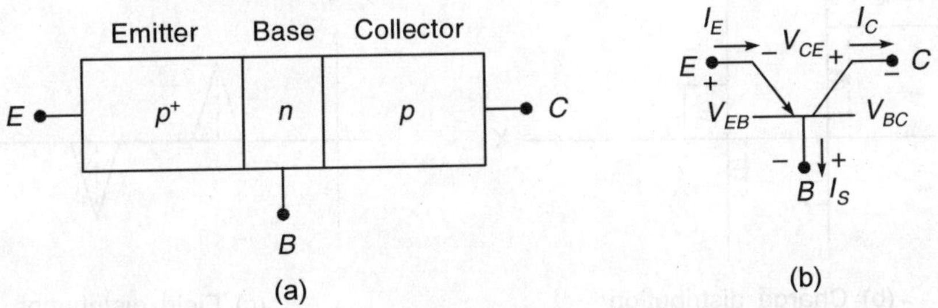

FIGURE 9.4 One dimensional view and symbol of *p-n-p* transistor.

From Fig. 9.4, it is clear that the current flow and voltage polarity are all reversed for a *p-n-p* transistor.

9.3 OPERATION IN THE ACTIVE MODE

Figure 9.5 shows an idealized *n-p-n* trasistor at thermal equilibrium, i.e., no bias is applied at any leads. The ideal transistor behaviour is based on the ideal *p-n* junction diode. As we have studied in Chapter 4, a depletion region will form near the two junctions as shown in Fig. 9.5(a). The depletion region is more wider in the base than the emitter at emitter base junction because of heavily doped emitter region. Similarly, at the collector base junction depletion region is more wider in collector region than the base. The depletion regions at the two junctions are formed due to diffusion of majority carriers to other side. The carrier distribution, electric field near the junction and equilibrium band diagram of the transistor are shown in Figs. 9.5(b), (c) and (d). Since, at thermal equilibrium, net current flow is zero, i.e., the probability of getting charges is same everywhere, therefore, Fermi level is constant in the regions as shown in Fig. 9.5(d). The results obtained for a *p-n* junction in Chapter 4 are equally applicable at the two junctions where emitter-base junction (n^+-p) behaves as one sided junction and collector base junction (p-n) as a normal diode.

In Fig. 9.5(a), W_E, W_B and W_C represent the width of the emitter, base and collector regions.

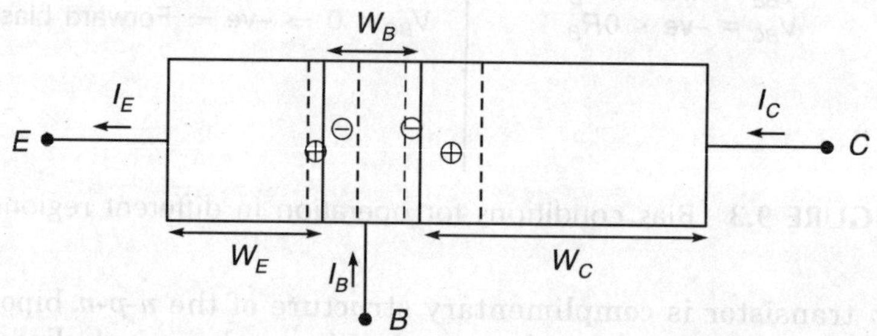

(a) Schematic representation of *n-p-n* transistor

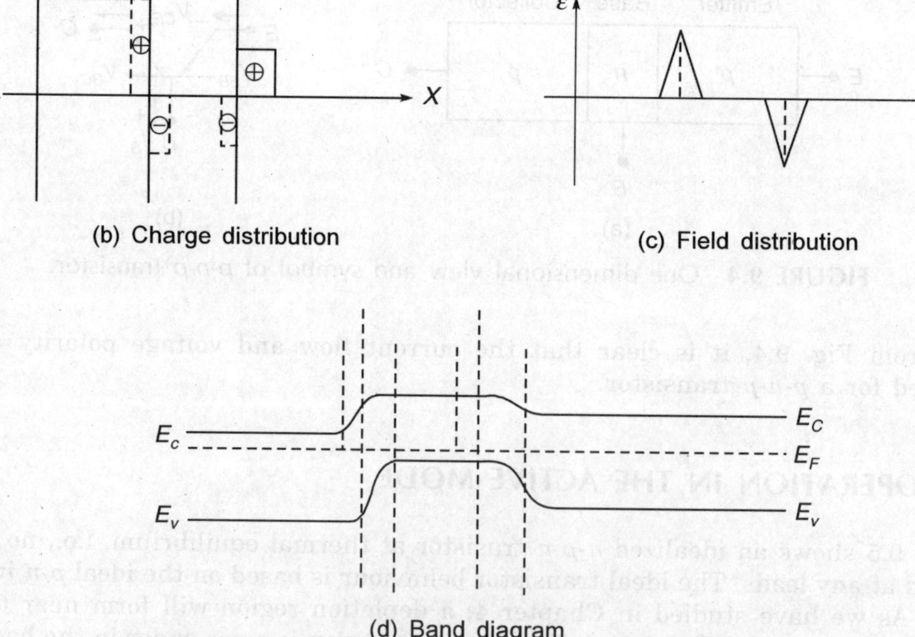

(b) Charge distribution   (c) Field distribution

(d) Band diagram

FIGURE 9.5 *n-p-n* transistor under no bias condition.

Consider n-p-n transistor is biased in forward active mode, i.e., the emitter-base junction (input junction) becomes forward biased and collector-base junction (output junction) becomes reversed biased as shown in Fig. 9.6(a). Due to forward bias of input p-n junction, electrons injected (or emitted) into the base region from n^+-emitter region and majority carrier holes are injected from base to emitter (due to diffusion of carriers). Assuming ideal situation, where there is no generation-recombination in the depletion width, means all the injected carriers move towords the respective contacts and constitute respective currents namely, emitter current I_E and base current I_B. Due to forward biasing, the depletion width at the emitter-base junction shrinks. The output junction is reversed biased means collector is at more positive potential than base, therefore, due to positive potential at the collector, the injected electrons into the base (from emitter) get attracted towards the collector. If the base region is very narrow or thin then almost all the injected electrons can cross the base region (without any recombination with majority carrier holes in the base) and move towards the collector contact and result in collector current I_C. Few minority electrons from base also enter into the collector region. Therefore, we can see that the majority carrier electrons injected from emitter or pushed away from emitter due to negative potential on the emitter travel through the base into the collector region.

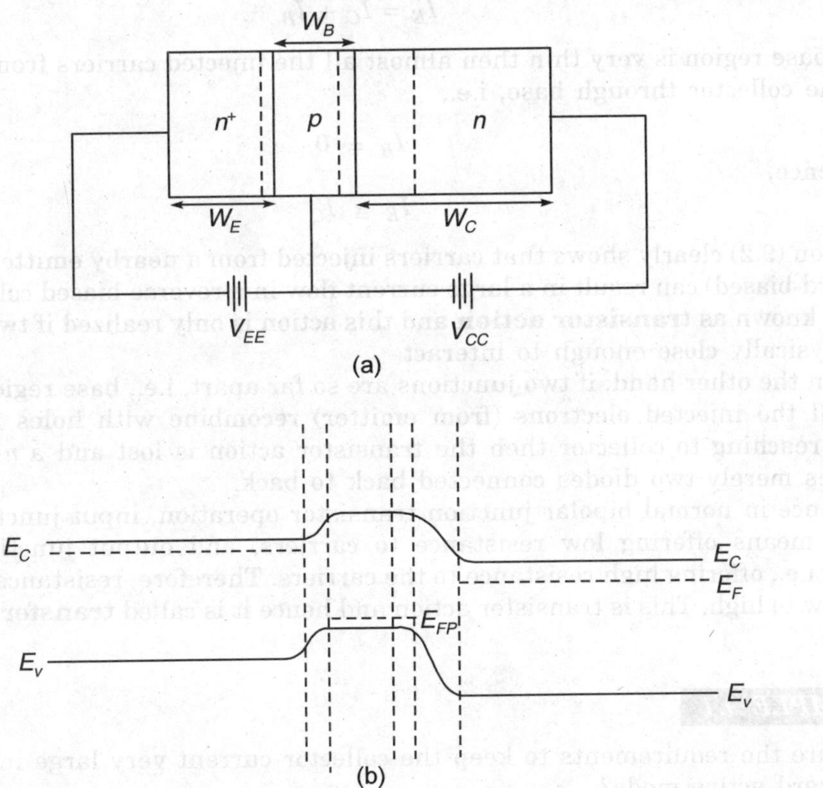

FIGURE 9.6 n-p-n transistor in active mode of operation and band diagram.

If the base region is very wide then most of the injected electrons will stay in base region for long time (because they take more time to cross base), and the probability of their recombination with majority carrier holes increases. Therefore, less number of electrons reach at the collector, reducing the collector current. Since large carriers are collected into the collector region, therefore, to dissipate the generated heat, the collector region should be wide.

The energy band diagram of a n-p-n transistor in normal active region is shown in Fig. 9.6(b).

Here V_{EE} and V_{CC} are power supplies connected to emitter-base junction and collector-base junction. From Fig. 9.6(b), it is clear that the electron from emitter side can easily cross the barrier height and reach into the collector region. The field distribution under such biasing scheme can be drawn in same way as in Chapter 4.

Since electrons are pushed away from n^+ region or injected into the base region or emitted into the base region. Therefore, n^+ region is called **emitter**. The injected carriers from emitter get collected at the n region and hence this region is called **collector region**. Since, collector-base junction is reverse biased, therefore, a small reverse current flows across the junction.

The total emitter current I_E is equal to current in base region plus collector current, i.e.,

$$I_E = I_C + I_B$$

If the base region is very thin then almost all the injected carriers from emitter move into the collector through base, i.e.,

$$I_B \simeq 0$$

and hence,

$$I_E \cong I_C \tag{9.2}$$

Equation (9.2) clearly shows that carriers injected from a nearby emitter-base junction (forward-biased) can result in a large current flow in a reverse-biased collector junction. This is known as **transistor action** and this action is only realized if two p-n junctions are physically close enough to interact.

On the other hand, if two junctions are so far apart, i.e., base region is very wide that all the injected electrons (from emitter) recombine with holes in base region before reaching to collector then the transistor action is lost and a n-p-n transistor becomes merely two diodes connected back to back.

Since in normal bipolar junction transistor operation, input junction is forward biased means offering low resistance to carriers, and output junction is reverse-biased, i.e., offering high resistance to the carriers. Therefore, resistance is transferred from low to high. This is transistor action and hence it is called **transfer of resistance device**.

EXAMPLE 9.1

What are the requirements to keep the collector current very large in a BJT biased in forward active mode?

Solution: Requirements are to make the base width as thin as possible. The important condition is $W_B \ll L_n$ where L_n is diffusion length of electron in the base. If this requirement is satisfied then an average electron injected into base will diffuse to the depletion region of the collector junction without recombination in the base.

Second requirement is that the electron life time should be long; which also satisfies the condition $W_B \ll L_n$.

Other important requirement is to dope base region lightly compared to the emitter so that the probability of recombination in the base should be as small as possible.

9.3.1 Current Amplification (Current Gain)

The base current in a transistor is due to three important mechanisms:

1. Although $W_B \ll L_n$, still there is finite probability that some of the injected electrons recombine with majority carrier holes in the base region. The loss of holes in the base due to recombination must be resupplied through the base current. In ideal situation, some 99% of the carriers injected from emitter into base are swept to the collector.
2. Due to forward biased junction of emitter-base, some of the holes are injected into the emitter from base, even if the emitter is heavily doped compared to the base. These lost holes must also be supplied by I_B.
3. Some of the minority carrier holes are swept into the base region from the collector due to reverse biased collector junction because the positive applied potential on the collector pushes the holes into the base region. This small current reduces I_B by supplying holes to the base.

    Out of these three, the first two are the dominant processes in contributing base current I_B.

The various current components in an ideal *n-p-n* transistor is shown in Fig. 9.7 under the forward active mode. It is assumed that no generation-recombination is taking place in the depletion region. As seen from Fig. 9.7 the electrons injected from emitter of the transistor gives emitter current I_{En}, this is the largest current-component in a well designed transistor. If the recombination of carriers in the base is negligible then most of the injected electrons reach to collector and give rise to a collector I_{Cn}. Due to forward bias of emitter-base junction, holes are also injected from base to emitter and gives current I_{Ep}. There are three factors contributing the base current as discussed earlier. The current I_{Ep} is undesirable and can be minimized by heavily doped emitter region or heterojunction. As clear from Fig. 9.7, the total emitter current is given by

$$I_E = I_{En} + I_{Ep} \tag{9.3a}$$

i.e, sum of the number of electrons entered from emitter to collector and number of holes entered from base to emitter.

Similarly, the terminal current I_C is:

$$I_C = I_{Cn} + I_{Cp} \tag{9.3b}$$

i.e, sum of the number of electrons injected from emitter moves into collector through base and the number of thermally generated holes in the collector region.

From Eq. (9.2), the base terminal current is:

$$I_B = I_E - I_C \qquad (9.3c)$$

Using Eqs. (9.3a) and (9.3b) into Eq. (9.3c), we have

$$I_B = I_{En} + I_{Ep} - I_{Cn} - I_{Cp}$$

$$= I_{En} + (I_{Ep} - I_{Cn}) - I_{Cp} \qquad (9.3d)$$

An important parameter in the bipolar junction transistor is defined as the common base current gain and is given as:

$$\alpha = \frac{I_{Cn}}{I_E} = \frac{I_{Cn}}{I_{En} + I_{Ep}} \quad \text{[Using Eq. (9.3a)]} \qquad (9.3e)$$

Multiplying numerator and denominator by I_{En} of Eq. (9.3e), we have

$$\alpha = \left(\frac{I_{En}}{I_{En} + I_{Ep}} \right) \left(\frac{I_{Cn}}{I_{En}} \right) \qquad (9.3f)$$

where $(I_{En}/I_{En} + I_{Ep})$ is defined as the emitter injection efficiency and represented by γ and the term (I_{Cn}/I_{En}) is represented by α_T and called **base transport factor**. Therefore, Eq. (9.3f) reduces to

$$\alpha = \gamma . \alpha_T \qquad (9.4a)$$

For an efficient transistor,

$$\gamma \cong 1 \quad \text{and} \quad \alpha_T \simeq 1 \qquad (9.4b)$$

means

$$\frac{I_{En}}{I_{En} + I_{Ep}} = 1$$

or

$$I_{En} = I_{En} + I_{Ep} \quad \text{or} \quad I_{Ep} = 0$$

This clearly shows that the total emitter current is entirely due to injected electrons from emitter.

$$\alpha_T \cong 1$$

gives

$$\frac{I_{Cn}}{I_{En}} \simeq 1 \quad \text{or} \quad I_{Cn} \simeq I_{En}$$

This shows that most of the injected electrons should eventually participate in the collector current. From Eqs. (9.4a) and (9.4b), it is clear that

$$\alpha \simeq 1$$

But in real diode, α is less than unity and actually does not reflect any amplification.

From Eq. (9.3b) and using $\alpha_T = I_{Cn}/I_{En}$, we have

$$I_C = \alpha_T I_{En} + I_{Cp}$$

$$= \left(\frac{\alpha_T}{\gamma}\right)(\gamma I_{En}) + I_{Cp}$$

or

$$I_C = (\alpha_T\gamma)\left(\frac{I_{En}}{\gamma}\right) + I_{Cp}$$

or

$$I_C = \alpha\left(\frac{I_{En}}{\gamma}\right) + I_{Cp}$$

or

$$I_C = \alpha I_E + I_{Cp} \tag{9.4c}$$

To get Eq. (9.4c), we used

$$\frac{I_{En}}{\gamma} = I_{En} + I_{Ep} = I_E \qquad \text{[from Eq. (9.3a)]}$$

where I_{Cp} corresponds to the collector-base current flowing with the emitter open-circuited, i.e., $I_{Cp} = I_C$ for $I_E = 0$ and also represented as I_{CBO}. Therefore, Eq. (9.4c) can be further written as:

$$I_C = \alpha I_E + I_{CBO} \tag{9.4d}$$

where I_{CBO} is leakage current between the collector and base with emitter-base junction open. Eq. (9.4d) represents the collector current for common-base configuration.

Let α_T represent the number of injected electron from emitter enter the collector region without recombining in the base, then the number of electrons recombining into the base region is $(1 - \alpha_T)$. Therefore, the total base current is:

$$I_B = I_{Ep} + (1 - \alpha_T)\,I_{En} \tag{9.5}$$

In Eq. (9.5), we have neglected the contribution of collector saturation current.

Using Eqs. (9.4c) and (9.5), one has

$$\frac{I_C}{I_B} = \frac{\alpha I_E}{I_{Ep} + (1 - \alpha_T)\,I_{En}}$$

Neglecting I_{Cp},

$$\frac{I_C}{I_B} = \frac{\alpha I_E}{(I_E - I_{En})\,I_{En} + (1 - \alpha_T)\,I_{En}}$$

$$= \frac{\alpha I_E}{I_E + (1 - \alpha_T - 1)\,I_{En}} = \frac{\alpha I_E}{I_E - \alpha_T I_{En}}$$

$$= \frac{\alpha}{1 - \alpha_T \left(\dfrac{I_{En}}{I_E} \right)} = \frac{\alpha}{1 - \alpha_T \gamma} = \left(\frac{\alpha}{1 - \alpha} \right)$$

or

$$\frac{I_C}{I_B} = \left(\frac{\alpha}{1 - \alpha} \right) = \beta \tag{9.6}$$

where, β relates the collector current to the base current and called **base-to-collector current amplification factor**.

If $\alpha \simeq 1$ then $\beta = 1/0 = \infty$ means, for a good amplifier action β should be as large as possible, and $I_C \gg I_B$.

Means the base current should be negligible. This is only possible if base region is as thin as possible.

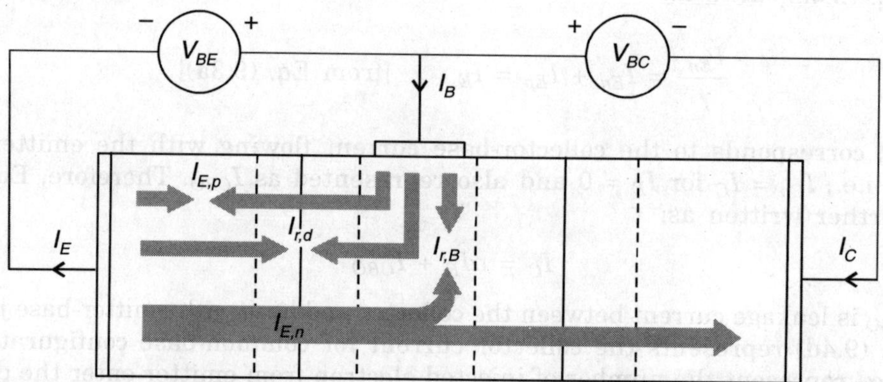

FIGURE 9.7 Various current components in *n-p-n* transistor in active mode of operation.

The minority carrier distribution in the three regions of the transistor when it is operating in forward active mode is shown in Figure 9.8.

EXAMPLE 9.2

For an ideal *n-p-n* transistor, $I_{En} = 3$ mA, $I_{Ep} = 0.01$ mA, $I_{Cn} = 2.99$ mA and $I_{Cp} = 0.001$ mA. Determine

  (a) Total emitter current
  (b) Emitter injection efficiency
  (c) Total collector terminal current
  (d) Common base current gain
  (e) Base transport factor
  (f) Leakage current between the collector and the base with emitter-base junction
     open

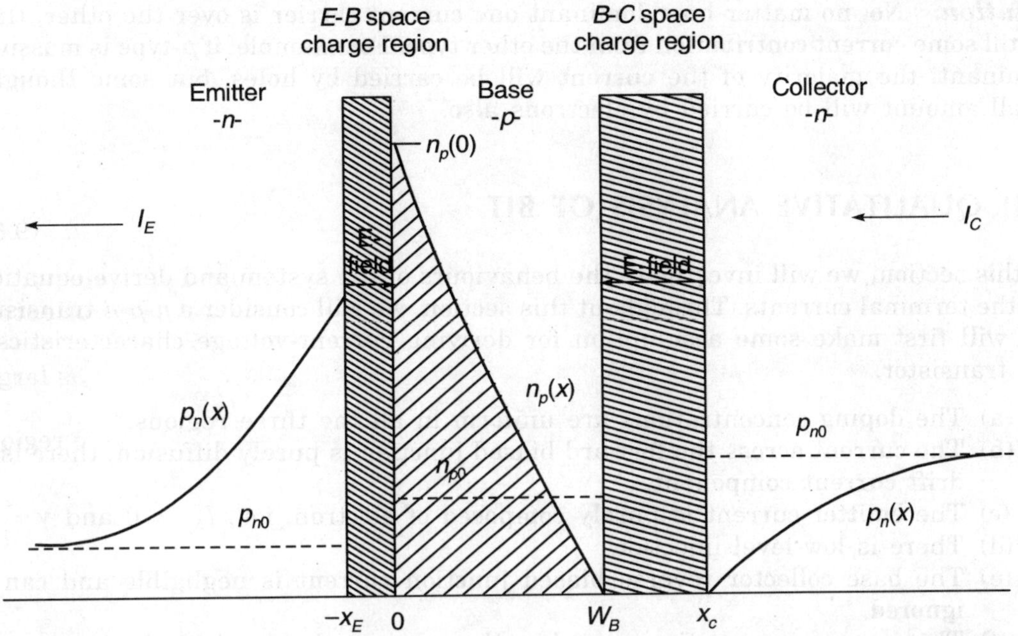

FIGURE 9.8 Carrier concentration in *n-p-n* transistor in active mode operation.

Solution: (a) $I_E = I_{En} + I_{Ep} = 3.01$ mA

(b) $\gamma = \dfrac{I_{En}}{I_{En} + I_{Ep}} = \dfrac{I_{En}}{I_E} = \dfrac{3}{3.01} = 0.99967$

(c) $I_C = I_{Cn} + I_{Cp} = 2.991$ mA

(d) $\alpha = \dfrac{I_{Cn}}{I_{En} + I_{Ep}} = \dfrac{2.99}{3.01} = 0.9934$

(e) Since $\alpha = \gamma\alpha_T$,

or

$$\alpha_T = \frac{\alpha}{\gamma} = \frac{0.9934}{0.9967} = 0.9967$$

(f) $I_C = \alpha I_E + I_{CBO}$

or

$$I_{CBO} = I_C - \alpha I_E$$
$$2.991 - 0.9934 \times 3.01 = 0.87 \ \mu A$$

EXAMPLE 9.3

Will the forward bias current in the transistor (BJT) due to majority carriers only?

Solution: No, no matter how dominant one current carrier is over the other, there is still some current contribution from the other type. For example, if p-type is massively dominant, the majority of the current will be carried by holes, but some though a small amount will be carried by electrons also.

9.4   QUALITATIVE ANALYSIS OF BJT

In this section, we will investigate the behaviours of the system and derive equations for the terminal currents. Throughout this section, we will consider a n-p-n transistor. We will first make some assumption for deriving current-voltage characteristics of the transistor.

 (a) The doping concentrations are uniform in all the three regions.
 (b) The current across the forward biased junction is purely diffusion, there is no drift current component.
 (c) The emitter current is purely composed of electron, i.e., $I_{Ep} = 0$ and $\gamma = 1$.
 (d) There is low-level injection.
 (e) The base collector reverse biased junction current is negligible and can be ignored.
 (f) There are no generation-recombination currents in the depletion region.
 (g) There are no series resistances in the device so that all currents and voltages are stable and constant.

To obtain the current from the minority-carrier gradient, it is required to find out the minority-carrier distribution in the base region because the majority carriers are injected into the base under forward bias condition where these carriers become minority carriers, and then diffuse across the base region and reach the collector junction.

The minority-carrier (electron), distribution in the neutral base region is given by the continuity equation,

$$D_n \left(\frac{\partial^2 n_p}{\partial x^2} \right) - \frac{n_p - n_{po}}{\tau_n} = 0 \tag{9.7}$$

where D_n and τ_n are the diffusion constant and the life time of the minority carriers respectively. The solution of this second order differential equation is given as:

$$n_p(x) = n_p + C_1 e^{x/L_{nB}} + C_2 e^{-x/L_{nB}} \tag{9.8a}$$

where $L_{nB} = \sqrt{D_n \tau_{nB}}$ is the diffusion length of electrons in base region and C_1 and C_2 are constants, can be determined by boundary conditions for the active mode:

$$n_p(0) = n_{p0} e^{qV_{BE}/kT} \tag{9.8b}$$

and

$$n_p(W_B) = 0 \tag{9.8c}$$

where n_{p0} is the equilibrium minority-carrier concentration in the base and given by n_i^2/N_a^- where N_a^- is uniform acceptor concentration in the base. The boundary condition given by Eq. 9.8(b) states that the minority carrier concentration at the edge of emitter-base junction varies exponentially with V_{BE} for forward bias, whereas Fig. 9.8 shows that carrier concentration at the edge of emitter-base junction is almost linear. This is due to fact that base region is very narrow. Second boundary condition given by the Eq. (9.8c) states that the minority carrier distribution at the edge $(x = W_B)$ of reverse biased base-collector junction is zero.

Substituting Eq. (9.8b) for $x = 0$ in Eq. (9.8a) we have,

$$n_{p0}\, \exp(qV^{BE/kT}) = n_p + C_1 + C_2 \tag{9.9a}$$

and for $x = W_B$, we have

$$0 = n_p + C_1 e^{W_B/L_{nB}} + C_2\, e^{-W_B/L_{nB}} \tag{9.9b}$$

On solving Eqs. (9.9a) and (9.9b), we get

$$C_1 = \frac{n_p(e^{-W_B/L_{nB}} - 1) - n_{p0}e^{qV_{BE}/kT}e^{-W_B/L_{nB}}}{2\sinh(W_B/L_{nB})} \tag{9.10a}$$

and

$$C_2 = \frac{n_p(1 - e^{W_B/L_{nB}}) - n_{p0}e^{qV_{BE}/kT}e^{W_B/L_{nB}}}{2\sinh(W_B/L_{nB})} \tag{9.10b}$$

Substituting Eqs. (9.10a) and (9.10b) in Eq. (9.8a), one has

$$n_p(x) = n_{p0}(e^{qv_{BE}/kT} - 1)\left\{\frac{\sinh(W_B - x)/L_{nB}}{\sinh W_B/L_{nB}}\right\} + P_{n0}\left\{1 - \frac{\sinh(x/L_{nB})}{\sinh(W_B/L_{nB})}\right\} \tag{9.11}$$

For $(W_B/L_n) \ll 1$, Eq. (9.11) can be rewritten as:

$$n_p(x) = n_{p0}(e^{qv_{BE}/kT})(1 - x/W_B) \tag{9.12a}$$

$$= n_p(0)\left(1 - \frac{x}{W_B}\right) \tag{9.12b}$$

For narrow width of the base, the minority carrier distribution at any point x approaches to straight line as shown in Fig. 9.8. If the base layer thickness is not small compared to the diffusion length, the minority-carrier distribution is given by hyperbolic functions [Eq. (9.11)]. The holes injected into emitter from base is represented by $p_n(x)$ and decreases as $\exp(-x/L_p)$ when the thickness of the emitter region is larger than the diffusion length. In an integrated-circuit transistor, the emitter-layer thickness could be less than the diffusion length $(W_E < 0.11\ p)$ and $p_n(x)$ would have a linear variation in the emitter. The minority carriers in the collector vary as

$[1 - \exp(-x/L_p)]$ for the reverse biased collector base junction (as discussed in Chapter 4) or a linear variation of $p_n(x)$ if the collector layer thickness is less than the diffusion length ($W_C < 0.1L_p$).

The stored charge in the base region Q_B is represented by the shaded area in Fig. 9.8 and can be given as:

$$Q_B = qA \int_0^{W_B} n_p(x)\, dx \tag{9.13a}$$

Substituting Eq. (9.12a) and after simplification, we have

$$Q_B = qA\, np_0\, \exp\left(\frac{qv_{BE}}{kT}\right) \frac{W_B}{2} \tag{9.13b}$$

where A is cross-sectional area of the base and q is electronic charge. The base stored charge Q_B will be used in several expression for various currents in the bipolar transistor.

The excess electron concentration at the edge of the emitter depletion region is:

$$\Delta n_E = n_p[e^{qv_{BE}/kT} - 1]$$

and excess electron concentration on the collector side of the base is:

$$\Delta n_C = n_p[e^{qV_{BC}/kT} - 1]$$

If emitter-base junction is strongly forward-biased, i.e., $\dfrac{V_{BE}}{(kT/q)} \gg 1$ and hence

$$\Delta n_E \simeq n_p\, e^{qv_{BE}/kT}$$

If base-collector junction is strongly reverse-biased, i.e.,

$$v_{BC} \ll 0$$

Then

$$\Delta n_c \simeq -\, n_p$$

9.4.1 Ideal Transistor Currents for Active Mode Operation

The electron current injected from emitter at $x = 0$ is proportional to area as well as the gradient of the minority carrier concentration, i.e.,

$$I_{En} = qAD_{nB} \cdot \left.\frac{dn_p(x)}{dx}\right|_{x=0} \tag{9.14a}$$

Using Eq. (9.12a) in Eq. (9.14a), we get

$$I_{En} = -\frac{qA}{W_B} Dn_B n_{po} e^{qv_{BE}/kT} \qquad (9.14b)$$

Assuming $(W_B/L_{nB}) \ll 1$. If base-region is not thin then Eq. (9.14b) cannot be used. In this case, one can get the expression for emitter-current under forward biased by substituting Eq. (9.11) into Eq. (9.14a). Negative sign in Eq. (9.14b) only indicates that current always flows opposite to the electron motion.

Similarly, the electron current collected by the collector at $x = W_B$ is:

$$I_{Cn} = \frac{qAD_n}{W_B} n_{po} e^{qv_{BE}/kT} \qquad (9.14c)$$

Again assuming $(W_B/L_{nB}) \ll 1$. From Eq. (9.14b) and (9.14c), it is clear that for a thin base region such that $(W_B/L_{nB}) \ll 1$, the current I_{En} must be equal to I_{Cn}. But it is not true when base region is not thin then some carriers will recombine in the base and result in a base current.

The current due to movement of holes from base into emitter is:

$$I_{Ep} = A\left[-qD_{pE} \left.\frac{dp_E(x)}{dx}\right|_{x=-x_E} \right]$$

and simplified as:

$$I_{Ep} = -\frac{qAD_{pE}p_{E0}}{L_E}(e^{qV_{BE}/kT} - 1) \qquad (9.15a)$$

Therefore, the total emitter terminal current is given by the sum of Eqs. (9.14b) and (9.15a), i.e.,

$$I_E = I_{En} + I_{Ep}$$

$$= -\frac{qA}{W_B} D_{nB} n_{p0}\, e^{qV_{BE}/kT} - \frac{qA}{L_E} D_{pE}\, p_{E0}(e^{qV_{BE}/kT} - 1) \qquad (9.15b)$$

where D_{pE} is diffusion constant in the emitter region for hole. Equation (9.15b) can be simplified further as:

$$I = -\left(\frac{qA}{W_B} D_{nB} n_{p0} + \frac{qA}{L_E} D_{pE}\, p_{E0}\right) e^{qV_{BE}/kT} + \frac{qA}{L_E} D_{pE}\, p_{E0} + \frac{qA}{W_B} D_{nB} n_{p0} - \frac{qA}{W_B} D_{nB} n_{p0}$$

$$= -qA\left(\frac{D_{nB} n_{p0}}{W_B} + \frac{D_{pE}\, p_{E0}}{L_E}\right)(e^{qV_{BE}/kT} - 1) - \frac{qA}{W_B} D_{nB} n p_0 \qquad (9.15c)$$

$$= -\left[a_{11}(e^{qV_{BE}/kT} - 1) + a_{12}\right]$$

where
$$a_{11} = qA\left(\frac{D_{nB}n_{p0}}{W_B} + \frac{D_pE\,p_{E0}}{L_E}\right)$$

and
$$a_{12} = qA\left(\frac{D_{nB}n_{p0}}{W_B}\right)$$

Similarly, the current I_{Cp} due to flow of holes from collector to base is:

$$I_{Cp} = \frac{qAD_{pC}\,p_{C0}}{L_C} \tag{9.15d}$$

where D_{pC} is the diffusion constant in the collector for holes and L_C is diffusion length of carriers in the collector. Total collector current is given as:

$$I_C = I_{Cn} + I_{Cp}$$

From Eqs. (9.14c) and (9.15d), we have

$$I_C = \frac{qAD_{nB}}{W_B}\, n_{p0}\, e^{qv_{BE}/kT} + \frac{qAD_{pc}\,p_{C0}}{L_C}$$

$$= \frac{qAD_{nB}}{W_B}\, n_{p0}\, (e^{qV_{BE}/kT} - 1) + qA\left(\frac{D_{pc}\,p_{C0}}{L_C} + \frac{D_{nB}n_{p0}}{W_B}\right) \tag{9.15e}$$

or

$$I_C = a_{21}(e^{qv_{BE}/kT} - 1) + a_{22} \tag{9.15f}$$

where

$$a_{21} = \frac{qAD_{nB}}{W_B}\, n_{p0} \quad \text{and} \quad a_{22} = qA\left(\frac{D_{nB}}{W_B}\, n_{p0} + \frac{D_{pC}\,p_{C0}}{L_C}\right)$$

Again, these two Eqs. (9.15c) and (9.15e) are true when base region is thin, i.e., true for short-diode. p_{E0} and p_{C0} are the thermal equilibrium concentration of minority carriers in the emitter and collector regions respectively.

For an ideal transistor,

$$I_B = I_E - I_C$$

Therefore, using Eqs. (9.15c) and (9.15e), we have

$$I_B = -(a_{11} + a_{21})\,(e^{qv_{BE}/kT} - 1) - (a_{12} + a_{22}) \tag{9.16}$$

From these discussion, it is clear that the transistor's terminal currents are mainly determined by the minority carrier distribution in the base region.

Note: These three-terminal currents equations are true for a thin base region. For wide base region, one should use Eq. (9.11) for the derivation of terminal currents.

Assuming collector is more lightly doped than the base region, means $p_{CO} \ll 1$ and hence neglecting the second term from Eq. (9.15e), one has

$$I_C = \frac{qAD_{nB}}{W_B}\, n_{p0}\, e^{qV_{BE/kT}} \tag{9.17a}$$

Using $n_{p_0} = \dfrac{n_i^2}{N_a^-}$ where N_a^- is acceptor impurity dopant in the base and n_i is intrinsic carrier concentration.

Equation (9.17a) can be further rewritten as:

$$I_C = \frac{qAD_{nB}}{W_B}\left(\frac{n_i^2}{N_a^-}\right)\exp\left(\frac{qV_{BE}}{kT}\right) \tag{9.17b}$$

Rewriting Eq. (9.17b) as:

$$I_C = \left[qAW_B\,\frac{n_i^2}{2N_a^-}\exp\left(\frac{qV_{BE}}{kT}\right)\right]\left[\frac{2D_{nB}}{W_B^2}\right] \tag{9.17c}$$

where first term of Eq. (9.17c) gives the charge stored in the base as given by Eq. (9.13b) and hence Eq. (9.17c) reduces to

$$I_C \simeq Q_B\left(\frac{2D_n}{W_B^2}\right) \simeq \frac{Q_B}{\tau_D} \tag{9.18}$$

where τ_D is the base transist time. Equation 9.18 shows that the collector current is directly proportional to the stored base charge.

Gumel number

The doping in the base region per unit area is calculated as:

$$G_N = \int_0^{W_B} N_B(x)\, dx = N_a W_B \tag{9.19}$$

Using Eq. (9.19) in Eq. (9.17b), we have

$$I_C \simeq \frac{qAD_n n_i^2}{G_N}\exp\left(\frac{qv_{BE}}{kT}\right) \tag{9.20}$$

where G_N is termed as the *Gumel number* and $N_B(x)$ represents the impurity profile in the base region. From Eq. (9.20) it is clear that in absence of any doping in the base region, the collector current is inversely proportional to Gumel number. The typical values of G_N in a high performance bipolar transistor range from $10^{12}/cm^2$ to $10^{13}/cm^2$.

Gumel number is often defined as the total integrated base dose. All the effects in the base region, such as band gap narrowing, bandgap non-uniformity, and dopant-distribution are contained in the parameter G_N.

Current transfer ratio

For the case $(W/L_{nB}) << 1$, the emitter efficiency γ is obtained by substituting Eqs. (9.14b) and (9.15a) in the following expression:

$$\gamma = \frac{I_{En}}{I_{En} + I_{Ep}}$$

and hence,

$$\gamma = \frac{(qA/W_B)\,D_{nB}n_{p0}\,e^{qv_{BE}/kT}}{\left[(qA/W_B)\,D_{nB}n_{p0}\,e^{qv_{BE}/kT} + \dfrac{qAD_{pE}\,p_{E0}}{L_E}\,(e^{qv_{BE}/kT} - 1)\right]} \tag{9.21a}$$

Assuming $(v_{BE}/kT) >> 1$, i.e., more forward bias, Eq. (9.21a) reduces to

$$\gamma = \frac{(qA/W_B)\,D_{nB}n_{p0}\,\exp(qv_{BE}/kT)}{\left[(qA/W_B)\,D_{nB}n_{p0} + \dfrac{qAD_{pE}\,p_{E0}}{L_E}\right]\exp(qv_{BE}/kT)}$$

$$\gamma = \frac{(D_{nB}/W_B)\,n_{p0}}{(D_{nB}/W_B)\,n_{p0} + (D_{pE}/L_E)\,p_{E0}}$$

$$\gamma = \frac{1}{1 + (D_{pE}/D_{nB})\,(W_B/L_E)\,(p_{E0}/n_{p0})}$$

or

$$\gamma = \left[1 + \left(\frac{D_{pE}}{D_{nB}}\right)\left(\frac{W_B}{L_E}\right)\left(\frac{p_{E0}}{n_{p0}}\right)\right]^{-1}$$

or

$$\gamma = \left[1 + \left(\frac{D_{pE}}{D_{nB}}\right)\left(\frac{W_B}{L_E}\right)\left(\frac{N_B}{N_E}\right)\right]^{-1} \tag{9.21b}$$

where $N_B = (n_i^2/n_{p0})$ is the impurity in base region and $N_E = (n_i^2/p_{EO})$ is the impurity in the emitter region. Equation (9.21b) shows that by reducing the ratio (N_B/N_E) one can increase the value of γ, i.e., emitter should be heavily doped than the base region. This is the main reason for heavy doping of emitter.

The base transport efficiency α_T (often called **current transport factor**) is defined as the fraction of electrons injected from the emitter into the base that reach the collector, i.e.,

$$\alpha_T = \frac{I_{nC}}{I_{nE}} = \frac{I_{nE} - I_{rec}}{I_{nE}} = 1 - \left(\frac{I_{rec}}{I_{nE}}\right) \qquad (9.21c)$$

For ideal case, if there is no recombination in the base region, $I_{rec} = 0$ and then

$$\alpha_T \simeq \frac{I_{nC}}{I_{nE}} = \frac{I_{nE}}{I_{nE}} \simeq 1$$

and

$$I_{rec} = \frac{qAW_B n_i^2}{2\tau_n N_a} \exp\left[\frac{qV_{BE}}{kT}\right] = \frac{Q_B}{\tau_n} \qquad (9.21d)$$

Substituting Eqs. (9.17b) and (9.21d) into Eq. (9.21c), and $D_{nB} = D_n$, we have

$$\alpha_T = 1 - \left(\frac{qAW_B n_i^2}{2\tau_n N_a^-}\right)\left(\frac{W_B N_a^-}{qAD_n n_i^2}\right)$$

$$= 1 - \frac{W_B^2}{2\tau_n D_n} \qquad (9.21e)$$

Using $D_n\tau_n = L^2 n$, Eq. (9.21e) can be further rewritten as:

$$\alpha_T = 1 - \frac{W_B^2}{2L_{nB}^2} \qquad (9.21f)$$

Equation (9.21f) shows that the base transport factor approaches to unity for $W_B/L_{nB} \ll 1$. Using γ and α_T given by Eqs. (9.21b) and (9.21f), one can get the value of common base current gain α.

EXAMPLE 9.4

Why base region for any bipolar junction transistor should be thin?

Solution: For an ideal transistor behaviour, it is required

$$\gamma = 1$$

$$\alpha_T = 1$$

From Eq. (9.21f), it is clear that $\alpha_T = 1$ when $(W_B/L_{nB}) \ll 1$. This means either L_{nB} should be as large as possible, but this is not desirable, or $W_B \ll 1$, means as thin as possible. This is the best way to achieve the unity base transport factor.

EXAMPLE 9.5

Find the β for a *n-p-n* junction transistor with a non-degenerate emitter. Assume that the emitter, base and collector are non-compensated and

$$N_{dE} = 2 \times 10^{18}/\text{cm}^3, \; N_{AB} = 10^{16}/\text{cm}^3$$

$$N_{dC} = 10^{15}/\text{cm}^3, \; L_E = W_E = 0.2 \; \mu\text{m}, \; W_B = 0.1 \; \mu\text{m}$$

$$D_E = 5.5 \; \text{cm}^2/\text{sec}, \; D_n = 32 \; \text{cm}^2/\text{sec}$$

$$L_{nB} = 400 \; \mu\text{m}$$

Solution: Using Eq. (9.21b) and substituting all values, we get

$$\gamma = \left[1 + \frac{5.5}{32} \times \frac{0.1}{0.2} \times \frac{10^{16}}{2 \times 10^{18}} \right]^{-1} = 0.9996$$

Substituting values in Eq. (9.21f),

$$\alpha_T \simeq 1 - \left(\frac{0.1}{400} \right)^2 \times \frac{1}{2}$$

$$\alpha_T = 0.9999 \simeq 1$$

Hence

$$\alpha = \gamma \alpha_T$$

$$= 0.996$$

Therefore,

$$\beta = \frac{\alpha}{1 - \alpha} = \frac{0.996}{1 - 0.996} \simeq 2500$$

EXAMPLE 9.6

An ideal *n-p-n* transistor has impurity concentration of 10^{18}, 10^{16} and $5 \times 10^{15}/\text{cm}^3$ in the emitter, base and collector region respectively. The corresponding lifetimes are 10^{-8}, 10^{-7} and 10^{-6} sec in respective regions. Let effective cross-section area A is 0.05 mm^2 and emitter-base junction is forward biased with 0.6 V. Find the common-base current gain of the transistor. Other parameters are: $D_E = 1 \; \text{cm}^2/\text{sec}$, $D_{pn} = 40 \; \text{cm}^2/\text{sec}$, $D_c = 2 \; \text{cm}^2/\text{sec}$ and $W_B = 0.5 \; \mu\text{m}$.

Solution: In the base region,

$$L_n = \sqrt{D_n \tau_n}$$

$$= \sqrt{40 \times 10^{-7}} = 2 \times 10^{-3} \text{ cm}$$

$$n_{p0} = \frac{n_i^2}{N_B}$$

$$= \frac{2.25 \times 10^{20}}{10^{16}}$$

$$= 2.25 \times 10^4/\text{cm}^3$$

In the emitter region,

$$L_p = \sqrt{D_E \tau_E}$$

$$= \sqrt{1 \times 10^{-8}} = 10^{-4} \text{ cm}$$

$$p_{E0} = \frac{2.25 \times 10^{20}}{10^{18}} = 2.25 \times 10^2/\text{cm}^3$$

and

$$\frac{W_B}{L_n} = \frac{0.5 \times 10^{-4}}{2 \times 10^{-3}} = \frac{0.5}{2} \times 10^{-1} = 0.025 \ll 1$$

Hence, current components are:

$$|I_{En}| = \frac{1.6 \times 10^{-19} \times (0.5)^2 \times 10^{-2}}{0.5 \times 10^{-4}} \times 40 \times 2.25 \times 10^4 \times e^{0.6/26 \times 10^3}$$

$$= \frac{0.36}{5} \times 10^{-21} = 7.5 \times 10^{-3} \text{ A}$$

From Eq. (9.15a),

$$|I_{Ep}| = \frac{1.6 \times 10^{-19} \times (0.5)^2 \times 10^{-2} \times 1 \times 2.25 \times 10^2}{10^{-4}} (e^{0.6 \times 1000/26} - 1)$$

$$= 9.472 \times 10^{-8} \text{ A}$$

$$I_{Cn} = \frac{qAD_{nB}}{W_B} n_{p0} \, e^{qv_{BE}/kT}$$

$$= 7.5 \times 10^{-3} \text{ A}$$

Hence

$$\alpha_T = \frac{I_{Cn}}{I_{En} + I_{Ep}} = 0.99998 \simeq 1$$

9.4.2 Modes of Operation

The four modes of operation for the *n-p-n* transistor are shown in Fig. 9.3. These four modes of operation depend on the voltage polarities of the emitter-base junction and collector-base junction. The polarities are:

- Forward active mode: base-emitter junction: forward biased
  collector-base junction: reverse biased
  i.e., V_{BE} is +ve and V_{CB} is +ve.
- Cut-off mode: both junctions should be reverse biased.
  i.e., V_{EB} is +ve and V_{CB} is +ve.
- Saturation region: both junctions are forward biased.
  i.e., V_{EB} is −ve and $V_{CB} = $ −ve.
- Reverse active mode: emitter-base junction is reverse biased and collector-base junction is forward biased.
  i.e., V_{EB} is +ve and V_{CB} is −ve.

Since both junctions are forward biased in saturation region, therefore, the contribution of minority-carriers are almost-negligible. As seen from Fig. 9.9(a), there is non-zero minority-carrier distribution at the edge of the each depletion region. The saturation mode of transistor corresponds to small biasing voltage and large output current, i.e., transistor is in conducting mode and acts as a closed (or ON) switch.

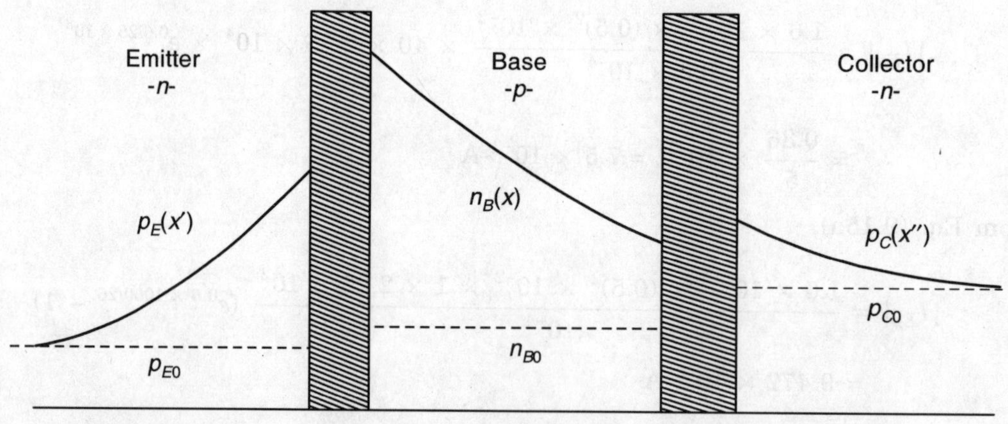

FIGURE 9.9(a) Carrier distribution in *n-p-n* transistor in saturation.

Both junctions are reverse biased in the cut-off region and hence, $n_p(0) = n_p(W_B) = 0$ as shown in Fig. 9.9(b). The cut-off mode corresponds to open (or OFF) state of the transistor as a switch.

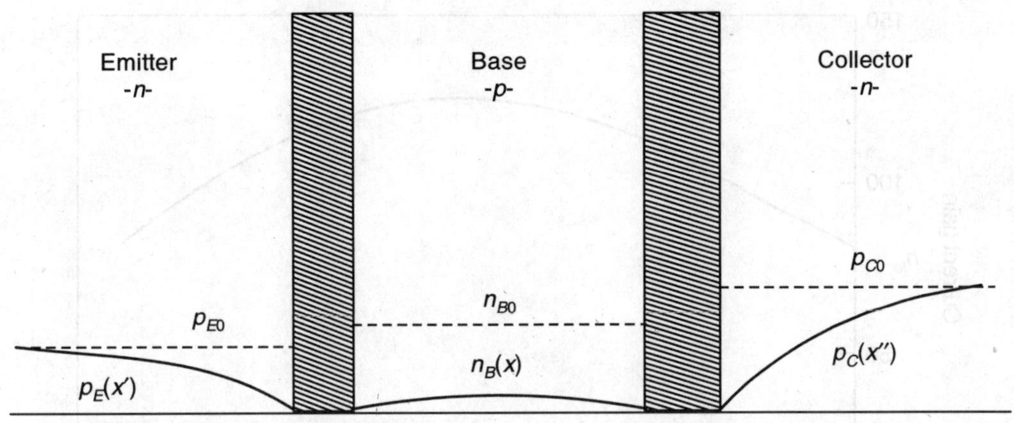

FIGURE 9.9(b)   Carrier distribution in *n-p-n* transistor in cut-off region.

In the fourth mode where emitter-base junction is reversed-biased and collector-base junction is forward biased. For this condition, the collector serves to inject electrons from the collector into the base. However, the injection efficiency is poor because collector is generally more lightly doped than the base. The carrier distribution is shown in Fig. 9.9(c).

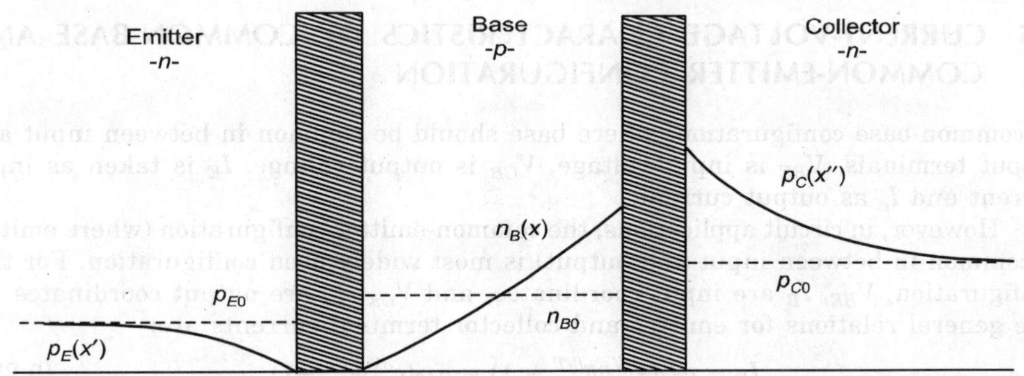

FIGURE 9.9(c)   Carrier distribution in *n-p-n* transistor in reverse active region.

Note: The main characteristic of a bipolar junction transistor to be its current gain value. In practice, this value is not a *universal constant*, but depends on many factors, e.g. the transistor's temperature, the size and shape of its base region, etc. The variation of the current gain with collector current is shown in Fig. 9.10. From Fig. 9.10, it is clear that the proportion of electrons (injected from emitter junction) *caught* by hole in the base region whilst trying to cross the base region does vary a bit depending on the current level.

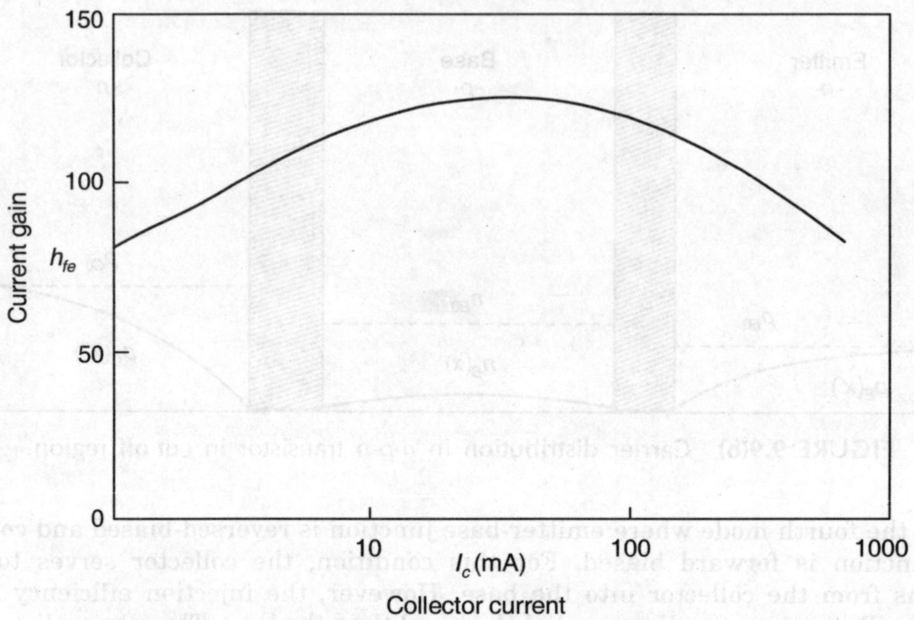

FIGURE 9.10   Variation of current gain with collector current.

9.5   CURRENT-VOLTAGE CHARACTERISTICS OF COMMON-BASE AND COMMON-EMITTER CONFIGURATION

In common-base configuration, where base should be common in between input and output terminals, V_{BE} is input voltage, V_{CB} is output voltage, I_E is taken as input current and I_c as output current.

However, in circuit applications, the common-emitter configuration (where emitter is common in between input and output) is most widely used configuration. For this configuration, V_{BE}, I_B are input coordinates, and V_{CE}, I_C are output coordinates. The general relations for emitter and collector terminal currents are:

$$I_E = a_{11}(e^{qV_{BE}/kT} - 1) - a_{12}(e^{qV_{BC}/kT} - 1) \qquad (9.22a)$$

$$I_E = a_{21}(e^{qV_{BE}/kT} - 1) - a_{22}(e^{qV_{BC}/kT} - 1) \qquad (9.22b)$$

where a_{11}, a_{12} and a_{22}, a_{21} are defined earlier.

By taking the proper sign for different voltages, one can easily find out the terminal currents.

In this section, we will only discuss the common emitter junction.

9.5.1   Common-Emitter Configuration

The common emitter configuration for a *n-p-n* transistor is shown in Fig. 9.11(a). The output current is given as:

$$I_C = \alpha I_E + I_{CBO}$$

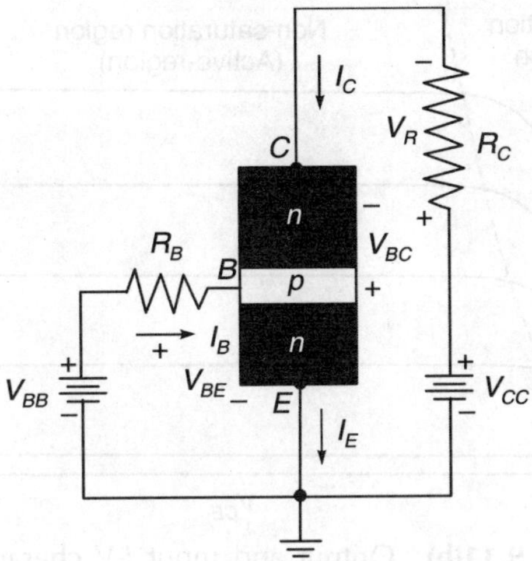

FIGURE 9.11(a)   Common emitter configuration of *n-p-n* transistor.

Substituting $I_E = I_C + I_B$, we have

$$I_C = \alpha(I_C + I_B) + I_{CBO}$$

or

$$I_C = \left(\frac{\alpha}{1 - \alpha}\right) I_B + \left(\frac{1}{1 - \alpha}\right) I_{CBO} \qquad (9.23)$$

Using Eq. (9.6), Eq. (9.23) can be written as:

$$I_C = \beta I_B + \left(\frac{\beta}{1 + \beta}\right) I_{CBO} \qquad (9.24a)$$

Expressing $\left(\dfrac{\beta}{1 + \beta}\right) I_{CBO} = I_{CEO}$, Eq. (9.24a) reduces to

$$I_C = \beta I_B + I_{CEO} \qquad (9.24b)$$

If $I_B = 0$,

$$I_{CEO} = I_C \qquad (9.25)$$

This shows that I_{CEO} is a collector-emitter leakage current for $I_B = 0$. The output characteristics of a common-emitter configuration is shown in Fig. 9.11(b) for an ideal transistor. The three regions of operations are also shown. Since, the value of α is very close to unity. β is very large quantity. It is defined as $\beta = \Delta I_C/\Delta I_B$ means a small change in base current can give rise to a large change in the collector current.

The characteristics in Fig. 9.11(b) also known as output characteristics for ideal BJT, where collector current is independent of V_{CE} in active region for any base current.

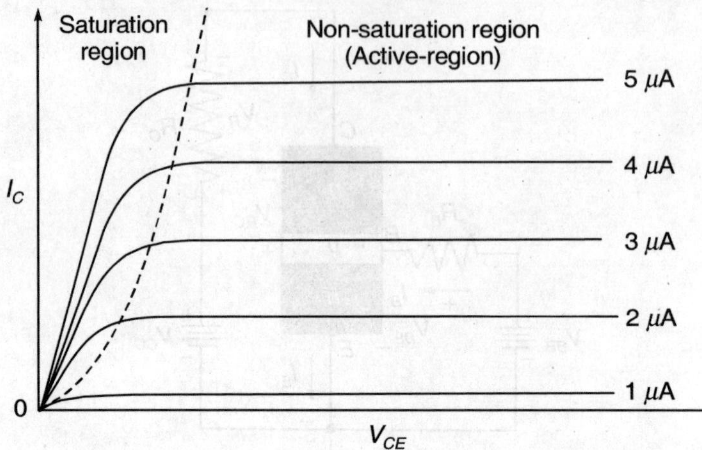

FIGURE 9.11(b) Output and input *I-V* characteristics.

Curves below $I_B = 0$, gives cut-off region.

In an ideal transistor the collector current for a given base-current I_B, in common emitter configuration, is expected to be independent of V_{CE}. This is true when it is assumed that the neutral base width (W_B) is constant.

Main advantage of using a transistor in a common-emitter configuration is that a small-variation in base current causes a larger collector current.

The input characteristics is shown in Fig. 9.11(c).

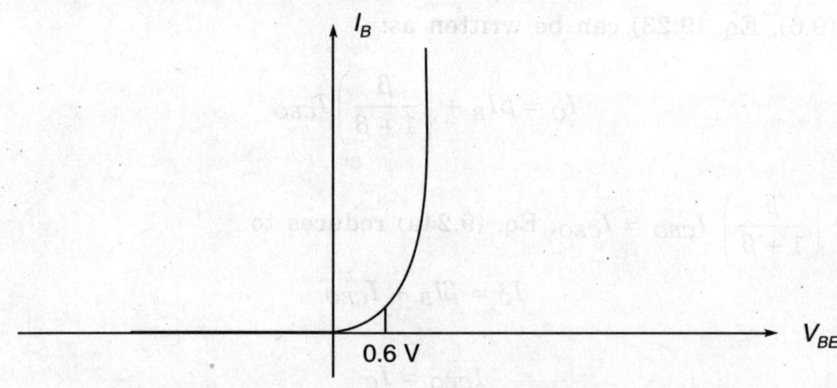

FIGURE 9.11(c)   Output and input *I-V* characteristics.

For Si transistor, required voltage for saturation is:

$$V_{CE(\text{sat.})} \simeq 0.2 \text{ V}$$

Also,
$$V_{CE} = V_{Cb} + V_{EB}$$

or
$$(V_{CE} - V_{EB}) = V_{CB}$$

For *n-p-n* transistor, if

$$V_{CE} < V_{EB}, \ V_{CB} = -\text{ve}$$

means collector is at lower potential than the base, i.e., collector base junction is forward biased.

Saturation results in large stored-base charge, and therefore, this condition must be avoided for high-speed operation.

9.6 EBERS-MOLL MODEL

The most common dc and large-signal, non-linear model (and the one used in SPICE) is the Ebers-Moll model. This model is used to relate the large-signal behaviour to the commonly used small-signal parameters. The Ebers-Moll model for a *n-p-n* transistor is shown in Fig. 9.12. This model describes an *n-p-n* transistor as two *p-n* junctions connected in back-to-back in series such that base should be common. When a voltage V_{BE} is applied to emitter-base diode then a forward current I_F flows in the emitter-base diode. Due to this forward biased current, a collector current $I_C = \alpha_F I_F$ flows where α_F is common-base current gain in the forward direction. When a voltage V_{BC} is applied across the base collector junction in such a way so that the base-collector junction is reverse-biased then a reverse current I_R flows in the collector-base diode, causing a current $\alpha_R I_R$ to flow in the emitter, where α_R is the common-base current gain in the reverse direction. These currents are shown in Figure 9.12.

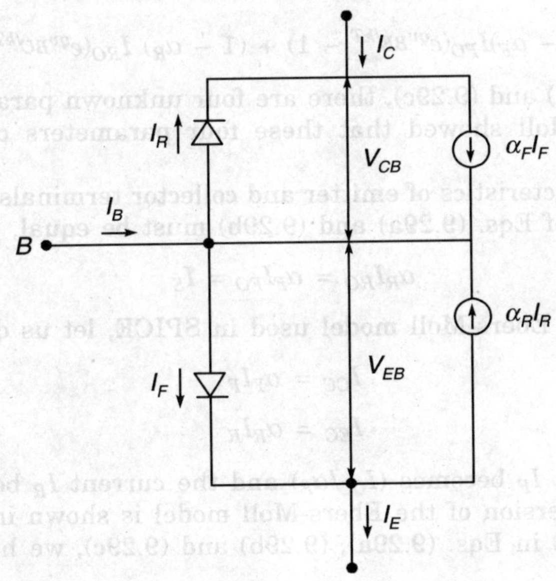

FIGURE 9.12 Ebers-Moll model for *n-p-n* transistor.

For *n-p-n* transistor, the forward current is given as:

$$I_F = I_{FO}[e^{qV_{BE}/kT} - 1] \tag{9.26a}$$

and the reverse current I_R is:

$$I_R = I_{RO}[e^{qV_{BC}/kT} - 1] \tag{9.26b}$$

where I_{FO} is the emitter-base junction saturation current and I_{RO} is the collector-base saturation current.

From Fig. 9.12, it is seen that

$$I_E = I_F - \alpha_R I_R \tag{9.27a}$$

and

$$I_C = -I_R + \alpha_F I_F \tag{9.27b}$$

and base current $I_B = (I_E - I_C)$

Substituting Eqs. (9.27a) and (9.27b), one has

$$I_B = I_F - \alpha_R I_R + I_R - \alpha_F I_F$$

$$= I_F(1 - \alpha_F) + I_R(1 - \alpha_R)$$

$$I_B = [I_F(1 - \alpha_F) + I_R(1 - \alpha_R)] \tag{9.28}$$

Equations (9.27a, b) and (9.28) give the Ebers-Moll model for the *n-p-n* transistor.

Substituting Eqs. (9.26a) and (9.26b) into Eqs. (9.27a), (9.27b) and 9.28, we have

$$I_E = I_{FO}(e^{qv_{BE}/kT} - 1) - \alpha_R(e^{qv_{BC}/kT} - 1)I_{RO} \tag{9.29a}$$

and

$$I_C = \alpha_F I_{FO}(e^{qv_{BE}/kT} - 1) - I_{RO}(e^{qv_{BC}/kT} - 1) \tag{9.29b}$$

and

$$I_B = (1 - \alpha_F)I_{FO}(e^{qv_{BE}/kT} - 1) + (1 - \alpha_R) I_{RO}(e^{qv_{BC}/kT} - 1) \tag{9.29c}$$

In Eqs. (9.29a), (9.29b) and (9.29c), there are four unknown parameters, I_{FO}, I_{RO}, α_R and α_F. Ebers and Moll showed that these four parameters can reduce to three parameters.

Reciprocity characteristics of emitter and collector terminals require that the off diagonal coefficients of Eqs. (9.29a) and (9.29b) must be equal, i.e.,

$$\alpha_R I_{RO} = \alpha_F I_{FO} = I_S \tag{9.30a}$$

To obtain the form of Ebers-Moll model used in SPICE, let us define

$$I_{CC} = \alpha_F I_F \tag{9.30b}$$

$$I_{EC} = \alpha_R I_R \tag{9.30c}$$

Therefore, the current I_F becomes (I_{CC}/α_F) and the current I_R becomes (I_{EC}/α_R) and hence the transport version of the Ebers-Moll model is shown in Fig. 9.13.

Using Eq. (9.30a) in Eqs. (9.29a), (9.29b) and (9.29c), we have

$$I_E = \frac{I_S}{\alpha_F} [e^{qv_{BE}/kT} - 1] - I_S[e^{qv_{BC}/kT} - 1] \tag{9.31a}$$

$$I_C = I_S[e^{qv_{BE}/kT} - 1] - \frac{I_S}{\alpha_F} [e^{qv_{BC}/kT} - 1] \tag{9.31b}$$

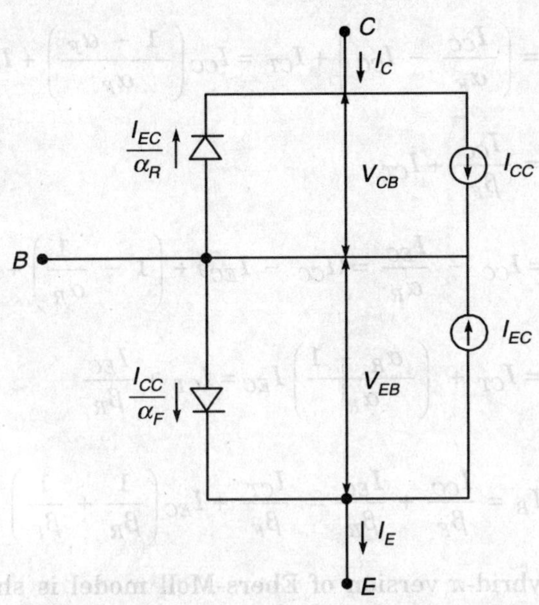

FIGURE 9.13 Transport version of Ebers-Moll model.

and

$$I_B = \frac{I_S}{\beta_F}\,[\exp(qv_{BE}/kT) - 1] + \frac{I_S}{\beta_R}\,[\exp(qv_{BE}/kT) - 1] \qquad (9.31c)$$

where $\beta_F = (\alpha_F/1 - \alpha_F)$ and $\beta_R = (\alpha_R/1 - \alpha_R)$ called **common-emitter current gains** in the forward and reverse directions respectively.

Using Eqs. (9.30b) and (9.30c) into Eq. (9.27a), we have

$$I_E = \frac{I_{CC}}{\alpha_F} - I_{EC} \qquad (9.32a)$$

and

$$I_C = I_{CC} - \frac{I_{EC}}{\alpha_R} \qquad (9.32b)$$

Next, introducing a single current source to replace the two reference currents I_{CC} and I_{EC} as

$$I_{CT} = I_{CC} - I_{EC} \qquad (9.32c)$$

Rearranging Eqs. (9.32a) and (9.32b) and using Eq. (9.32c), the emitter and collector currents can be redefined as:

$$I_E = \frac{I_{CC}}{\alpha_F} - I_{EC} + I_{CC} - I_{CC}$$

$$= \left(\frac{I_{CC}}{\alpha_F} - I_{CC} \right) + I_{CT} = I_{CC} \left(\frac{1 - \alpha_F}{\alpha_F} \right) + I_{CT}$$

$$= \frac{I_{CC}}{\beta_F} + I_{CT} \tag{9.33a}$$

$$I_C = I_{CC} - \frac{I_{EC}}{\alpha_R} = (I_{CC} - I_{EC}) + \left(1 - \frac{1}{\alpha_R} \right) + I_{EC}$$

$$= I_{CT} + \left(\frac{\alpha_R - 1}{\alpha_R} \right) I_{EC} = I_{CT} + \frac{I_{EC}}{\beta_R} \tag{9.33b}$$

and

$$I_B = \frac{I_{CC}}{\beta_F} + \frac{I_{EC}}{\beta_R} = \frac{I_{CT}}{\beta_F} + I_{EC} \left(\frac{1}{\beta_R} + \frac{1}{\beta_F} \right) \tag{9.33c}$$

The non-linear hybrid-π version of Ebers-Moll model is shown in Fig. 9.14.

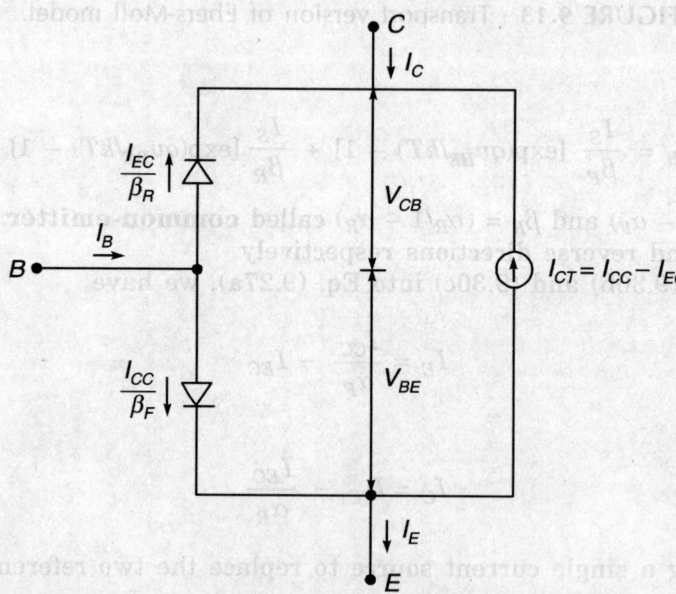

FIGURE 9.14 Non-linear hybrid-π version of Ebers-Moll model.

EXAMPLE 9.7

Find the terminal currents for a *n-p-n* BJT which is operating in active region.

Solution: Since, in active mode, emitter-base junction is forward biased and collector-base junction is reverse biased, i.e.,

$$e^{qv_{BE}/kT} >> 1 \quad \text{and} \quad e^{qv_{BC}/kT} << 1$$

then Eqs. (9.31a, b and c) reduce to

$$I_E = \frac{I_S}{\alpha_F} e^{qV_{BE}/kT} \implies \frac{I_C}{\alpha_R} = I_C\left(1 + \frac{1}{\beta_F}\right)$$

$$I_C = I_S \, e^{qV_{BE}/kT}$$

and

$$I_B = \frac{I_S}{\beta_F} e^{qV_{BE}/kT} = \frac{I_C}{\beta_F}$$

EXAMPLE 9.8

Find the common-emitter current gain and express I_{CEO} in terms of β and I_{CBO}. Take $\alpha = 0.9934$.

Solution: (a) For $\alpha = 0.9934$,

$$\beta = \frac{\alpha}{1 - \alpha} = \frac{0.9934}{1 - 0.9934} = 150.5$$

and Eq. (9.24a) can be rewritten as:

$$I_{CEO} = \left(\frac{\alpha}{1 - \alpha} + 1\right) I_{CBO}$$

(b) Find I_{CEO} if $I_{CBO} = 0.87$ μA and hence

$$I_{CEO} = (\beta + 1) \, I_{CBO} = (150.5 + 1) \times 0.87 \text{ μA}$$

$$= 0.132 \text{ mA}$$

This shows that the leakage current is very small.

EXAMPLE 9.9

Why collector current of a bipolar junction transistor remain constant (for ideal case) in forward active mode?

Solution: In active forward mode, the collector base junction is reverse biased. Hence all the injected electrons from emitter into base are collected by the collector. There is no electron injection from the collector into base. Therefore, the collector current is constant and independent of V_{CE}. The current gain is also constant and constant I_B curves are spaced apart by an amount determined by the base-current step.

Note: The measured current-voltage characteristics of a typical bipolar devices are not ideal. The degree of deviation from ideal characteristics depends on the device structure, the device design, the device fabrication process and on the bias condition of the transistor.

The typical measured current gain (I_C/I_B) as a function of collector current is shown in Fig. 9.15.

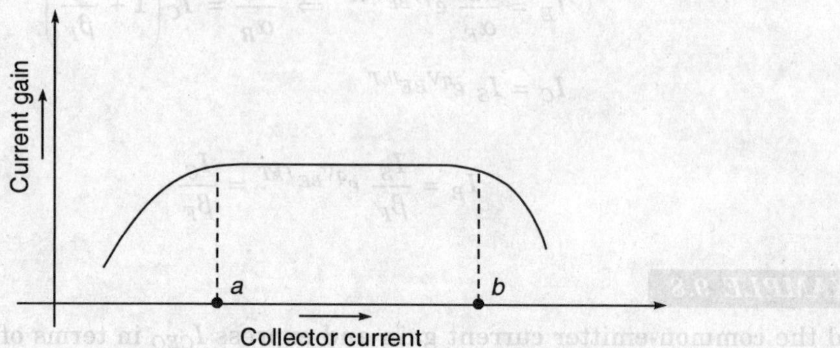

FIGURE 9.15 Characteristic of current gain versus collector gain.

For current range from a to b, the transistor behaves as ideal ones and hence, current gain is constant. At low currents, the current gain is less than the ideal one because the base current is larger than its ideal value. At high collector current, the current gain rolls off because the percentage by which the collector current is smaller than its ideal value is larger than the percentage by which the base current is smaller than its ideal value.

9.7 NON-IDEAL BEHAVIOUR OF THE BIPOLAR JUNCTION TRANSISTOR

9.7.1 Base Resistance and Emitter Crowding

Since, one of the basic requirements for the design of a bipolar junction transistor is that the base should be moderately doped and thin. This requires to avoid any recombination of carriers and to increase current gain. But this thin base region has a high resistance which degrades the device performance. Therefore, a base current flows from points within the base region towards each end and this causes a voltage drop across the resistance. This drop along the emitter edge means that the base-emitter voltage V_{BE} is not uniform, but instead varies with position. The potential decreases from the edge of the emitter towards the centre. The base resistance consists of two parts, the intrinsic-base resistance R_{Bi}, which is determined by the design of the intrinsic-base region and the other part is extrinsic-base resistance R_{Bx}, which includes all other resistances associated with the base terminals.

The emitter-base diode voltage drop due to the flow of base current is:

$$\Delta V_{BE} = I_B R_B$$

$$= I_B(R_{Bi} + R_{Bx}) \tag{9.34a}$$

Since, emitter is heavily doped, so as a first approximation it is treated as an equipotential region. The relation between V_{BE} (applied between the emitter and base terminals) and voltage V'_{BE} (appearing across the intermediate emitter-base junction due to base series resistance R_B) is:

$$V'_{BE} = V_{BE} - \Delta V_{BE} \tag{9.34b}$$

Since, as we know that the emitter current I_E is strong (exponential) function of V_{BE},

$$I_E = I_{EO} \, (e^{qV_{BE}/kT} - 1) \tag{9.34c}$$

in absence of base series resistance R_B and hence, the modified emitter current due to presence of series base resistance is given as:

$$I'_E = I_{EO} \, (e^{qV'_{BE}/kT} - 1) \tag{9.34d}$$

Since, V'_{BE} lateral dependent, therefore, Eq. (9.34d) gives that a small lateral voltage drop in the base causes a large spatial dependence on the emitter current. Since, the polarity of this lateral voltage drop is such that V_{BE} is maximum at the emitter edge nearest to the base contact, therefore, according to Eq. (9.34d), emitter current is maximum at the emitter edge. Hence, we can say that injection of electrons is greatest there. This effect is called **emitter crowding** or **current crowding**.

Note: Base saturation current is a function of the emitter parameters only which does not vary with the minority-carrier injection level because emitter is heavily doped. Therefore, at high currents, deviation of the base current from its ideal behaviour is due to ΔV_{BE} alone.

Figure 9.16 shows the base current distribution and resulting lateral potential drop in the base region.

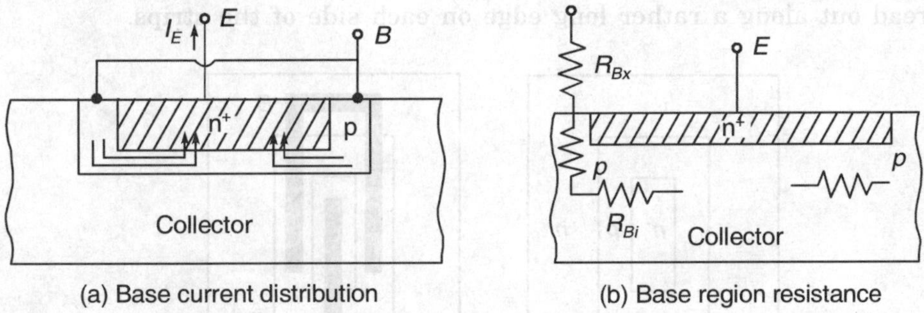

(a) Base current distribution   (b) Base region resistance

FIGURE 9.16   Non-ideal behaviour of base region.

This current crowding effect is schematically shown in Fig. 9.17. The current crowding increases as base current increases. This means a larger fraction of the current is flowing into the emitter at the end near the base contact. The larger current near the emitter edge may cause localized heating effects as well as localized high-injection effects.

The most effective way to handle the emitter crowding is to distribute the emitter current along a relatively large emitter edges, i.e., increasing area A.

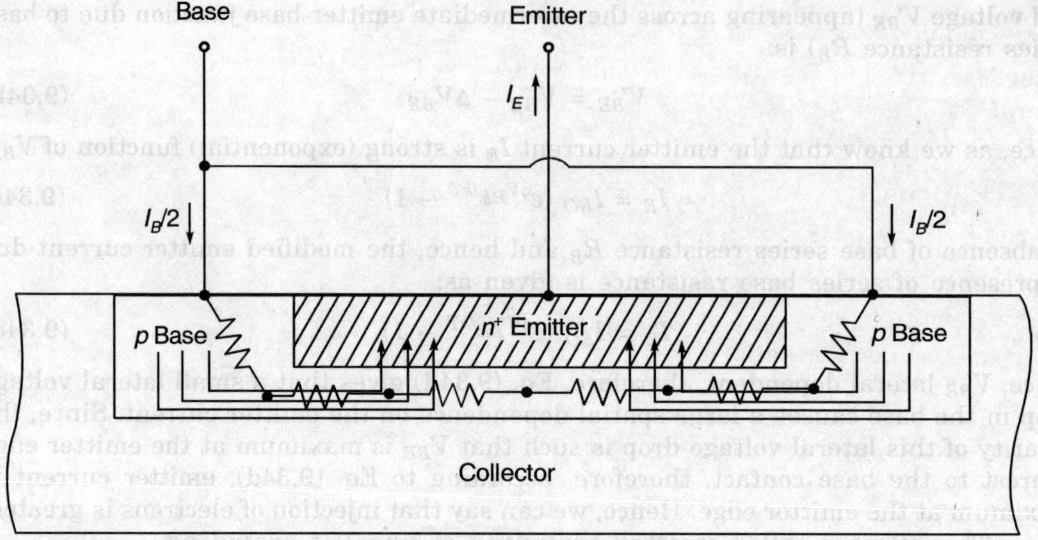

FIGURE 9.17 Schematic representation of current crowding effect in transistor.

EXAMPLE 9.10

In modern IC BJTs, a long thin stripe geometry for emitter with base contacts on each side is preferred, why?

Solution: Because with this geometry as shown in Fig. 9.18, the total emitter current I_E is spread out along a rather long edge on each side of the strips.

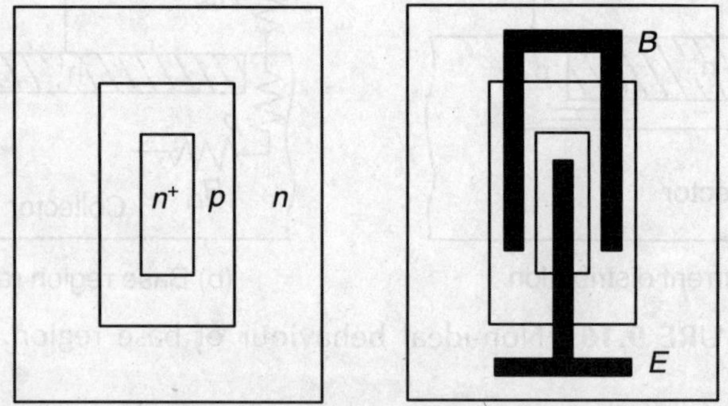

FIGURE 9.18 Stripe geometry representation for emitter.

EXAMPLE 9.11

What is an interdigitated geometry?

Solution: An even better geometry is to use several emitter stripes which are connected electrically by the metallization and separated by interspersing base contacts as shown in Fig. 9.19. Many such thin emitter and base contact *fingers* can be interlaced to provide for handling a large current in a power transistor. Such a design is called **interdigitated geometry**. Current density reduces at that point which reduces the lateral voltage drop. Therefore, a preferable design for BJT is to make emitter region with a larger perimeter compared with its area.

Note: The high current density at the emitter-edge reduces the current gain β.

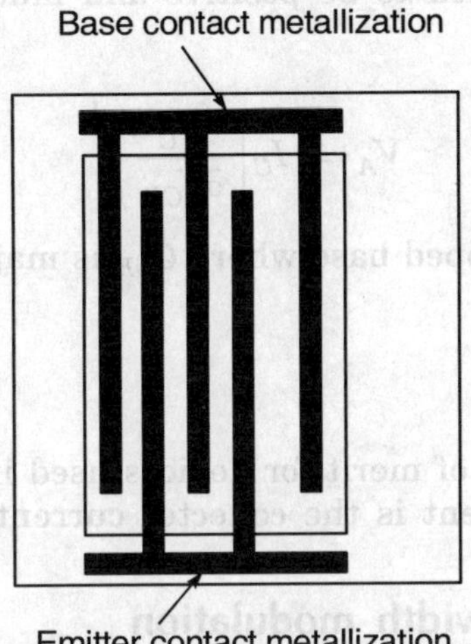

FIGURE 9.19 Interdigitated geometry.

9.7.2 Base Width Modulation and Early Effect (Base Narrowing)

In an ideal transistor, it is assumed that the base width W_B remains constant. But in real bipolar junction transistor, the base-collector region is reverse biased, therefore, as the reverse bias across the base-collector junction is increased, the base-collector junction depletion-layer width increases and the quasineutral base width W_B decreases. In other words, we can say that in real (practical) bipolar transistor the base width W_B is a function of the base-collector voltage. This in turn causes the collector current to increase because the reduced base width causes the gradient in the minority-carrier concentration to increase. Due to increase in minority-carrier concentration gradient, the diffusion current increases and hence β also. This effect is known as **base-width modulation**. It is also called **Early effect**. Therefore, in real transistor, the collector current increases with collector voltage as shown in Fig. 9.21(b). The early effect is demonstrated in Fig. 9.21(a). By extrapolating to zero collector current and the point where these current curves intersect, the V_{CE} axis, one can get the early voltage V_A as shown in Fig. 9.21(b). The typical value of V_A is ranging from (50 to 100). There is an additional term to account the early voltage. Non-zero slope means the output resistance is not infinite and given as:

$$r_0 \simeq \frac{V_A}{I_C}$$

(9.35a)

and collector current $I_C = I_S\, e^{qv_{BE}/kT}\left(1 + \dfrac{V_{CE}}{V_A}\right)$ $\hspace{2em}$ (9.35b)

where $I_s = \dfrac{qAD_n n_i^2}{N_a^- W_B}$ is reverse saturation current.

In practice, V_A is assumed to be positive and much larger than the operation range of V_{CE} and is given as:

$$V_A \simeq I_C \left(\frac{\partial I_C}{\partial V_{CE}}\right)^{-1} \hspace{3em} (9.35c)$$

and $V_A = Q_{pB}$ for uniformly doped base where Q_{pB} is majority-carrier hole charge per

unit area $= q \displaystyle\int_0^{W_B} p_p(x)\, dx$.

Early voltage is a figure of merit for devices used in analog circuits. Larger the early voltage, more independent is the collector current on the collector voltage.

How to reduce the base-width modulation

The base width modulation effect can be reduced if we

    (a) increase the base width.
        Consequence: probability of recombination increases
    (b) increase the base doping concentration N_B.
        Consequence: more recombination
    (c) decrease the collector doping concentration.

Note: Larger V_A, i.e, larger r_0 is desirable for proper operation. Since V_A is function of W_B, and W_B is function of collector voltage, therefore, strictly speaking early voltage is function of the collector voltage at which the slope is used for extrapolating $I_C = 0$. In otherwords, I_C does not increase linearly with V_{CE}, but the linear dependence is a good approximation for circuit analysis and modelling.

    The product of current gain and early voltage (βV_A) is also treated as a figure of merit and depends on base doping profile only.

EXAMPLE 9.12

For the geometry shown in Fig. 9.20, the base doping $= 10^{16}/\text{cm}^3$ and neutral base width $x_B = 0.8$ μm, the emitter width $W_E = 10$ μm and the emitter length is $L = 10$ μm.

    (a) Determine the potential difference between $x = 0$ to $x = W_E/2$. Take $\mu_p = 400$ cm^2/V-sec and assume uniform base current given by $I_B/2 = 5$ μA.
    (b) Find the ratio of emitter current density at $x = 0$ and $x = W_E/2$.

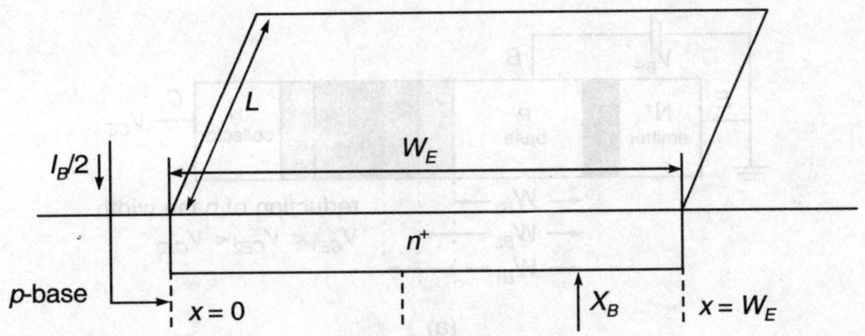

FIGURE 9.20 Geometry of base region.

Solution: (a) $R = \rho \dfrac{L}{A} = \left(\dfrac{1}{q\mu_p N_B} \right) \left(\dfrac{W_E/2}{x_B x_L} \right) = 9.8 \text{ k}\Omega$

$$\Delta V = \left(\frac{I_B}{2} \right) R = 5 \times 10^{-6} \times 9.8 \times 10^{3} = 49 \times 10^{-3} = 49 \text{ mV}$$

(b) $\dfrac{I_E(x = 0)}{I_E(x = W_E/2)} = \exp\left(\dfrac{q\Delta V}{kT} \right) = \exp\left(\dfrac{0.0049}{0.0259} \right) \simeq 6.59$

EXAMPLE 9.13

Explain, why slope of the curves as shown in Fig. 9.21(b) intercepts the V_{CE} axis at the same value V_A?

Solution: The early voltage is given by Eq. (9.35c) for larger V_A,

where
$$I_C = A J_{CO} \exp(q V_{BE}/kT)$$

$$= \frac{qA \exp(q V_{BE}/kT)}{F(W_B)} \tag{9.36a}$$

where
$$F(W_B) \cong \frac{q}{J_{CO}} \int_0^{W_B} \frac{p_p(x)}{D_{nB}(x)\, n_{ieB(x)}^2}\, dx$$

Substituting all these parameters and after simplification, we have

$$V_A \cong \frac{q D_{nB}(W) n_{ieB(W)}^2}{C_d B_C} \int_0^{W_B} \frac{p_p(x)}{D_{nB}(x)\, n_{ieB(x)}^2}\, dx \tag{9.36b}$$

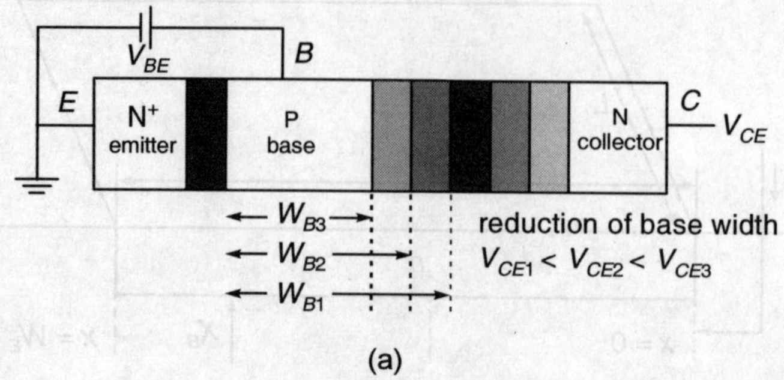

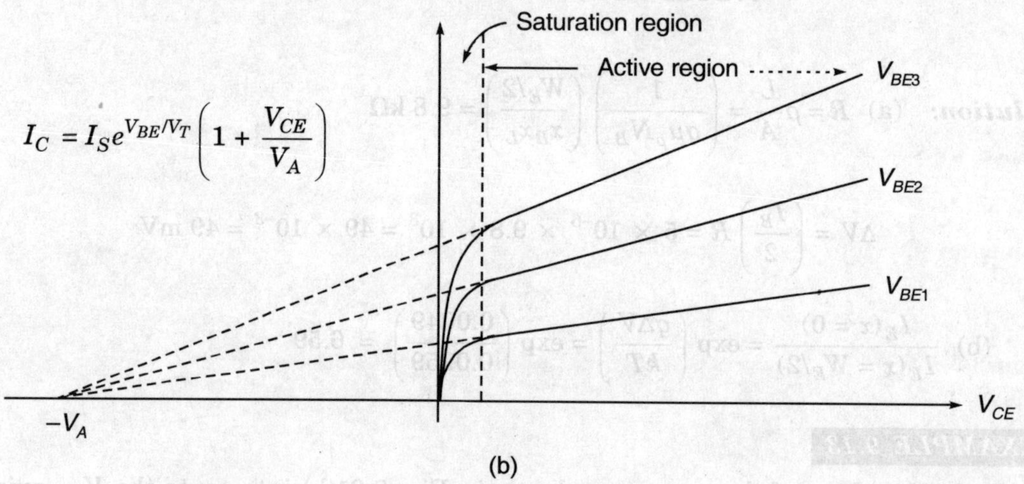

$$I_C = I_S e^{V_{BE}/V_T}\left(1 + \frac{V_{CE}}{V_A}\right)$$

FIGURE 9.21 Base width modulation and early effect.

Hence Eq. (9.36b) can be rewritten as:

$$V_A = \frac{qAD_{nB}(W_B)\,n_{ieB}^2(W_B)}{C_dB_c}\int_0^{W_B}\frac{N_B(x)}{D_{nB}(x)n_{ieB}^2(x)}\,dx \qquad (9.36c)$$

where C_dB_c is the base collector junction depletion-layer capacitance, and is given as:

$$C_dB_c \cong -\frac{dQ_{pB}}{dV_{CB}}$$

Equation (9.36c) shows that V_A is independent of base current, therefore, the slope of all the curves intercept the V_{CE} axis at the same point.

At sufficiently low collector current,

$$p_p \simeq p_{p_n} \simeq N_B$$

Emitter-collector punchthrough

The emitter collector punchthrough occurs when base width becomes so small that the collector-base voltage starts to affect the potential barrier height of the emitter base junction. This process is similar to DIBL effect in MOSFET.

When transistor is operated near punchthrough, collector current is no longer controlled by V_{BE}.

At collector emitter punchthrough, the collector current becomes excessively large being limited only by the emitter and collector series resistances. Punchthrough must be avoided for normal operation of the device. To avoid punchthrough, design a device which has sufficiently large majority carrier base charge N_B because

$$N_B \ \alpha \ \frac{1}{W_B^2}$$

Neglecting the built-in potential the reverse-biased base-collector voltage at punchthrough

is $V_{pt} = \dfrac{qW_B^2}{2\varepsilon_{Si}} \dfrac{N_B(N_B + N_C)}{N_C}$ where N_B and N_C are uniform base and collector concentrations.

Injection level, Thermal effects

In an ideal bipolar transistor analysis, we have assumed that α and β are independent of level of injection. But in practical transistor these parameters may vary considerably with injection level. The level of injection is determined by the currents I_E or I_C. The assumption of negligible recombination in the junction depletion regions is not valid if the level of injection is low. Any recombination in the emitter junction reduces the emitter junction efficiency paramater γ which would decrease the value of α and β for low value of I_C. This causes the curves of the collector characteristics to spread more closely for low currents than for higher currents.

As current I_C is increased beyond the low level injection, α and β first increases and then again fall off at very high injection. Figure 9.22(b) shows a typical common-emitter current gain versus collector curent curve and Fig. 9.22(a) shows the minority and majority carrier concentrations in base under low and high level injections.

Large collector current I_C and the voltage drop across the collector base, i.e., V_{CB} determines the power dissipated on the BJT transistor. This power dissipation generates heat. If heat cannot dissipate quickly enough then the silicon warms up. Therefore, it is very important that the transistor be operated in a range such that $I_C V_{CB}$ does not exceed the maximum power rating of the device. If silicon warms up then β increases and hence collector current increases. This will produce more power because of I_C^2 R and again temperature of transistor will rise further. This further rise in temperature causes further increase in β and this further increases I_C and again more heat generates. If this process is continuous then so much heat is produced that the device fails.

The process of variation in β value due to change in temperature is called **thermal runaway**.

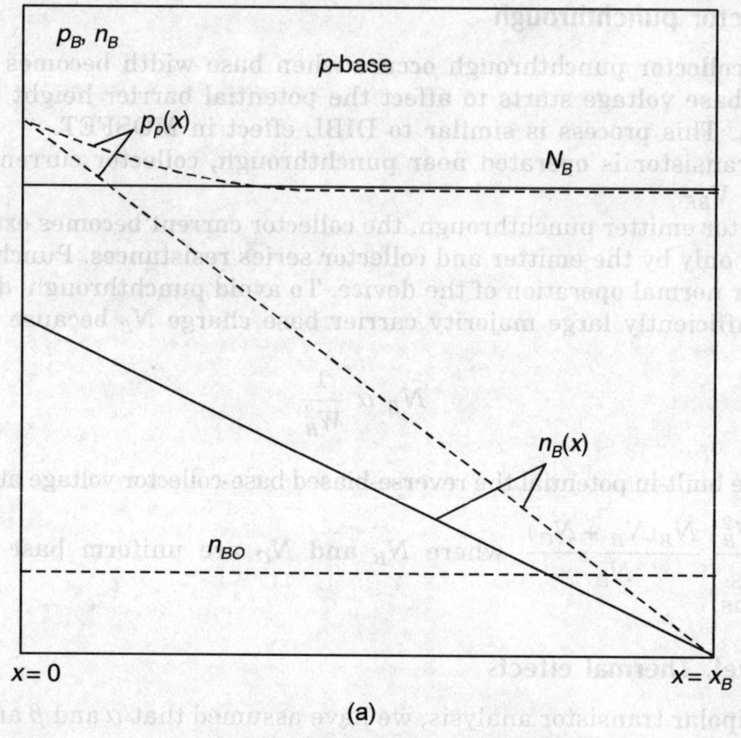

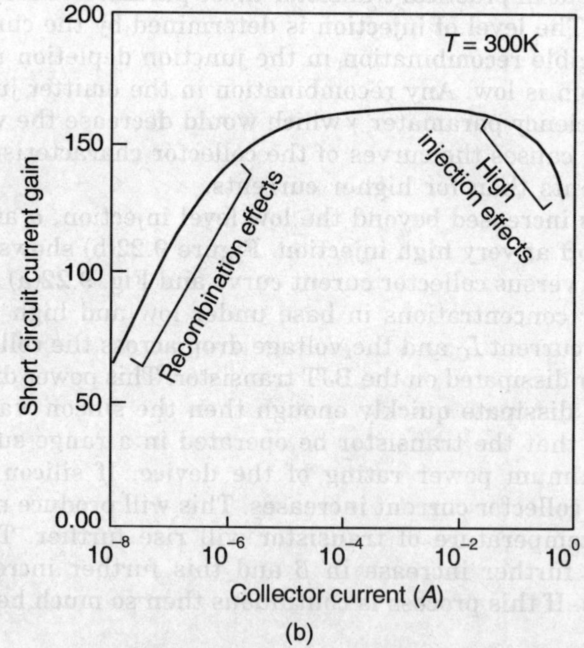

FIGURE 9.22 (a) Majority and minority carrier concentration in base region under low and high level injection, (b) Typical common-emitter current gain versus collector current.

Circuit is properly designed to avoid thermal runway. Emitter degenerated common emitter circuits are immune to this.

Note: In the device designed for high power capability the transistor is mounted on an efficient heat sink to transfer the thermal energy away from the junction.

Most important parameters of Si and Ge depends on temperature are the carrier lifetime and diffusion coefficient. In Si and Ge devices

$$\text{Temperature } T \uparrow \to \tau_n \uparrow \to \beta \uparrow \to I_C \to \text{More power dissipation}$$

$$\text{Temperature } T \uparrow \to \text{Mobility of carrier } \downarrow$$

EXAMPLE 9.14

How increase in temperature due to heating effect, causes β to increase?

Solution: Most important parameter of Ge and Si depends on temperature are carrier's lifetime (τ_n or τ_p) and mobility. Due to increase in temperature τ_n or τ_p increases for most cases. This increase is due to thermal re-excitation from recombination centre. This increase in τ_n or τ_p tends to increase β of the transistor.

9.7.3 Drift in Base Region (Non-uniform Base Doping)

During the analysis of current-voltage characteristics of an ideal bipolar junction transistor, it is assumed that regions are uniformly doped. But in practical transistor uniform doping rarely occurs because of implanted junction transistor. The doping profile in the three different regions of a transistor is shown in Fig. 9.23. The most likely doping distribution in the base is a portion of Gaussian. Due to the graded base region a built-in-electric field develops from collector-to-emitter in an n-p-n transistor as discussed in Chapter 4, i.e., in the negative x-direction. This built-in-field adds a drift component to the transport of electrons across the base. For a p-type base region in thermal equilibrium, we can write

$$J_p = N_a q \mu_p E_x - q D_p \frac{dN_a}{dx} = 0 \tag{9.37a}$$

or
$$E_x = \text{Built-in electric field}$$

$$= \left(\frac{kT}{q}\right) \frac{1}{N_a(x)} \frac{dN_a(x)}{dx} \tag{9.37b}$$

Since, from Fig. 9.23, it is clear that dN_a/dx is negative and hence, induced electric field is in the negative x-direction.

Let us represent an exponential distribution in the base region as

$$N_a(x) = N_0\, e^{-\alpha x_p / W_B} \tag{9.38a}$$

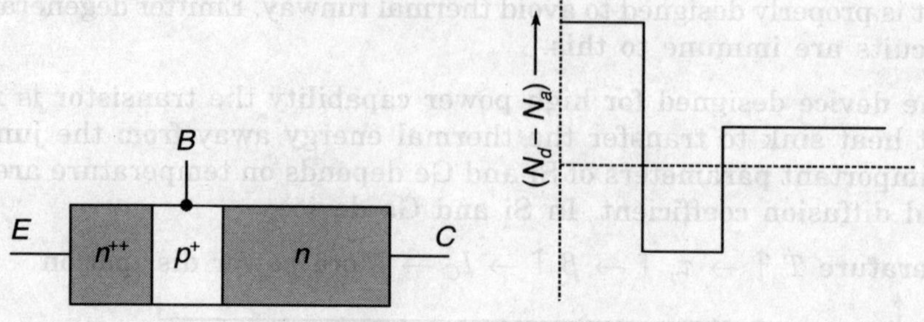

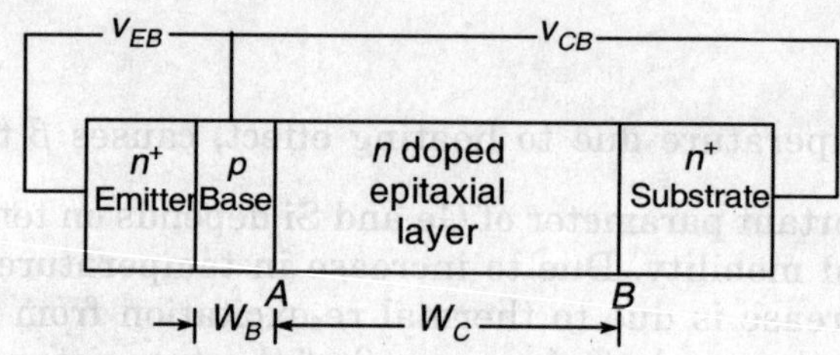

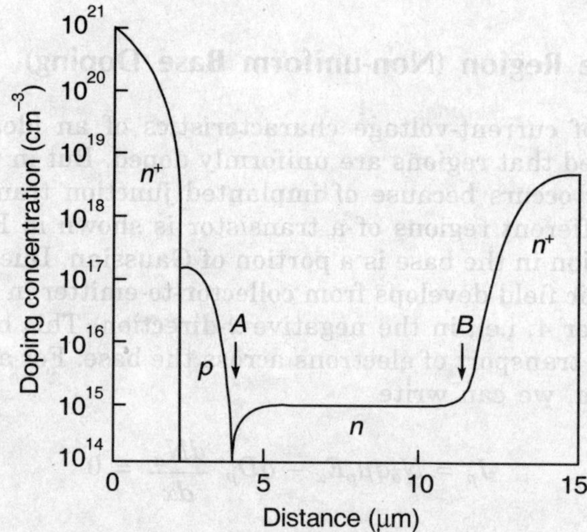

FIGURE 9.23 Non-uniform base doping effect in transistor.

where

$$\alpha = \ln \frac{N_0}{N(W_B)}$$

Using Eq. (9.38a) into Eq. (9.37b), we have

$$E_x = \frac{kT}{q} \cdot \frac{1}{N_0 e^{-\alpha x_p/W_B}} \times N_0(\alpha/W_B)\, e^{-\alpha x_p/W_B}$$

$$E_x = -\frac{kT}{q}\left(\frac{\alpha}{W_B}\right) \tag{9.38b}$$

where W_B is the base region width. The induced electric field in the base exerts an accelerating force on the electrons in the direction towards the collector or in other words, this electric field aids the movement of injected electrons from emitter to collector across the base region. Therefore, the electric field is called **accelerating field**. Since, the minority carrier electron concentration varies across the base and hence, the drift-current density will not be constant.

EXAMPLE 9.15

Why a real BJT is said to be doubly compensated transistor?

Solution: We have to take a uniformly doped semiconductor (substrate) as a starting material for the fabrication of transistor. The substrate is then diffused with an acceptor impurity from the surface to form a compensated p-type base. We further diffuse donor atoms from the surface to form a doubly compensated n^+-emitter region.

EXAMPLE 9.16

As we know that due to non-uniform base doping, the drift current density is not constant, but at the thermal equilibrium total current density (drift component plus diffusion component) remains constant. How?

Solution: The non-uniform doping in the base region will produce an induced electric field that aids the flow of electrons from emitter to collector. This induced electric field will also alter the minority carrier concentration through the base so that the total current density, i.e., sum of diffusion current density and drift current density will be a constant.

Note: Transist time τ_t is defined as the time required for minority carriers to diffuse across the base. It is given as:

$$\tau_t = \frac{W_B^2}{2D_B}$$

This time will also influence the switching speed of the BJT.

EXAMPLE 9.17

Why transist time of minority carriers to diffuse across the base is smaller for non-uniformly doped base than the uniformly doped base?

Solution: Since, due to non-uniform doping of base, an induced electric field exists in base which aids the movement of injected carriers from emitter to collector through base. In other words, the time taken for injected minority carriers to cross the base reduces. But in case of uniformly doped base, there is no induced electric field.

Note: 1. Another approach for obtaining the built-in-field is to vary the alloy composition, which will affect the bandgap.
- Fix the bandgap and decrease N_B from emitter to collector.
- Fix N_B and reduce the band gap from emitter to collector.

2. Shortening of the transist time can be very important in high frequency devices.

9.7.4 Avalanche Breakdown

There are mainly two breakdown mechanisms in the bipolar junction transistor, namely, punchthrough and avalanche.

For an *n-p-n* transistor the avalanche breakdown process is shown in Fig. 9.24(a). In normal *n-p-n* BJT operation, the electrons are injected into base from emitter and then collected by the reverse-biased base-collector junction. Some of the injected electrons in base-collector depletion region has enough energy to cause impact-ionization and generate electron-hole pairs (EHPs). The generated secondary electrons are swept into the collector terminal, adding to the measured collector current. The secondary generated holes will flow towards the base terminal and reduces the measured base current.

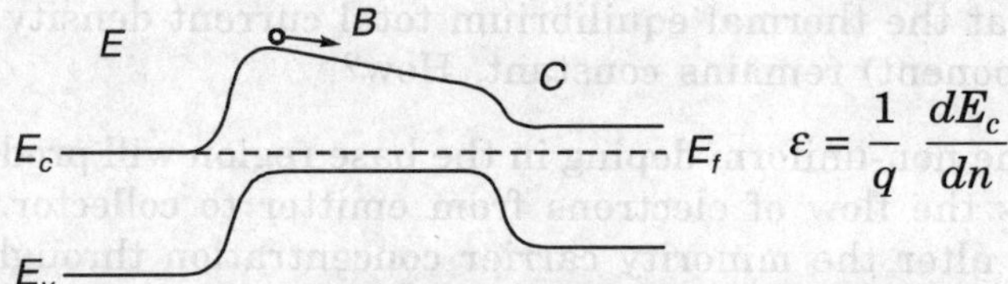

FIGURE 9.24(a) Avalanche breakdown mechanism.

If the generated secondary hole carriers are very large, they cause a negative base current. The increase in collector current due to generation of secondary electron is called **avalanche breakdown**.

In the forward-bias, a current I_{CEO} produces mainly due to the injection of electrons from emitter into base and these injected electrons diffuse into the collector and current $\alpha\,I_{CEO}$ flows. Assuming recombination in the base region,

$$I_{CEO} = \alpha I_{CEO} + I_{CBO} \tag{9.39a}$$

where I_{CBO} is the normal reverse biased *B-C* junction current.

$$I_{CEO} \simeq \frac{I_{CBO}}{(1 - \alpha)} \simeq \alpha\beta I_{CBO} \tag{9.39b}$$

where α is common base current gain, β is common emitter current gain and O represents the open circuit case. For a transistor biased in open emitter configuration, the current I_{CBO} at breakdown becomes

$$I_{CBO} \rightarrow MI_{CBO}$$

where M is the multiplication factor and for that empirical relation is

$$M = \frac{1}{1 - (V_{CB}/BV_{CBO})^n} \tag{9.40}$$

where n is the empirical constant, usually between 3 and 6 and BV_{CBO} is base-collector breakdown voltage with emitter left open.

Now, consider a case, when transistor is biased in such a way so that the base is open circuited as shown in Fig. 9.24(b).

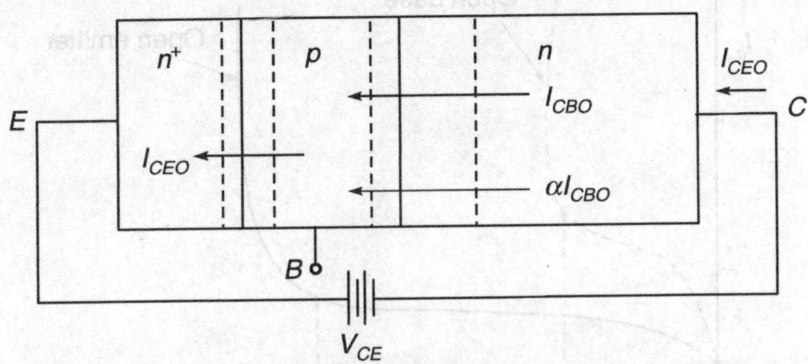

FIGURE 9.24(b)　Avalanche breakdown in transistor.

The current in the base-collector junction at breakdown is given by multiplying Eq. (9.39a) with the multiplication factor M,

$$I_{CEO} = M(\alpha I_{CEO} + I_{CBO}) \tag{9.41a}$$

and on solving,

$$I_{CEO} = \frac{MI_{CBO}}{(1 - \alpha M)} \tag{9.41b}$$

From Eq. (9.41b), it is clear that the current increased indefinitely when

$$M\alpha = 1 \tag{9.41c}$$

This is also called **breakdown condition**.

Using Eq. (9.40) and replacing $V_{CB} \simeq V_{CE}$, Eq. (9.41c) reduces to

$$\frac{\alpha}{1 - \left(\dfrac{BV_{CEO}}{BV_{CBO}}\right)^n} = 1$$

or

$$BV_{CEO} = BV_{CBO} \sqrt[n]{1 - \alpha}$$

or
$$BV_{CEO} = \frac{BV_{CBO}}{\sqrt[n]{\beta}}$$
(9.41d)

where BV_{CEO} is collector-emitter voltage at breakdown when base-collector junction is open junction.

From Eq. (9.41d), it is observed that the breakdown voltage in open-base configuration is smaller by factor $\sqrt[n]{\beta}$ than the actual avalanche junction breakdown voltage as shown in Fig. 9.24(c).

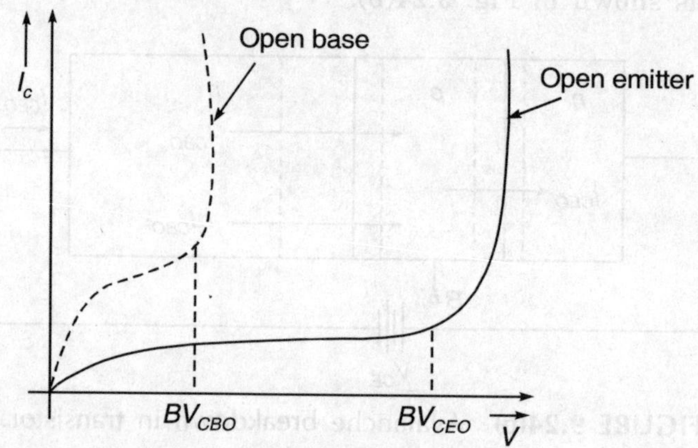

FIGURE 9.24(c)   Current-voltage characteristic after breakdown.

Note: From Fig. 9.24(c), it is clear that there is a strong influence of carrier multiplication over a fairly broad range of collector voltage.

EXAMPLE 9.18

Why multiplication is so important in the common-emitter case? Explain physically.

Solution: When an ionizing collision occurs in the collector junction depletion region, secondary electron-holes pairs (EHPs) are created. The primary and secondary electrons are swept into the collector of an *n-p-n* transistor, whereas holes swept into the base by junction field. This enhances the supply of holes in the base and from charge control analysis it is clear that to maintain the charge neutrality condition, electron injection at the emitter must increase. This is a regenerative process, in which an increased injection of electrons from the emitter causes an increased multiplication current at the collector junction. This increases the rate at which the secondary holes are swept into base and hence requiring more injection of electrons. Because of regenerative in nature it is required that the multiplication factor M should be slightly greater than unity to start the avalanching process.

EXAMPLE 9.19

What is base-conductivity modulation effect?

Solution: As the electrons are injected into the p-type base, the hole concentration in the base increases in order to maintain the charge neutrality condition. If this increase in hole concentration is appreciable then the collector saturation current density decreases. At the same time, as the injected electrons reach the base-collector junction, they add to space charge in the base-collector space-charge region. This results in widening of the quasineutral base layer. As W_B increases, the collector saturation current density also decreases. This is known as **base-conductivity modulation effect**.

Note: Base-conductivity modulation and base-widening effects are not really separate and do not act independently.

9.7.5 Base Widening or Kirk Effect

At high current densities (> 0.5 mA/μm^2), the charge concentration is comparable to or larger than the fixed ionized impurity concentration. Therefore, the depletion approximation is not valid. Furthermore, the excess electrons in collector of n-p-n transistor can produce a substantial electric field in the collector and hence, the classical concept of junction boundary in base-collector junction is no longer valid. It is also seen that to maintain the quasineutrality, the excess electrons in the collector induce an excess of holes in the collector of n-p-n transistor. This excess holes in the collector region appears to extend the base region because in n-p-n transistor, base region is p-type. In other words, we can say that the base region extends into the collector region. Due to widening of the base, the excess minority-carrier charge Q_n increases which causes base-collector depletion-region transit time τ_{BC} to increase and hence the cut off frequency decreases. At sufficiently high collector current densities, base widening can push the base-collector junction deep into the collector region. This effect is known as **base-widening** or **Kirk effect**; observed by Kirk in 1962. This Kirk effect also causes the decrease in saturation collector current density J_{CO}.

Note: At low collector currents, the maximum electron concentration in n-type collector region is equal to the collector doping concentration N_C and maximum electron current density that can be supported by N_C is:

$$J_{max} = qv_{sat}N_C$$

where v_{sat} is saturated electron velocity $\simeq 1 \times 10^7$ cm/s.

When injected electron current density approaches to J_{max} there is an increase in electron concentration in excess of N_C to support this injected electron current flow. As the excess electrons build-up, there is a build-up, of excess holes in order to maintain the quasineutrality condition. Due to this the high-field region shifts. It was observed that the base widening starts at a collector current density of approximately $0.3J_{max}$.

Note: To avoid the base widening, a bipolar transistor could be avoided to operate at collector current density approaching J_{max}.

EXAMPLE 9.20

Find out the maximum collector-current density I_C for an n collector doping of $N_{dc} = 5 \times 10^{16}/\text{cm}^3$ to avoid base widening.

Solution: $N_{dc} = 5 \times 10^{16}/\text{cm}^3$

$$J_{max} = 1.6 \times 10^{-19} \times 10^7 \times 5 \times 10^{16}$$

$$= 8 \times 10^4 \text{ A/cm}^2$$

To avoid Kirk effect,

$$\simeq 0.3 \times J_{max}$$

$$\simeq 2.4 \times 10^4 \text{ A/cm}^2$$

$$\simeq 2.4 \times 10^4 \times 10^{-8} \text{ A/\mu m}^2$$

$$\simeq 2.4 \times 10^{-4} \text{ A/\mu m}^2$$

$$\simeq 0.24 \text{ mA/\mu m}^2$$

9.7.6 Non-ideal Base Current at Low Currents

The base current is larger than the ideal value at low emitter-base voltages. The main reasons for this excess base current are:

(a) The generation-recombination current in the emitter-base junction depletion region

(b) The tunnelling current in the emitter-base junction

Base current due to generation recombination

In modern VLSI devices, the generation-recombination current due to defect centres in silicon is negligible because of device design and fabrication process. The generation recombination hole current adds to the base current and hence degrades the current gain. This further reduces the value of β. Figure 9.25 shows the plot of I_C as a function of V_{BE} and this plot is known as **Gummel plot**.

Since, the collector curent I_C is essentially equal to I_{NO} which is due to injected electrons from the base into the emitter and proportional to exp (qv_{BE}/kT). Thus, on semilog scale, as shown in Fig. 9.25, the $I_C - V_{BE}$ plot is a straight line of slope $(\log e^{qV_{BE}/kT})$. The base current has two components: (a) Injection current due to injected holes from the base into the emitter and varies as $\exp(qv_{BE}/kT)$, (b) The recombination current results from the base supplying the holes to support the electron-hole pair recombination in the emitter-base transition region. This current vaires as exp $(qv_{BE}/\eta kT)$ where $\eta = 2$. Since, the injection current varies more rapidly than the recombination current with V_{BE}. Therefore, the injection current predominates at higher base-emitter voltage while the recombination current predominates at lower voltages.

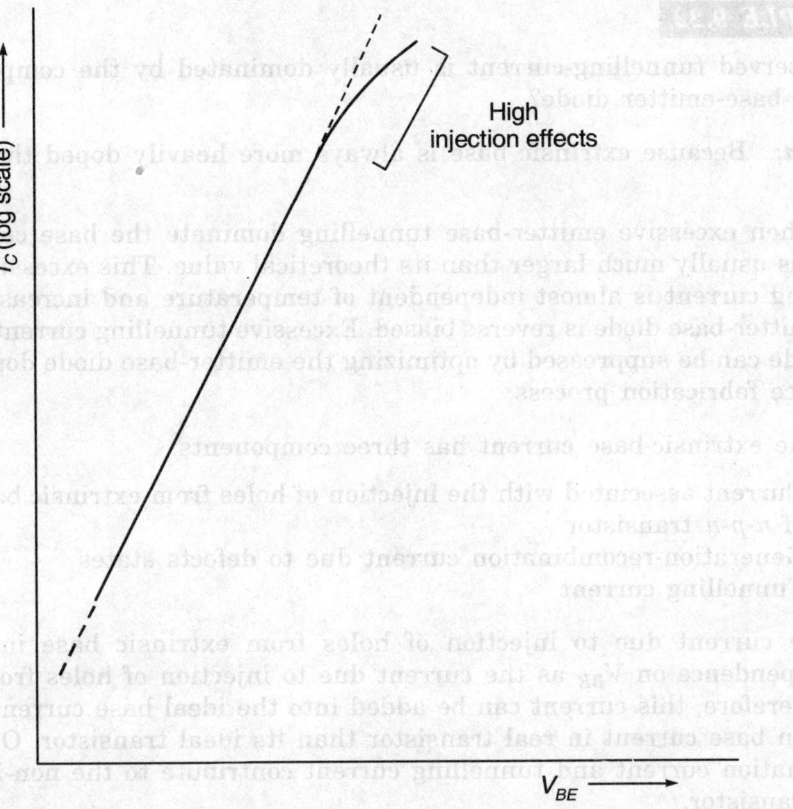

FIGURE 9.25   Gummel plot.

Base current due to tunnelling

As the transistor dimension is scaled down, the tunnelling current in the emitter-base junction is expected to increase. The emitter-base junction is an abrupt-junction where emitter is much heavily doped and base doping is moderate. Due to scaling of transistor dimension the typical peak base doping concentration is increased. This results in enhanced tunnelling current in the emitter-base diode.

Note: The base region directly underneath the emitter is referred to as the *intrinsic base* and remaining parts of the base are collectively referred to as the *extrinsic base*. To minimize the parasitic resistance and electron injection from the emitter into the extrinsic base region, the extrinsic base region is usually doped much heavily than the intrinsic base.

EXAMPLE 9.21

Why the contribution of extrinsic base to the collector current is very small than the intrinsic base?

Solution: Due to large width and high doping concentration in the extrinsic base region.

EXAMPLE 9.22

Why observed tunnelling-current is usually dominated by the component from the extrinsic-base-emitter diode?

Solution: Because extrinsic base is always more heavily doped than the intrinsic base.

Note: When excessive emitter-base tunnelling dominate the base current, the base current is usually much larger than its theoretical value. This excessive-emitter-base tunnelling current is almost independent of temperature and increases with voltage when emitter-base diode is reverse biased. Excessive tunnelling current in the emitter-base diode can be suppressed by optimizing the emitter-base diode doping profile and the device fabrication process.

Note: The extrinsic-base current has three components:

1. Current associated with the injection of holes from extrinsic base into emitter of n-p-n transistor
2. Generation-recombination current due to defects states
3. Tunnelling current

The current due to injection of holes from extrinsic base into emitter has same dependence on V_{BE} as the current due to injection of holes from the intrinsic base. Therefore, this current can be added into the ideal base current and shows no change in base current in real transistor than its ideal transistor. Only generation-recombination current and tunnelling current contribute to the non-ideal behaviour of the transistor.

Note: BJT performance requirements,

(a) High gain, i.e., $\beta \gg 1$. This can be achieved by one sided emitter junction, i.e., emitter should be heavily doped so that $\gamma = 1$. Other requirement is that the base should be narrow, so base transport factor $\alpha_T = 1$, i..e, quasineutral base width << minority-carrier diffusion length.

(b) The current I_C should be determined only by I_B, i.e., $I_C \neq$ function of V_{CE} and V_{CB}. This requirement is possible if the collector junction is one-sided junction and so quasineutral base width W_B does not change drastically with changes in V_{CE} or V_{CB} and base should be doped more heavily than the collector.

9.7.7 Emitter Bandgap Narrowing

Since, it is already seen that the emitter injection efficiency factor would continue to increase and approach to unity if the ratio of doping concentration of emitter to base continued to increase. In other words, we can say that the emitter region becomes more and more heavily doped. As the silicon becomes heavily doped, the discrete donor energy level in an n-type emitter splits into a band of energies and, therefore, the energy gap reduces. This effect is known as **bandgap narrowing**.

A reduction in the bandgap energy increases the intrinsic carrier concentration in the emitter as:

$$n_{iE}^2 = n_i^2 \; e^{\Delta E_{gE}/kT} \tag{9.42}$$

where $\Delta E_{gE} = E_g - E_{go}$ is the bandgap narrowing factor.

From Eq. (9.21b), we have

$$p_{EO} = \frac{n_{iE}^2}{N_E}$$

which is the thermal equilibrium minority hole concentration in the emitter of n-p-n transistor. Using Eq. (9.42), this expression can be further rewritten as:

$$p_{EO} = \frac{n_i^2}{N_E} \; e^{\Delta E_{gE}/kT} \tag{9.43}$$

As the emitter doping increases and ΔE_{gE} increases, p_{EO} does not continue to decrease with increased emitter doping. If p_{EO} starts to increase because of bandgap narrow then according to Eq. (9.21b), γ begins to fall off instead of continuing to increase with increased emitter doping.

Note: ΔE_g is negligible if $N_E < 10^{18}/\text{cm}^3$. But

$$\Delta E_g = 35 \text{ meV at } N_E = 10^{18}/\text{cm}^3$$
$$75 \text{ meV at } N_E = 10^{19}/\text{cm}^3$$

EXAMPLE 9.23

Assume $D_B = 3D_E$, $W_E = 3W_B$, $N_B = 10^{18}/\text{cm}^3$ and $n_{iB}^2 = n_2^i$. What is β for
(a) $N_E = 10^{19}/\text{cm}^3$; (b) $N_E = 10^{20}/\text{cm}^3$; (c) $N_E = 10^{19}/\text{cm}^3$ and a SiGe base with $\Delta E_{gB} = 60$ meV.

Solution: (a) At $N_E = 10^{19}/\text{cm}^3$, $\Delta E_{gE} = 35$ meV

$$\beta = \frac{D_B W_E}{D_E W_B} \cdot \frac{N_E n_i^2}{N_B n_{iE}^2}$$

$$n_{iE}^2 = n_i^2 \; \exp\left(\Delta E_g/kT\right) = 38 \; n_i^2$$

hence,
$$\beta = \frac{9 \times 10^{19} \times n_i^2}{10^{18} \times 3.8 \times n_i^2} = 23.6$$

(b) $N_E = 10^{20}/\text{cm}^3$

$$n_{iE}^2 = 320 \; n_i^2$$

and
$$\beta = 3$$

This shows that increasing the doping concentration in the emitter, does not increase the gain β, but it reduce. This is due to emitter bandgap narrowing.

(c) $n_{iB}^2 = n_i^2 \exp\left(\Delta E_{gB}/kT\right) = 10\ n_i^2$

Hence
$$\beta = 236$$

This suggests that the emitter bandgap narrowing effect is reduced by using SiGe base.

9.8 GUMMEL-POON MODEL

Several important second order effects like low base current region where space charge recombination and/or surface recombination can have an effect on the current gain, high level injection and base-width modulation, which are present in actual bipolar transistor, but are not included in the Ebers-Moll model. These effects are included in Gummel-Poon model. Gummel-Poon model is a charge control model. With the total charge of one type (either electrons or holes) in the active part of the BJT, we have

$$Q_b = q \int_{\text{emitter}}^{\text{collector}} A_E \cdot p \cdot dx \tag{9.44a}$$

where A_E is cross-sectional area of emitter region. At zero bias, the Q_b is given as Q_{b0}. Taking I_s as the collector saturation current, the general expression for the collector current becomes

$$I_c = \frac{Q_{b0}}{Q_b}\ I_s\left[e^{V_{BE}/V_t} - e^{V_{BC}/V_t}\right] \tag{9.44b}$$

where $V_t = kT/q$ is the thermal voltage.

Splitting up this current into forward and reverse current, we have

$$I_C = I_f - I_r \tag{9.44c}$$

where

$$I_f = \frac{Q_{b0}}{Q_b}\ I_s\left(e^{V_{BE}/V_t} - 1\right) \tag{9.44d}$$

and

$$I_r = \frac{Q_{b0}}{Q_b}\ I_s\left(e^{V_{BC}/V_t} - 1\right) \tag{9.44e}$$

The total base charge Q_b consists of

- The zero bias base charge Q_{b0}
- The depletion charges of emitter and collector junctions (Q_{Tc}, Q_{Te})
- The stored charges of the minority carrier (Q_{be}, Q_{bc})

i.e.,

$$Q_b = Q_{b0} + Q_{Te} + Q_{Tc} + Q_{be} + Q_{bc} \tag{9.45}$$

The charges Q_{be} and Q_{bc} are related to forward current and reverse current as:

$$Q_{be} = \tau_f I_f \tag{9.46a}$$

$$Q_{bc} = \tau_r I_r \tag{9.46b}$$

where τ_f and τ_r are the forward and reverse transit times.

Neglecting the junction charges, and substituting Eqs. (9.46a) and (9.46b) into Eq. (9.45), we get

$$Q_b = Q_{b0} + \tau_f I_f + \tau_r I_r \tag{9.47}$$

Equation (9.47) gives the stored charges of injected minority carriers.

But the depletion charges of the emitter and collector junctions cause the increase in collector current with the collector-emitter voltage in the saturation region and these charges depend on their respective junction voltages. This effect is called **early effect**. These charges are given as:

$$Q_{Te} = \int_0^{V_{BE}} C_{Te}\, dV \tag{9.48a}$$

and

$$Q_{Tc} = \int_0^{V_{BC}} C_{Tc}\, dV \tag{9.48b}$$

where C_{Te} and C_{Tc} are the respective capacitances. Equations (9.48a) and (9.48b) reduce as:

$$Q_{Te} = C_{Te}\, V_{BE} = Q_{b0}\, \frac{V_{BE}}{V_{\text{early}f}} \tag{9.49a}$$

and

$$Q_{Tc} = C_{Tc}^{\sim}\, V_{BC} = Q_{b0}\, \frac{V_{BC}}{V_{\text{early}r}} \tag{9.49b}$$

Hence

$$Q_b = Q_{b0} + Q_{TC} + Q_{Tc}$$

$$= Q_{b0} + Q_{b0}\, \frac{V_{BE}}{V_{\text{early}f}} + Q_{b0}\, \frac{V_{BC}}{V_{\text{early}r}}$$

$$= Q_{b0} + Q_{b0}\left[\frac{V_{BE}}{V_{\text{early}f}} + \frac{V_{BC}}{V_{\text{early}r}} \right]$$

$$= Q_{b0} \left[1 + \left\{ \frac{V_{BE}}{V_{\text{early}\,f}} + \frac{V_{BC}}{V_{\text{early}\,r}} \right\} \right]$$

$$Q_b = Q_{b0}[1 + q_1] \tag{9.50}$$

where the term $(1 + q_1)$ is called the **early factor**. Therefore, the whole base charge Q_b is:

$$Q_b = Q_{b0}\,(1 + q_1) + \tau_f \left(\frac{Q_{b0}}{Q_b} \right) i_f + \tau_r \left(\frac{Q_{b0}}{Q_b} \right) i_r \tag{9.51}$$

On solving Eq. (9.51), we have

$$\frac{Q_b}{Q_{b0}} = \left(\frac{1 + q_1}{2} \right) \left[1 + \left\{ 1 + \frac{4}{1 + q_1} \left(\frac{\tau_f}{Q_{b0}} i_f + \frac{\tau_r}{Q_{b0}} i_r \right) \right\} \right] \tag{9.52}$$

The term Q_{b0}/τ_f is replaced by a quantity I_k. This parameter is called **knee-voltage** in the forward mode and defined as the current at which high injection in the base starts. Similarly, $I_{kr} = Q_{b0}/\tau_r$ is called **knee-voltage** in reverse mode.

The base current is built by the ideal components:

$$I_{b_1 f} = \frac{i_f}{\beta_f} \tag{9.53a}$$

and
$$I_{b_1 r} = \frac{i_r}{\beta_r} \tag{9.53b}$$

Non-ideal base current components are given by Eqs. (9.44a) and (9.44b). These currents arise due to recombination in the depletion region of forward biased junction, i.e., leakage currents; assuming $\eta = 1$.

The model parameters for the current equations are:

$$I_s, \ \beta_f, \ \beta_r \quad V_{\text{early}\,f}, \ V_{\text{early}\,r}, \ I_{kr}, \ I_{kf}$$

The space charge layer capacitance are:

$$C_{Te} = \frac{C_{je}}{[1 - V_{BE}/V_{bie}]^{me}} \tag{9.54a}$$

$$C_{Tc} = X_{cje} \frac{C_{jc}}{[1 - V_{BE}/V_{bie}]^{me}} \tag{9.54b}$$

$$C_{Bc'} = (1 - X_{cje}) \frac{C_{jc}}{[1 - V_{BC}/V_{bic}]^{me}} \tag{9.54c}$$

and

$$C_{c's} = \frac{C_{js}}{[1 - V_{Cs}/V_{je}]^{mj}} \tag{9.54d}$$

where C_{Te} is emitter-base depletion capacitance, C_{je} is zero-bias capacitance, V_{bie}, V_{bic} are built-in potentials of the respective junctions, m_e is the emitter-base grading coefficient, C_{Tc} is capacitance between internal base and collector and C_{Bc} is the capacitance between the external and the collector. This capacitance is generally important at high frequencies.

The resistance (series) R_e and R_c are linear and constant assuming the resistance of the wire as zero, i.e., $R_{CW} = 0$. The base resistance has a constant part R_{be} for the inactive region outside the emitter and a variable part R_{bv} for the base region under the emitter as shown in Fig. 9.26. The resistance R_{bv} depends on the emitter-base crowding.

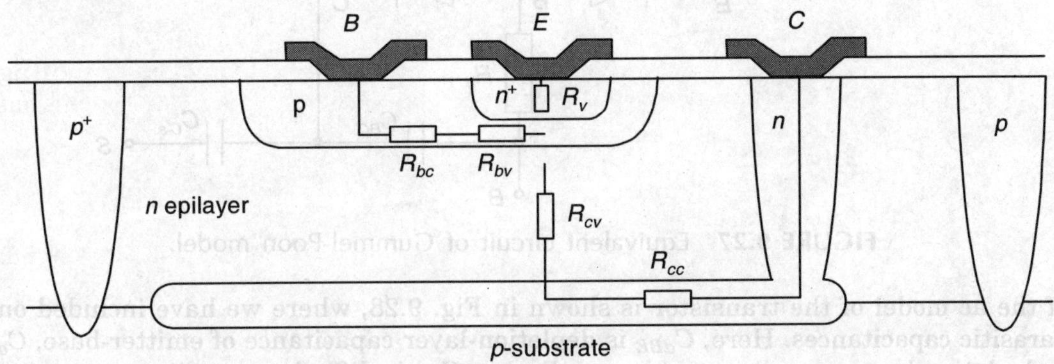

FIGURE 9.26 Emitter-base crowding effect in transistor.

The equivalent circuit diagram of the Gummel-Poon model is shown in Fig. 9.27, where I_{b2f} and I_{b2r} respresent the non-ideal base currents. The C, B, E terminals are the external connections to the transistor while C', B', E' points are the idealized internal collector base and emitter regions.

The complete parameters of the spice Gummel-Poon model are listed in Table 9.1.

9.9 BASIC AC MODEL

It is necessary to include the parasitic internal capacitors and resistances of the bipolar transistor (n-p-n or p-n-p) to model the ac behaviour of the transistor. By taking larger device areas and device layout techniques, one can minimize the effect of internal parasitic resistances effect. But the internal parasitic capacitors increase due to larger device areas. As a result, the basic behaviour of a transistor is determined by its parasitic capacitances rather than by its parasitic resistance. Let first ignore the effect of parasitic resistance, the common emitter-equivalent-circuit representation

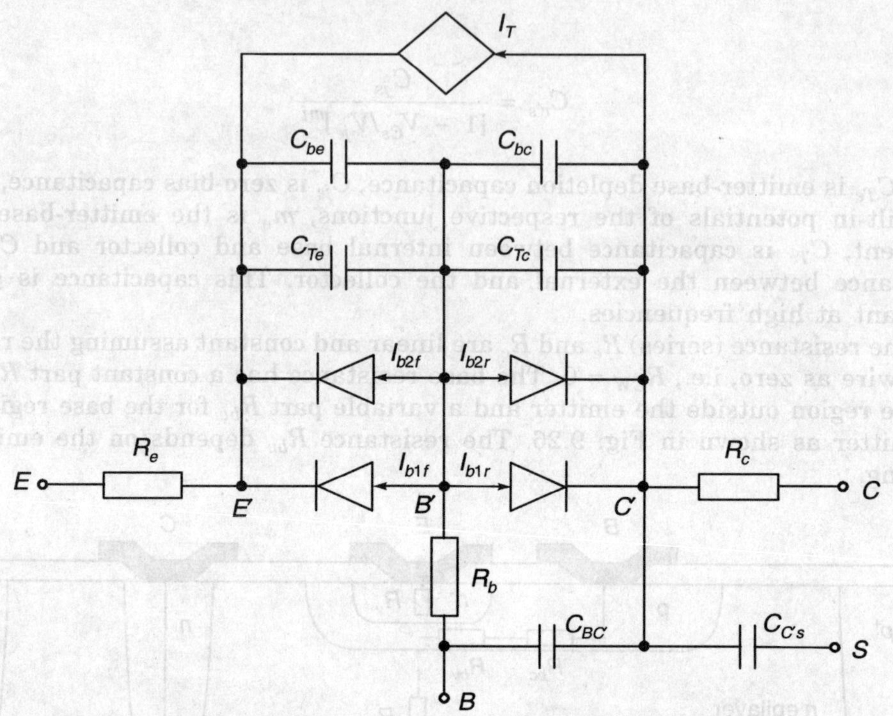

FIGURE 9.27 Equivalent circuit of Gummel-Poon model.

of the ac model of the transistor is shown in Fig. 9.28, where we have included only parasitic capacitances. Here, C_{dBE} is depletion-layer capacitance of emitter-base, C_{dBc} is depletion layer capacitance of base-collector, C_{DE} is diffusion capacitance associated with forward-biasing the emitter-base junction and C_{dcs} is the depletion layer capacitance of the collector-substrate diode.

This model given in Fig. 9.28, represents the ac model of an *n-p-n* transistor biased in active forward mode. The equivalent ac model, including parasitic resistances, is shown in Fig. 9.29.

Here, r_{bi} is the intrinsic base resistance and r_{bx} is the extrinsic part of the base resistance.

The depletion layer capacitance of the base-collector junction has two components: intrinsic part C_{dBci}, and extrinsic part C_{dBCx}.

Figure 9.29 is equivalent-circuit representation of ac Moll model of an *n-p-n* transistor including both parasitic resistances as well as capacitances in an active mode of operation.

9.9.1 Small-Signal ac Model

Equivalent circuit

Small signal ac means that the ac voltages and currents are small as compared to the dc values, and the ac output varies linearly with the ac input. The ac signal should

Table 9.1 Various Spice Model Parameters of Transistor

	Name	Symbol	Parameter	Units	Default value
1	IS	I_{sO}	transport saturation current	A	1.00
2	BF	β_{fO}	ideal maximum forward beta	–	100
3	NF	n_f	forward current emission coefficient	–	1
4	VAF	$V_{early\,f}$	forward early voltage	V	infinity
5	IKF	I_{kf}	corner for forward beta high current roll-off	A	infinity
6	ISE	I_{sEo}	B-E leakage saturation current	A	0
7	NE	n_E	B-E leakage emission coefficient	–	1.5
8	BR	β_{ro}	ideal maximum reverse beta	–	1
9	NR	n_r	reverse current emission coefficient	–	1
10	VAR	$V_{early\,r}$	reverse early voltage	V	infinity
11	IKR	I_{kr}	corner for reverse beta high current roll-off	A	infinity
12	ISC	I_{sco}	B-C leakage saturation current	A	0
13	NC	n_c	B-C leakage emission coefficient	–	2
14	RB	R_B	zero-bias base resistance	Ω	0
15	IRB	I_{RB}	current where base resistance falls halfway to its minimum	A	infinity
16	RBM	R_{BM}	minimum base resistance at high currents	Ω	RB
17	RE	R_e	emitter resistance	Ω	0
18	RC	R_c	collector resistane	Ω	0
19	CJE	C_{je}	B-E zero-bias depletion capacitance	F	0
20	VJE	V_{je}	B-E built-in potential	V	0.75
21	MJE	m_{je}	B-E junction exponential factor	–	0.35
22	TF	τ_f	ideal forward transit time	s	0
23	XTF	$X_{\tau f}$	coefficient for bias-dependence of TF	–	0
24	VTF	$V_{\tau f}$	voltage describing VBC dependence of TF	V	infinity
25	ITF	$I_{\tau f}$	high-current parameter for effect on TF	A	0
26	PTF	$\varphi_{\tau f}$	excess phase at freq = 1.0/(TF*2PI) Hz	deg	0
27	CJC	C_{jc}	B-C zero-bias depletion capacitance	F	0
28	VJC	V_{jc}	B-C built-in potential	V	0.75
29	MJC	m_{jc}	B-C junction exponential factor	–	0.35
30	XCJC	X_{cjc}	fraction of B-C depletion capacitance conducted to internal base node	–	1
31	TR	τ_r	ideal reverse transit time	s	0

(Contd.)

Table 9.1 Various Spice Model Parameters of Transistor (Contd.)

	Name	Symbol	Parameter	Units	Default value
32	CJS	C_{js}	zero-bias collector-substrate capacitance	F	0
33	VJS	V_{jc}	substrate junction built-in potential	V	0.75
34	MJS	m_{js}	substrate junction exponential factor	–	0
35	XTB	X_{tb}	forward and reverse beta temperature exponent	–	0
36	EG	W_g	energy gap for temperature effect on IS	eV	1.12
37	XTI		temperature exponent for effect on IS	–	3
38	KF		flicker-noise coefficient	–	0
39	AF		flicker-noise exponent	–	1
40	FC		coefficient for forward-bias depletion capacitance formula	–	0.5
41	TNOM	T_{nom} in K	parameter measurement temperature	°C	27

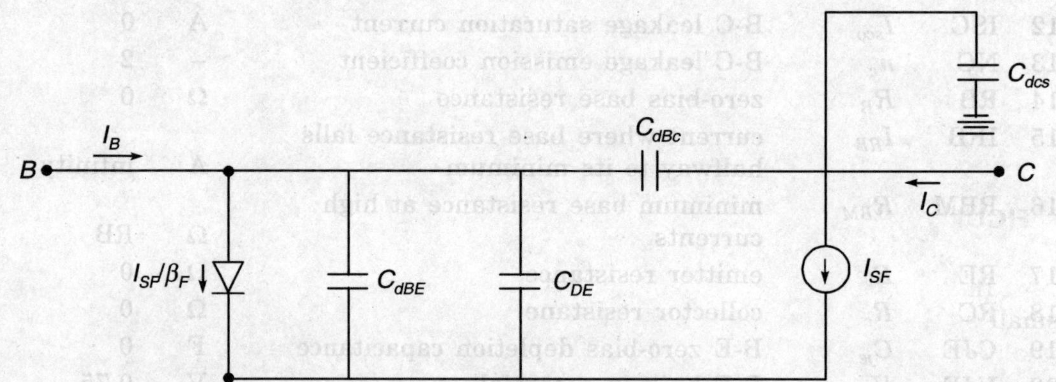

FIGURE 9.28 ac model of *n-p-n* transistor biased in active region.

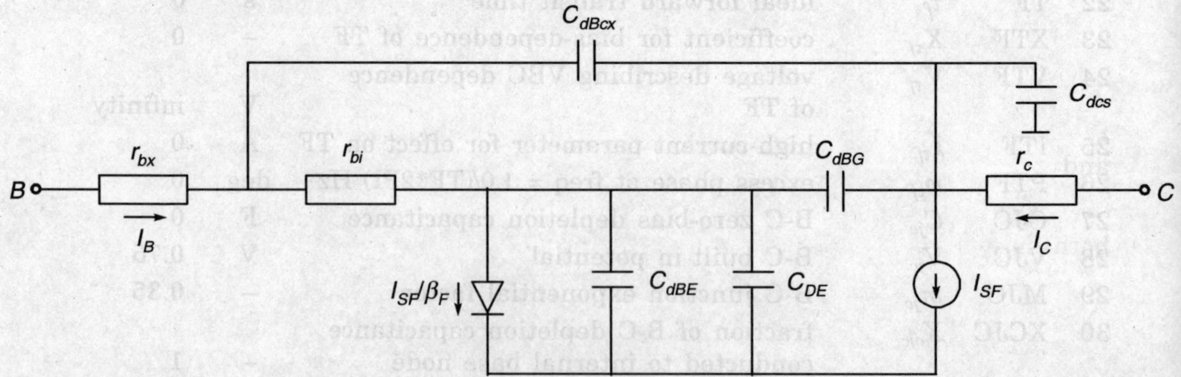

FIGURE 9.29 The equivalent ac model of *n-p-n* transistor in active mode including parasitic resistances.

be small enough so that the region of dc characteristics can be approximated by a linear representation. dc sources are required for proper biasing of the transistor so that it operates in the required region.

A BJT can be used as an amplifier as well as switch. A common-emitter circuit of an *n-p-n* transistor is biased in such a way that it operates in forward-active mode and small ac signal is superimposed on it as shown in Fig. 9.30.

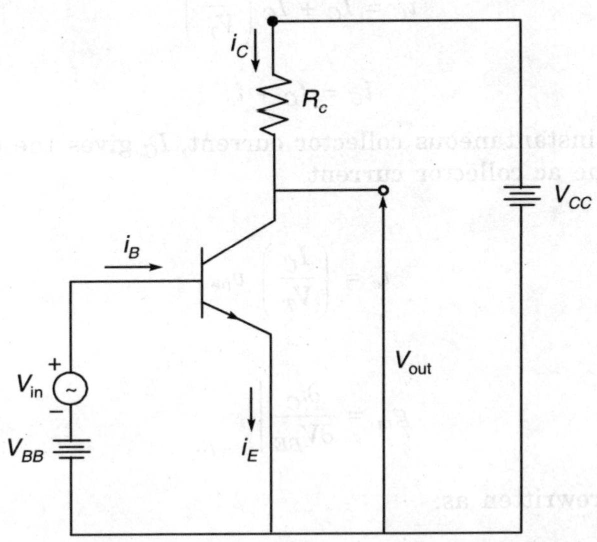

FIGURE 9.30   *n-p-n* transistor biasing in active mode in presence of small ac signal.

In Fig. 9.30, V_{BB} and V_{CC} represent the two dc power supplies, V_{in} represents the small sinusoidal signal waves. Considering only dc bias,

$$I_c = I_s \, e^{qV_{BE}/kt} \tag{9.55a}$$

$$I_E = \frac{I_c}{\alpha} \tag{9.55b}$$

$$I_B = \frac{I_c}{\beta} \tag{9.55c}$$

and

$$V_{CE} = V_C - V_E = V_{CC} - I_C R_C \tag{9.55d}$$

The dc bias adjust the operating point Q of the transistor. Now, in presence of both dc and small ac signal,

$$i_c = I_s \, e^{q(V_{BE} + v_{be})/kt}$$

$$i_C = I_C \, e^{qv_{be}/kt} \tag{9.56a}$$

where v_{be} is small-applied sinusoidal signal and kT/q is the thermal voltage represented by V_T.

If $v_{be} \ll V_T$ then Eq. (9.56a) reduces to

$$i_c = I_C \left(1 + \frac{v_{be}}{V_T} \right) \tag{9.56b}$$

or

or

$$i_C = I_C + I_C \left(\frac{v_{be}}{V_T} \right)$$

$$i_C = I_C + i_c \tag{9.56c}$$

where i_C gives the instantaneous collector current, I_C gives the dc collector current and i_c represents the ac collector current.

Since

$$i_c = \left(\frac{I_C}{V_T} \right) v_{be} \tag{9.56d}$$

and defining

$$g_m = \left. \frac{\partial i_C}{\partial V_{BE}} \right|_{i_c = I_C} \tag{9.56e}$$

Eq. (9.56d) can be rewritten as:

$$i_c = g_m v_{be} \tag{9.57a}$$

where g_m is the transconductance of the transistor. Similarly, the total base current is given as:

$$i_B = I_B + i_b \tag{9.57b}$$

where i_B is instantaneous base current and $i_b = (1/\beta) g_m v_{be}$ is ac base current.

The small-signal resistance looking into the base is denoted by r_π and is defined as:

$$r_\pi = \frac{v_{be}}{i_b} = \frac{v_{be} \times \beta}{g_m v_{be}} = \frac{\beta}{g_m} \tag{9.57c}$$

The hybrid-π model of the bipolar junction transistor is shown in Fig. 9.31.

Sometimes r_π is also called as **input resistance**. We can represent the small-signal model for the transistor as a voltage controlled current source or a current controlled current source.

Looking into the emitter, the effective small signal resistance between base and emitter r_e is:

$$r_e = \frac{v_{be}}{i_e} = \frac{V_T}{I_E} = \frac{\alpha}{g_m} \simeq \frac{1}{g_m} \tag{9.57d}$$

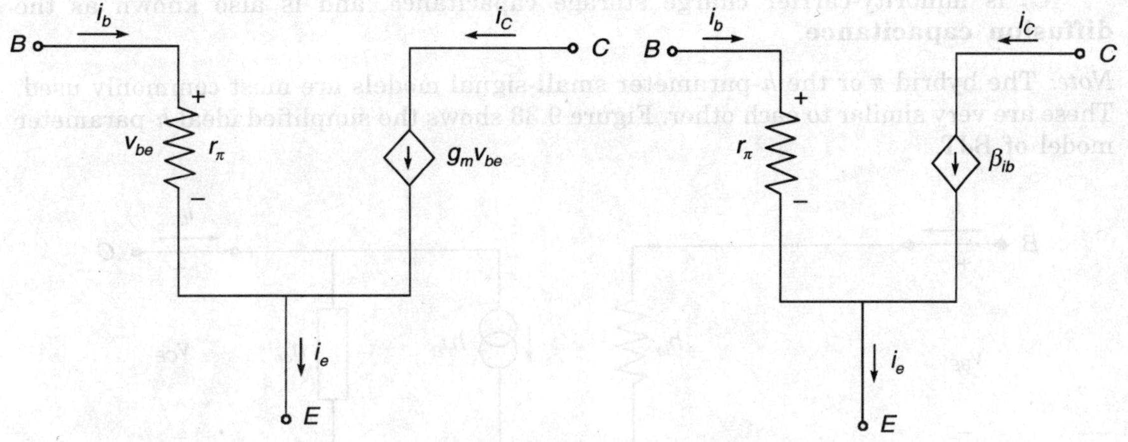

FIGURE 9.31 Hybrid-π model.

Therefore,

$$r_\pi = (\beta + 1)\, r_e \tag{9.57e}$$

The output resistance r_0 relates i_c to v_{ce} as:

$$r_0 = \left(\frac{\partial I_c}{\partial V_{CE}}\right)^{-1} = \frac{V_A}{I_C} \tag{9.58a}$$

and base-collector depletion capacitance

$$c_\mu = C_{dBC} \tag{9.58b}$$

and capacitance of emitter-base diode is

$$C_\pi = C_{dBE} + C_{DE} \tag{9.58c}$$

The complete small-signal hybrid π-model of a BJT is shown in Fig. 9.32 where all the parasitic resistances are neglected.

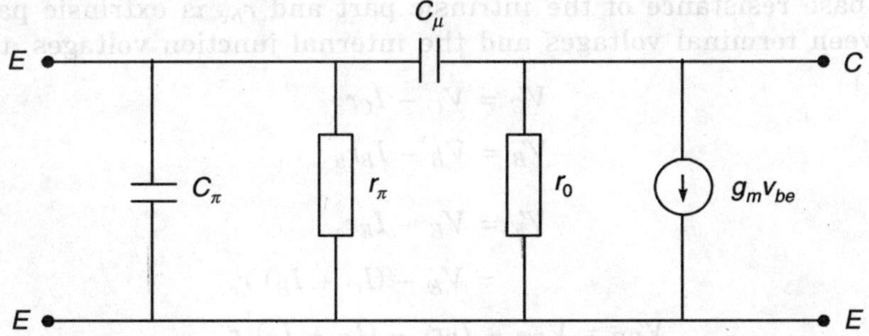

FIGURE 9.32 Complete small signal hybrid-π model of *n-p-n* transistor.

C_π is minority-carrier charge storage capacitance, and is also known as the **diffusion capacitance**.

Note: The hybrid π or the h-parameter small-signal models are most commonly used. These are very similar to each other. Figure 9.33 shows the simplified ideal h-parameter model of BJT.

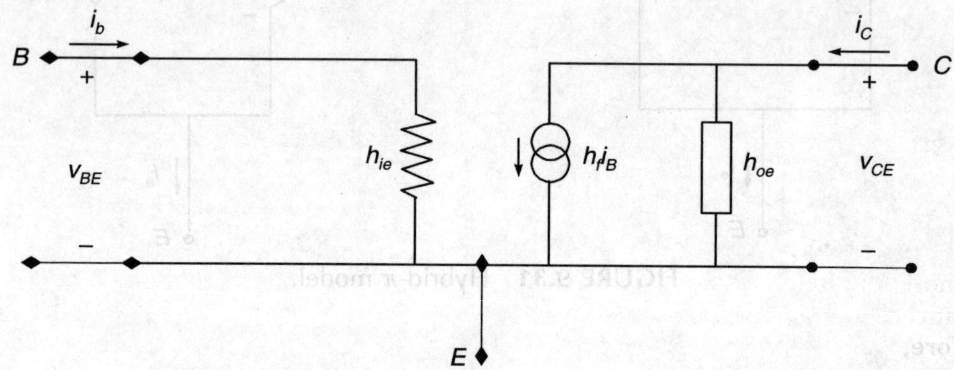

FIGURE 9.33 Simplified ideal h-parameter model of BJT.

Here,

$h_{ie} \Rightarrow r_{in} \Rightarrow$ The equivalent resistance across the forward biased junction and also the base and emitter bulk resistances. It is very low.

$h_{oe} \Rightarrow r_0 \Rightarrow$ The equivalent resistance across the reverse biased junction and also the collector, base and emitter bulk resistances. It is very high.

$h_{fe} \Rightarrow \beta \Rightarrow$ It is heavily dependent on the emitter efficiency and base transport factor. Ideally it should be high.

Small signal model including parasitic resistances

In the presence of parasitic resistance, the device terminal voltages V_C, V_E and V_B are no longer equal to the internal junction voltages V'_C, V'_E and V'_B due to potential drop. IR in the parsitic resistance r_e, r_c and r_b. For simplicity, r_{bi} and r_{bx} lumped into r_b where r_{bi} is base resistance of the intrinsic part and r_{bx} is extrinsic part. And the relation between terminal voltages and the internal junction voltages are given as:

$$V'_C = V_C - I_C r_c \tag{9.59a}$$

$$V'_B = V_B - I_B r_b \tag{9.59b}$$

and

$$V'_E = V_E - I_E r_e$$

$$= V_E - (I_C + I_B)\, r_e \tag{9.59c}$$

$$V'_{BE} = V_{BE} - I_B r_b - (I_C + I_B)\, r_e$$

$$= V_{BE} - I_C r_e - I_B(r_b + r_e) \tag{9.59d}$$

$$V'_{CE} = V_{CE} - I_C r_e - (I_C + I_B)\, r_e$$
$$\simeq V_{CE} - I_C(r_c + r_e) \tag{9.59e}$$

Assuming I_B is very small.

Using Eqs. (9.59d) and (9.59e), the transconductance g_m in the presence of parasitic resistances is:

$$g'_m \cong \frac{g_m}{1 + g_m r_e} \tag{9.60a}$$

where g_m is tranconductance in absence of parasitic resistance. Similarly,

$$r'_\pi = r_\pi(1 + g_m r_e) + (r_b + r_e) \tag{9.60b}$$

and
$$r'_0 = r_0 + (r_e + r_c) \tag{9.60c}$$

The small signal ac model in the presence of parasitic capacitances as well as parasitic resistances is shown in Fig. 9.34.

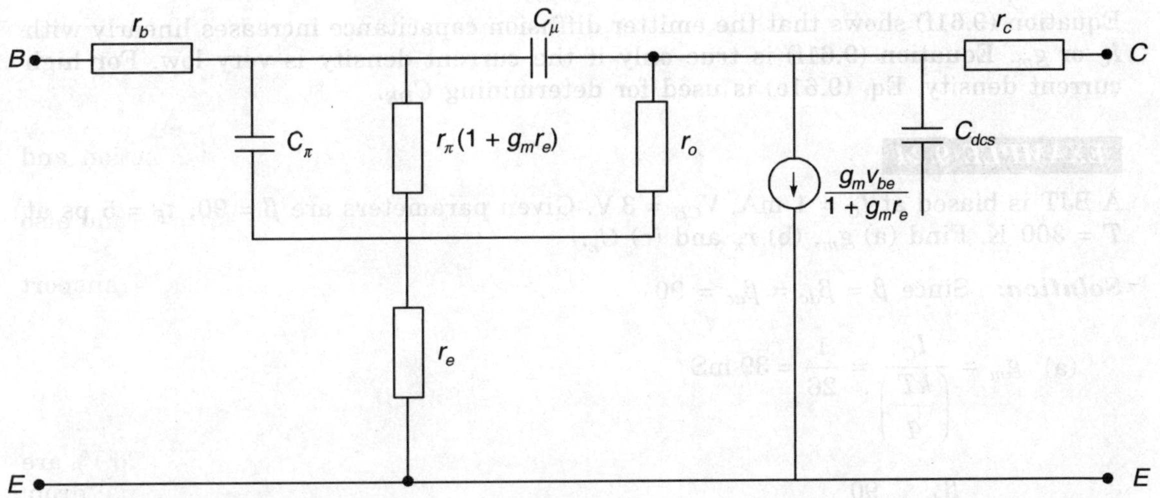

FIGURE 9.34 Small signal model including all the parasitic resistances and capacitances.

Note: The total minority-carrier charge can be written as:

$$Q_{DE} = |Q_E| + |Q_B| + |Q_{BE}| + |Q_{BC}| \tag{9.61a}$$

where Q_E, Q_B, Q_{BE} and Q_{BC} represent the minority-carrier charge in the emitter, base, emitter-base space charge region and base-collector space-charge region. From Eq. (9.61a), it is clear that Q_{DE} is sum of the absolute values of the individual minority-charge carrier components, not the sum of net charges.

Equation (9.61a) can be further written as:

$$Q_{DE} = (\tau_E + \tau_B + \tau_{BE} + \tau_{BC})\, I_C \tag{9.61b}$$

where

$$\tau_F = \tau_E + \tau_B + \tau_{BE} + \tau_{BC} \tag{9.61c}$$

called **forward transit time**. τ_E is emitter delay time, τ_B is base transit time, τ_{BE} is emitter-base depletion region transit time, and τ_{BC} is base-collector depletion-region transit time.

From Eq. (9.61c), it is clear that to reduce τ_F, the emitter as well as the depletion layers must be kept thin and hence Eq. (9.61b) reduces to

$$Q_{DE} = \tau_E I_c \tag{9.61d}$$

and the emitter diffusion capacitance is:

$$C_{DE} = \frac{\partial Q_{DE}}{\partial V'_{BE}} \tag{9.61e}$$

Assuming τ_F is constant, independent of I_C, we get

$$C_{DE} = \tau_F \, g_m \tag{9.61f}$$

Equation (9.61f) shows that the emitter diffusion capacitance increases linearly with I_C or g_m. Equation (9.61f) is true only if the current density is very low. For high current density, Eq. (9.61e) is used for determining C_{DE}.

EXAMPLE 9.24

A BJT is biased at $I_C = 1$ mA, $V_{CE} = 3$ V. Given parameters are $\beta = 90$, $\tau_F = 5$ ps at $T = 300$ K. Find (a) g_m, (b) r_π and (c) C_π.

Solution: Since $\beta = \beta_{dc} = \beta_{ac} = 90$

(a) $g_m = \dfrac{I_C}{\left(\dfrac{kT}{q}\right)} = \dfrac{1}{26} = 39$ mS

(b) $r_\pi = \dfrac{\beta_{dc}}{g_m} = \dfrac{90}{39} = 2.3$ kΩ

(c) $C_\pi = \tau_F g_m = 5 \times 10^{-12} \times 39 = 19$ pF

Note: 1. Sometimes other small signal models can be used more conveniently as shown in Figs. 9.35(a) and 9.35(b).

These two models representation are known as **T-model**.

2. Once the small-signal model parameters have been determined, the circuit can be analyzed with arbitrary source and load impedance. The parameters are recently determined through comprehensive measurement of the BJT ac and dc characteristics.

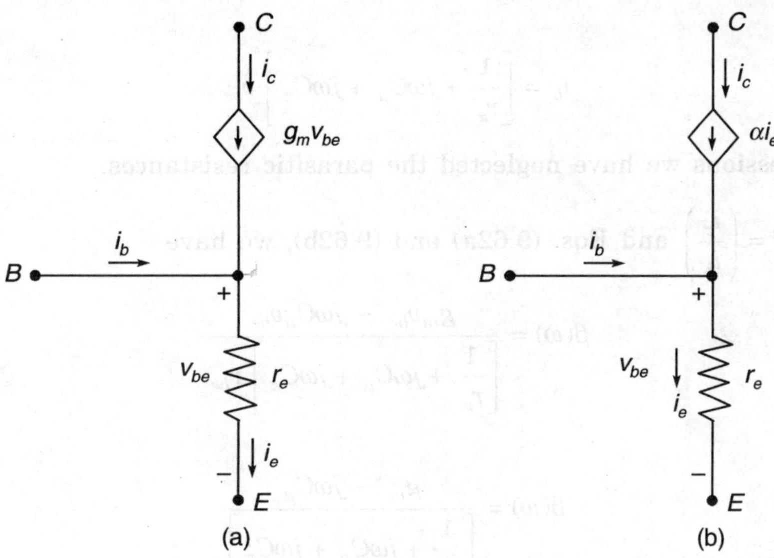

FIGURE 9.35   T-model of BJT.

9.9.2   Cutoff Frequency f_T

The two most important figures of merit that describe the high frequency behaviour of a BJT are the cutoff frequency f_T and the maximum frequency of oscillation f_{max}. The cutoff frequency f_T is defined as the frequency at which the common emitter short-circuit current gain is equal to unity. The maximum frequency is defined as the frequency at which the unilateral power gain of the transistor approaches unity.

Short the output terminal of the small signal circuit of Fig. 9.32 as shown in Fig. 9.36.

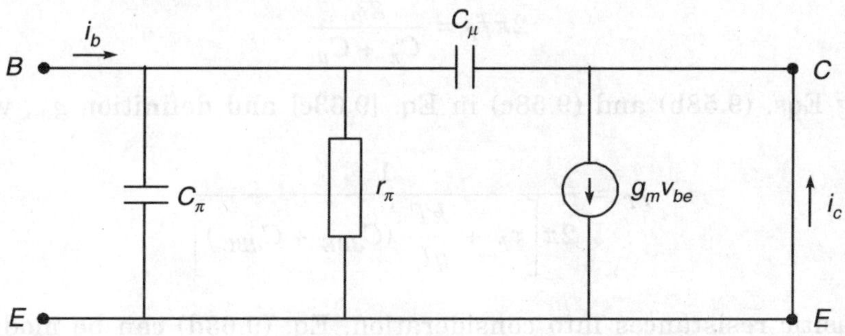

FIGURE 9.36   Small signal circuit with output shorted.

From Fig. 9.36,

$$i_C = g_m v_{be} - j\omega C_\mu v_{be} \tag{9.62a}$$

and

$$i_b = \left[\frac{1}{r_\pi} + j\omega C_\mu + j\omega C_\pi\right] v_{be} \tag{9.62b}$$

in both expressions we have neglected the parasitic resistances.

Using $\beta = \left(\dfrac{i_c}{i_b}\right)$ and Eqs. (9.62a) and (9.62b), we have

$$\beta(\omega) = \frac{g_m v_{be} - j\omega C_\mu v_{be}}{\left[\dfrac{1}{r_\pi} + j\omega C_\mu + j\omega C_\pi\right] v_{be}}$$

or

$$\beta(\omega) = \frac{g_m - j\omega C_\mu}{\left[\dfrac{1}{r_\pi} + j\omega C_\mu + j\omega C_\pi\right]} \tag{9.63a}$$

and Eq. (9.63a) reduces to

$$\beta(\omega) \cong \frac{g_m r_\pi}{1 + j\omega r_\pi (C_\mu + C_\pi)}$$

At high frequency the imaginary part of denominator dominates and hence

$$\beta(\omega) \approx \frac{g_m}{\omega(C_\mu + C_\pi)} \tag{9.63b}$$

Defining $\beta(\omega) \to 1$, for $f = f_T$, Eq. 9.63(b) reduces to

$$2\pi f_T = \frac{g_m}{C_\pi + C_\mu} \tag{9.63c}$$

Substituting Eqs. (9.58b) and (9.58c) in Eq. [9.63c] and definition g_m, we have

$$f_T = \frac{1}{2\pi\left[\tau_F + \dfrac{kT}{qI_c}(C_{dBE} + C_{dBC})\right]} \tag{9.63d}$$

Taking parasitic resistances into consideration, Eq. (9.63d) can be modified as:

$$f_T = \frac{1}{2\pi\left[\tau_F + (C_{dBE} + C_{dBC})\dfrac{kT}{qJe} + C_{dBC}(r_e + r_c)\right]} \tag{9.63e}$$

f_T is commonly used as a metric for the speed of transistor.

Alternatively, the cutoff frequency can be written in terms of the base transit time τ_B as:

$$f_T = \frac{1}{2\pi\tau_B} \tag{9.63f}$$

Equation (9.63f) provides only an estimate of the cutoff frequency since it is based on the approximate hybrid-pi model.

The maximum frequency of operation of a BJT is:

$$f_{\max} = \sqrt{\frac{f_T}{8\pi r_b C_\mu}} \tag{9.63g}$$

Equation (9.63g) shows that $f_{\max}$ increases with inceasing cutoff frequency and reduction in base parasitic resistance. Since, base resistance depends on the base doping concentration, therefore, to maximize $f_{\max}$, it is required to use a very heavily doped base region, but this will affect the overall performance of the BJT.

The cutoff frequency f_T can be maximized by

- Increasing collector current I_C
- Minimizing C_{dBE}, C_{dBC}
- Minimizing r_e and r_c
- Minimizing τ_F

Note: In bipolar junction transistor, the parasitics (capacitances) are smaller and less than the MOSFETs.

9.9.3 BJT Structure for High Speed

To improve the frequency response, the transit time of minority carriers across the base must be short and that can be achieved by making narrow base. Since the electron diffusion constant in silicon is about three times larger than that of holes, therefore, all the high frequency silicon transistors are of n-p-n type, i.e, in the base minority carriers are electrons. Another way to reduce base transit time is to use a graded base with a built-in-field which will aid the motion of the carriers towards the collector and reduce the base transit time.

Other requirements for high speed BJT structure are:

- n^+ poly-Si emitter
- Self-aligned p^+-poly-Si base contacts
- Lightly doped collector
- Heavily doped epitaxial subcollector
- Shallow trenches and deep trenches filled with SiO_2 for electrical isolation

The high-speed BJT circuit is shown in Fig. 9.37 which is most widely used in modern bipolar transistor.

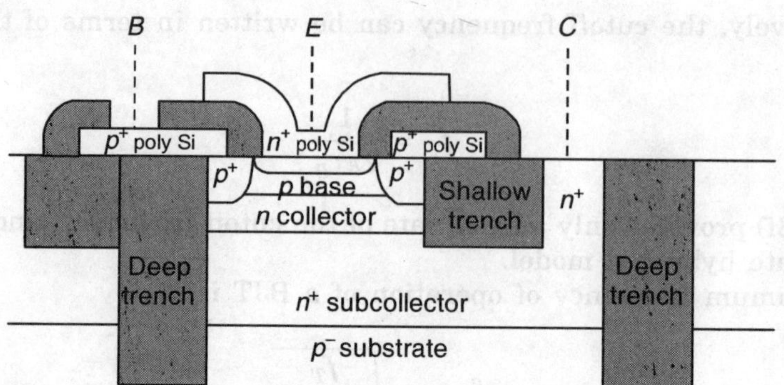

FIGURE 9.37 BJT structure for high speed.

The salient features of the transistor, shown in Fig. 9.37, are:

(a) Polysilicon emitter allows extremely small emitter-junction depth without the large base current associated with metal-contacted shallow emitter.

(b) Polysilicon base contact is self-aligned to the emitter. This allows the extrinsic base to be formed independently of the intrinsic base.

(c) The isolation region must be deep enough to isolate the subcollector of adjacent transistors. This technique reduces significantly the area taken up by isolation.

(d) A pedestal collector, i.e., collector region directly underneath the emitter is more heavily doped than its surrounding regions. This minimizes the base-widening effect and reduces the total base-collector junction capacitance.

Note: For high-performance logic applications, the most commonly used bipolar circuit is the emitter-coupled logic or ECL circuit.

9.9.4 Switching Behaviour of a BJT

For digital application, a transistor is designed to function as a switch, i.e., either operates in saturation region or cutoff region. In the saturation mode both junctions (emitter-base and collector-base) should be forward biased. In other words, both the *p-n* junctions are in forward biased mode and, therefore, a large current flows. Large current in the circuit means low resistance and hence low voltage drop. The saturation mode corresponds to the ON state of the transistor in digital application. Since, the voltage of the device operated in the saturation mode is low, therefore, the saturation mode corresponds to logic 0 in digital binary logic. Conversely, in the cutoff mode, both the *p-n* junctions are reverse biased and hence a low current flows in the circuit. The cutoff mode corresponds to OFF state of the transistor. Due to small current in the circuit, a large voltge drop occurs. Since, the voltage drop is high in the cutoff mode, therefore, cutoff represents the logic high or logic 1. When the device switches from 1 to 0 means it switches from *cutoff* to staturation and conversely if it switches from 0 to 1 means transistor moves from *saturation region* to *cutoff region*. The speed of switching behaviour of transistor is decided by the fact that how fast it crosses the active region.

In brief, we can say that the transistor behaves as short circuit in the saturation mode, i.e, switch is ON and open-circuited in cutoff mode, i.e, switch is OFF.

The switching time is the time required for a transistor to switch from the OFF state to ON state or vice versa. The turn ON time depends on how fast one can add the minority carriers (i.e., electrons in *n-p-n* transistor) to the base of the transistor. The turnoff time depends on how fast one can remove the minority carriers by recombination. The most important parameter for switching action of a transistor is its minority carrier lifetime τ_n. For faster switching action it is required to reduce the value of τ_n (for *n-p-n* transistor) and that can be achieved by introducing efficient generation-recombination centres near the midgap.

EXAMPLE 9.25

Find the base transit time for electrons in the *n-p-n* prototype transistor of base doping of $10^{17}/cm^3$ and base width of $W_B = 0.1$ μm. Find the cutoff frequency also. Take $D_n = 20$ cm^2.

Solution: Using

$$\tau_B = \frac{W_B^2}{2D_n}$$

$$= \frac{(10-5)^2}{2 \times 20} = 2.5 \text{ ps}$$

This time is more than 5 orders of magnitude smaller than the electron lifetime in the base using Eq. (9.63e).

$$f_T = \frac{1}{2\pi\tau_B} = 63.6 \text{ GHz}$$

Note: At low collector currents (i.e., low collector current densities) the base widening is negligible and hence total transit time τ_F is contant, independent of I_C. But once base widening becomes appreciable, τ_F increases with I_C. Therefore, the transit time delay component of ECL is independent of I_C for low current density and increases with current once the current densities exceed the base widening threshold.

EXAMPLE 9.26

One has to reduce the emitter and collector areas to reduce junction capacitance. This results in higher speed of the device. Explain for that what price one should pay?

Solution: For higher speed of the device one should keep the size of the device as small as possible and that can be achieved by narrowing the base and reducing the emitter, collector area. Unfortunately, the requirement of small size will affect the power rating of the device because there is generally a trade-off between frequency and power.

EXAMPLE 9.27

What is the application of interdigitation method?

Solution: This method provides a means of increasing the emitter edge length while keeping the overall emitter area to minimum. Therefore, some form of interdigitation is generally used in designing a transistor for high frequency and reasonable power requirements.

EXAMPLE 9.28

What are the other methods to achieve a high speed device?

Solution: Since, the resistance associated with collector, base and emitter affects the various R_C charging time, therefore, it is required to keep these parasitic resistances as small as possible. Therefore, the metallization patterns contacting the emitter and base regions must not present significant series resistance. Also, the semiconductor region must be designed to reduce resistance. For example, the parasitic base series resistance r_b can be reduced by performing p^+ diffusion between the contact area on the surface and the active part of the base region. The reduction of base resistance by heavy doping requires the use of heterojunction to maintain r at an acceptable value.

The heavily doped substrate provides a low-resistance contact to the collector region, while maintaining low doping in the epitaxial collector material to ensure a high breakdown voltage of the collector junction. It is required to keep the collector depletion region small so that the carrier can drift through the collector junction very fast, i.e, to reduce the transit time. Narrow collector depletion region can be achieved by making collector lightly doped and narrow. By choice of proper packaging one can reduce the parasitic resistance, inductance or capacitance at high frequencies.

9.10 DESIGN CONSIDERATION OF BJT

As we have already seen that the emitter parameters only affect the base current. It has no effect on the collector current. Theoretically, we can say that a designer can vary the emitter design to vary the base current. But practically it is rarely done due to two reasons:

1. In digital-current applications, if the current gain is not very low or base current is usually high then the performance of the bipolar transistor is not affected by its base current. For many analog-circuit applications, once the current gain is adequate then the reproducibility of the base current is more important than its magnitude. Therefore, it is not required to tune the base current by tuning the emitter design.

2. The emitter is formed at the end of the device fabrication process. Any change to the emitter design to tune the base current could affect the doping profile of the other regions and hence could affect the other device parameters.

As a result, once the bipolar transistor is ready for manufacturing, its emitter fabrication process is usually fixed and the designer can only alter the device and circuit characteristics by changing the base and collector designs. This change can be accomplished independently of the emitter process and hence no affect on the base current.

During the design of an emitter region, designer should consider the requirements of low and reproducible base current and minimum emitter series resistance. Commonly used technique for designing emitter is either a diffused emitter or a polysilicon emitter. But diffused emitter is not suitable for base widths less than 100 nm.

Practically all modern high performance BJTs with base widths of 100 nm or smaller are fabricated by using a polysilicon emitter. In this case, emitter is formed by a doping of polysilicon layer heavily and then activating the doped polysilicon layer to obtain reproducible base current and low series emitter resistance.

Since it is experimentally verified that as long as transistor remains in the active region its performance is determined by the collector current which is affected by the base-region parameters. The design of the base of a bipolar transistor can be very complex due to the tradeoffs between the ac and dc characteristics. In this book we will not consider the design consideration of the base region.

Since, one of the most important trends in VLSI technology is the continued miniaturization of devices while simultaneously increasing their operating current densities. Now, in modern technology, a transistor with emitter areas of much less than 1 μm^2 can be easily fabricated while device currents of 0.5 mA and larger are desired on many bipolar circuits. This means that in many modern bipolar transistors the collector current densities can easily exceed the value of 0.5 mA/μm^2. At these high current densities, base widening can easily occur, therefore, a designer should pay special attention to design the collector and subcollector regions such that base widening effects can be minimized.

Note: The base transit time for a bipolar transistor is:

$$\tau_B = \frac{|Q_B|}{|J_C|}$$

where Q_B is the minority carrier charge per unit area stored in the base region and J_C is the collector current density. For *n-p-n* transistor,

$$Q_B = -q \int_0^{W_B} n_p(x)\, dx$$

At low currents,

$$\tau_B \simeq \frac{W_B^2}{2 D_{nB}}$$

Assuming uniform-bandgap approximation.

EXAMPLE 9.29

How can base transit time be reduced?

Solution: By reducing the intrinsic-base width one can reduce the base transit time. But as the base width is reduced, the base doping concentration must be increased to avoid the emitter-collect or punch through or the early voltage becoming unacceptably small.

Note: To maintain negligible base widening, the collector current density J_C should be small as compared to the maximum current density, i.e,

$$|J_C| << J_{max} = qv_{sat} N_C$$

where N_C is collector doping concentration and v_{sat} is saturated velocity for electrons in silicon. In many literature it is mentioned that the base widening becomes appreciable in modern n-p-n transistor when $J_C > 0.3 J_{max}$.

For typical value of $N_C = 10^{16}/cm^3$, $J_{max} = 0.16$ mA/μm^2 and the allowed J_C is of order of 0.05 mA/μm^2, which is much smaller. To increase J_C without increasing base widening effect, it is required to increase N_C proportionately. But due to increase in N_C, the base collector junction capacitance increases which adversely affect the base-collector junction avalanche and early voltage V_A decreases.

9.10.1 Base-Collector Junction Avalanche Effect

Base-collector junction avalanche occurs when the electric field in the junction space-charge region becomes too large. The junction avalanche can cause the base and collector currents to increase out of control and hence can affect the functionality of the circuits. The avalanche can also cause the device breakdown. In modern bipolar devices, the probability of base-collector junction avalanche effect is more due to sufficient high voltages. Therefore, the designer should take extra care to minimize this effect.

One way to minimize avalanche effect is to reduce N_C. But reduced N_C will reduce the allowed collector current density. Other method is the base and/or the collector doping profiles, at or near the base-collector junction, can be designed to reduce the electric field on the junction.

The Gaussian base doping profile is one of the best choice to reduce the field at the collector-base junction. To reduce the electric field in the base-collector junction, the collector doping profile should be retrograded, i.e., graded with its concentration increasing with distance into the silicon. Retrograding of the collector doping profile can be achieved readily by high-energy ion implantation. By all these methods one can also reduce the base-collector-junction depletion-layer capacitance which will improve the device and circuit performance. But reducing base-collector junction avalanche can reduce the current density capability and maximum speed of the bipolar transistor, i.e., there is a tradeoff between base-junction avalanche effect and device and circuit speed.

Note: Base widening readily occurs in modern bipolar devices, when a bipolar transistor is operated with significant base widening. The diffusion capacitance of emitter limits the speed and cutoff frequency of the transistor. Therefore, to minimize the emitter diffusion capacitance, the total excess minority carriers stored in the collector should be minimized. This can be achieved by retrograding profile of collector and minimizing the total collector volume available for minority-carrier storage, i.e., the thickness of the collector layer should be minimized. But the thin collector can lead to an increase in the base-collector junction depletion-layer capacitance. At low current density,

Speed of thinner collector < Speed of thicker collector

But at higher current density,

Speed of thinner collector transistor > Speed of thicker collector transistor

Therefore, as the designer one should consider all these tradeoffs during designing of collector region of a bipolar transistor.

9.10.2 SiGe-Base Bipolar Transistors

The SiGe-base transistor is particularly useful in analog circuit applications. The bandgap energy of Ge (~ 0.67 eV) is significantly lower than the silicon (~ 1.12 eV) bandgap. Incorporation of Ge into Si reduces the bandgap of the alloy. If Ge is introduced into the base region of Si bipolar transistor, the bandgap will reduce which influences the device characteristics. It is required larger concentration of Ge near the base-collector junction and minimum near the emitter-base junction. Figure 9.38 shows an ideal uniform boron doping concentration in the p-type base and a linear Ge concentration profile.

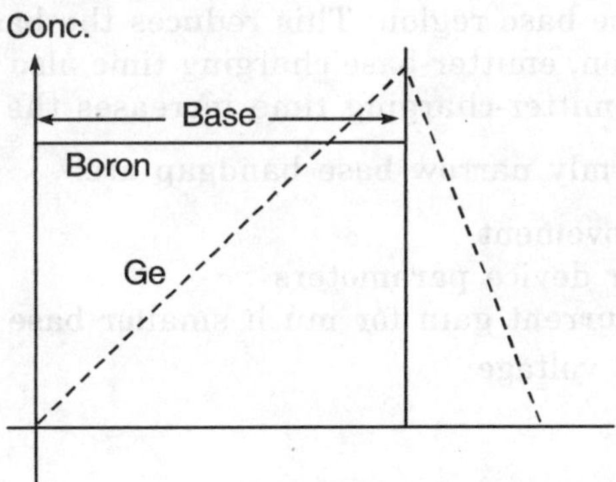

FIGURE 9.38 Concentration distribution in SiGe-base bipolar transistor.

The energy band diagram of a bipolar transistor with and without Ge in base region is shown in Fig. 9.39. The emitter-base junctions of two transistors are identical because the doping concentration of Ge is very low, but the energy bandgap of the SiGe base transistor near the collector-base junction is smaller than Si base transistor.

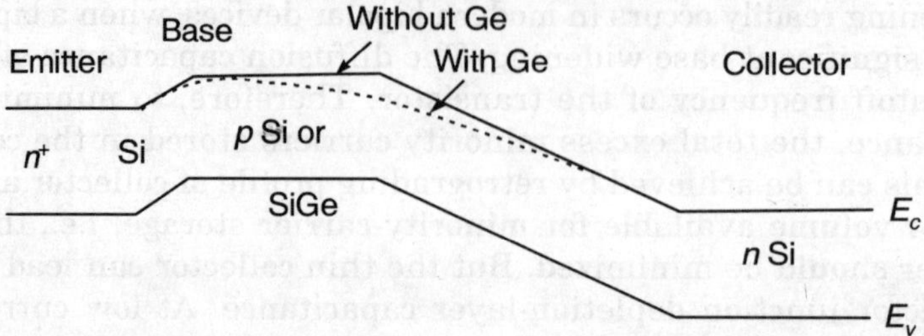

FIGURE 9.39   Energy-band diagram of SiGe-base bipolar transistor.

Since the base current is determined by the base-emitter junction parameters, therefore, for two transistors, base currents remain same, whereas collector currents for both transistors will be different.

The intrinsic carrier concentration in both transistors are related as:

$$\frac{n_i^2(\text{SiGe})}{n_i^2(\text{Si})} = \exp\left(\frac{\Delta E_g}{kT}\right) \tag{9.64}$$

where ΔE_g is change in the bandgap of SiGe material with respect to silicon material. Since, barrier for electron injection into p^+ SiGe is lowered by the ΔE_g in SiGe which increases collector current. But barrier for hole injection into n^+ Si has not changed, therefore, base current remains same. Increase in collector current in SiGe transistor implies that the current gain in the SiGe base transistor is larger than Si base transistor.

Also it was found that early voltage of SiGe base transistor is larger than Si base transistor. The decrease in bandgap energy from the base-emitter junction to collector-base junction induces an electric field in the base that helps electron to accelerate across the p-type base region. This reduces the base-transit time. Due to Ge doping in the base region, emitter-base charging time also reduces. The reduction in base-transit time and emitter-charging time increases the cutoff frequency.

Note: The merits of uniformly narrow base bandgap are:

- Current gain improvement
- No change in other device parameters
- Tradeoff in large current gain for much smaller base resistance
  * Increase in early voltage
  * Increase in C_{dBE}

9.11   HETEROJUNCTION BIPOLAR TRANSISTOR

The first idea for using heterojunctions in BJTs was directed at increasing β by making the emitter defects smaller. This possibility was implied by Schockley and proposed explicitly by Herb Kroemer in 1958. He reproduced the idea in 1980 when it was technologically possible to explore this idea and that is when heterojunction bipolar transistor (HBTs) took off. Heterojunction bipolar transistors are those which are composed of at least two different semiconductor materials. As a result, the energy bandgap as well as other material properties can be different in the emitter, base and collector.

An HBT device consists of a heterostructure emitter-base junction typically produced by a wide bandgap emitter contact with a narrow bandgap base region. The heterojunction can be either abrupt, in which the materials composition changes suddenly or grad͏ͅ ͏herein a much more gradual change from one material to another occurs vi͏ͅ ͏uous composition grading. Figure 9.40(a) shows a schematic structure of a bas͏ͅ HBT. In this device, the n-type emitter is performed in the wide bandgap Alx͏ͅ s, whereas the p-type base is formed in the lower bandgap GaAs. The n-type collector and n-type subcollector are formed in GaAs with light doping and heavy doping respectively.

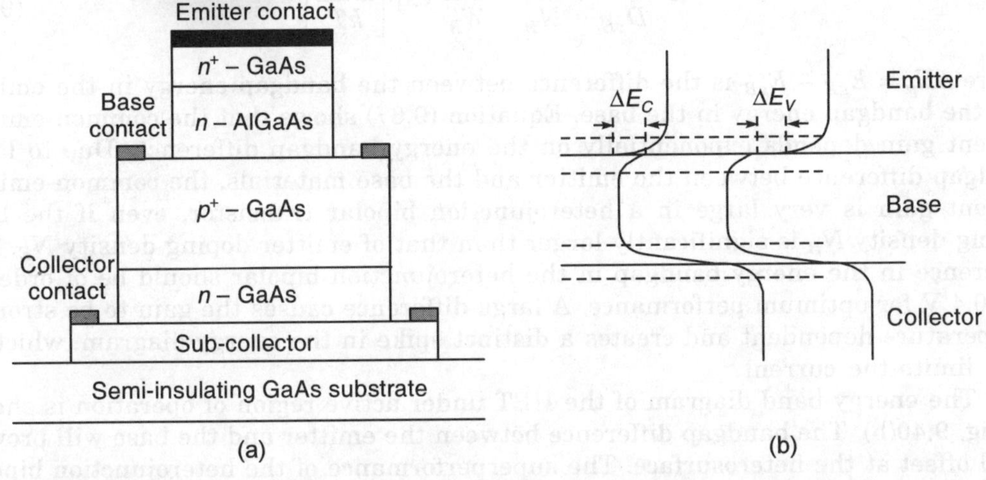

FIGURE 9.40 Structure and band diagram of heterojunction transistor.

Assuming base transport factor $\alpha_T \simeq 1$ then the common emitter current gain can be given as:

$$\beta = \frac{\gamma}{1 - \gamma} \tag{9.65a}$$

Substituting the value of γ, for n-p-n transistor,

$$\beta \simeq \frac{n_{p0}}{p_{E0}} \tag{9.65b}$$

where the minority carrier concentration in the emitter and base regions are given as:

$$p_{E0} = \frac{n_i^2(\text{emitter})}{N_E} = \frac{n_{iE}^2}{N_E} = \frac{N_C N_v e^{E_{gE}/kT}}{N_E} \tag{9.66a}$$

and

$$n_{p0} = \frac{n_i^2(\text{base})}{N_B} = \frac{n_{iB}^2}{N_B} = \frac{N_C' N_v' e^{-E_{gB}/kT}}{N_B} \tag{9.66b}$$

Therefore, Eq. (9.65b) can be expressed as:

$$\beta \simeq \frac{N_C' N_v' e^{-E_{gB}/kT}}{N_B} \times \frac{N_E}{N_E N_v e^{-E_{gE}/kT}}$$

$$\simeq \frac{D_{nB}}{D_{pE}} \cdot \frac{N_E}{N_B} \cdot \frac{W_E'}{W_B'} \exp\left[\frac{\Delta E_g}{kT}\right] \tag{9.67}$$

where $\Delta E_g = E_{gE} - E_{gB}$ is the difference between the bandgap energy in the emitter and the bandgap energy in the base. Equation (9.67) shows that the common-emitter current gain depends exponentially on the energy bandgap difference. Due to large bandgap difference between the emitter and the base materials, the common-emitter current gain is very large in a heterojunction bipolar transistor, even if the base doping density N_B is significantly larger than that of emitter doping density N_E. The difference in the energy bandgap in the heterojunction bipolar should be of order of 0.2–0.4 V for optimum performance. A large difference causes the gain to be strongly temperature dependent and creates a distinct spike in the energy diagram, which in turn limits the current.

The energy band diagram of the HBT under active region of operation is shown in Fig. 9.40(b). The bandgap difference between the emitter and the base will provide band offset at the heterosurface. The superperformance of the heterojunction bipolar transistor results from the valence-band discontinuity ΔE_v at the heterointerface. The presence of ΔE_v increases the valence-band barrier height in the emitter-base heterojunction which reduces the injection of the holes from the base to the emitter. This effect in the HBT allows the use of a heavily doped base region while maintaining the higher emitter-efficiency and current gain. Since the base is heavily doped, therefore, the base sheet resistance can be reduced. Base can also be made very thin without concern about the punch-through effect in narrow-base region. A thin base is more desirable because it reduces the base transit time and increase the cutoff frequency which is given as:

$$f_T = \frac{1}{2\pi\tau}$$

where $\tau = \tau_B + \tau_E + \tau_C$ is the total transit time. The transit frequency or cutoff frequency can be further improved by using materials with a higher mobility for the base layer and higher saturation velocity for the collector layer.

An HBT can be further improved by grading the composition of the base layer such that the bandgap energy of the material is gradually reduced throughout the base. The grading causes electric field which reduces the transit time.

The maximum frequency f_{max} of the operation is one of the most important figure of merit that characterizes RF performance. Since, f_{max} is directly proportional to f_T and inversely proportional to base resistance, therefore, with increase of f_T or reduce of base resistance in the heterojunction bipolar, the f_{max} increases.

HBT advantages

- Low base spreading resistance
- High electron mobility and ballistic and overshoot effects
- Can be integrated with HEMTs, MESFETs, Schotlky diodes
- Low $1/f$ noise

HBT applications in systems

- Analog and data conversion circuits
  * Differential amplifiers
  * Logarithmic amplifiers
  * Direct-coupled wide band-widths amplifiers
- Microwave ICs
  * Low phase noise high frequency oscillators, mixers, etc. (lower $1/f$ noise than for MESFETs)

Digital applications

- A to D converters
- Track and hold circuits
- Quantizers
- Voltage comparators
- HBT-diode logic
- Microprocessors

Note: 1. The primary advantage of heterojunction BJT over homojunction BJT is its high emitter efficiency; and higher speed and higher frequency capability in circuit operation.

2. The fundamental restriction apply to fabricate a HBTs is that the materials must have a similar lattice constant so that they can be grown without reducing the quality of the material.

9.11.1 Materials Used for HBTs

HBTs are important devices for power amplifiers. Several different material systems are used to fabricate HBT structures. The major heterostructures utilized in HBT technology are GaAs-AlGaAs, SiGe and InP based heterostructures, primarily

InGaAs-InP and InGaAs-AlInAs. Main advantages of using InGaP emitters rather than AlGaAs in HBTs are:

- Higher yield
- Better reliability
- Higher etch selectivity
- Large valence band discontinuity
- Low surface recombination

And also because of higher electron mobility in InGaAs than GaAs superior high-frequency performance is expected. The InP cellector region has higher drift velocity at high field than that in the GaAs collector. The InP collector breakdown voltage is also higher than that in the GaAs.

InP based HBT are potentially better candidates for high power at higher frequency operation.

Another heterojunction is in the Si/SiGe material system. The SiGe heterostructure is not lattice matched and thus is strained upon growth. The major advantages of the use of SiGe for HBT devices are:

- Low cost alternative to compound semiconductor devices
- High speed capability
- Minimum surface recombination
- High current gain even at low collector current
- Compatibility with the standard silicon

But SiGe HBT has lower cutoff frequency than the GaAs and InP based HBTs because of low mobility in Si.

9.11.2 Design Aspects of HBTs

The conduction band discontinuity ΔE_c as shown in Fig. 9.40(b) is not desirable, since the discontinuity will make it necessary for the carriers in the heterojunction to transport by means of thermionic emission across a barrier or by tunnelling through it. Due to that the emitter efficiency and collector current will degrade. This problem can be avoided by using graded-layer and the graded-base heterojunctions. Figure 9.41 shows the energy band diagram in which ΔE_c is removed by a graded layer placed between emitter and base heterojunction where W_g is the thickness of the graded layer.

The base region should also be graded profile which results in a reduction of the bandgap from the emitter side to collector side. In Fig. 9.41, dotted line represents the energy band diagram of the graded base HBT. Due to graded profile, there is an induced electric field which reduces the base transit time and increases the common-emitter current gain and cutoff frequency also increases.

During the design of collector layer of HBT, designer should consider the breakdown voltage requirement and collector transit time delay. A thicker collector layer will improve the breakdown voltage, but for that we have to pay price in terms of increase transit time.

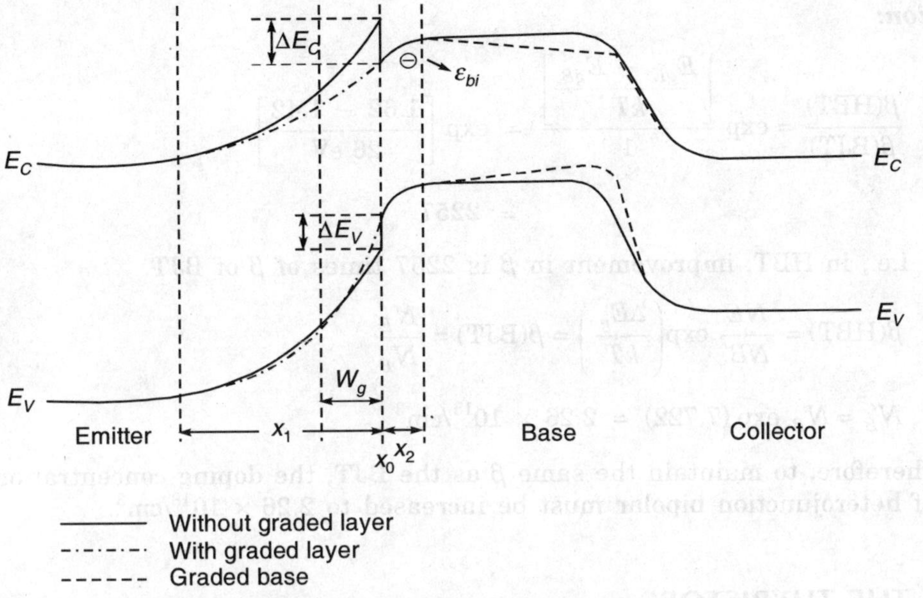

FIGURE 9.41   Energy band diagram of heterojunction transistor to remove the discontinuity at the collector.

EXAMPLE 9.30

What is ballistic collector transistor (BCT)?

Solution: In most HBT devices for high-power applications, the carriers move through the collector at their saturation velocities because very large electric fields are maintained in this layer. The velocities can be increased by lowering the electric field with certain doping profile in the collector layer. One way is to use p^- collector with p^+ pulse doped layer near the subcollector of n-p-n HBT. Therefore, electrons entering the collector layer can maintain their higher mobility of the lower valley during most of collector transit time. Such a device is called a **ballistic collector transistor**.

EXAMPLE 9.31

A HBT has a bandgap of 1.62 eV for the emitter and a bandgap of 1.42 eV for the base. A BJT has a bandgap of 1.42 eV for both the emitter and base materials. It has an emitter doping of $10^{18}/cm^3$ and base doping of $10^{15}/cm^3$.

(a) If the HBT has the same doping as the BJT, find the improvement of β.
(b) If the HBT has the same emitter doping and same β as the BJT, how much one can increase the base doping of the HBT. Assume all the device parameters are same.

Solution:

(a) $\dfrac{\beta(\text{HBT})}{\beta(\text{BJT})} = \exp\left[\dfrac{\dfrac{E_{gE} - E_{gB}}{kT}}{1}\right] \simeq \exp\left[\dfrac{1.62 - 1.42}{26 \text{ eV}}\right]$

$$\simeq 2257$$

i.e., in HBT, improvement in β is 2257 times of β of BJT.

(b) $\beta(\text{HBT}) = \dfrac{NE}{NB'} \exp\left(\dfrac{\Delta E_g}{kT}\right) = \beta(\text{BJT}) = \dfrac{N_E}{N_B}$

$$N'_B = N_B \exp(7.722) \simeq 2.26 \times 10^{18}/\text{cm}^3$$

Therefore, to maintain the same β as the BJT, the doping concentration of the base of heterojunction bipolar must be increased to $2.26 \times 10^{18}/\text{cm}^3$.

9.12   THE THYRISTORS

The thyristor is an important power device that is designed to handle the high voltage and large currents. The thyristor is mainly used for switching application that requires the device to change from ON (conducting state) to OFF (blocking state) or vice versa. We have seen that normal bipolar transistor consists of two closely spaced and electrically interacting p-n junctions, whereas the thyristor consists of three interacting junctions. The switching mechanism of thyristor is different from the normal BJT. Also because of device construction, thyristors have a much wider range of current and voltage-handling capabilities.

9.12.1   Basic Structure

The schematic cross-sectional view of a thyristor is shown in Fig. 9.42 which has four layers p-n-p-n and three p-n junctions in the series. The contact electrode to the outer p-layer is called the **anode** and that to the outer n-layer is called the **cathode**. This structure without any additional electrode is a two-terminal device and is called the ***p-n-p-n* diode**. If an additional electrode, called the **gate electrode**, is connected to the inner p-layer (p_2), the resulting three terminal-device is commonly known as **semiconductor controlled rectifier (SCR)** or **thyristor**. A typical doping profile of a thyristor is shown in Fig. 9.42(b) and the energy band diagram at the equilibrium is shown in Fig. 9.43. At each junction there is built-in-potential which is determined by the impurity doping profile. The I_A–V_{AK} characteristics of such a diode is shown in Fig. 9.44. The different regions on the current voltage characteristics are:

- 0–1:   In this region, device is in the forward-blocking or offstate. This region has very high impedance. At point 1 where $dV/dI = 0$, the voltage V_{BF} is called **forward-break over voltage** and I_S is the switching current.

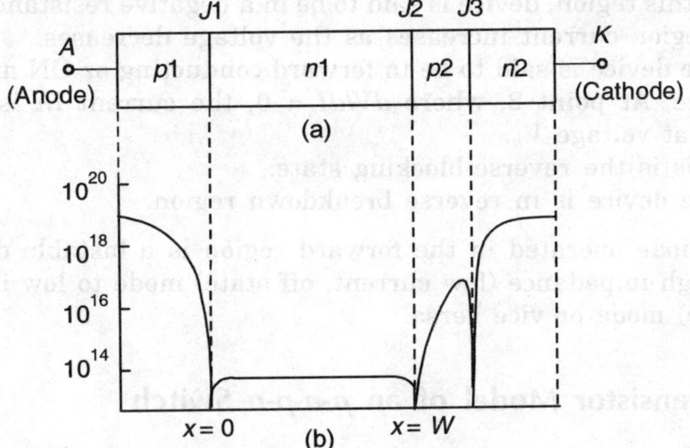

FIGURE 9.42   Structure and carrier distribution in thyristor.

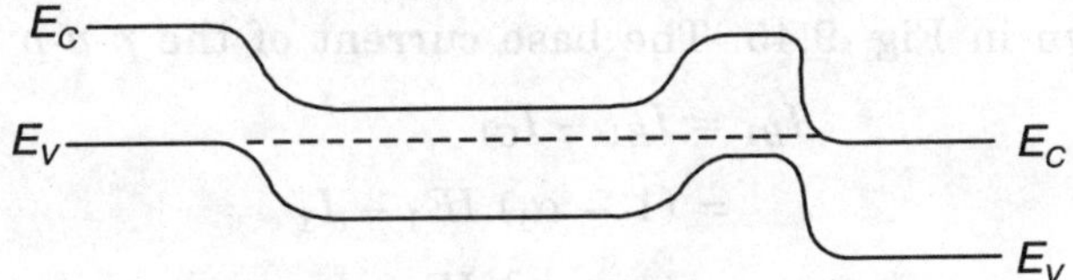

FIGURE 9.43   Band diagram of thyristor.

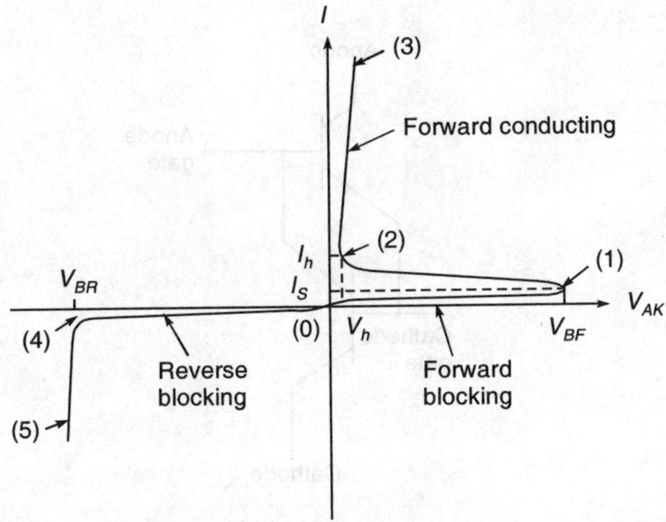

FIGURE 9.44   *I-V* characteristic of thyristor.

- 1–2: In this region, device is said to be in a negative resistance region because in this region current increases as the voltage decreases.
- 2–3: The device is said to be in forward conducting or ON mode and has low impedance. At point 2, where $dV/dI = 0$, the current Ih is called **holding current** at voltage V_h.
- 0–4: This is the reverse-blocking state.
- 4–5: The device is in reverse breakdown region.

A *p-n-p-n* diode operated in the forward region is a bistable device which can switch from a high impedance (low current, off state) mode to low impedance (high-current, on state) mode or vice versa.

9.12.2 Two-transistor Model of an *p-n-p-n* Switch

The two transistor representation of thyristor is shown in Fig. 9.45. The thyristor is considered as two bipolar transistors, *p-n-p* transistor and an *n-p-n* transistor. They are connected with the base of one transistor attached to the collector of the other and vice versa as shown in Fig. 9.45. The base current of the *p-n-p* transistor is:

$$I_{B1} = I_{E1} - I_{C1}$$

$$= (1 - \alpha_1)\, IE_1 - I_1$$

$$= (1 - \alpha_1)\, IE - I_1 \tag{9.68a}$$

where I_1 is leakage current $(I_C B_O)$ of the transistor 1 and α_1 is common emitter current gain. I is total current in the device.

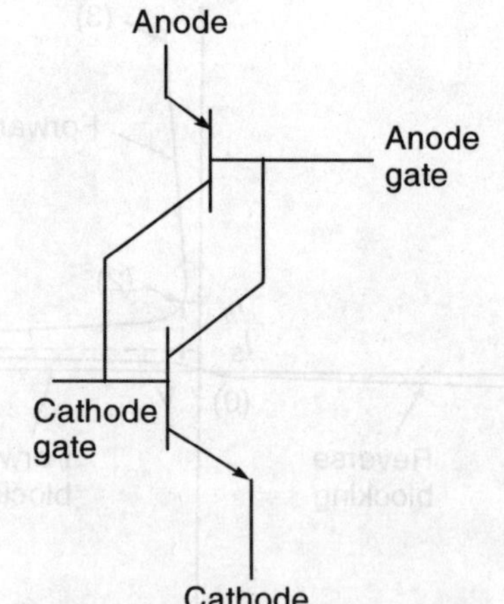

FIGURE 9.45 Two transistor model of a *p-n-p-n* switch.

The collector current in the transistor n-p-n is:

$$I_{C2} = \alpha_2 I_{E2} + I_2$$

$$= \alpha_2 I + I_2 \qquad (9.68b)$$

where I_2 is leakage current (I_{CBO}) from the transition n-p-n. As clear from Fig. 9.45 that

$$I_{B1} = I_{C2}$$

and hence equating Eqs. (9.68a) & (9.68b), we have

$$I = \frac{I_1 + I_2}{1 - (\alpha_1 + \alpha_2)} \qquad (9.68c)$$

In the forward active region, α_1 and α_2 are close to unity, but transistors never get to that mode operation. If $(\alpha_1 + \alpha_2) = 1$, $I = \infty$, but this current is limited by the voltage drop across the external resistance R in the circuit.

Note: A bidirectional thyristor is a switching device that has ON and OFF-states for positive and negative anode voltages and, therefore, it is very useful in ac applications. The bidirectional p-n-p-n diode switch is called a **diac** (diode ac switch). It behaves like two conventional p-n-p-n diodes where anode of one is connected to cathode of other and vice versa. The symmetry of this structure will result in identical performance for either polarity of applied voltage.

A bidirectional three-terminal thyristor is called a **triac** (triode ac switch).

EXAMPLE 9.32

Consider a thyristor in which the leakage currents are 0.4 and 0.6 mA respectively. Explain the forward blocking characteristics when $(\alpha_1 + \alpha_2)$ is 0.01 and 0.9999.

Solution:

$$I = \frac{I_1 + I_2}{1 - (\alpha_1 + \alpha_2)}$$

When

$$I_1 = 0.4 \text{ mA}$$

$$I_2 = 0.6 \text{ mA}$$

$$\alpha_1 + \alpha_2 = 0.01$$

Then

$$I = \frac{0.4 \times 10^{-3} + 0.6 \times 10^{-3}}{1 - 0.01} = 1.01 \text{ mA}$$

In this case, current flowing through the device is the sum of the leakage currents I_1 and I_2, i.e, $I_1 + I_2 = 1.0$ mA. As the applied voltage increases, the anode current I also increases, as do α_1 and α_2. This in turn causes I to increase further which again increases α_1 and α_2, i.e., a regenerative behaviour.

When $\alpha_1 + \alpha_2 = 0.9999$,

$$I = \frac{1}{1.999} \simeq 10 \text{ A}$$

This value is 10,000 times larger than $(I_1 + I_2)$. Therefore, as $(\alpha_1 + \alpha_2)$ approaches to 1, the current I increases without limit, i.e., the device is in forward breakdown mode.

Note: 1. The thyristor model is simulated as resistor R_{ON}, an inductor L_{ON} and a dc voltage source, representing the forward voltage V_f, connected in series with a switch as shown in Fig. 9.46(a). The switch is controlled by a logical signal depending on the voltage V_{ak}, the current I_{ak} and gate signal V_g. When a small current flows into the gate g, this allows a larger current to flow from anode to cathode. Even when the current into the gate stops, the thyristor continues to allow current to flow from anode to cathode. The symbol representation of SCR is shown in Fig. 9.46(b).

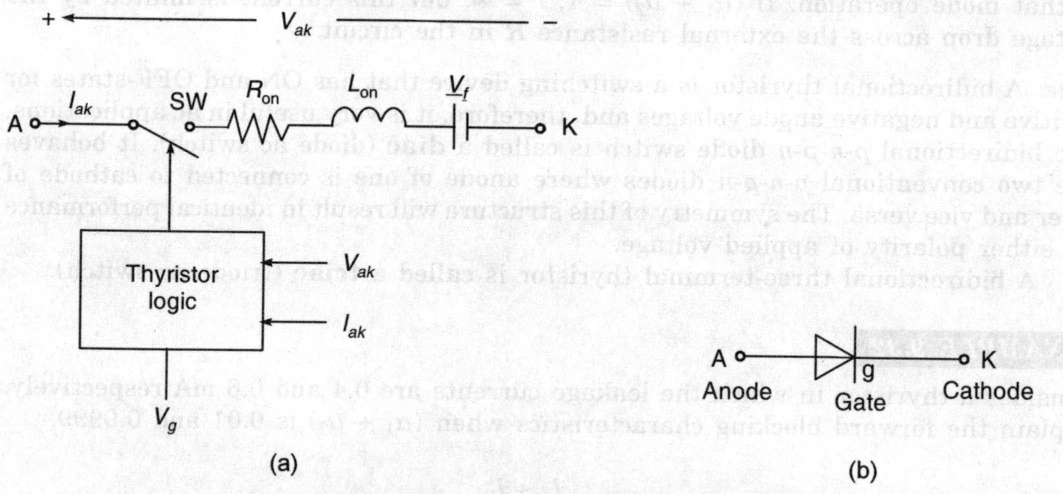

FIGURE 9.46   Thyristor model and symbol.

2. The effect of gate currents on the current-voltage characteristics of a thyristor is shown in Fig. 9.47. From the figure, it is observed that as the gate current increases, the forward breakover voltage decreases.

9.13   THE UNIJUNCTION TRANSISTOR

It is one of the oldest semiconductor device in use. As its name suggests, this is a three-terminal device which nevertheless has only one *p-n* junction. It cannot amplify the signal, but it nevertheless can be used as the active element in an oscillator circuit.

The UJT consists of a bar of *n*-type silicon with electrical connection at either end plus an aluminium wire bonded to a point along the length of the silicon bar. At the bonding point, the aluminium creates a *p*-type region in the silicon bar, thus

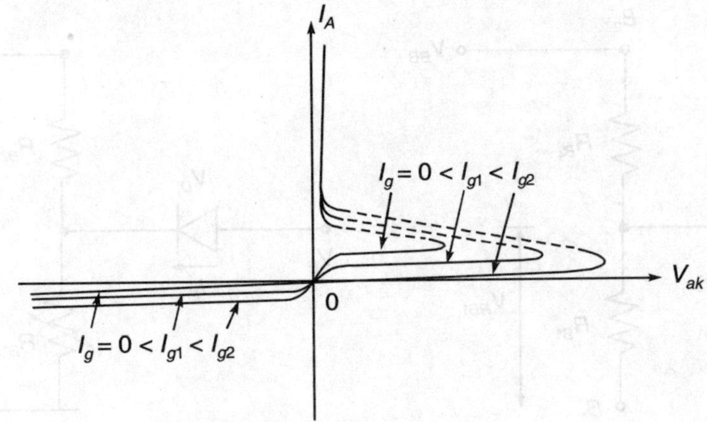

FIGURE 9.47 *I-V* characteristic of thyristor under different gate currents.

forming a *p-n* junction. The physical construction of a typical UJT is shown in Fig. 9.48(a). Since there is only one junction, therefore, it is not reasonable to use the terms *anode* or *cathode*. But designation of UJT is taken from transistor notation. The *p*-type connection is known as the **Emitter** and the two ends of the *n*-type bar are designated as **Base 1** (B_1) and **Base 2** (B_2). For this reason the device is sometimes called a **Double base diode**. The schematic symbol of UJT is shown in Fig. 9.48(b). The Base 1 is connected to zero volts and Base 2 to the positive supply. The resistance between the two base (i.e. interbase resistance) is typically 10 kΩ due to that a relatively small current will flow through it and applied voltage will be distributed evenly along its length. If emitter is grounded, the junction will be reverse biased and the bar acts as a potential divider and about 0.5 V appears at the emitter. If a voltage is connected

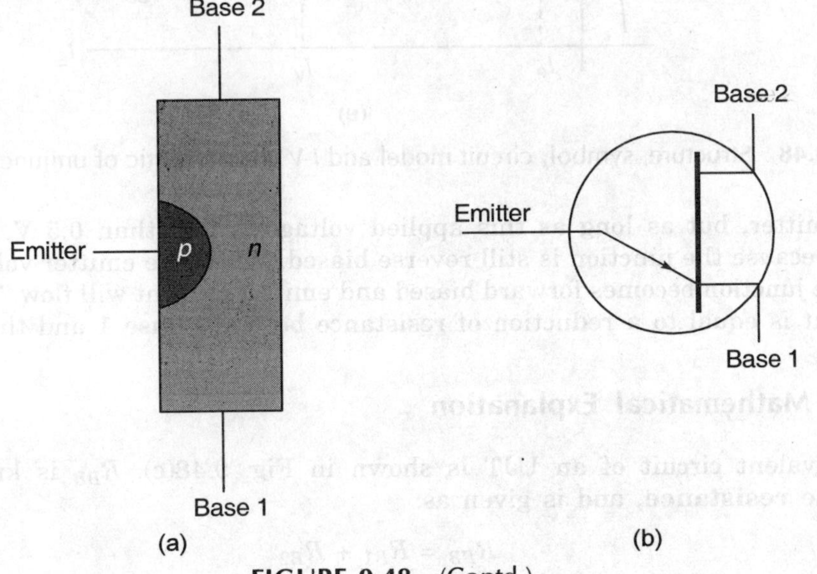

FIGURE 9.48 (Contd.)

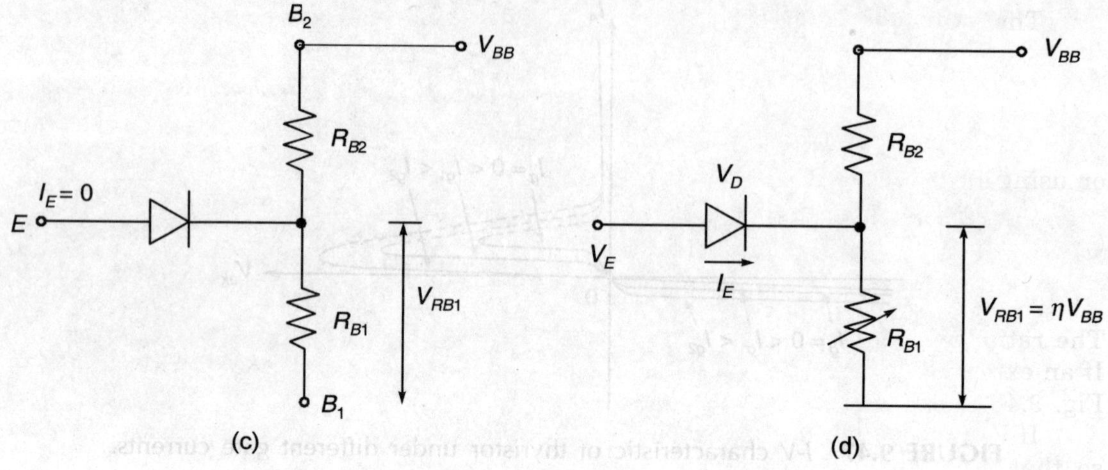

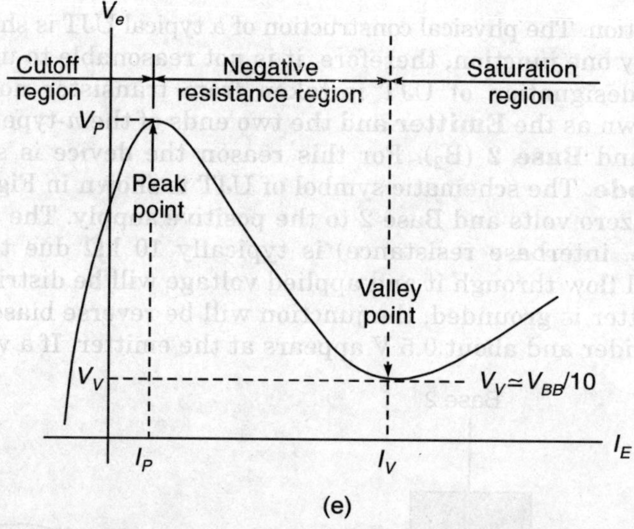

FIGURE 9.48 Structure, symbol, circuit model and *I-V* characteristic of unijunction transistor.

to the emitter, but as long as this applied voltage is less than 0.5 V, nothing will happen because the junction is still reverse biased. When the emitter voltage exceeds 0.5 V, the junction becomes forward biased and emitter current will flow. This increase in current is equal to a reduction of resistance between Base 1 and the emitter.

9.13.1 Mathematical Explanation

The equivalent circuit of an UJT is shown in Fig. 9.48(c). R_{BB} is known as the **interbase resistance**, and is given as:

$$R_{BB} = R_{B1} + R_{B2} \qquad (9.69a)$$

(This formula is true only when emitter is open circuit.)

The voltage drop across R_{B1}, i.e, resistance of Base 1 is:

$$V_{RB1} = \left(\frac{R_{B1}}{R_{B1} + R_{B2}} \right) V_{BB} \qquad (9.69b)$$

or using Eqs. (9.69a) & (9.69b),

$$V_{RB1} = \frac{R_{B1}}{R_{BB}} \times V_{BB} \qquad (9.69c)$$

The ratio (R_{B1}/R_{BB}) is known as the **intrinsic standoff ratio** and is denoted by η. If an external voltage V_E is connected across emitter, the equivalent circuit shown in Fig. 9.48(c) can be redrawn as Fig. 9.48(d).

If V_E is less than V_{RB1}, the junction is reverse biased. However, if V_F is increased so that it exceeds V_{RB1} by at least 0.7 V, the diode becomes forward biased and emitter current I_E flows into the Base 1 region. Because of this, RB1 value decreases. At this point it is thought that the emitter infects holes into the silicon bar, greatly reducing the effective resistance of the bar between emitter and Base 1. This will lower the emitter voltage required to keep the junction forward biased and will sustain a heavy emitter current. In other words, further increase in V_E causes the emitter current to increase which in turn reduces R_{B1} and this causes further increase in current. This run away effect is termed as **regeneration**. The value of emitter voltage at which this occurs is known as the **peak voltage V_p**, and is given as:

$$V_p = \eta \, V_{BB} + V_D \qquad (9.69d)$$

The current-voltage characteristics of a UJT is shown in Fig. 9.48(e). It is seen from figure that when the current rises sharply with emitter voltage up to the peak voltage V_p, after that current falls rapidly with rise in voltage V_E. This region is called the **negative resistance region**. After the valley point, the device runs into saturation. At this point, R_{B1} is at its lowest value which is known as **saturation resistance**. Simplest application of a UJT is a relaxation oscillator.

Note: The emitter terminal does not inject current into the base region until its voltage reaches V_p. Once V_p is reached the base circuit conducts and a positive pulse appears at B1 and negative pulse at B2. The UJT incorporates a negative resistance region. The very basic specifications of UJT are:

- $V_{BB(\text{max})}$: Maximum interbase voltage that can be applied to UJT
- R_{BB} : The interbase resistance of the UJT
- η : The intrinsic standoff ratio which defines V_p. η is a function of emitter location between B1 and B2
- I_p : The peak point emitter current

A UJT could be built either with a p-type region in an n-type substrate (bar), called an **n-channel UJT** or with an n-type region in p-type substrate (or bar), called **p-channel UJT**. In practice, only n-channel UJTs are to be manufactured.

9.14   SCHOTTKY-CLAMPED TRANSISTOR

When the input of a saturated transistor is changed, the output does not change immediately, but it takes some extra time to come out the saturation. This extra time is called **storage time** which accounts for a significant portion of the propogation delay in the original TTL logic family. Storage time can be eliminated and propagation delay can be reduced by ensuring that transistors do not saturate in normal operation. This can be achieved by using a new type of transistor, called **Schottky clamped transistor** or simply **Schottky transistor**. The schmatic representation of such transistor is shown in Fig. 9.49(a) and the physical structure is shown in Fig. 9.49(b). As seen in Fig. 9.49(b), a schottky diode is placed in between the base and the collector of the transistor. Schottky-clamped transistor does not saturate. In the Sckottky-clamped transistor, the built-in-voltage for the Schottky barrier diode from base to collector is less than that of base-collector *p-n* junction. When Schottky diode is forward biased, its potential drop is much less than that of a standard *p-n* junction diode so for a given forward voltage the current through the Schottky barrier is greater than that through the *p-n* junction. The Schottky current is carried out by majority carriers, whereas *p-n* junction current is carried by minority carriers. This means the injection of minority carriers across the base-collector junction as well as the storage minority charges in Schottky transistor are greatly reduced compared with that of conventional bipolar transistor with reduced storage charge. The time required to remove the charges is also decreased.

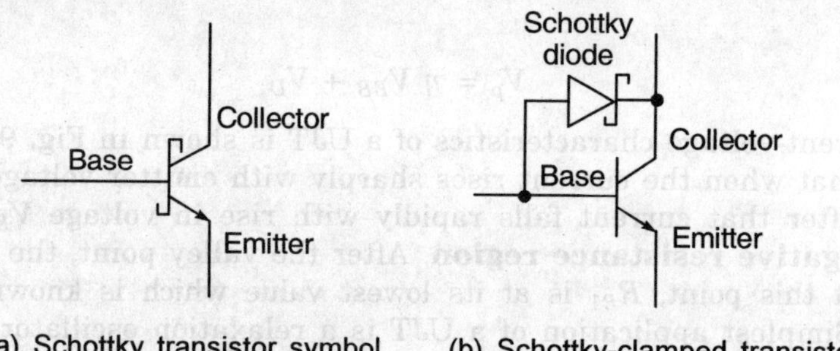

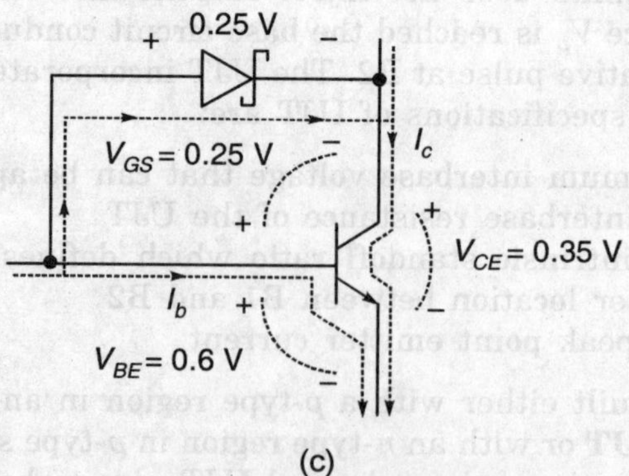

FIGURE 9.49   Schottky-clamped transistor.

It is also seen that in Schottky transistor, the Schottky diode shunts current from base into collector before the transistor goes into saturation as shown in Fig. 9.49(c).

9.15 BiCMOS TECHNOLOGY

When it is necessary to drive long interconnects or chip output pads, CMOS circuits are slower than bipolar circuits. Bipolar transistors can provide large driving currents, but have high power dissipation, whereas the power dissipation of CMOS is lower. Therefore, to retain the benefits of bipolar and CMOS, a new technology emerges which combines bipolar and CMOS transistor in a single integrated circuits, called **BiCMOS technology**. BiCMOS is able to achieve VLSI circuits with speed, power, density, performance previously unattainable with either technology individually.

Advantages in BiCMOS technology

Design uses CMOS gates along with bipolar totem-pole stage where deriving of high capacitance loads is required.

Resulting benefits of BiCMOS technology over CMOS or bipolar are:

- Improved speed over purely CMOS-technology
- Lower power dissipation than purely bipolar technology
- Flexible I/OS BiCMOS is well suited for I/O intensive applications
- High performance analogue
- Latchup immunity

Note: BiCMOS can provide applications with CMOS power and densities at speeds which were previously the exclusive domain of bipolar. This has been demonstrated in applications ranging from SRAMS to gate arrays to microprocessors.

Further advantages of BiCMOS technology

High impedance CMOS transistors may be used for the input circuitry while the remaining stages and output drivers are realized using bipolar transistors.

In general, BiCMOS devices offer many advantages where high load current sinking and sourcing are required. The high current gain of the *n-p-n* transistor greatly improves the output drive capability of the conventional CMOS device.

Compared to CMOS, BiCMOS reduced dependence on capacitive load and the multiple circuit. I/Os configurations greatly enhance the design flexibility and can lead to reduced cycle time, i.e., faster devices. BiCMOS is inherently robust with respect to temperature and process variations, resulting in less variability in final electrical parameters, resulting in higher yield, an important economic considerations. Large circuits can impose severe penalties due to simultaneous switching noise, internal clock skews and high nodal capacitances in critical paths, but BiCMOS has demonstrated superiority over CMOS in all these factors.

Disadvantages with BiCMOS technology

- Greater process complexity compared to CMOS
- Increase in die costs over conventional CMOS
- BiCMOS cannot be scaled as aggressively as CMOS

Applications

Most early BiCMOS applications were analogue. BiCMOS operational amplifiers were introduced in mid-70s. Introduction of digital LSI BiCMOS were motivated by high power dissipation of bipolar and speed limitations of MOS devices and need of high I/O throughput.

Development of BiCMOS resulted in very high performance memories, gate arrays and microprocessors.

BiCMOS has been established as the technology of choice for high speed VLSI.

Advantages of CMOS over bipolar

- Lower static power dissipation
- Higher noise margins
- Higher package density
- Lower manufacturing cost per device
- High yield with large integrated complex functions

Advantages of bipolar over CMOS

- Higher switching speed
- Higher current drive per unit area, higher gain, i.e., higher g_m value
- Better noise performance and high frequency characteristics
- Better analogue capability
- Improved I/O speed

Other properties of bipolar

- Higher power dissipation
- Lower input impedance, i.e, high drive current
- Low voltage swing
- Low packing density
- Low delay sensitivity to load
- High unity gain bandwidth at low current
- Essentially unidirectional

9.16 SUMMARY

The bipolar junction transistor is basically a three-terminal device, called **emitter, collector** and **base**. Emitter should be heavily doped and collector should be lightly doped. The doping of base is moderate. Current in bipolar device is due to both majority carriers as well as minority carriers. Bipolar is of two types: *n-p-n* and *p-n-p*. In modern technology, usually *n-p-n* transistor is more preferred because of high mobility of electron. Bipolar junction transistor consists of two *p-n* junctions, called **emitter-base junction** and **collector-base junction**. The vast majority of BJT applications are in analog electronics, but can be employed in digital electronics also. Deep emitter gives higher current gain, but difficult to control thin base width. The base should be as thin as possible to avoid the recombination as well as to increase the base transit time. Thin base means higher current gain and smaller stored charge and hence higher performance. Doping must be such that to keep base series resistance low and avoid *E-C* punch through. The collector region should be very wide to allow the power dissipation. Higher collector doping reduces base widening and gives higher base-collector capacitance. It is required that subcollector region should be highly doped to minimize resistance. Since, there are only three leads for a BJT, therefore, one of the leads must be common to both the junctions, i.e., common to input and output terminals. Therefore, the cofigurations are called **common emitter configuration** (CE), **common base configuration** (CB) or **common collector configuration** (CC). The most commonly employed configuration is common emitter. Bipolar transistor is used as an amplifier in analog elecronics and the preferred region of operation is active region (forward mode) where input side is forward biased and output terminal is reverse biased. In digital electronics, BJT is used as a switch, i.e., either it operates in cutoff mode (where both terminals are reverse biased) or in saturation region (where input terminal as well as output terminal both are forward biased). The base current I_B is due to:

- Recombination of injected electrons from the emitter (*n-p-n* transistor) with holes in the base. The lost holes due to recombination must be restored due to space-charge neutrality requirements, I_B restores the holes lost in the base from recombination through the base contact.
- Holes injection into the emitter from the base in the forward biased emitter-base junction. I_B must supply the holes lost.
- The reverse biased collector-base junction sweeps holes into the base. These holes reduce the magnitude of the base current I_B.

In the CE configuration, the input base current is amplified by a factor β, called **common-emitter current gain** and defined as the ratio of collector current to the base current. In the BJT, total current I_E is given by the sum of the collector current and the base current. The requirement for the high performance BJT device is that the base width should be small compared with the minority carrier diffusion length, i.e., base transit time is much less than that of minority carrier recombination lifetime in the base. In the ideal transistor, one has to neglect the base recombination-generation process. Gummel number is defined as the total integrated base dose. Emitter-collector punchthrough occurs when base width becomes so small that V_{BC} starts to affect the potential barrier of emitter-base diode. This situation is like DIBL effect in MOSFET. When transistor operated near the punchthrough, collector current is no longer controlled by V_{BE}. To avoid punchthrough the base doping N_B should be increased. Saturation current depends on temperature due to n_i^2 term.

BJTs are not symmetric devices because doping and physical dimensions are different for emitter and collector.

The most common DC and large-signal non-linear model as used in SPICE, is the Ebers-Moll model. This model is used to relate the large-signal behaviour to the commonly used small-signal parameters.

The measured current-voltage characteristics of a typical bipolar transistors is not ideal. The degree of deviation from the ideal characteristics depend on the device structure, the device design and the device fabrication process and bias conditions.

The built-in field due to inhomogeneity of base aids the minority carriers transport through the base and hence reduces the base transit time. The reduction in the base transit time acts to increase the cutoff frequency of the device. Thus, for high frequency operation, it is desirable to use a non-uniform base doping concentration to produce field aided diffusion.

The two most important figures of merits that describe the high frequency behaviour of a BJT are the cutoff frequency and the maximum frequency of oscillation.

Several important second order effects which are present in actual BJT are not included in the Ebers-Moll model and, therefore, a new SPICE model is developed to take care of all these effects. This model is given by Gummel-Poon.

For high-performance logic applications, the most commonly used bipolar circuit is emitter coupled logic (ECL) circuit. The Si-Ge base transistor is particularly useful in analog circuit applications. Incorporation of Ge into Si reduces the bandgap of the alloy. If Ge is introduced into the base region of Si BJT, the bandgap will reduce which influences the device characteristics. Heterojunction BJTs are those which are composed of at least two different semiconductor materials. As a result, the energy bandgap as well as other material properties can be different in the emitter, base and the collector. In traditional heterojunction emitter, emitter bandgap is larger than the base bandgap at emitter base junction. Base current is suppressed by the large hole injection barrier. Collector current is not affected.

The thyristor is an important power device that is designed to handle the high voltage and large currents, mainly used for switching applications. In thyristor, there are mainly three interacting *p-n* junctions. Its switching behaviour is different from normal BJT. It is also called **silicon controlled rectifier (SCR)** or **semiconductor controlled rectifier**.

For normal bipolar transistor,

Forward active region: $V_{BE} > 0.6$ V

$V_{CE} > 0.2$ V

i.e., $V_{BC} < 0.4$ V

Cutoff: $V_{BE} < 0.6$ V

Saturation: $V_{CE} < 0.2$ V

REVIEW QUESTIONS

1. What are the basic differences between bipolar junction transistor and field effect transistor?

2. Explain, why MOS devices is more preferable in VLSI design than the bipolar.

3. Why bipolar transistor has larger g_m value than the MOS transistor?

4. Why common emitter configuration of BJT is preferred more? Explain briefly.

5. Why bipolar junction transistor is not a symmetric device?

6. As a designer, what are your considerations during the design of a BJT?

7. Discuss the current flow mechanism in the BJT.

8. What is the requirement of base region in the BJT although we want to make it as thin as possible?

9. What are the differences between ideal BJT and real BJT?

10. Why it is required to keep the base width as thin as possible?

11. What will happen if the base region has non-uniform doping profile?

12. How can designer achieve a unity value of base transport efficiency?

13. Is the base width affect the base transport of efficiency? Explain.

14. On which factors, the current of bipolar transistor depend?

15. What is the significance of Ebers-Moll and Gummel-Poon models?

16. What is punchthrough? Discuss its effect on the performance of the bipolar device.

17. Explain why common emitter breakdown voltage is always lower than the common base breakdown voltage.

18. Discuss the curent crowding effect in BJT in your own words.

19. As a designer, how you overcome the problem of base-width modulation and for that what price will you pay?

20. What is the main cause of thermal runaway in the bipolar devices?

21. Shortening of the transit time can be very important in high frequency devices. Explain this statement with proper argument.

22. Discuss the consequences of Kirk effect.

23. Explain how a small varition in I_B produces a large, proportional variation in I_C.

24. As a designer, how do you minimize the parasitic base resistance in a BJT?

25. Explain the operation of Schottky clamped transistor.

26. Why in modern technology BiCMOS is an emerging technology?

27. Explain how bandgap narrowing in the degenerate emitter affects the curent gain β. What is the physics behind this?

28. Explain in your own words, why an increasing V_{CE} causes the collector current to increase even in the active region.

29. Show that the base transport factor α_T can be simplified to $(1 - W^2/2L_p^2)$.

30. Explain why turn-ON transient of a BJT is faster when the device is driven into oversaturation?

31. The variation of β with V_{EB} for a transistor is shown in Fig. 9.50. Explain what will happen to γ and α_T in the region marked A.

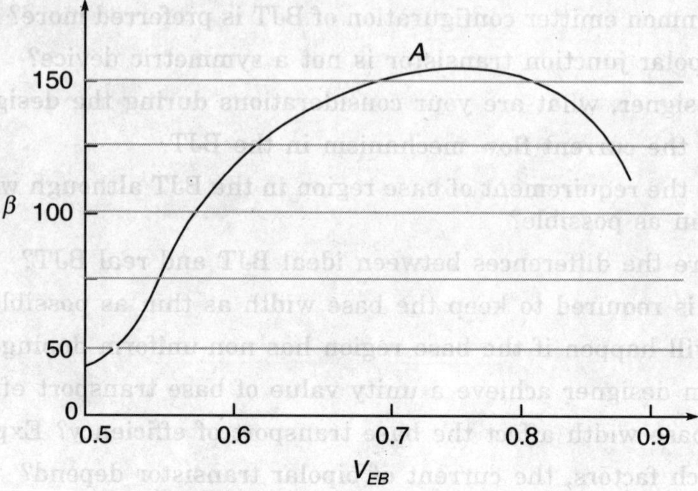

FIGURE 9.50 Variation of β with V_{EB}.

32. Draw the Gummel-Poon model for *p-n-p* transistor and level the current components.

33. Describe the advantages of polysilicon emitter BJT.

34. How UJT is different from BJT? Discuss its applications.

35. Explain qualitatively the effect of gate current on the voltage-current characteristics of a thyristor shown in Fig. 9.47.

36. In the circuit of Fig. 9.51,
    (a) Explain the role of *p-n* junction diode.
    (b) What type of transistor is it?
    (c) What will happen if the polarity of diode change?
    (d) What is the need of this transistor?

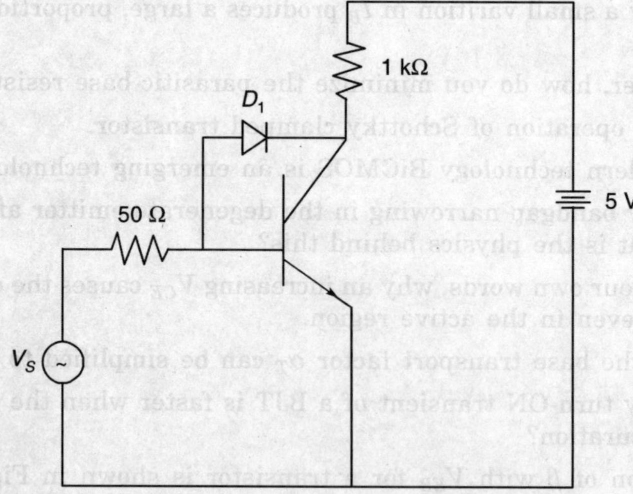

FIGURE 9.51 Circuit for Problem 36.

37. What is the effect of introducing a material having an energy bandgap of $E_g = 0.72$ eV into the Si base? Draw the resultant band diagram. Discuss its advantages.

38. For a *p-n-p* transistor in the CE active mode, the ratio of the hole current injected into the base from the emitter to the space charge recombination current decreases by 25%. (a) Does γ change or remain constant? Explain your answers (b) Does α_T change or remain constant? Explain.

39. Why should we reduce back injection? What can be done to control it?

NUMERICAL PROBLEMS

1. For a Si *n-p-n* transistor biased in forward active mode, the emitter-base electron-diffusion current is 2.5 mA, the base recombination current is 0.016 mA, the emitter-base hole diffusion current is 0.061 mA, and the space charge recombination is 0.033 mA. Find

    (a) Base transport factor
    (b) Emitter efficiency
    (c) Common-emitter current gain

2. In Problem 9.1, the emitter-base bias is 0.55 V and if the bias voltage is increased to 0.65 V, will the common-emitter current gain β increase or decrease? Explain.

3. By using the Ebers-Moll model of a BJT determine the emitter collector voltage V_{EC}. Consider the operation of a BJT when the collector-base voltage is zero.

4. Consider a silicon *p-n-p* BJT in which the base width is 0.5 µm, the base doping is 5×10^{16}/cm^3, the collector doping concentration is 10^{15}/cm^3 and the collector-base bias is $- 10$ V. Determine whether it is in punchthrough conditions.

5. Given that an ideal transistor has an emitter efficiency of 0.999 and the collector-base leakage curent is 10 µA. Calculate the active region emitter current due to holes if $I_B = 0$.

6. A symmetrical *p-n-p* BJT has the following parameters:

$A = 10^{-4}$ cm^2
Base width = 1 µm
Emitter: $N_a \ 10^{17}$/cm^3, $\tau_n = 0.1$ µs, $\mu_p = 200$ cm^2/V sec $\mu_n = 7$
Base: $N_d = 10^{15}$/cm^3, $\tau_p = 10$ µs, $\mu_n = 1300$, $\mu_p = 450$

    (a) Calculate the saturation current.
    (b) Calculate I_B for $V_{EB} = 0.3$ V and $V_{CB} = - 40$ V. Assume $\gamma =$
    (c) Calculate γ and β.

7. A prototype *n-p-n* transistor has its emitter doped to $N_d = 5 \times 10^{19}$/cm^3. The base is doped with $N_a = 2 \times 10^{17}$/cm^3. Find ΔE_g accounting for narrowing of the emitter conduction band. Find β if $W_E = 0.12$ µ and $W_B = 0.07$ µm.

8. For step junction n-p-n transistor,

$N_{dE} = 5 \times 10^{19}/cm^3 \qquad N_{aB} = 4 \times 10^{17}/cm^3 \qquad N_{dC} = 10^{17}/cm^3$

$W_{EM} = 0.13$ μm $\qquad W_{BM} = 0.20$ μm

$V_{BE} = 0.8$ V $\qquad V_{CB} = 2$ V

Find

(a) The minority carrier lifetimes
(b) The built-in voltage of the collector-base junction
(c) The built-in voltage of the emitter base junction neglecting bandgap narrowing
(d) W_B
(e) γ
(f) α assuming $M = 1$
(g) β

9. If the built-in field in the base of a BJT is made via doping concentration gradient, determine the doping concentration profile needed to produce a field of 100 kV/cm in a silicon p-n-p device. Assume that the doping concentration within the base region has an exponential dependence and that the base width is 0.1 μm.

10. Consider a silicon p-n-p BJT. Neglect I_{cn} with respect to I_{cp}. Calculate the cutoff frequency if the total capacitance of the structure is 1.0×10^{-8} F. Given $A = 1$ μm^2, $D_B = 11.6$ cm^2/s, $L_B = 0.15$ μm, $W_B = 0.1$ μm, $N_{dB} = 10^{17}/cm$, $V_{EB} = 0.7$ V, $V_{CB} = -10.0$ V and $(kT/q) = 0.0259$ V.

11. A silicon n-p-n BJT has impurity concentration of 3×10^{18}, 2×10^{16} and $5 \times 10^{15}/cm^3$ in the respective regions. Determine the current component in each region with $V_{BE} = 0.6$ V.

Take A = 0.01 mm^2, neutral-base width = 0.5 μm. Assume $\mu_n = 1000$ cm^2/V sec. and minority carrier lifetime in each region is same and equal to 10^{-6} sec.

12. Determine the base transport factor for a silicon p-n-p BJT given that the hole lifetime in the base is 10^{-6} sec, the hole mobility is 450 cm^2/V sec and $kT = 0.0259$ eV. Assume that the base width is:

(a) $W_B = 0.1$ μm
(b) $W_B = 5.0$ μm

13. For an ion-implanted n-p-n transistor, the net impurity doping in the neutral base is given by $N(x) = N_o e^{-x/l}$ where $N_o = 2 \times 10^{18}/cm^3$, $l = 0.3$ μm. Find

(a) Base region width
(b) Induced electric field
(c) The total number of impurities in the neutral-base region per unit area

14. Consider a silicon n-p-n bipolar transistor with a common emitter current gain of 100 and a base doping concentration of $10^{17}/cm^3$. The minimum open-base breakdown voltage is to be 15 V. Find BV_{CBO}. Also find the multiplication factor for avalanche breakdown.

15. Consider a uniformly doped silicon bipolar transistor with base doping of $5 \times 10^{16}/\text{cm}^3$ and a collector doping of 2×10^{15} cm^{-3}. Assume the metallurgic base width is 0.70 μm. Calculate the change in the neutral base width as the C_B voltage changes from 2 to 10 V. Discuss this effect qualitatively on the performance of the device.

16. The current amplification factor β of a BJT is very sensitive to the base width as well as the ratio of the base doping to the emitter doping. Calculate and plot β for an n-p-n transistor with $L_n^p = L_p^n$ for:

    (a) $n_n = p_p$, $W_B/L_n^p = 0.01$ to 1
    (b) $W_B = L_n^p$, $p_p/n_n = 0.01$ to 1

    Assume constant mobility.

17. Find out the collector doping concentration to avoid the base widening for $J_{max} = 8 \times 10^4$ A/cm^2 and assume $v_{sat} = 10^7$ cm/sec. Find the current density to avoid Kirk effect.

18. Calculate the change in p_{EO} when the emitter doping of a silicon BJT increases from $10^{18}/\text{cm}^3$ to $10^{19}/\text{cm}^3$. Take $T = 300$ K.

19. The base width of a bipolar transistor is normally small to provide a large current gain and increased speed. The base width also affects the early voltage. In a silicon n-p-n transistor, $N_B = 10^{16}/\text{cm}^3$ $N_C = 5 \times 10^{15}/\text{cm}^3$. Assume $D_B = 20$ cm^2/sec and $\tau_{BO} = 5 \times 10^{-7}$ sec. Let $V_{BE} = 0.70$ V. Using voltages $V_{CB} = 5$ V and $V_{CB} = 10$ V, calculate the early voltage for the metallurgical base widths of

    (a) 1 μm   (b) 0.80 μm   (c) 0.40 μm

    Interpret the result.

20. A Si transistor has D_p of 10 cm^2/sec and W_B of 0.5 μm. Find the cutoff frequencies for the transistor with the common-base current gain of 0.998. Neglect the emitter and collector delays.

21. Three n-p-n transistors are shown in Fig. 9.52. Assume the transistors are identical in terms of emitter region, collector region. The doping concentrations of the emitter and the collector are same.

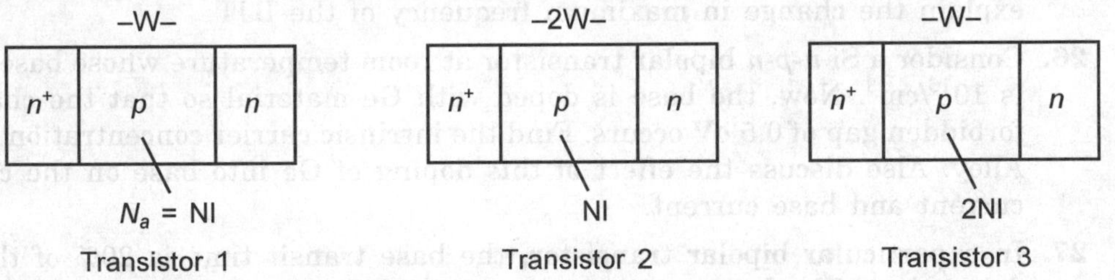

FIGURE 9.52 *n-p-n* transistors.

Give mathematical reasons to know which transistors have the largest and lowest value of each parameter.

(a) γ

(b) α_T

(c) Punchthrough voltage

(d) β

(e) Collector junction capacitance with V_{CB} reverse biased at 15 V.

22. A Si n-p-n transistor has the following parameters at room temperature:

Emitter	*Base*	*Collector*
$N_d = 10^{18}/\text{cm}^3$	$N_a = 10^{16}/\text{cm}^3$	$N_d = 10^{15}$
$\tau_p = 1\ \mu\text{s}$	$\tau_n = 2.5\ \mu\text{s}$	$\tau_p = 2\ \mu\text{s}$
$\mu_n = 500\ \text{cm}^2/\text{V-sec}$	$\mu_n = 100$	$\mu_n = 1000$
$\mu_p = 300\ \text{cm}^2/\text{V-sec}$	$\mu_p = 400$	$\mu_p = 450$
$W_B = 0.1\ \mu\text{m}$		

Area $= 10^{-4}\ \text{cm}^2$

Calculate β and γ. Also calculate the charge stored in the base when $V_{CB} = 0$ and $V_{BE} = 0.7$ V. If the base transit time is the dominant delay component for this BJT, calculate f_T.

23. A n-p-n Si transistor has an emitter doping of $5 \times 10^{18}/\text{cm}^3$ and a base doping of $10^{16}/\text{cm}^3$. Calculate the forward emitter-base junction voltage at which high level injection occurs.

24. For an n-p-n silicon transistor, the collector current at $V_{CE} = 5$ V is obtained as 1 mA. Given parameters are $\beta = 90$, $\tau_F = 5p_{\text{sat}}$ $T = 300$ K.

Find

(a) g_m (b) Input resistance (c) r_π (d) c_π

Draw the small signal model without parasitic capacitances at $\tau_b = 1\ \mu\text{A}$.

25. In an n-p-n silicon BJT, the collector current becomes twice of its initial value. The transit time of minority carriers in the base is reduced to half its earlier value. The parasitic capacitances and resistances remain unaffected. Numerically explain the change in maximum frequency of the BJT.

26. Consider a Si n-p-n bipolar transistor at room temperature whose base doping is $10^{16}/\text{cm}^3$. Now, the base is doped with Ge material so that the change in forbidden gap of 0.5 eV occurs. Find the intrinsic carrier concentration in SiGe Alloy. Also discuss the effect of this doping of Ge into base on the collector current and base current.

27. In a particular bipolar transistor, the base transit time is 20% of the total delay time. The base width is 0.5 μm and the base diffusion coefficient is $D_B = 20\ \text{cm}^2/\text{S}$. Determine the f_T.

28. One condition for switching a thyristor is that $\alpha_1 + \alpha_2 = 1$. Show that this condition corresponds to $\beta_1 + \beta_2 = 1$, where β_1 and β_2 are the common-emitter current gains of the p-n-p and n-p-n BJTs in the equivalent circuit of thyristor.

29. An HBT device doped in the base at $10^{16}/cm^3$ has f_{max} of 20 GHz. If the base doping concentration is increased by a factor of 100 to $10^{18}/cm^3$ and all other parameters are held constant, what is the new value of f_{max}? Assume it is an n-p-n device.

30. For the doping profile shown in Fig. 9.42, find the width W of n_1-region so that the thyristor has a reverse blocking voltage of 120 V. If the current gain α_2 for the n_1-p_2-n_2 transistor is 0.4, and α_1 of the p_1-n_1-p_2 transistor can be expressed as $0.5 \sqrt{L_p/W}$ $\ln (J/J_0)$ where $L_p = 25$ μm and $J_0 = 5 \times 10^{-6}$ A/cm^2, find the cross-sectional area of the thyristor that will switch at a current I_s of 1 mA.

31. Find the intrinsic standoff ratio if the resistance of the Base 1 is 1 kΩ and interbase resistance is 3 kΩ. If the built-in-potential of the diode is 0.7 V in forward direction, find the peak voltage for $V_{BB} = 5$ V.

32. An n-p-n transistor has $N'_{DE} = 10^{20}$, $N'_{AB} = 5 \times 10^{17}$ and $N'_{DC} = 10^{16}$. For $V_{BE} = 0.75$ V, how small will be the base width to keep the punchthrough voltage above 12 V?

CHAPTER 10

Optoelectronic Devices and Power Devices

- Radiative transition
- LEDs
- LASER
- Photodetector
- Photoconductor
- Photodiode
- *p-i-n* photodiode
- Solar cell
- Microwave diode
- Tunnel diode
- IMPATT diode
- GUNN diode
- Hot electron devices
- Fibre transmission

10.1 INTRODUCTION

So far we have only considered the basic physics of the electronic devices. Semiconductor devices can also be designed and fabricated to detect and generate optical signals. These devices are called **photonic** or **optoelectronic devices**. In these devices, the basic particle of light—the photon—plays a major role. The primary optoelectronic devices which are useful in lightwave communication systems are emitters, amplifiers, modulators and detectors. Emitters are the front-end components of a lightwave communication systems. The most important emitter are light emitting diodes (LEDs) and lasers. In this chapter, we will mainly concentrate on light emitting diodes, lasers, photodetectors, photodiodes and photoresistors.

Several important devices for high-frequency applications use the instabilities that occur in semiconductors. The important type of instabilities involves negative conductance. We will concentrate mainly on three important negative conductance devices: microwave devices, tunnel diode, IMPATT diode and gunn diode. These are mainly two-terminal devices that can be operated in a negative conductance mode to provide amplification or oscillation at microwave frequencies in a proper circuit.

10.2 RADIATIVE TRANSITIONS AND OPTICAL ABSORPTION

As we know that light can be treated as a particle as well as wave. The particle nature of light can be explained on the basis of photon whose energy is given as $E = h\nu$, where h is Planck's constant and ν is frequency of incident light, generally ultraviolet-rays (UV).

Wavelength of the incident light and energy can be related as:

$$\lambda = \frac{c}{\nu} = \frac{hc}{h\nu} = \frac{hc}{E} = \frac{1.24}{E} \, \mu m \tag{10.1}$$

where E is the energy in eV and c is speed of light. There are several possible photon-semiconductor interaction mechanisms. Photon can interact with the semiconductor's crystal lattice and on this interaction photon energy is converted into heat energy. Photon can also interact with the impurity atoms (either donor or acceptor) and with defects in the semiconductor material. But main interest is to see the interaction of light (photon) with the valence band of semiconductor. When the photon interacts with the valence band electrons and energy $h\nu$ is assumed to be sufficient and greater than bandgap E_g then some amount is utilized in removing the valence electron and rest is used in exciting the valence electron to conduction band. In this process, one electron in conduction band and one hole in valence band are created. This process is known as **electron-hole pair generation**. The process of excitation of valence electron to conduction band is called **optical absorption**. Any photon, whose energy is less than the energy gap of semiconductor material, i.e., $h\nu < E_g$, is not able to excite valence electron to conduction band. The requirement for optical absorption is $h\nu \geq E_g$.

The direct excitation of valence electron to conduction band after absorbing energy equal to forbidden gap, i.e., $h\nu \geq E_g$ is called **intrinsic transitions** or **band-to-band transitions** as shown in Fig. 10.1(a).

If $h\nu$ is less than E_g, a photon will be absorbed if only there are available energy states in the forbidden gap (due to chemical impurities or physical defects). The transition is called **extrinsic absorption** as shown in Fig. 10.1(b).

The reverse of the process is also true, i.e., an electron from conduction band edge jumps into V_B edge, recombine with hole there and produces a photon of energy $h\nu \, (= E_g)$.

Let us consider a semiconductor material which is illuminated from a light source, having energy $h\nu$ greater than E_g. It is also assumed that the number of photons per unit area, defined as **photon flux**, falling on the material is ϕ_0. Assume

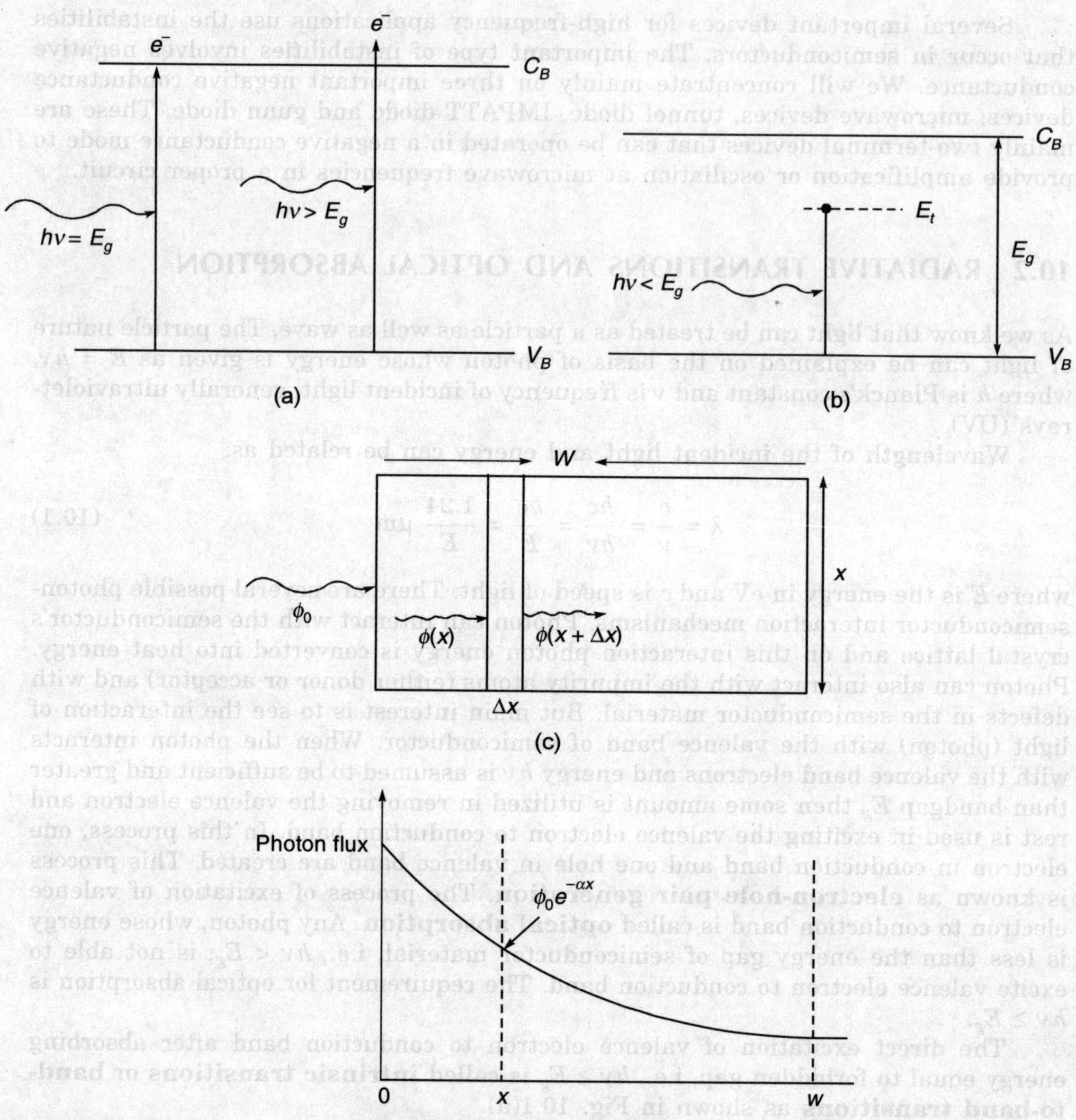

FIGURE 10.1 Schematic representation of radiative transitions and optical absorption.

the thickness of the semiconductor material is x. As the photon flux travels through the material, some of the photons are absorbed. The fraction of photons absorbed is proportional to the intensity of the flux. Consider an incremental thickness of the material is Δx as shown in Fig. 10.1(c). The number of photons absorbed within this incremental thickness is $\alpha\phi(x)\Delta x$, where α is a proportionality constant and defined as the absorption coefficient from the continuity of flux,

$$\phi(x + \Delta x) - \phi(x) = \frac{d\phi(x)\Delta x}{dx} = -\alpha\phi(x)\Delta x$$

or
$$\frac{d\phi(x)}{dx} = -\alpha\phi(x) \tag{10.2a}$$

Negative sign only indicates that photon flux intensity decreases due to absorption. Using boundary condition, $\phi(x) = \phi_0$ at $x = 0$, the solution of differential Eq. 10.2(a) is given as:

$$\phi(x) = \phi_0\, e^{-\alpha x} \tag{10.2b}$$

Hence, the fraction of photon flux that exists at the other end of the material, i.e,. at $x = W$, is given as:

$$\phi(W) = \phi_0\, e^{-\alpha W} \tag{10.2c}$$

Equation 10.2(c) shows that the intensity of the photon flux decreases exponentially with distance through the semiconductor material as shown in Fig. 10.1(d). The absorption coefficient decreases rapidly at the cutoff wavelength λ_c, i.e.,

$$\lambda_c = \frac{1.24}{E_g}\ \mu m \tag{10.2d}$$

because the optical band-to-band absorption becomes negligible for $h\nu < E_g$ or $\lambda > \lambda_c$. Since, the absorption coefficient is very small for $h\nu < E_g$, therefore, the material appears to be transparent to photon in this energy range.

EXAMPLE 10.1

A single-crystal of silicon sample having a width of 0.3 µm, illuminated with a monochromatic light of energy 3 eV. The incident power is 10 MW. Find: (a) the total energy absorbed by the material per second. (b) the rate of excess thermal energy dissipated to the lattice (c) number of photon per second given off from recombination by intrinsic transitions. Take $\alpha = 4 \times 10^4/cm$.

Solution: (a) Energy absorbed per second is:

$$\phi_0(1 - e^{\alpha W}) = 10^{-2}[1 - e^{4 \times 10^4 \times 0.3 \times 10^{-4}}]$$

$$= 10^{-2}[1 - e^{1.2}] \simeq 6.3\ mW$$

(b) The portion of each photon energy that is converted into heat is:

$$\frac{h\nu - E_g}{h\nu} = \frac{3 - 1.12}{3} = 62\%$$

Therefore, the amount of energy dissipated per second to lattice is $62\% \times 6.3 \simeq 3.9$ mW.

(c) Number of photons per second from recombination is:

$$\frac{2.4}{1.6 \times 10^{-19} \times 1.12} = 1.3 \times 10^{16} \text{ photons/sec}$$

Recombination radiation = $(6.3 - 3.9) \simeq 2.4$ mW

10.2.1 Radiative Transitions

A photon interacts with electron in a solid material by three ways: absorption, spontaneous emission and stimulated emission. Consider Fig 10.2(a) where the two energy levels E_1 and E_2 represent the ground energy level and excited energy level. Any movement of electron from E_1 to E_2 after absorbing photon is called **absorption**. And if electron falls from E_1 to E_2 then process is called **emission** of photon. In both the processes, frequency of photon is given as:

$$v_{12} = \frac{E_2 - E_1}{h} \tag{10.3a}$$

Actually, at room temperature, most of the atoms in a solid occupies their ground state. If a photon of energy hv_{12} falls on the system then atom in the energy state E_1 absorbs this energy and jumps to higher state E_2. In other words, we can say that the probability of jumping atoms from states E_1 to state E_2 increases if the photon energy is equal to or greater than $(E_2 - E_1)$. Since the excited state E_2 is unstable, therefore, after finite time interval, these excited atoms again jump into the lower state or ground state E_1 which is a stable energy state. During this transition atoms give off the energy equivalent to $E_2 - E_1$ in the form of photon. This process is called **spontaneous emission** as shown in Fig. 10.2(b). Consider another situation where atoms are already in the excited state E_2. Now, a photon of energy hv_{12} impinges on this excited atom. Then the atom can be stimulated to make a transition from the excited state to ground state and gives off photon of energy hv_{12}, which is in the phase with the incident radiation. This process is called **stimulated emission** as shown in Fig 10.2(c). The radiation from the stimulated emission is monochromatic because each photon has precised an energy hv_{12} and is coherent because all emitted photons are in phase.

The stimulated emission process is a dominating process in laser diodes, whereas for photodetectors and the solar cell the dominant process is absorption.

Under a thermal equilibrium condition and for $(E_2 - E_1) > 3\,kT$, the population ratio in both the excited states is given as:

$$\frac{n_2}{n_1} = e^{-(E_2 - E_1)/kT}$$

$$= e^{-hv_{12}/kT} \tag{10.3b}$$

where n_2 is the instantaneous population in state E_2 and n_1 is instantaneous population of state E_1.

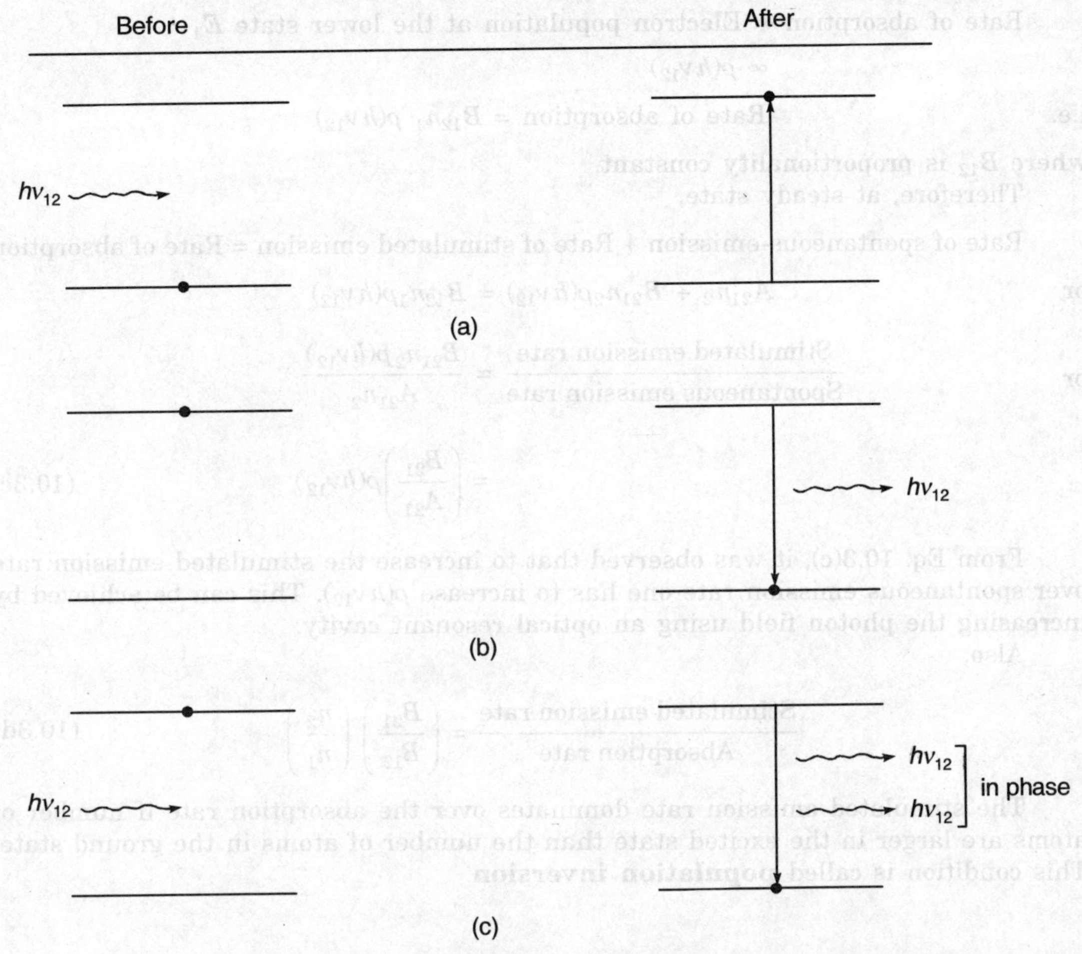

FIGURE 10.2 Radiative transitions in solids.

The negative sign in Eq. (10.3b) indicates that the excited state E_2 is less populated than the ground state E_1.

In steady state, the stimulated rate and spontaneous-emission rate must be balanced by the rate of absorption to maintain the population in the respective energy states.

Stimulated-emission rate is directly proportional to photon-field energy density $\rho(h\nu_{12})$ which is defined as the total energy in the radiation field per unit volume per unit frequency.

Spontaneous-emission rate is proportional to the population of the upper level, i.e.,

Stimulated-emission rate = $B_{21} n_2 \, \rho(h\nu_{12})$

Spontaneous-emission rate = $A_{21} n_2$

where B_{21} and A_{21} are proportionality constants.

$$\text{Rate of absorption} \propto \text{Electron population at the lower state } E_1$$

$$\propto \rho(h\nu_{12})$$

i.e.
$$\text{Rate of absorption} = B_{12}n_1\,\rho(h\nu_{12})$$

where B_{12} is proportionality constant.

Therefore, at steady state,

$$\text{Rate of spontaneous-emission} + \text{Rate of stimulated emission} = \text{Rate of absorption}$$

or
$$A_{21}n_2 + B_{21}n_2\rho(h\nu_{12}) = B_{12}n_1\rho(h\nu_{12})$$

or
$$\frac{\text{Stimulated emission rate}}{\text{Spontaneous emission rate}} = \frac{B_{21}n_2 p(h\nu_{12})}{A_{21}n_2}$$

$$= \left(\frac{B_{21}}{A_{21}}\right)\rho(h\nu_{12}) \tag{10.3c}$$

From Eq. 10.3(c), it was observed that to increase the stimulated emission rate over spontaneous emission rate one has to increase $\rho(h\nu_{12})$. This can be achieved by increasing the photon field using an optical resonant cavity.

Also,

$$\frac{\text{Stimulated emission rate}}{\text{Absorption rate}} = \left(\frac{B_{21}}{B_{12}}\right)\left(\frac{n_2}{n_1}\right) \tag{10.3d}$$

The stimulated emission rate dominates over the absorption rate if number of atoms are larger in the excited state than the number of atoms in the ground state. This condition is called **population inversion**.

10.3 FLUORESCENCE AND PHOSPHORESCENCE

A fluorescent material glows under a ultraviolet lamp, but stop glowing as soon as lamp is turned off. A phosphorescent material keeps glowing for a while.

Phosphorescent materials have the ability to store up the light and release it gradually. This behaviour can be explained by the metastable consideration. If molecules of a substance can be excited from ground state to metastable state and if the metastable state can decay slowly back to the ground state via photon emission, the one can get the phosphorescence.

10.4 LEDs

The LED has become one of the most ubiquitous compound semiconductor devices. It is commonly employed in clocks, appliances, calculators, commercial sign boards and traffic lights. LED is a very simple device, essentially it is just a forward biased *p-n* junction diode that can emit spontaneous radiation in ultraviolet, visible or infrared regions. The device is designed in such a way that the most probable recombination mechanism is radiative recombination. The visible LED has applications as an information link between electronic instruments and their users. The infrared LEDs are useful in opto-isolators and for optical fibre communications. As we have already studied that there are mainly two types of semiconductors: direct semiconductor and indirect semiconductor. In direct semiconductor, both energy and momentum remain conserved, whereas in indirect semiconductor, there is violation of conservation of momentum. Si and Ge are the examples of indirect semiconductor. All compound semiconductor materials fall in the category of direct semiconductor.

The radiative transition mechanisms are found predominately on direct bandgap semiconductors such as GaAs and $GaAs_{1-y}P_y$ due to conservation of momentum. The photon energy is approximately equal to the bandgap energy of the semiconductor. The choice of materials for an LED is dictated primarily by the requirements of colour depending on the applications, the nature of material itself and technological maturity. The most important materials for LEDs are GaAs and GaP. GaP is indirect semiconductor, but impurities are added to introduce special recombination-centres to enchance the radiative processes. GaP LEDs emit light in the green portion of the visible spectrum. AlGaInP, $GaAs_{1-x}P_x$ changes from direct to indirect as the proportion of phosphorous increases. Red colour emission can be achieved by using direct bandgap emission with a 40% phosphorous composition. Orange and yellow colours can also be achieved by inserting impurities within the indirect compositions of GaAsP. Blue LEDs can be made using the ternary compound $In_{1-x}Ga_xN$.

10.4.1 Advantages of LEDs

- LEDs have higher efficiency than incandescent lighting and can ultimately or possibly exceed fluorescence in efficiency.
- LEDs have a very long lifetime.
- LEDs are highly durable and reliable.
- LED white light sources can provide significant energy saving compared with the conventional lighting.

10.4.2 Structure of LEDs

An LED must be constructed such that the light emitted by the radiative recombination events can escape the structure. LEDs can be designed as either surface or edge emitters, as shown in Fig. 10.3. Surface emitting LEDs can be tailored such that the bottom edge reflects light back toward the top surface to enhance the output intensity. Edge emitters in contrast emit light out of the side of the device. The main advantage of edge emitter LED is that the emitted radiation is relatively direct.

The three major issues that influence the commercial success of an LED are:

1. Luminous performance
2. Cost
3. Reliability

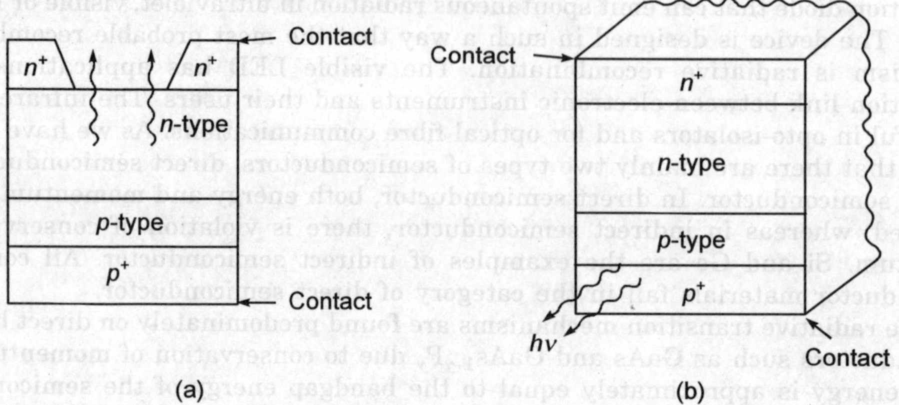

FIGURE 10.3   Structure of LED.

Luminous performance is related to the optical yield of the device. The optical yield measures the efficiency of LED to convert input electrical current into output photon. The luminous performance is given as the ratio of total output flux in lumens to the input power in watts. The lumen is the unit for measuring photometric flux.

There are three main factors that reduce the quality of the emitted photons:

1. Absorption within the LED material
2. Reflection loss when light passes from semiconductor to air due to differences in refractive index
3. Total internal reflection of a light at an angle greater than the critical angle θ_C

In LEDs, the applied voltage across the *p-n* junction will result in a diode current, which in turn can produce photons and light output. This inverse mechanism is called **injection electroluminescence**.

In brief, main working theory of LED is when a voltage is applied across a *p-n* junction, electrons and holes are injected across the space charge region, where they became excess minority carriers. These excess minority carriers diffuse into the neutral semiconductor regions and recombine with majority carriers. If this recombination process is a direct band-to-band process then photons are emitted. The output photon intensity is directly proportional to the ideal diode diffusion current.

The forward current-voltage characteristics of LED is same as the GaAs *p-n* junction as shown in Fig. 10.4. At low forward voltages, the diode current is dominated by the non-radiative recombination current mainly due to surface recombination near the perimeter of the LED chip. At higher voltages, the diode current is dominated by the radiative diffusion current. At even higher voltages, the series resistance will limit the diode current. The total diode current can be written as:

$$I = I_d \, \exp\left[\frac{q(V - IR_s)}{kT}\right] + I_r \, \exp\left[\frac{q(V - IR_s)}{2\,kT}\right] \tag{10.4}$$

where R_s is the device series resistance, I_d is saturation current due to diffusion and I_r is saturation current due to recombination.

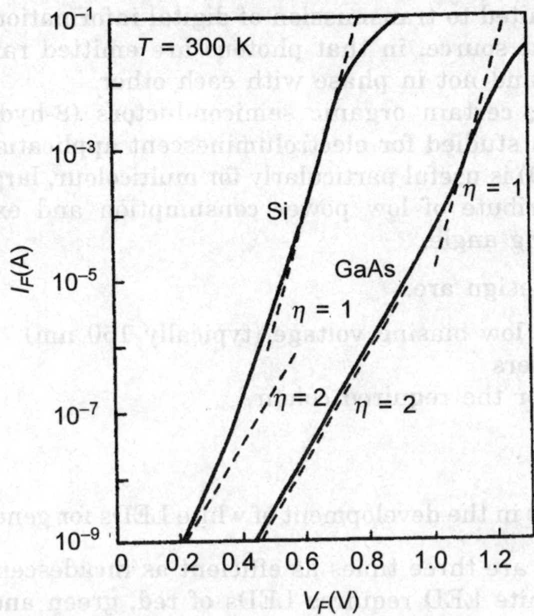

FIGURE 10.4   Forward bias current-voltage characteristic of LED.

10.4.3   Internal Quantum Efficiency and External Quantum Efficiency

The internal quantum efficiency of an LED is the fraction of diode current that will produce luminescence. The internal quantum efficiency is a function of injection efficiency and function of percentage of radiative recombination events compared with total number of recombination events.

The factors that most influence the internal quantum efficiency of a homojunction LED are: (1) the quality and purity of the material and the substrate (2) the existence of a direct energy gap spectrally matched to the desired wavelength. The presence of traps and impurities reduce the internal quantum efficiency.

The external quantum efficiency is defined as the ratio of the number of photons emitted from the device per incident injected electron.

The relation between internal quantum efficiency and external quantum efficiency is given as:

$$\eta_{\text{ext}} = \eta_{\text{int}}\, C_{ox} \tag{10.5}$$

where C_{ox} is the extraction efficiency defined as the fraction of generated photons that escape from the device package and emerge from the chip.

There are several issues that determine the extraction efficiency. These are:

(1) Absorption coefficient of the material at the emission wavelength
(2) Radiation geometry of the LED
(3) Substrate

Note: 1. LEDs are less suited to transmission of digital information than lasers. An LED is an incoherent light source, in that photons are emitted randomly from the junction on all directions and not in phase with each other.

2. In recent years, certain organic semiconductors (8-hydroxy-quinolinato aluminium etc.) have been studied for electroluminescent applications. The organic-light emitting diode (OLED) is useful particularly for multicolour, large area flat-panel display because of its attribute of low power consumption and excellent emissive quality with a wide viewing angle.

Criteria for OLEDs design are:

(i) Ultrathin layer for low biasing voltage (typically 150 nm)
(ii) Low injection barriers
(iii) Proper bandgaps for the required colour.

EXAMPLE 10.2

Why there has been interest in the development of white LEDs for general illumination?

Solution: Because LEDs are three times as efficient as incandescent lamps and can last 10 times longer. A white LED requires LEDs of red, green and blue colours.

EXAMPLE 10.3

Discuss the type of OLEDs, its advantages and limitations.

Solution: There are two general types of OLEDs devices. These devices are made with small organic molecules and polymers. In either case, organic material is highly disordered and thus can be readily deposited at very low manufacturing cost. The main advantage of OLEDs over standard LEDs is their cost.
Since, OLEDs are non-crystalline materials, therefore, they can be easily deposited in a relatively low cost manner. However, OLEDs have some limitations. One problem faced by OLED is lifetime. The organic materials used in LEDs are very sensitive to water and oxygen. Thus, they must be deposited on glass substrates and covered with a second sheet of glass. Unfortunately the usage of glass makes the cost of manufacturing higher.

EXAMPLE 10.4

What is the purpose of using heterojunction LEDs?

Solution: Since, a photon emitted due to radiative recombination process of direct bandgap material in one part of the device and reabsorbed before it reaches the surface, therefore, one obvious solution is to put the junction close to the surface to avoid reabsorption. But the presence of surface states, producing non-radiative recombination channels, can result in a reduction in the radiative efficiency of the device. Therefore, the solution of this problem is to use heterojunction instead of avoiding reabsorption. A commonly used heterojunction is that of GaAs and AlGaAs. The heterojunction LED is designed such that the recombination occurs on the narrow gap semiconductor. The light is subsequently emitted through an AlGaAs window into the external environment. Since bandgap of AlGaAs is greater than that of GaAs and hence, most of the light will transmit through the AlGaAs without being absorbed.

EXAMPLE 10.5

How can one improve the extraction efficiency of LEDs?

Solution: The extraction efficiency of LEDs can be improved by choosing a transparent substance. In such a device the light that is internally reflected, but is reabsorbed will eventually escape after being randomly scattered by rough spots on the chip surface. Therefore, the extraction efficiency can be quite high (of order of 0.3 in most structure).

EXAMPLE 10.6

What are the importance of infrared LEDs?

Solution: Infrared LEDs include GaAs LEDs which can emit light near 0.9 μm and many III–IV compounds such as quantenary $Ga_xIn_{1-x}As_yP_{1-y}$ LEDs, which emit light from 1.1 to 1.6 μm.

An important application of infrared LEDs is in optoisolators where an input or control signal is decoupled from the output. Another important application is for transmission of an optical signal through an optical fibre.

10.5 LASER OPERATION

In an LED the emitted light results from spontaneous emission wherever in laser the light is emitted through stimulated emission. A semiconductor exhibits a net stimulated emission rate when the difference in the quasi-Fermi levels exceeds the bandgap. This condition is called **population inversion** and can be realized by using a forward biased, degenerately doped *p-n* junction diode. In order for a system to lase, the optical gain must exceed all optical losses such that the optical field grows with distance and the field must be reasonantly amplified.

Note: Semiconductor lasers are similar to the solid-state ruby laser and helium-neon gas laser as the emitted radiation is highly monochromatic and produces a highly directional beam of light. The semiconductor laser is different from other lasers in that it is small (of order of 0.1 nm long) and easily modulated at high frequencies. Due to these features, the semiconductor laser is one of the most important light sources for optical fibre communications. Also used in video recording, optical reading and high-speed laser printing.

10.5.1 Semiconductor Materials

All lasing semiconductors have direct bandgaps. GaAs was the first material to emit laser radiation and is related III–V compound alloys are the most extensively studied and developed. The three most important III–V compound alloy systems are Ga_xIn_{1-x} AS_yP_{1-y}, Ga_xIn_{1-x} As_ySb_{1-y} and Al_xGa_{1-x} As_ySb_{1-y} solid solutions. To achieve a heterostructure with negligible onto face traps, the lattices between the two semiconductors must be matched closely.

10.5.2 Population Inversion

To enhance the stimulated emission for laser operation we need population inversion. To achieve population inversion in a semiconductor laser, it is required to form a *p-n* junction or heterojunction between two degenerated semiconductors. This mean Fermi level on *p*-side is below the valence band edge and Fermi level on *n*-side is above the conduction band edge as shown in Fig. 10.5(a). This will enhance the carrier injection across the junction when sufficiently large bias is applied, i.e., large concentration of electrons and holes are injected into the transition region. As a result the transition region *d* contains a large concentration of electrons in the conduction band and a large concentration of holes in the valence band as shown in Fig. 10.5(b). This is the required condition for population inversion. For band-to-band minimum required energy is the bandgap E_g. Therefore, the condition required for population inversion is:

$$(E_{FC} - E_{Fv}) > E_g \qquad (10.6)$$

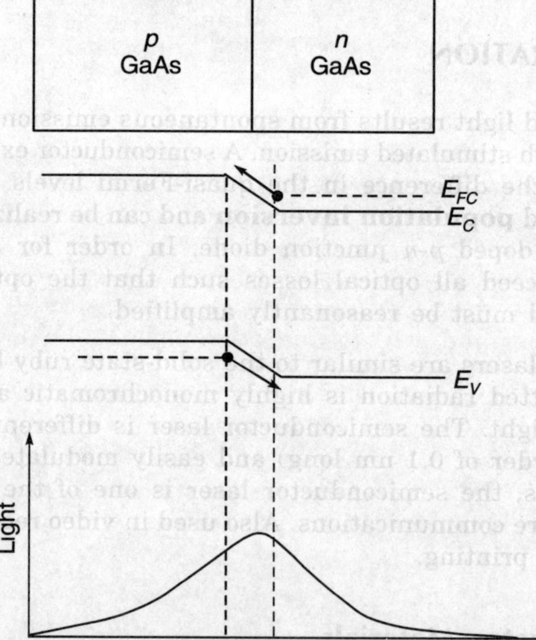

FIGURE 10.5 Schematic representation of laser action.

10.5.3 Resonant Amplification

Resonant amplification can be achieved by confining the emitted light in a resonant cavity. The simplest design of a cavity is to use two reflective surfaces bounding the active region in which stimulated emission occurs. This region is also called the **gain medium**. The overall structure is similar to waveguide on which light is confined to bounce back and forth through the gain medium between two reflective surfaces. In gas laser, reflective surfaces are simply mirror, whereas in a semiconductor laser, cleaving the ends of waveguide from the reflective surfaces. By confining the radiation to a resonant cavity we can oscillate only one frequency and thus, it is resonantly amplified. As a result we get a monochromatic light at the output. This is often referred to as a condition for oscillation. The spacing of reflective surfaces decide which wavelength of light is going to resonantly enhanced. Constructive interference of light occurs when the light is coherently reflected back and froth within the cavity. The condition for the constructive interference is given as:

$$L = \frac{m\lambda}{2} \tag{10.7}$$

where L is length of cavity, λ is wavelength of incident light and $m = 1, 2, 3, \ldots$

Generally, the optical field within the resonant cavity of laser experiences multiple reflections between the two reflective surfaces and passing through the gain medium several times. Each time the field passes through the gain medium is amplified by the addition of photons produced through stimulated emission events. The additional photons produced during each pass time from stimulated emission add coherently to the optical field. A fraction of light is transmitted through second reflective surface. This fraction of light E_t provides the output beam, and is given as:

$$E_t = \frac{E_i t_1 t_2 e^{-\Gamma L}}{1 - r_1 r_2 e^{-2\Gamma L}} \tag{10.8a}$$

where E_i is amplitude of incident wave, t_1 and t_2 are the transmission coefficients of reflective surface, r_1 and r_2 are the reflection coefficients of respective surface and Γ is a complex number and gives propagation constant.

For resonance, the required condition is:

$$E \to \infty$$

i.e.,
$$1 - r_1 r_2 e^{-2\Gamma L} = 0$$

or
$$r_1 r_2 \, e^{-2\Gamma L} = 1 \tag{10.8b}$$

Defining
$$\Gamma = j\beta k_0 - \alpha \tag{10.8c}$$

where the first term leads to a propogation wave. The negative figure of the real part indicates that the wave attenuates with the distance travelled. Substituting Eq. 10.8(c) into Eq. 10.8(b) and after simplification, we have

$$1 = r_1 r_2 \, e^{-2j\beta k_0 L} \, e^{2\alpha L} \tag{10.8d}$$

The phase term $e^{-2j\beta k_0 L}$ does not change the magnitude of the field because it has unity magnitude. Therefore, the amplitude of the field depends upon the product of the reflectivities and the real exponential term

$$1 = r_1 r_2 \ e^{2\alpha L} \tag{10.8e}$$

The parameter α can be either positive or negative.
For

$$\alpha = \begin{cases} \text{negative, field must attenuate with distance and medium must be lossy.} \\ \text{positive, field grows with distance and propagation medium provides gain.} \end{cases}$$

In a laser, α has both positive and negative real components and can be expressed as:

$$\alpha = g - \alpha_i \tag{10.9a}$$

where g represents the gain and α_i the losses within the propagation medium. From Eq. (10.9a) it is seen that if $g > \alpha_i$ then α is a positive quantity. This means medium provides gain when sum of all the losses must be less than the gain g. In this case system will lase. On the other hand, if $g < \alpha_i$ means α is a negative. In other words, we can say the medium behaves as lossy medium if sum of all loses exceeds the gain. This system will not lase.

From Eq. (10.9a) we have

$$g = \alpha + \alpha_i$$

and substituting α from Eq. (10.8e), we get

$$g = \alpha_i + \frac{1}{2L} \ln\left(\frac{1}{r_1 r_2}\right) \tag{10.9b}$$

where α_i represents the various loss factors like scattering loss, non-radiative loss, coupling loss and mirror loss.

When the gain is equal to the loss term, then system can begin to lase. The corresponding gain is known as the **threshold gain**. For threshold gain, there should be a minimum current density injected into the active region of the laser which is called **threshold current density** J_{Th}.

In steady state, the continuity equation is given as:

$$\frac{dn}{dt} = \frac{J}{qd} = \frac{n}{\tau} = 0 \tag{10.9c}$$

where τ is recombination time, d is the width of the active region and n is electron concentration within the active region.

Using Eq. (10.9c), the threshold current density is expressed as:

$$J_{\text{Th}} = \frac{q n_{\text{Th}} d}{\tau} \tag{10.9d}$$

where n_{Th} is concentration of electron in active region at threshold. Equation (10.9d) shows that a minimum current density and corresponding minimum electron concentration must be injected into active region to achieve the threshold and lasing action.

EXAMPLE 10.7

The reflection coefficients for a semiconductor laser are equal and have the value 0.5. The length of cavity is 1 μm. Assume that there are no additional loss terms. Calculate the gain of the laser.

Solution: No internal loss means $a_i = 0$. Using Eq. (10.9b),

$$g = \frac{1}{2L} \ln \left(\frac{1}{r_1 r_2} \right)$$

$$= \frac{1}{2 \times 1} \ln \left(\frac{1}{0.5 \times 0.5} \right)$$

$$= 6.93 \times 10^3 / \text{cm}$$

10.5.4 Types of Semiconductor Laser

In a simple homojunction semiconductor laser, one can achieve relatively high threshold current densities because both carriers and optical confinement within the active region of device are minimum. Carriers injected into active region of device pass through it without any recombination and do not contribute to laser action. It is also seen that the optical fields leak into the surrounding inactive layers. Due to these two reasons, the threshold current density increases which reduces the efficiency of laser action.

Heterostructure provides a mean of confining both the optical fields and carriers within the active region and reducing the threshold current density. For heterostructure it is required that the lattice constants of two materials should match, e.g., GaAs and AlGaAs. The structure of a simple GaAs–AlGa is shown in Fig. 10.6 which is called a **double heterostructure laser** because it contains two GaAs–AlGaAs interface. Since the energy bandgap of AlGaAs is greater than GaAs, therefore, there is an energy gap discontinuity between two materials. GaAs and AlGaAs are said to form a type I heterojunction in which the energy gap discontinuity is equal to the sum of the conduction and valence band edge discontinuities.

The presence of conduction and valence band energy discontinuities in the two heterostructures act to greatly confine the injected carriers within the narrow gap GaAs layer as shown in Fig. 10.7 which ultimately can result in a radiative recombination event with the subsequent emission of a photon. Heterostructure has lower threshold current density than the homojunction laser. The double heterostructure also acts to confine the optical fields within the GaAs layer due to different refractive indices of the used semiconductor materials.

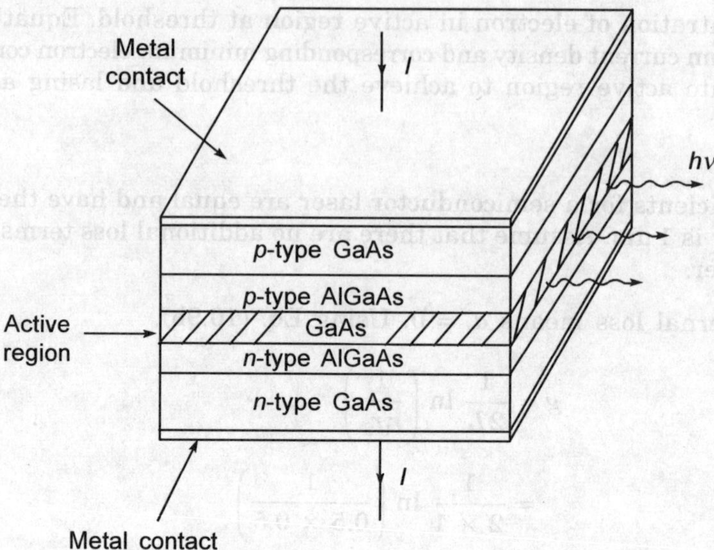

FIGURE 10.6   Structure of GaAs-AlGa laser.

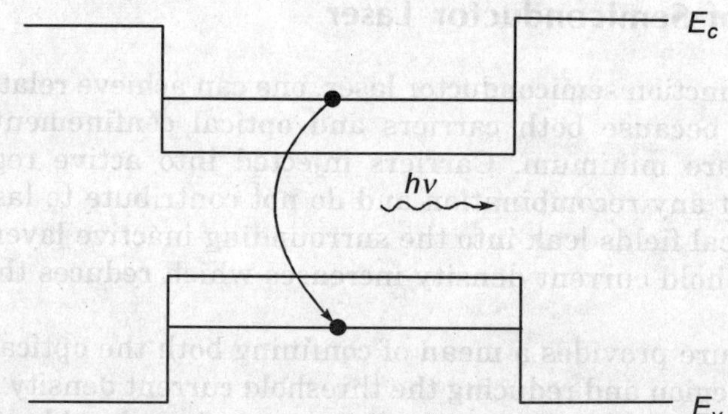

FIGURE 10.7   Radiative event in GaAs layer.

In the laser action, photon is emitted by stimulated emission. The light bounces back and forth due to reflection from the boundaries of the device. The reflecting surfaces are produced by cleaving the crystal such that the cleaved surface provides for nearly complete reflection. Stimulated emission within the GaAs layer is triggered by reflected light. The emitted photon has the same phase as that of the stimulating light, i.e., coherence is maintained. In double heterostructure laser, the emitted light has an energy close to that of the energy gap of narrow gap material. The energy of the emitted photon thus depends on the intrinsic properties of the active medium.

Quantum well lasers

Quantum well lasers offer an alternative to double heterostructures. In quantum well lasers, the energy of emitted photon can be made subsequently higher than that of the bandgap. By proper choice of well width, the energy of the emitted photon can be adjusted so that lasing can be achieved at various wavelengths using the same material. A single quantum well laser is similar in design to a double heterostructure laser except that the narrow gap. This means active region is intentionally made thin so that the spatial quantization effects should occur. As we know that when the dimension of the confining region is equal to that of electron de Broglie wavelength, quantum mechanical effects occur. According to quantum mechanical theory, only certain discrete energy levels are allowed in the atom and electrons/holes can occupy only these allowed energy levels. Spatial quantization results in the production of energy levels above the conduction band minimum in a quantum well. The minimum electron and hole energies within the well are higher than the conduction band and valence band minima respectively. If electron and hole recombine, the energy of the emitted photon is greater than that of the energy gap. It is found that the energy of quantum states is a strong function of the well width. The narrower the width, the greater the quantum state energy. This means in a very narrow width device, the energy of emitted photon is greater than that of a wider well width device. By adjusting the well width, the energy of the emitted photon can be tuned.

Limitations of quantum well lasers. The cross-section for the capture of carriers in the well is relatively small. In order to obtain a radiative recombination event, it is necessary for the injected electrons and holes to be captured within the well. To confine the carriers within the well, it is required that they must undergo an inelastic scattering event in order to lose sufficient energy. If there is no scattering events within the well width, the injected carriers will traverse the well without being captured. In other words, we can say that the injected carriers can pass directly over the well without becoming trapped and do not contribute to the optical output. As a result, threshold current density can be relatively high.

Multiple quantum well lasers. The capture probability of a single quantum well laser can be improved by using multiple quantum wells as shown in Fig. 10.8. As can be seen from Fig. 10.8, the single quantum well within the active region is replaced by a series of quantum wells. Since, more quantum wells are present in the structure, therefore, there is a higher probability that the injected carrier will suffer an inelastic scattering event and be captured within a well before it can completely traverse the device. Main drawback of multiple quantum well device is that the injected electrons

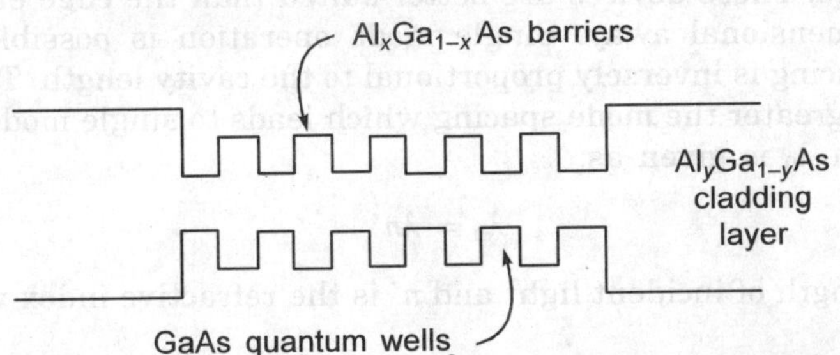

FIGURE 10.8 Structure of multiple quantum well laser.

and holes enter the active region from opposite ends. Therefore, the electrons and holes are not necessarily captured within the same well which acts to reduce the radiative recombination rate. For this reason the multiple quantum well devices do not have the lowest threshold current density of semiconductor laser.

GRINSCH laser. The acronym stands for **graded** index separate confinement heterostructures. This has the lowest threshold current density of existing semiconductor lasers. The structure of this laser is shown in Fig. 10.9.

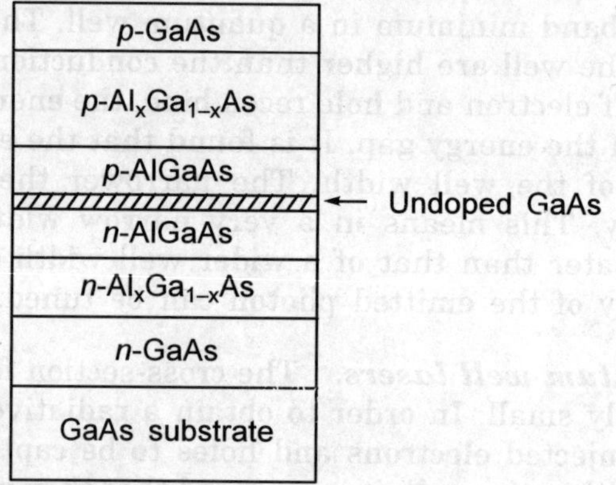

FIGURE 10.9 Structure of GRINSCH laser.

In the structure as shown in Fig. 10.9, a single quantum well is embedded within a graded funlike region. The graded region surrounding the quantum well provide a means by which the injected carriers can be tunnelled into the well. The bandgap grading induces an electric field that directs the electrons and holes into the same quantum well where they can recombine radiatively. As a result fewer electrons and holes can traverse the quantum well without being trapped. Thus, the threshold current density of GRINSCH laser is relatively lower than any other lasers.

Vertical cavity surface emitting lasers (VCSELs). In this structure, the light is emitted normal to the surface as shown in Fig. 10.10. The primary advantages of VCSELs is that the single mode operation can be achieved due to the short cavity length of the device. These devices are better suited than the edge emitting lasers to forming a two-dimensional away. Single mode operation is possible with VCSELs since the mode spacing is inversely proportional to the cavity length. Thus, the smaller the cavity length, greater the mode spacing which leads to single mode operation. The output wavelength λ_0 is given as:

$$\lambda_0 = \lambda n' \tag{10.10a}$$

where λ is wavelength of incident light and n' is the refractive index within the laser.

$$\text{Mode index } m = \frac{2Ln'}{\lambda_0} \tag{10.10b}$$

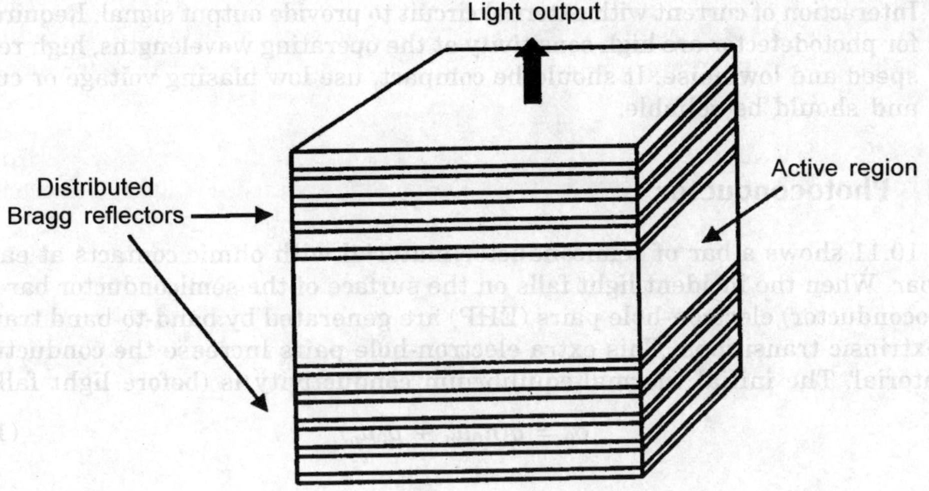

FIGURE 10.10   Structure of VCSELs laser.

Hence

$$\frac{dm}{d\lambda_0} = \frac{-2\,Ln'}{\lambda_0^2} + \frac{2L}{\lambda_0}\,\frac{dn'}{d\lambda_0} \qquad (10.10c)$$

where $dm/d\lambda_0$ defines the mode spacing with wavelength. Solving Eq. (10.10c), we have

$$d\lambda_0 = \frac{(\lambda_0^2/2\,Ln')\,dm}{\left[1 - \dfrac{\lambda_0}{n'}\,\dfrac{dn'}{d\lambda_0}\right]} \qquad (10.10d)$$

The value of dm is 1 for adjacent modes.

From Eq. (10.10d) it is clear that the wavelength separation between adjacent modes is inversely proportional to the cavity length.

The major disadvantage of VCSELs is that the cavity length is short as the round trip gain of the device is low. The threshold current density of VCSELs is significantly higher than that of an edge emitting laser.

10.6 PHOTODETECTOR

Photoconductors are semiconductor devices that can convert optical signals into electrical current. The operation of photodetector involves three steps:

- Carrier generation, i.e., the absorption of photon create electron-hole pairs
- Carrier transport

- Interaction of current with external circuit to provide output signal. Requirements for photodetector are high sensitivity at the operating wavelengths, high response speed and low noise. It should be compact, use low biasing voltage or currents and should be reliable.

10.6.1 Photoconductor

Figure 10.11 shows a bar of semiconductor material with ohmic contacts at each end of the bar. When the incident light falls on the surface of the semiconductor bar (which is photoconductor) electron-hole pairs (EHP) are generated by band-to-band transition or by extrinsic transitions. This extra electron-hole pairs increase the conductivity of the material. The initial thermal-equilibrium conductivity is (before light falls):

$$\sigma_0 = q(n_0\mu_n + p_0\mu_p) \tag{10.11a}$$

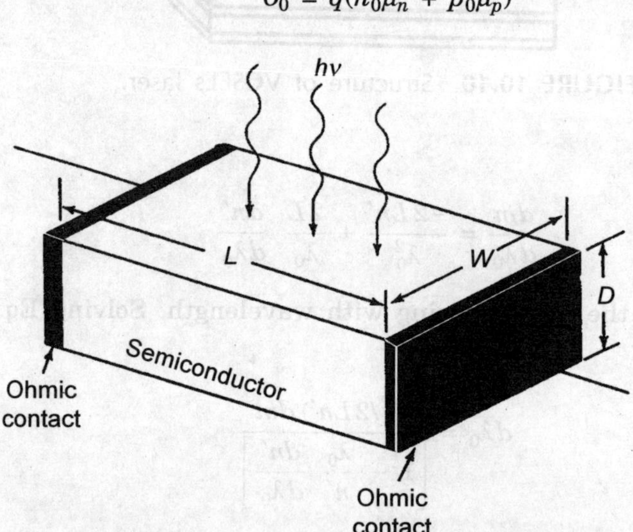

FIGURE 10.11 A Semiconductor bar exposed with light.

The increase in conductivity under illumination is mainly due to the increase in the number of carriers, and is given as:

$$\Delta\sigma = q\,\delta p(\mu_n + \mu_p) \tag{10.11b}$$

This change in conductivity due to optical excitation is known as the **photoconductivity**.

When a voltage is applied across the bar of length L then the electric field produces current in the bar. The current density is written as:

$$J = J_0 + J_L = (\sigma_0 + \Delta\sigma)E \tag{10.11c}$$

where J_0 is the current density (when no optical excitation, i.e., dark condition) and J_L is photocurrent density.

Let us assume electrons and holes are generated uniformly throughout the bar then the current due to the optical excitation is:

$$I_L = J_L A = \Delta \sigma A E$$

$$= \delta_p (\mu_n + \mu_p) A E q$$

$$= G_L \tau_p (\mu_n + \mu_p) A E q \qquad (10.11d)$$

where A is the cross-sectional area of the device. Equation (10.11d) shows that photocurrent is directly proportional to the area of the device and the rate at which excess-carriers are generated.

The time required by electron to flow through the bar of the photoconductor is known as **transit time**, and is given as:

$$t_n = \frac{L}{\mu_n E}$$

Substituting this value in Eq. (10.11d), we have

$$I_L = G_L \tau_p (\mu_n + \mu_p) A q \, \frac{L}{\mu_n t_n}$$

$$= q A G_L \left(\frac{\tau_p}{\tau_n} \right) (\mu_n + \mu_p) \frac{L}{\mu_n}$$

$$= q A G_L L \left(\frac{\tau_p}{t_n} \right) \left(1 + \frac{\mu_p}{\mu_n} \right) \qquad (10.11e)$$

The photocurrent gain is defined as the ratio of the rate at which charges are collected by the contacts to the rate at which charges are generated within the photoconductor, i.e.,

$$\text{Gain} = \frac{I_L}{q A G_L L} \qquad (10.11f)$$

Substituting Eq. (10.11e) into Eq. (10.11f), we get

$$\text{Gain} = \frac{\tau_p}{t_n} \left(1 + \frac{\mu_p}{\mu_n} \right) \qquad (10.11g)$$

If $\mu_p / \mu_n << 1$,

$$\text{Gain} \simeq \frac{\tau_p}{t_n} \qquad (10.11h)$$

EXAMPLE 10.8

Calculate gain of a photoconductor device of length 100 μm and cross-sectional area of 10^{-7} cm^2. Assume the bar is of n-type with minority carrier lifetime is 10^{-6} sec. Take $\mu_n = 1350$ cm^2/V-sec and applied voltage is 10 V.

Solution: $\quad t_n = \dfrac{L}{\mu_n E} = \dfrac{L^2}{\mu_n V} = 7.41 \times 10^{-9}$ sec

$$\text{Gain} = \frac{\tau_p}{t_n}\left(1 + \frac{\mu_p}{\mu_n}\right) \simeq 1.83 \times 10^2$$

10.6.2 Photodiode

A photodiode is basically a p-n junction or metal semiconductor contact operates under reverse bias. A cross-section of a typical silicon photodiode is shown in Fig. 10.12. The starting material is n-type silicon. A thin p layer is formed on the front surface of the device by thermal diffusion or ion implantation of the appropriate doping. Small metal contacts are applied to the front surface of the device and entire back is coated with a contact metal. The back contact is cathode and front contact is anode. The active area is coated either with silicon nitride, silicon monoxide or silicon oxide for protection and to serve as the anti-reflection coating. The thickness of the top p layer is determined by the wavelength of the radiation to be detected. The depth of the depletion region can be varied by applying a reverse bias voltage across the junction. When the depletion region reaches the back of the photodiode then it is said to be *fully depleted*. The depletion region is important to photodiode performance since most of the sensitivity to radiation originates these.

When a light is absorbed in the active region, an electron-hole pair is created. The electrons are collected in the n-region and holes in the p-region. The migration of electrons and holes to their respective region is called photovoltaic effect. Photodiodes are most useful as current generators although a voltage is also generated by illumination. Also, the thermally generated minority carriers within the diffusion length of each side of the junction diffuse to the depletion region, and are swept to the other side by the electric field.

When a reverse bias is applied, some current will flow without illumination. The Dark current is specified for every device. In cases where a very low bias voltage is applied, shunt resistance is specified. This is determined by measuring dark current with +/− 0.010 V applied bias.

Let us assume that the junction is uniformly illuminated by photons of energy $h\nu > E_g$ and due to that FHPs are created at the rate of g_{op} (EHP/cm^3-sec) which take part in the conductivity. The number of holes created per second within the diffusion length of transition region on n side is $AL_p g_{op}$, where A is the cross-sectional area. Similarly, $AL_n g_{op}$ represents the number of electrons created per second within the diffusion length of the transition region on p-side. AWg_{op} is the number of generated

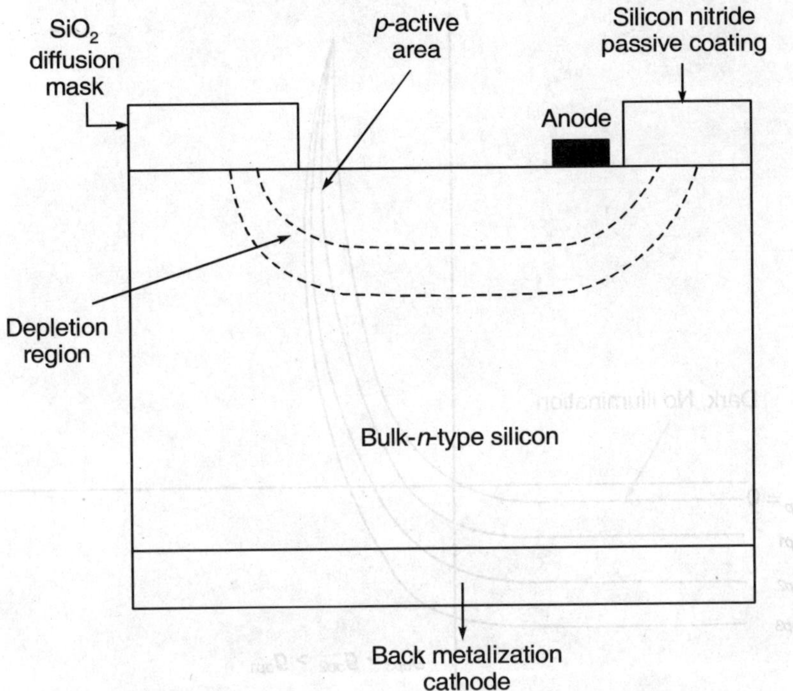

FIGURE 10.12 Cross-section of silicon photodiode.

carriers within the transition region of width W. These all optically generated carriers take part in conduction and hence resulting current is:

$$I_{op} = qAg_{op}(L_n + L_p + W) \qquad (10.12a)$$

Adding the thermally generated current and optically generated current, the total reverse bias diode current is given as:

$$I = I_{Th}(e^{qV/kT} - 1) - I_{op}$$

$$= qA\left(\frac{L_p}{\tau_p}\,p_n + \frac{L_n}{\tau_n}\,n_p\right)(e^{qV/kT} - 1) - qAg_{op}(L_n + L_p + W) \qquad (10.12b)$$

Equation (10.12b) clearly shows that the total diode current reduces due to the generation rate. The first term of the RHS of Eq. (10.12b) represents the usual diode current equation and the second term gives current due to optical generation. The I–V characteristics in the presence of optical generation is shown in Fig. 10.13.

For $V = 0$, means device is short circuited then Eq. (10.12b) reduces to

$$I = -I_{op}$$

This shows that the short circuit current from n to p is equal to optical generation current in magnitude, i.e.,

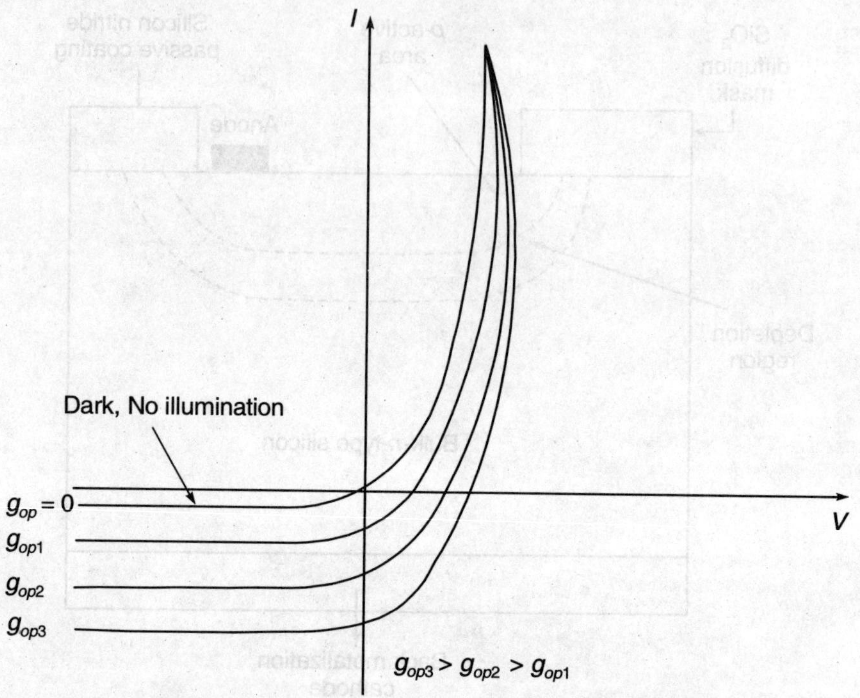

FIGURE 10.13 *I-V* characteristic of photodiode.

$$I = -qAg_{op}(L_n + L_p + W) \tag{10.12c}$$

or for constant values of A, L_n, L_p and W,

$$I \propto g_{op}$$

Therefore, the short circuit current of the device is proportional to the optical generation rate.

Let us assume V_{OC} represents the open circuit voltage for $I = 0$ and substituting $I = 0$ in Eq. (10.12b), we have

$$qA\left(\frac{L_p}{\tau_p} p_n + \frac{L_n}{\tau_n} n_p\right)(e^{qV_{oc}/kT} - 1) = qAg_{op}(L_n + L_p + W)$$

or

$$(e^{qV_{oc}/kT} - 1) = g_{op}\left(\frac{L_n + L_p + W}{(L_p/\tau_p)p_n + (L_n/\tau_n)n_p}\right)$$

or

$$V_{oc} = \frac{kT}{q} \ln\left(\frac{L_n + L_p + W}{(L_p/\tau_p)p_n + (L_n/\tau_n)n_p} \cdot g_{op} - 1\right) \tag{10.12d}$$

Consider a special case, when junction is symmetrical,

$$\tau_p = \tau_n \text{ and } n_p = p_n$$

Equation (10.12d) can be further written as:

$$V_{oc} = \frac{kT}{q} \ln\left(\frac{L_n + L_p + W}{(L_n + L_p)\,p_n/\tau_n} \cdot g_{op} + 1 \right) \tag{10.12e}$$

Defining $p_n/\tau_n = g_{Th}$ = Thermal generation rate, Eq. (10.12e) can be further simplified as:

$$V_{oc} \simeq \frac{kT}{q} \ln\left(\frac{L_n + L_p + W}{L_n + L_p} \cdot \frac{g_{op}}{g_{Th}} + 1 \right) \tag{10.12f}$$

Assume $L_n + L_p \gg W$ and $(g_{op}/g_{Th}) \gg 1$. Then Eq. (10.12g) reduces to

$$V_{OC} \simeq \frac{kT}{q} \ln\left(\frac{g_{op}}{g_{Th}} \right) \tag{10.12g}$$

Note: V_{OC} cannot increase indefinitely with increased g_{op} and limit on V_{OC} is that it can take maximum value of equilibrium contact potential.

EXAMPLE 10.9

Why V_{OC} cannot increase indefinitely with g_{op}?

Solution: As the minority-carrier concentration increases due to optical generation rate, the lifetime τ_n becomes small and hence the term (p_n/τ_n) becomes larger which results in increase of g_{Th}. And hence the ratio (g_{op}/g_{Th}) cannot increase for indefinite period.

Note: The magnitude of the photocurrent generated by a photodiode is dependent upon the wavelength of the incident light. Silicon photodiodes exhibit a response from the UV through the visible and into the near infrared part of the spectrum.

Responsivity R_e is a measure of sensitivity which takes into account the active area of the photodiode chip. This parameter is obtained by dividing the short-circuit current by the energy of light per unit area.

Photodiodes have unity internal gain. In order to increase the sensitivity of photodiodes to light, one can either increase the active area of the photodiode chip itself or use lenses to increase the effective active area. There is a linear relationship between sensitivity and active area.

Note: Depending upon the required applications of the photodiode, it can be operated in either quadrants of its *I-V* characteristics. Since, the photodiode is reverse biased so the preferable quadrants are either third or fourth as shown in Fig. 10.14.

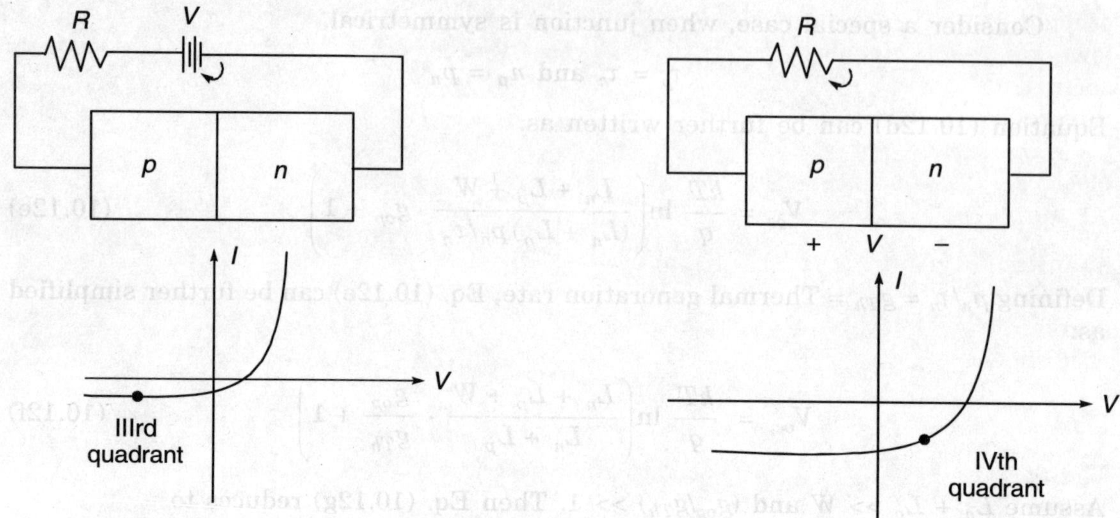

FIGURE 10.14 Schematic representation of operating quadrant in photodiode.

In third quadrant, both the current and junction voltage are either positive (or negative). In this quadrant power is delivered to the device from the external circuit because $P = VI = (-V)(-I)$. Power quantity is always positive.

In fourth quadrant, junction voltage is positive and current is negative and hence power is delivered from the junction to the external circuit.

If one is interested to extract the power from the device then fourth quadrant is used. But usually in photodetector we are using reverse bias and hence third quadrant will be chosen for operation.

Quantum efficiency

The quantum efficiency relates, as a percentage, the energy per photon and the quantum yields (electrons per photon). The mathematical expression for quantum efficiency is given as:

$$\eta = \left(\frac{I_{op}}{q}\right)\left(\frac{P_{op}}{h\nu}\right)^{-1} \qquad (10.13a)$$

where I_{op} is optically generated current and P_{op} is the optical power. It can also be given as:

$$\eta = \left(\frac{(124)\,S_R}{\lambda}\right) \qquad (10.13b)$$

where S_R is radiometric sensitivity and λ is the wavelength of incident light.

Increasing the operating temperature of a photodiode device results in two distinct changes in operating characteristics. The first change is a shift in the quantum efficiency due to changes in the radiation absorption of the device. QE value shifts lower in the UV region and higher in the IR region due to negative temperature coefficient at shorter wavelength of light and positive temperature coefficient in the infrared part of the spectrum.

The second change is caused by the exponential increase in the thermally excited electron-hole pairs resulting in increasing dark current.

Note: In ultraviolet and visible region, metal-semiconductor photodiodes show good quantum efficiencies. In the near-infrared region, silicon photodiodes can reach 100% quantum efficiency near the 0.8 to 0.9 µm region. In the 1.0 to 1.6 µm region, germanium photodiodes and group III–V photodiodes have shown high quantum efficiencies.

Dark current

The dark current is the leakage current that flows when the photodiode is in dark, i.e., no illumination of light and a reverse bias voltage is applied across the junction. The dark current may vary from pA to µA depending upon the junction area and process used. Dark current is always specified at a particular value of reverse applied voltage. The dark current is temperature dependent and becomes double for every 10°C increase in ambient temperature.

Other parameters

Shunt resistance R_{SH}. The shunt resistance or dynamic junction resistance at zero voltage is determined by applying a small voltage to the photodiode (~ mV) and measuring the resulting current. Values of shunt resistance may vary from 100 kΩ to 100 GΩ. Shunt resistance is voltage dependent and also depends on the active area of the diode chip and on the type of processing used. Shunt resistance decreases with increasing temperature.

Junction capacitance C_j. This is the capacitance associated with the depletion region which exists at the *p-n* junction. The response time of a photodiode is dependent upon the product of junction capacitance and the external load resistor. Junction capacitance increases with increasing junction area and decreases with reverse applied voltage.

Reverse breakdown voltage V_{BR}. This is the maximum reverse voltage that can safely be applied across the photodiode before breakdown occurs at the junction. The breakdown voltage is determined by process. Typical values for V_{BR} range from 5 V to over 100 V.

Open circuit voltage. V_{OC} is that voltage which is generated by photodiode when photocurrent is zero. V_{OC} varies logarithmically with light. Typical range of V_{OC} is 300 mV to 450 mV. Due to large temperature coefficient V_{OC} is not recommended as an accurate measure of light level.

Response time. A photodiode takes a certain amount of time to respond to a sudden change in light levels. It is expressed in terms of rise time t_r and fall time t_f.

Rise time t_r is the time required for the output to rise from 10% to 90% of its final value. t_f is the fall time, defined as the time required for the output to fall from 90% to 10% of its onstate value.

The response time of a photodiode depends on many factors, like wavelength of incident light, the value of applied voltage across the diode (affects the junction capacitor) and the load resistance.

Noise current. A photodiode will act as a source for electrical noise and generate a noise current J_N. The noise current will limit the usefulness of the photodiode at very low light levels where the magnitude of noise approached that of the signal photocurrent. The amount of noise generated depends on the characteristics of the photodiode and operating conditions.

There are three main components which contribute on the total noise of photodiode, namely, thermal noise, short noise and flicker noise. Thermal or Jhonson noise is inversely related to the shunt resistance of the photodiode. Thermal noise tends to be dominant noise component when the diode is operated under zero applied reverse bias. Shot noise is dependent on the leakage or dark current of the photodiode. It tends to dominate when the photodiode is used in the photoconductive mode. Flicker noise is unlike thermal or shot noise. It possesses a (1/f) spectral density. Flick noise may dominate when the bandwidth of the interest contains frequencies less than 1 kHz.

Noise equivalent power (NEP). In many design applications, the designer needs to know the minimum detectable light of the photodiode. The minimum incident power required on a photodiode to generate a photocurrent equal to the total photodiode noise current is defined as the noise equivalent power (NEP), i.e.,

$$\text{NEP} = \frac{\text{Noise current}}{\text{Responsivity}} \qquad (10.14a)$$

The NEP is dependent on the bandwidth of the measuring system. The NEP is always quoted at a particular wavelength and is non-linear over the wavelength range.

Note: The response speed of photodiode is limited by three factors:

1. Diffusion of carriers
2. Drift time in the depletion region
3. Capacitance of the depletion region.

Equivalent operating circuits

The equivalent circuit and symbol of a photodiode is shown in Fig. 10.15.

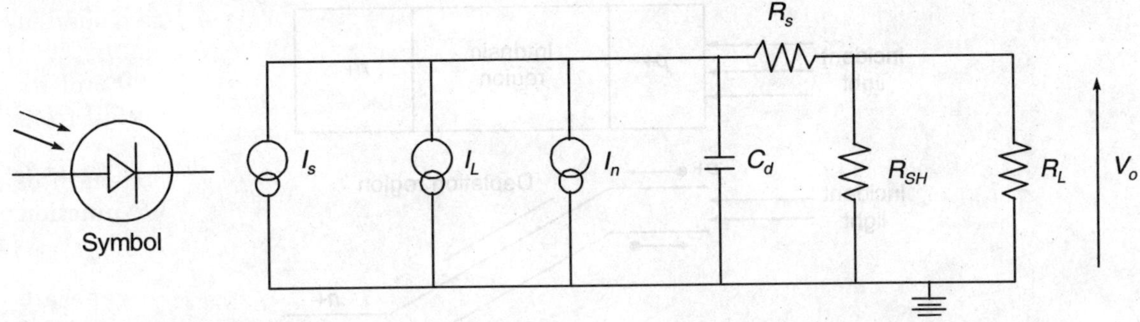

FIGURE 10.15 Equivalent operating circuit model of photodiode.

where I_s is signal current, I_L is leakage current, I_n is noise current, C_d is diode junction capacitance, R_s is diode series resistance, R_{SH} is diode shunt resistance and R_L is load resistance.

The output voltage V_o is given as:

$$V_o = (I_s + I_L + I_n)\left(\frac{R_L\,R_{SH}}{R_L + R_{SH} + R_s}\right) \qquad (10.14b)$$

Fundamentally a photodiode is a current generator. The junction capacitance of the photodiode depends on the depletion layer width and hence on bias voltage. Generally R_{SH} is very high of order of MΩ and R_s is very low.

Note: 1. Photodiodes are attractive for light wave communication system because they provide high quantum efficiency and very high bandwidth.

2. In order to fully absorb the incident radiation in the photodiode, it is necessary to utilize a wide depletion region. This is the reason why *p-i-n* junction diode is often used. However, as the width of the depletion region increases the transit time across the device increases and hence the bandwidth reduces. This is a tradeoff in photodiode.

p-i-n photodiode

The *p-i-n* photodiode is one of the most common photodetectors because of its high quantum efficiency and bandwidth. Typically, a *p-i-n* photodiode is illuminate from *p*-side as shown in Fig. 10.16. The depletion region, because device is reverse biased, of the device contains the intrinsic region and within the intrinsic region the electric field strength is contant. Holes generated near the edge of the depletion region on the *p*-side generally recombine at the contact or within the *p*-region. In general, the electrons can be photogenerated within the p^+ region as well as the intrinsic region of the diode. Most of the *p-i-n* photodiodes are designed such that the vast majority of carriers are photo generated within the intrinsic region. This intrinsic region acts as a depletion region because *n*-side as well as *p*-side is heavily doped. In normal photodiodes, some of the carriers are generated outside the depletion region.

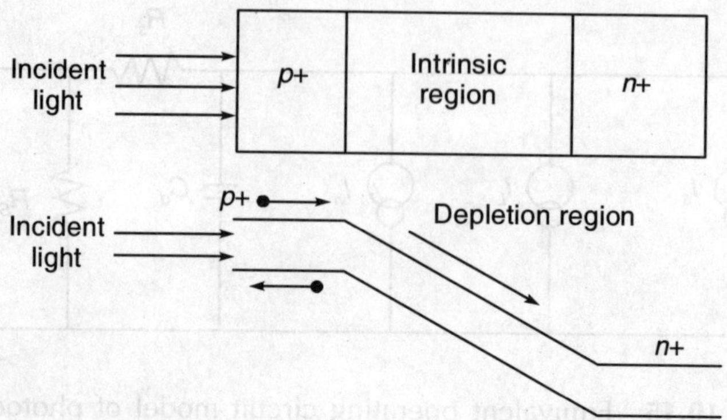

FIGURE 10.16 Structure and band diagram of *p-i-n* photodiode.

Once within the intrinsic region the current flows mainly from drift owing to the relatively high electric field. Figure 10.16 shows that illuminating *p*-side rather *n*-side electrons are the carrier species that drift through the depletion region. In most *p-i-n* photodiodes, the p^+ layer is relatively thin so that most of the photo-generated carriers are produced within the intrinsic region. Photodiodes can be operated basically in two different ways. If the applied voltage is such that the electric field within the intrinsic region is less than the critical field then relatively few events occur in the device.

Photodiode has unity gain, means there is no carrier multiplication or gain in the structure.

If magnitude of the reverse bias is increased to a point at which field within the intrinsic region becomes equal to or exceeds the critical field then carrier multiplication occurs due to impact ionization. These photodiodes are called **avalanche photodiodes** and they have gain.

Limitations of p-i-n photodiodes: The frequency response of a *p-i-n* photodiode is limited by transit time effects or circuit parameters. The most important parameter which is affecting the carrier transit time within *p-i-n* diode is width of the intrinsic region. As the width of the intrinsic region increases, the total distance travelled by the photogenerated carriers also increases and hence device bandwidth reduces. However, if the intrinsic width is made too small, the quantum efficiency is reduced. So, as a designer one has to select optimal intrinsic width.

Typically, a *p-i-n* photodiode is designed such that the transit time through the intrinsic region is equal to one-half of the modulation period of the optical signal.

$$f = \frac{1}{2\tau_t} \tag{10.15}$$

where τ_t is transit time within the intrinsic region.

10.6.3   Phototransistor

A second optoelectronic device that conducts current when exposed to light is the phototransistor. A phototransistor is much more sensitive to light and produces more output current for a given light intensity than does a photodiode. Figure 10.17 shows one type of phototransistor, which is made by placing a photodiode in the base circuit of a *n-p-n* transistor. Light falling on the photodiode changes the base current of the transistor which in turn changes or amplified the collector current. Phototransistor can be of *p-n-p* type also.

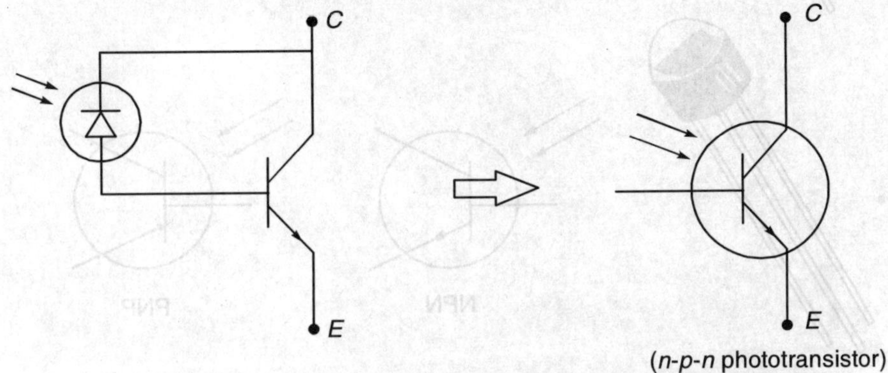

FIGURE 10.17   Schematic representation of phototransistor.

Phototransistors may be of the two-terminal type, in which the light intensity on the photodiode alone determines the amount of conduction. They may also be of the three-terminal type, which have an added base lead that allows an electrical bias to be applied to the base. Symbols for various types of phototransistors are shown in Fig. 10.18.

10.6.4   Solar Cell

Solar cell is a device which converts light energy into electrical energy. Solar cells are useful for both space as well as terrestrial applications. Main advantages of solar cells are: electrical energy conversion at very high efficiency, provide nearly permanent power at low operating cost and is virtually non-polluting.

p-n junction solar cell

A schematic representation of *p-n* junction solar cell is shown in Fig. 10.19. It consists of shallow *p-n* junction formed on the surface, a front ohmic contact stripe and fingers, a back ohmic contact that covers the entire back surface and an antireflection coating on the front surface. When the cell is exposed to the light then only those photon whose energy greater than forbidden gap contributes an energy E_g to the cell output. Energy greater than E_g is wasted as heat.

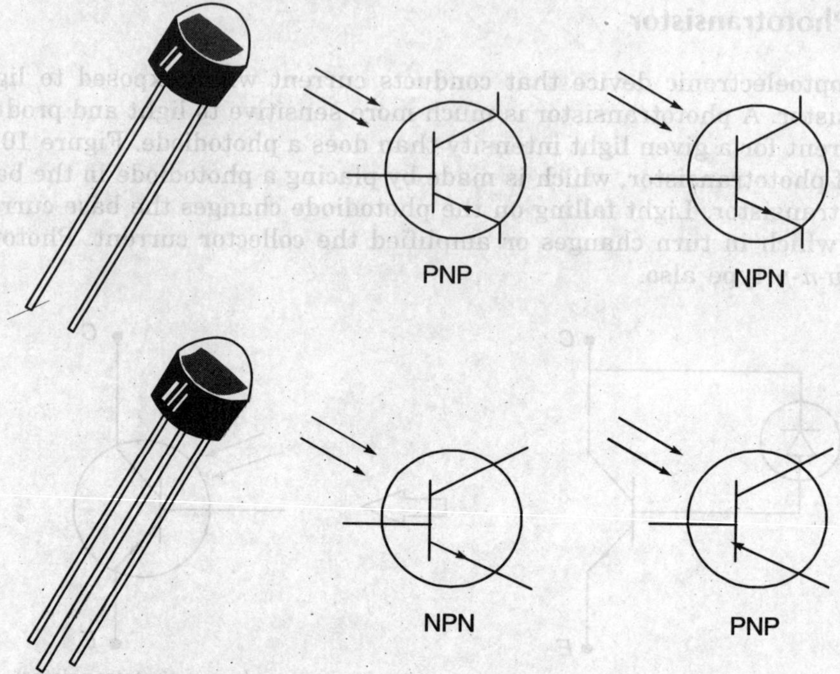

FIGURE 10.18 Various type of phototransistors.

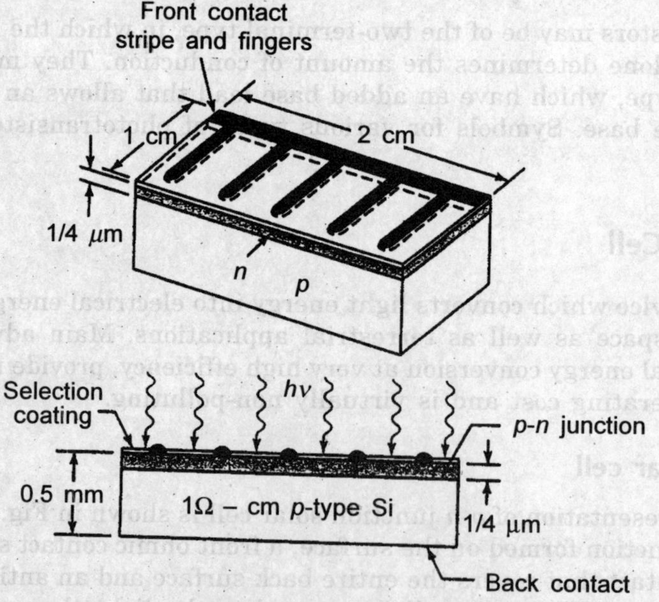

FIGURE 10.19 Schematic representation of *p-n* junction solar cell.

The ideal *I-V* characteristics of solar cell device is given by

$$I = \text{Net } p\text{-}n \text{ junction current} = I_s[e^{qV/kT} - 1] - I_L \qquad (10.16a)$$

where I_L is current due to photogenerated carriers.

and

$$J_s = \frac{I_s}{A} = qN_cN_V\left[\frac{1}{N_a}\sqrt{\frac{D_n}{\tau_n}} + \frac{1}{N_d}\sqrt{\frac{D_p}{\tau_p}}\right]e^{-E_g/kT} \qquad (10.16b)$$

where A is the cross-sectional area. A plot of Eq. (10.16b) is shown in Fig. 10.20(a).

The parameters values are $I_L = 100$ mA, $I_s = 1$ nA, $A = 4$ cm^2 and $T = 300$ K. As seen from Fig. 10.20 that the curve passes through the fourth quadrant and, therefore, power can be extracted from the device.

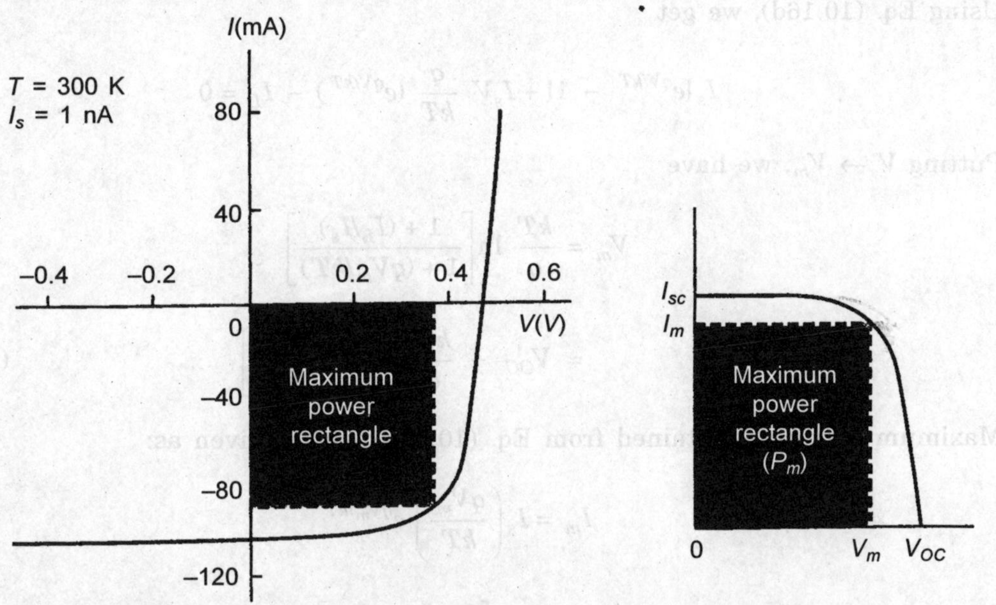

FIGURE 10.20(a)   *I-V* characteristic of *p-n* junction solar cell.

Now, let $I = 0$ in Eq. (10.16a), the voltage at which current becomes zero is called **open circuit voltage**. It is represented as V_{OC}, i.e.,

$$I_s[e^{qV_{OC}/kT} - 1] = I_L$$

or

$$V_{OC} = \frac{kT}{q}\ln\left(\frac{I_L}{I_s} + 1\right) \qquad (10.16c)$$

Hence, for a given I_L, V_{OC} increases logarithmically with decreasing I_s. The output power is:

$$P = IV$$

Using Eq. (10.16a), we have

$$P = I_s V[e^{qV/kT} - 1] = I_L V \tag{10.16d}$$

Maximum power can be obtained by differentiating Eq. (10.16d) w.r.t. V and setting it equal to zero.

$$\frac{dP}{dV} = 0$$

Using Eq. (10.16d), we get

$$I_s [e^{qV/kT} - 1] + I_s V \frac{q}{kT} (e^{qV/kT}) - I_L = 0$$

Putting $V \to V_m$, we have

$$V_m = \frac{kT}{q} \ln\left[\frac{1 + (I_L/I_s)}{1 + (qV_m/kT)}\right]$$

$$\simeq V_{OC} - \frac{kT}{q} \ln\left[1 + \frac{qV_m}{kT}\right] \tag{10.16e}$$

Maximum current is obtained from Eq. (10.16a), and is given as:

$$I_m = I_s \left(\frac{qV_m}{kT}\right) e^{qV_m/kT}$$

$$\simeq I_L \left[1 - \frac{1}{\left(\dfrac{qV_m}{kT}\right)}\right] \tag{10.16f}$$

The maximum power is obtained by multiplying Eqs. (10.16e) and (10.6f), i.e.,

$$P_m = I_m V_m = I_L \left[V_{OC} - \frac{kT}{q} \ln\left(1 + \frac{qV_m}{kT}\right) - \frac{kT}{q}\right] \tag{10.16g}$$

EXAMPLE 10.10

Calculate V_{OC} and output power at the voltage of 0.35 V for a solar cell. Take

$$I_L = 100 \text{ mA}, \quad I_s = 1 \text{ nA} \quad \text{and} \quad A = 4 \text{ cm}^2.$$

Solution: Using Eq. (10.16c),

$$V_{OC} = (0.026) \ln \left[\frac{10^{-1}}{10^{-9}} \right] \approx 0.48 \text{ V}$$

Using Eq. (10.16d),

$$P = (-10^{-9})(0.35)\ [e^{0.35/.026} - 1] - (-0.1)\ (0.35)$$

$$= 3.48 \times 10^{-2} \text{ W}$$

Since, I_s and I_L are reverse currents, so why negative sign is used in front of them.

Conversion efficiency

The conversion efficiency of a solar cell is defined as:

$$\eta = \frac{P_m}{P_{\text{in}}} \times 100$$

Substituting Eq. (10.16g), we have

$$\eta = \frac{I_c \left[V_{OC} - \dfrac{kT}{q} \ln \left(1 + \dfrac{qV_m}{kT} \right) - \dfrac{kT}{q} \right]}{P_{\text{in}}} \times 100$$

or

$$\eta = \frac{FF\ I_L\ V_{OC}}{P_{\text{in}}} \times 100 \tag{10.17a}$$

where P_{in} (incident power) FF (fill factor) is defined as:

$$FF = \frac{I_m V_m}{I_L V_{oc}} = 1 - \frac{kT}{qV_{oc}} \ln \left[1 + \frac{qV_m}{kT} \right] - \frac{k}{qV_{oc}} \tag{10.17b}$$

FF is a measure of the realizable power from a solar cell. Typically the fill factor is between 0.7 and 0.8. Nonideal factors, such as series resistance and reflection from the semiconductor surface will lower the conversion efficiency typically to the range of 10 to 15 percent.

The *I-V* characteristics due to series resistance is:

$$\ln \left[\frac{1 + I_L}{I_s} + 1 \right] = \frac{q}{kT}(V - IR_s) \tag{10.18a}$$

and

$$\text{Output current } I = I_s \left\{ \exp\left[\frac{q(V + IR_s)}{kT} - 1 \right] \right\} - I_L \qquad (10.18b)$$

and

$$\text{Output power } P = I\left[\frac{kT}{q} \ln\left(\frac{I + I_L}{I_s} \right) + IR_s \right] \qquad (10.18c)$$

The series resistance depends on the junction depth, the impurity concentrations of *p*-type and *n*-type regions and the arrangement of the front surface ohmic contacts. The band diagram and circuit representation is shown in Fig. 10.20(b).

Note: Heterojunction solar cell should have better characteristics than a homojunction cell, especially at shorter wavelengths.

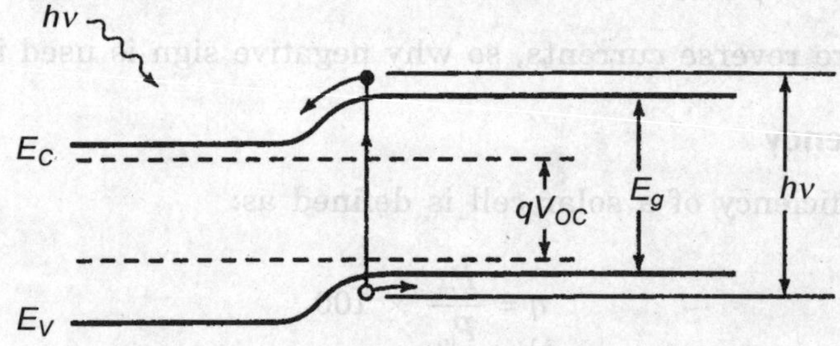

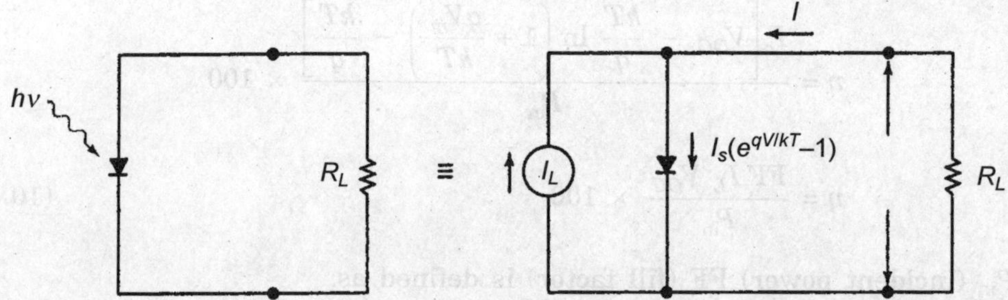

FIGURE 10.20(b)   Band diagram and circuit representation of *p-n* junction solar cell.

Amorphous Si solar cell

Single-crystal silicon solar cells are expensive and are limited to approximately 6 inch in diameter. A system powered by solar cells requires a very large area solar cell array to generate the required power. Amorphous silicon solar cells provide the possibility of fabricating large area and relatively inexpensive solar cell systems.

When silicon is deposited by CVD techniques at temperatures below 600°C, an amorphous film is formed regardless of the type of substrate. In amorphous silicon, there is only very short range order and no crystalline regions are observed. The optical absorption coefficient of amorphous silicon is very high and hence most of the sunlight is absorbed within approximately 1 µm of the surface. As a result only a very thin amorphous layer is needed for a solar cell conversion efficiency. This amorphous technique is cheaper.

Note: Another type of solar cell is Tandem solar cell. In this cell, a wide-bandgap p-n junction (GaInP with E_g = 1.9 eV) is on top of a smaller bandgap p-n junction (GaAs with E_g = 1.42 eV). Photons with energies greater than 1.9 eV will be absorbed in the p-n junction and photons with energies between $1.42 < h\nu < 1.92$ eV will pass through the top p-n junction, but will be absorbed in the bottom p-n junction.

10.6.5 Photocell

An older device that uses light in a way similar to the photodiode is the photo-conductive cell. Different types of photocells are shown in Fig. 10.21. Like the photodiode, the photocell is a light controlled variable resistor. However, a typical light-to-dark

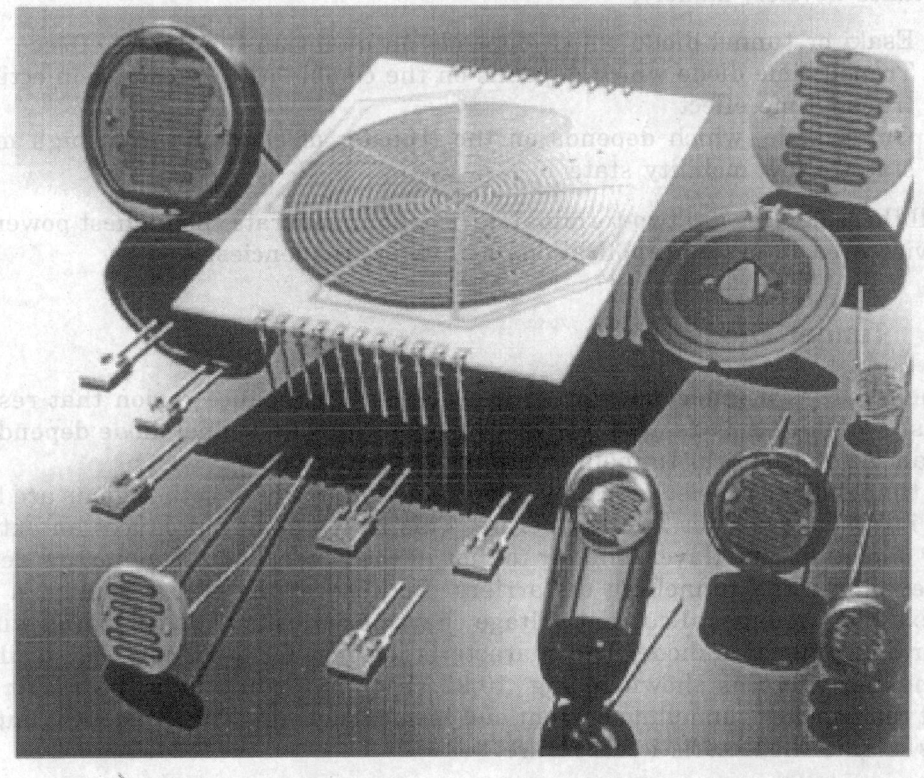

FIGURE 10.21 Different types of photocells.

resistance ratio for a photocell is 1:1000. This means that its resistance could range from 1000 ohms in the light to 1000 kΩ in the dark. Photocells are used in various types of control and timing circuits, e.g. street light controllers.

10.7 MICROWAVE DIODES

Ordinary electronic components behave in a different way at microwave frequencies (about 0.1 GHz to 3000 GHz) then they do at lower frequencies. The distributed effects must be considered at microwave frequencies because at these frequencies the wavelength is approximately equal to physical size of the components. At microwave frequencies the capacitors and inductors are realized by the transmission-line segments. Transmission lines are often used for microwave circuit interconnects as well. Several important devices use instabilities that occur in semiconductors for high frequency applications. The important instability that occurs is negative conductance. In the present topic we will concentrate on mainly three types of commonly used negative conductance devices, namely,

- Esaki or tunnel diode which depends on quantum-tunnelling effect
- Transit time diode which depends on the combination of carrier injection and transit time effect
- Gunn diode, which depends on the transfer of electrons from high mobility state to low mobility state

All these devices are two-terminal devices that generate the highest power levels per device area in system applications at higher frequencies.

10.7.1 Tunnel Diode

A tunnel diode is a semiconductor with a negative resistance region that results in very fast switching speeds upto 5 GHz. The operation of a tunnel diode depends upon the quantum mechanical tunnelling effect.

A tunnel diode consists of a simple p-n junction in which both regions are heavily doped, i.e., n- and p-sides are degenerate. This heavy doping produces an extremely narrow depletion zone (layer) similar to that in the zener diode. The narrow depletion width enhances the tunnelling of carriers. Because of the heavy doping, a tunnel diode exhibits an unusual current voltage characteristic curve as compared with that of an ordinary junction diode. The characteristic curve for a tunnel diode (called due to tunnelling effect) is shown in Fig. 10.22. The characteristic is the result of two current components: tunnelling current and thermal current. The three most important aspects of this characteristic curve are:

1. The forward current increases to a peak value given as I_p, with a small applied forward bias because the electrons can tunnel from n-side to p-side easily. When the applied bias voltage is equal to $(V_p + V_n)/3$, the forward current mainly tunnelling current reaches to its maximum value I_p and corresponding voltage V_p is called **peak voltage**.

2. The decreasing forward current with increasing forward voltage to a minimum valley current I_v because there are fewer unoccupied states on the p-side. The voltage corresponds to the valley condition is called valley voltage (V_v), i.e., for $V_p < V < V_v$, no tunnelling of carriers.

3. The normal increasing forward current with the further increase in bias voltage, i.e., after V_v. This current is mainly due to thermal current.

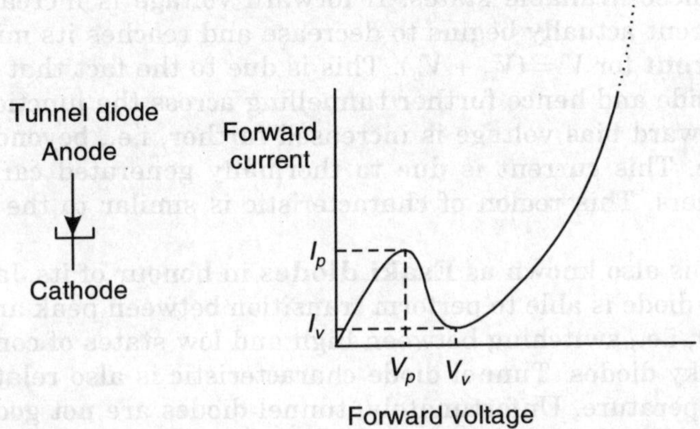

FIGURE 10.22   Symbol and forward characteristic of tunnel diode.

The portion of the characteristic curve between I_p and I_v is the region of the negative resistance and the value of negative resistance can be obtained by knowing the value of I_p and I_v. Therefore, the ratio (I_p/I_v) is used as figure of merit for tunnel diode. The most important specifications for the tunnel diode are:

- Peak current I_p
- Peak voltage V_p
- Valley current I_v
- Valley voltage V_v

The empirical relation for current and voltage in the tunnel diode is given as:

$$I = I_p\left(\frac{V}{V_p}\right)\exp\left(1 - \frac{V}{V_p}\right) + I_o\exp\left(\frac{qV}{kT}\right) \tag{10.19a}$$

where I is the total current, given as the sum of the tunnel current and thermal current, and V is the applied bias voltage.

Note: 1. In quantum-mechanical tunnelling, the carriers mainly electrons are crossing p-n junction without acquiring sufficient energy to cross junction barrier. Because of the heavy doping on both the sides, the width of the depletion region is only one-millionth of an inch, that enhances the tunnelling probability.

2. The major difference between tunnel diode and zener diode is that the tunnel diode is operated in forward bias, whereas zener diode is operated in reverse-biased or breakdown region.

Operation

Initially when a small forward bias is applied on the tunnel diode, it starts conducting. The current increases with applied forward bias, reaches to its maximum value, called **peak current**, for $V = (V_n + V_p)/3$. This is only due to fact that there exists a band of energy states that is occupied on the n-side and corresponding band of energy states is unoccupied on p-side. The electrons can easily tunnel from n-side to p-side to occupy these available states. If forward voltage is increased further, i.e., more than V_p, current actually begins to decrease and reaches its minimum value I_v, called **valley current** for $V = (V_n + V_p)$. This is due to the fact that no energy levels are free on the p-side and hence further tunnelling across the junction of electrons is stopped. If the forward bias voltage is increased further, i.e., beyond V_v, the current begins to increase. This current is due to thermally generated carriers not by the tunnelling of carriers. This region of characteristic is similar to the characteristic of normal diode.

Tunnel diode is also known as **Esaki diodes** in honour of its Japanese inventor Leo Esaki. Tunnel diode is able to perform transition between peak and valley current levels very quickly, i.e., switching between high and low states of conduction is much faster than Schottky diodes. Tunnel diode characteristic is also relatively unaffected by changes in temperature. Unfortunately, tunnel diodes are not good rectifiers, and they have relatively high leakage current when reverse biased. In order to exploit the tunnel effect, these diodes are operated at a bias voltage somewhere in between the peak and valley voltage levels and always in forward-bias polarity. The most common application of a tunnel diode is in simple high-frequency oscillator circuits, where they allow dc voltage source to contribute power to an LC tank circuit.

The reverse breakdown for tunnel diodes is very low, typically 200 mV. The tunnel diode conducts very heavily at the reverse breakdown voltage.

Note: 1. The negative differential resistance of the tunnel diode is given as:

$$R = \left(\frac{dI}{dV} \right)^{-1} \qquad (10.19b)$$

and can be obtained by using Eq. (10.19a).

2. A back diode is a tunnel diode with a suppressed I_p so that it has approximately a conventional diode characteristics.

10.7.2 IMPATT Diode

The name IMPATT stands for **impact ionization avalanche transit time**. IMPATT diode, as the name itself indicates, it employs impact ionization transit time properties of semiconductor devices to produce a negative resistance at mictowave frequencies. The IMPATT diode is one of the most powerful solid-state sources of microwave power. It is extensively used in radar systems and alarm sytems.

The structure, doping variation and field are shown in Fig. 10.23. The device essentially consists of two regions:

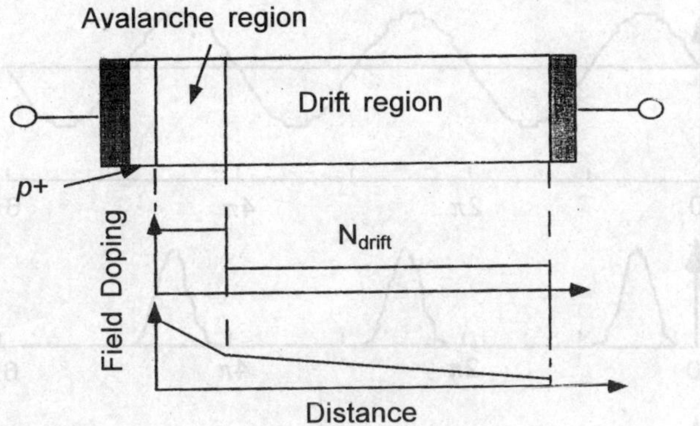

FIGURE 10.23   Impact ionization avalanche transit time diode

1. The region where avalanche multiplication occurs, is called **avalanche region**. It forms by n^+-p region.
2. The drift region through which generated holes must drift on moving to p^+ contact. This region is essentially intrinsic. The device operates in negative conductance mode. The negative conductance occurs due to two main processes causing current to lag behind the voltage in time.

1. A delay due to avalanche process
2. Further delay due to transit time of the carriers across the drift region

For negative conductance to occur, it is required that the sum of two delays must be approximately equal to one half cycle of the operating frequency. In the negative conductance mode, device can be used for oscillation and amplification.

The main purpose of introducing drift region is to create an additional 90°-phase delay. This delay occurs since the current flows even after the impact ionization stops during the transit time t_{tr} of the electrons across the drift region. The corresponding phase shift is equal to ωt_{tr}, where, ω is the an angular frequency of oscillation. If $\omega t_{tr} = \pi/2$, the total phase delay of the electron current w.r.t. the applied voltage is close to π.

Operation frequency

Since $\omega t_{tr} \cong \pi/2$, the drift transit time is equal to one half of the period of oscillations and operating frequency f of an IMPATT diode is:

$$f = \frac{V_{\text{drift}}}{2L} \tag{10.20}$$

where L is length of the drift region and V_{drift} is the drift velocity for holes.

Note: The rate of current increase is proportional to the generation rate. The generation rate G_i changes in phase with the ac voltage, and the phase shift between current density and rate of current increase is $\pi/2$ as shown in Fig. 10.24. Therefore, the current phase shift w.r. to ac voltage created by the IMPATT ionization is approximately $\pi/2$.

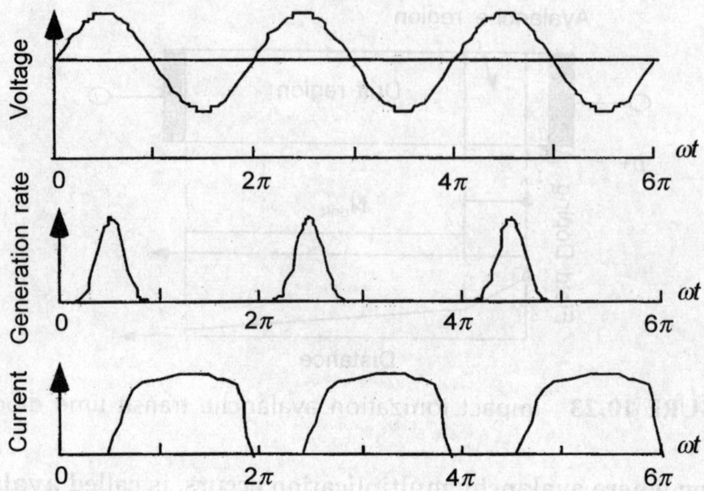

FIGURE 10.24 Variation of voltage, carrier generation rate and current with time for IMPATT diode.

10.7.3 Gunn Diode

In some materials (III–V compounds such as GaAs and InP), after an electric field in the material, reaches a threshold level, the mobility of electrons decreases as the electric field is increased, thereby producing negative resistance. This is called **Gunn effect**. A two-terminal device made from such a material can produce microwave oscillations, the frequency of which is primarily determined by the characteristics of the specimen of the material and not by any external circuit. The Gunn effect was discovered by J.B. Gunn of IBM in 1963. One form of a diode which makes use of Gunn effect consists of an epitaxial layer of n-type GaAs grown on a GaAs substrate. This diode is called Gunn diode. Electrons are transferred from one valley in the conduction band to another valley, i.e., transfer of conduction electrons from a high mobility energy valley to low mobility higher energy satellite valley.

Let us consider the electron drift velocity versus electric field (or current v_s voltage) relationship for GaAs which is shown in Fig. 10.25. Below the threshold field E_{Th} (of approximately 0.32 V/mw) the device acts as a passive resistance. However, above E_{Th} the electron velocity, i.e., current decreases as the field (voltage) increases producing a region of negative differential mobility (NDM) or negative differential resistance (NDR). This is the essential feature that leads to current instabilities and Gunn oscillations in an active device and is due to the spectral conductance band structure of direct bandgap semiconductors such as GaAs. For the transferred electron mechanism to give rise to the NDR, certain requirement must be met:

(a) The lattice temperature must be low enough that in absence of any field most of the lectrons remains in the lowest conduction valley, i.e., conduction band minima. In other words, $\Delta E > K$, where ΔE is the energy difference between two valleys.

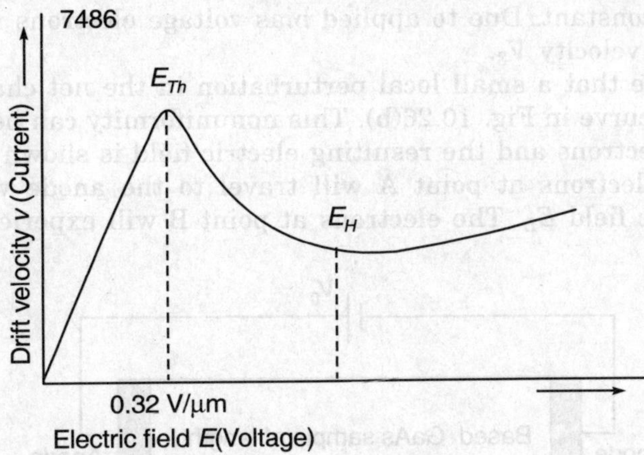

FIGURE 10.25 *I-V* characteristic of Gunn diode.

(b) In the lower valley the electrons have high mobility and low effective mass, whereas in the upper satellite valley the electron must have low mobility and higher effective mass.

(c) The energy separation between the two valleys must be smaller than the semiconductor bandgap so that the avalanche breakdown does not begin before the transfer of electrons into the upper valley.

These all requirements are satisfied by n-Gallium Arsenide (n-GaAs) whose energy momentum relationship contains two conduction band energy levels, Γ and L, also known as **valleys**. In the lower valley Γ, electrons exhibit low effective mass and higher mobility. In the satellite L valley, electrons exhibit a large effective mass and very low mobility. Two valleys are separated by a small energy gap of order of 0.31 eV.

Device operation

At thermal equilibrium, most electrons reside near the bottom of the Γ valley and because of their higher mobility ($\sim$ 8000 cm^{2}V^{-1}S^{-1}), they can readily be accelerated in a strong electric field to energies in the order of the Γ–L intervalley separation ΔE. Electrons are then able to scatter into satellite L valley, resulting in the decrease in the average electron mobility which is given as:

$$\mu = \frac{\mu_1 n_1 + \mu_2 n_2}{n_1 + n_2} \tag{10.21}$$

where n_1 and n_2 are the electron densities in Γ valley and L valley respectively.

Above the high electric field E_H most electrons reside in L valley and the device once again behaves as passive resistance.

In practical Gunn diode, electrons are transferred from the cathode by prevailing electric field when they have acquired sufficient energy, they begin to scatter into the low mobility satellite valley and slow down.

Let us consider a sample of uniformly doped n-GaAs of length L is biased with a constant voltage source V_0 as shown in Fig. 10.26(a). Therefore, the applied electric field $E_0(=V_0/L)$ is constant. Due to applied bias voltage electrons move from cathode to anode with the velocity V_3.

Let us assume that a small local perturbation in the net charge arise at $t = t_0$ as shown by solid curve in Fig. 10.26(b). This nonuniformity can be the result of local thermal drift of electrons and the resulting electric field is shown in Fig. 10.26(c) by solid curve. The electrons at point A will travel to the anode with velocity V_4 in presence of electric field E_L. The electrons at point B will experience a electric field

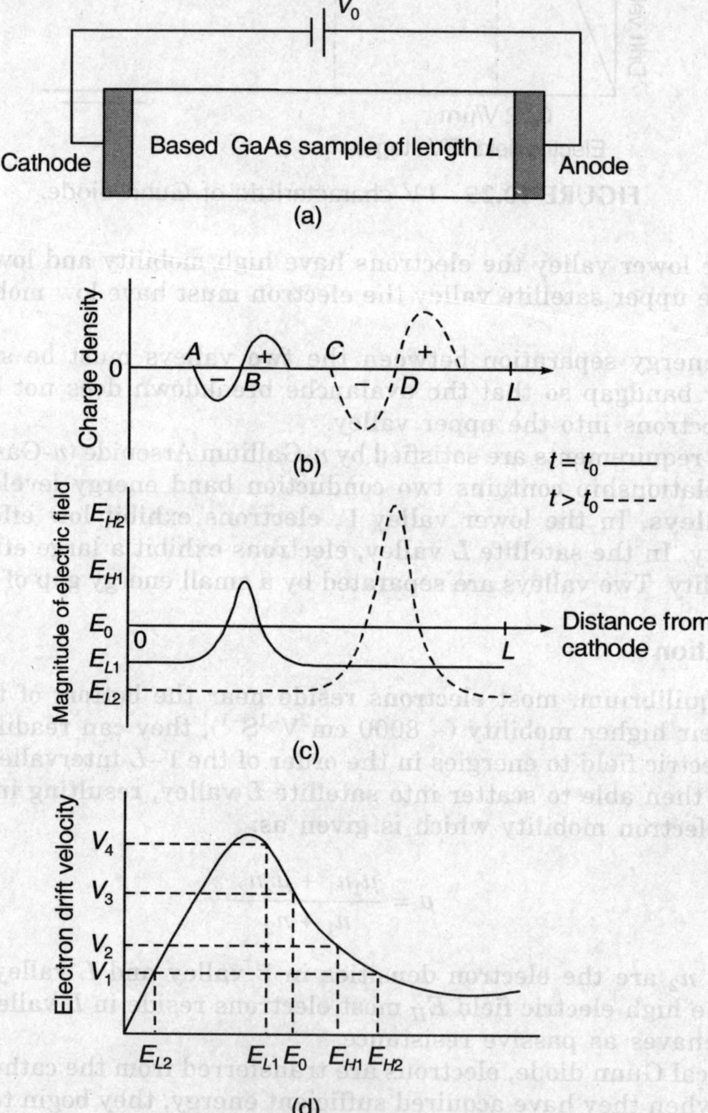

FIGURE 10.26 Biased condition of Gunn diode and variation of various electrical parameters of Gunn diode along the distance.

E_{H1} and drift towards the anode with velocity V_2 ($V_2 < V_4$). Due to that a pile up of electrons will occur between points A and B. This increases the net charge in that region. The region immediately to the right of B will become progressively more depleted of electrons due to their higher velocity towards the anode than those at B. The initial charge perturbation will grow into the dipole domain, commonly known as **Gunn domain**. Gunn domain will grow while propagating towards the anode until a stable domain has been formed. A stable Gunn domain is shown at time $t > t_0$, as shown by dashed curve in Fig. 10.26(b).

At this point in time, the domain has given sufficiently to ensure that electrons at both points C and D move at the same velocity V_1. It is important to note that the sample is biased in NDR region to produce a Gunn domain. Once the domain has formed, the electric field in the rest of the sample falls below the NDR region and will, therefore, inhibit the formation of second Gunn domain. As soon as the domain is absorbed by the anode contact region, the average electric field in the sample rises and domain formation can again take place. The successive formation and drift of Gunn domains through the sample leads to ac current oscillation observed at the contacts.

Note: The operational characteristics of a TED depends on five factors: doping concentration and doping uniformity in the device, length of active region, cathode contact characteristics, type of circuit and operating bias voltage.

EXAMPLE 10.11

A GaAs TED is 10 μm long and is operated in the transit-time domain mode. Find the minimum electron density n_o required and the time between current pulses.

Solution: Required condition

$$n_{oL} \geq 10^{12}/\text{cm}^2$$

or

$$n_o \geq \frac{10^{12}}{10 \times 10^{-4}} = 10^{15}/\text{cm}^3$$

The time between current pulses $t = \dfrac{L}{v} = \dfrac{10^{-3}}{10^7} = 10^{-10}$ sec

Applications

Gunn diodes are reliable, relatively easy to install and the lower output power levels fall well below the safety exposure limits. They are ideally suited for low noise sources such as local oscillators, locking oscillators, low and medium power transmitter applications and motion detection system.

Note: 1. The power output and efficiency of TEDs are generally lower than that of IMPATT diodes. TEDs are also used in detection systems, remote control and microwave test instruments.

2. **Trapped plasma avalanche triggered transit (TRAPATT):** In this regime, during a part of cycle, the drift region of the IMPATT diode is filled with high-density electron-hole plasma that is *trapped* within the device. In this state, the device impedance is very small and the voltage across the device is low. The electrons and holes are removed from the device via electron and hole transit to the contacts. This regime is called **trapped plasma avalanche triggered transit regime** and the diodes operating in this regime are called **TRAPATT diodes**. TRAPATT diodes typically operate at frequencies below 10 GHz and have high output power in pulsed regime of operation, in order to avoid excessive heating effect. Output powers higher than kilowatt for several TRAPATT diodes connected in series and operating at 1 GHz or so have been obtained.

10.7.4 Hot Electron Devices

Hot electrons are those electrons whose kinetic energies are greater than thermal energy (kT). As the dimensions of semiconductor shrink and applied supply remains same, then the internal field increases to a very high value (~10^5 V/cm) means kinetic energy of carriers increases tremendously. A large fraction of the carriers during their operation occupy the energy states of higher kinetic energy. At a given point in time and space, the velocity distribution of carriers may be narrowly peaked and one can speak about ballistic electron packets. May be at other times and locations the electron ensembles can have a broad velocity distribution. There are mainly two types of hot electron devices: hot electron heterojunction BJT (hot electron HBT) and real-space-transfer transistor.

Hot-electron injection is enabled in heterojunction bipolar transistors by designing structures with a wider-bandgap emitter. There are several advantages of hot-electron effect. The purpose of the ballistic injection is to shorten the base traversal time by replacing the relatively slow diffusion motion by faster ballistic propagation. The original real space transfer structure is a heterostructure with alternate doped wide gap (AlGaAs) and undoped narrow gap GaAs layers. If the power input into the structure exceeds the rate of energy loss by the system to the lattice then the *carriers heat up* and undergo partial transfer into the wide-gap layer where they may have a different mobility. If the mobility in the AlGaAs layer is much lower, negative differential resistance will occur in the two-terminal circuit, based on the momentum space intervalley transfer, so it is called **real space transfer**.

10.7.5 Fibre Transmission

There are several advantages to the fibre optic communication systems, namely

   (i) Smaller diameter, lighter weight
  (ii) Relatively low cost compared to copper cables
 (iii) Good isolation and cross talk immunity
 (iv) Low transmission loss and dispersion
  (v) High security transmission
 (vi) Tremendous capacity, i.e., large bandwidth

A typical optical fibre is shown in Fig. 10.27. The fibre consists of three different regions:

The core. The inner part of the fibre through which signal propagates. This region has larger refractive index.

Cladding. Outer region of the fibre, having refractive index lower than the core region.

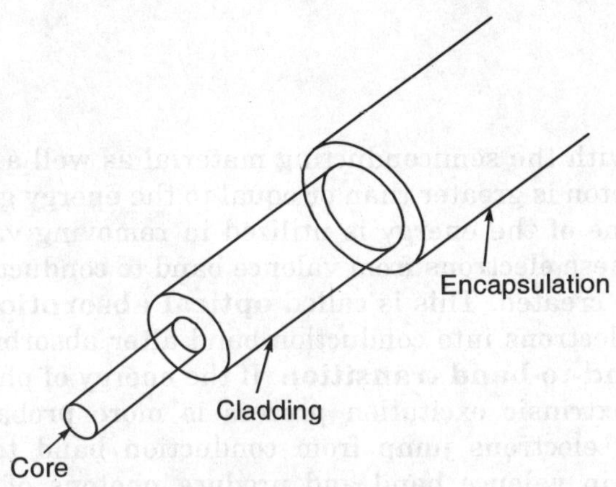

FIGURE 10.27 Optical fibre.

The signal is mainly confined within the core region by the total internal reflection. Since the core refractive index is larger than the cladding, the lightwave is totally internally reflected as it propagates through the fibre ensuring confinement.

The outermost part of the fibre is made up of glass to protect from environmental conditions. The important parameters to consider during the selection of fibre material are:

- Slight difference in the refractive indices of core and cladding
- Flexibility so that fibre can be bent to go around corners
- Ability to drawn into long fibres

Silica glass is most commonly used materials for fibre. Typically, a small concentration of impurities are added to alter the refractive index. These impurities or dopants also affect the attenuation and powerloss of the light in the fibre. There are two-loss transmission windows that are centered around 1310 nm and 1550 nm.

There are mainly two types of fibre: Single mode fibre where only one mode can propogate and multimode fibre in which more than one modes can propogate.

Attenuation generally varies linearly with distance. Thus, the longer a fibre optic cable is, more the signal will be attenuated at the end.

Multimodes are made available by designing the fibre to have a sufficiently large core diameter to accommodate the different modes. In multimode fibre, the light arrives at different times at each location leading to dispersion of the signal pulse. The problem is severe in multimode step index fibres. In multimode step index fibres the index of refraction changes abruptly from the core to the cladding regions. The graded-index multimode fibre overcome this limitation by gradually grading the index of refraction within the core from its centre to edge. The index of refraction is highest near the centre of the core and gradually decreases outwards. The light propogates with slower speed near the centre of the core, where the refractive index is higher, than near the cladding. In signal mode fibre, the width of the core is sufficiently small so that only one mode of the light can propagate and thus no dispersion.

10.8 SUMMARY

Photons can interact with the semiconducting material as well as the impurities also if the energy of the photon is greater than or equal to the energy gaps of the materials. In this interaction some of the energy is utilized in removing valence electrons and rest is used to excite these electrons from valence band to conduction band. Therefore, electron-hole pairs are created. This is called **optical absorption process**. A direct excitation of valence electrons into conduction band after absorbing optical energy is called **intrinsic** or **band-to-band transition**. If the energy of photon is less than the forbidden gap, then extrinsic excitation process is more probable. In the reverse process of absorption, electrons jump from conduction band to valence band and recombine with holes in valence band and produce photons of energy $h\nu$ $(=E_g)$. If atoms are already in the higher energy states or excited state and forced to jump into lower or stable energy states then an energy $h\nu_{12}$ is given off in this process, called **stimulated emission**. In spontaneous emission process, atoms from excited state jump to stable state without any stimulation process, i.e., by own. The radiation from the stimulated process is monochromatic. This process is dominant in LASERs. In solar cell dominant process is absorption.

A material that glows under an ultraviolet lamp, but stops glowing as soon as lamp is turned off, is called **fluorescence material**. But a phosphorescence material keeps glowing for a while even ultraviolet source is turned off.

A LED is just a forward biased *p-n* junction diode that can emit spontaneous radiation in UV, visible or infrared regions. The device is designed in such a way so that most probable recombination mechanism is radio-live recombination. The radiative transition mechanisms are found predominantly in direct bandgap semiconductors.

In LASER the light is emitted through the stimulated emission. A semiconductor exhibits a net stimulated emission rate when the difference in the quasi-Fermi levels exceeds the bandgap. This condition is called **population inversion** and can be realized by using a forward biased, degenerately doped *p-n* junction diode. In order for a system to lase, the optical gain must exceed all the optical losses.

Photodetectors are semiconductor devices that can convert optical signals into electrical currents. A photodiode is basically a *p-n* junction or metal semiconductor contact operated under reverse bias. The *p-i-n* photodiode is one of the most common photodetectors because of its high quantum efficiency and bandwidth. An optoelectronic device that conducts when exposed to light is the phototransistor. It is much more sensitive to light and produces more output current for a given light intensity than does a photodiode.

A solar cell is a device which converts light energy into electrical energy and are useful for both space as well as terrestrial applications.

A tunnel diode is a semiconductor with a negative resistance region that results in very fast switching speeds up to 5 GHz. The operation of a tunnel diode depends upon the quantum mechanical tunnelling effect. The IMPATT diode employs impact ionization and transit-time properties of the semiconductor devices to produce a negative resistance at microwave frequencies. It is one of the most powerful solid state devices in microwave power. In some materials like III–V compounds such as GaAs and when electric field reaches to its threshold value, the mobility of electrons decreases as the electric field increases, thereby producing negative resistance. This is called **Gunn theory**, and a form of diode which makes use of Gunn theory is called **Gunn diode**.

REVIEW QUESTIONS

1. Why direct bandgap materials are more preferred for optical devices than the indirect bandgap materials?

2. Can the indirect bandgap semiconductor materials be used for optical devices? If yes, explain how?

3. What is the intrinsic and extrinsic radiation?

4. Discuss the spontaneous and stimulated emission process.

5. Why radiation from stimulated emission is monochromatic?

6. What is population inversion process?

7. Why for laser action population inversion as well as stimulated emission processes are necessary?

8. How LEDs and LASERs are different? Discuss their applications.

9. For road signals which material will you prefer, fluorescence or phosphorescence? And why?

10. What are the criteria for choosing materials for LEDs? Discuss the advantages of LEDs.

11. What will happen if a light of energy $h\nu < E_g$ falls on the semiconductor substance? Explain in brief.

12. Why LEDs are less suited for transmission of digital information than lasers?

13. What is quantum efficiency in LEDs and what are the factors which influence the internal quantum efficiency of a homojunction LEDs?

14. What is the difference between semiconductor lasers and other types of lasers?

15. Discuss about quantum well lasers, its significance and limitations.

16. Why a photodiode is operated in reverse bias? What will happen if a forward bias is applied on that?

17. Why the magnitude of the photocurrent generated by a photodiode is dependent upon the wavelength of the incident light?

18. Why third quadrant is chosen for the photodetector operation? What will happen if one choses fourth quadrant?

19. What is the importance of dark current in photodiode?

20. Explain how the response speed of a photodiode is limited by the following three factors:

(a) Diffusion of carriers

(b) Drift time in the depletion region

(c) Capacitance of the depletion region

21. Why p-i-n photodiode is one of the most common photodetectors?

22. What is the purpose of introducing intrinsic region in p-i-n photodiode?

23. What is solar cell? Discuss its advantages and requirements for designing a solar cell.

24. Discuss the operation of a tunnel diode. How it is different from Gunn diode and IMPATT diode?

25. What will happen if tunnel diode is operated in reverse bias?

26. Why Gunn diode is also called transfer-electron devices? Discuss the factors on which operational characteristics of TED depends.

27. Discuss the significance of hot electron devices.

28. Why optical fibre is mainly used for the long range communication?

NUMERICAL PROBLEMS

1. A light of wavelength $\lambda = 1.5$ Å falls on the surface of the Si semiconductor material. Whether this light is strong enough to excite the valence electrons to conduction band or not?

2. A GaAs sample is illuminated with a light having a wavelength of 0.6 μm. The incident power is 15 mW. If one third of the incident power is reflected and another third exists from the other end of the sample, what is the thickness of the sample? Find the thermal energy dissipated per second to the lattice.

3. A single-crystal of silicon is illuminated with a light of energy 3 eV. The incident power is 15 mW. Find the width of the sample so that the energy absorbed per second by sample is 6.3 mW. Also, find the portion of each photon energy that is converted into heat. Assume $\alpha = 4 \times 10^4$/cm.

4. Calculate the thickness of a silicon semiconductor sample that will absorb 90% of the incident photon energy. Take $\lambda = 1.5$ μm, $\alpha = 10^2$/cm.

5. Find the population of ground state, if the excited state has 10^{16}/cm^3 population and difference in their energy states is 2.0 eV.

6. Determine the mirror reflectivity of a semiconductor laser if the system is at threshold. The mirrors have identical reflectivities, the cavity length is 10 μm, the collective losses are 10.0/cm; and the threshold gain is 10^3/cm. Assume that the optical confinement factor is unity.

7. Determine the threshold gain of a semiconductor laser with mirror reflectivities of 0.25 and 0.30, a cavity length of 20 μm and collective losses of Q_0/cm.

8. Determine the length of a semiconductor laser cavity if its collective losses are 50/cm. The collective losses are equal to that of the mirror loss. Assume that the two mirrors have identical reflectivities of 0.32.

9. For a GaAs laser diode operated at $\lambda_0 = 0.89$ μm, with $n' = 3.58$, $L = 300$ μm and $dn'/d\lambda_0 = 2.5/$μm. Find $d\lambda$ for adjacent modes.

10. A photoconductor with dimensions $L = 6$ mm, $W = 2$ mm and $D = 1$ mm is placed under uniform radiation. The absorption of light increases the current by 2.83 mA. A voltage of 10 V is applied across the device. As the radiation is suddenly cutoff, the current falls, initially at a rate of 23.6 A/S. The electron and hole mobilities are 3600 and 1700 cm^2/V-sec respectively. Find: (a) the equilibrium density of electron hole pairs generated under radiation (b) the minority-carrier life time (c) the excess density of electrons and holes remaining 1 ms after the radiation is cutoff.

11. Calculate the transit time through a bar of length $L = 100$ μm and cross-sectional area of 10^{-7} cm². The gain is 1800. Assume that the bar is of n-type with minority carrier life time is 10^{-6} sec; biased with the applied voltage of 10 V. Also, find the ratio of the mobilities of the carriers.

12. Calculate the gain and current generated when 1 μW of optical power with $h\nu = 3$ eV is shone onto a photoconductor of $\eta = 0.85$ and a minority carrier life time of 0.6 ns. The material has an electron mobility of 3000 cm²/V-sec., electric field is 5000 V/cm and $L = 10$ μm.

13. Show that the quantum efficiency η of a p-i-n photodetector is related to the responsivity $R(= I_p/I_{\text{opt}})$ at a wavelength λ by this equation $n\lambda/1.24$.

14. Consider a silicon p-n junction solar cell of area 2 cm². If the dopings of the solar cell are $N_a = 1.7 \times 10^{16}$/cm³, $N_d = 5 \times 10^{19}$/cm³, Given $\tau_n = 10$ μs, $\tau_p = 0.5$ μs, $D_n = 9.3$ cm²/s, $D_p = 2.5$ cm²/s and $I_L = 95$ mA then calculate:

    (a) Reverse saturation current I_s
    (b) Open circuit voltage
    (c) Output power for applied voltage of 10 V
    (d) Maximum output power of the solar cell at room temperature when $V = V_m$.

15. Use the values given in Problem 14, find the fill factor of the solar cell. If at room temperature, an ideal solar cell has $V_{OC} = 0.6$ V and $I_{SC} = 3$ A, find its efficiency for $P_{\text{in}} = 30$ ml.

16. A Si-solar cell has $I_{SC} = 100$ mA, $V_{OC} = 0.8$ V under full solar illumination. The fill factor is 0.7. What is the maximum power delivered to the load by this cell?

17. For the solar cell find the relative maximum power output for R_s of 0 and 5 Ω.

18. The current-voltage characteristics of a GaAs tunnel diode can be expressed by the empirical relation given by 10.19(a) with $I_p = 10$ mA, $V_p = 0.1$ V and

$I_0 = 0.1$ nA. Find the largest negative differential resistance and corresponding voltage.

19. A GaAs IMPATT diode is operated at 10 GHz with a dc bias of 100 V and an average biasing current of 100 mA. If the length of the drift region is 10 μm, find the drift velocity of holes.

20. An In P TED is 1 μm long with the cross-sectional area of 10^{-4} cm² and operated in the transit time mode (a) Find the minimum electron density (n_o) required for transit-time mode. (b) Find the time between current pulses.

21. What determines the peak tunnelling voltage V_p of a tunnel diode? Explain.

CHAPTER

11

Integrated Circuits and Fabrication

11.1 INTRODUCTION

The transistor is essentially a replacement for a vacuum tube because it is small, fast, reliable and effective. By replacing vacuum tubes from transistors in electronic devices same functions were realized, but the devices were now much smaller. This encouraged the engineers to build more functions into the same devices since there was now more space available. By the end of 1950's, electronic manufacturers saw the great possibilities of constructing more advanced circuits than before. However, as the complexity of the circuit grew problem started arising. When building a circuit, it is required that all the connections should be intact, if not, the current will not flow in the circuit means circuit would fail. Before invention of the integrated circuit (IC), the assembly workers had to construct circuits by hand, soldering each component at proper place and connecting them with metal wires. Then engineers soon realized that manually assembling the vast number of tiny electronic components (transistors, diodes, resistors and capacitors) needed for design of circuit would be impossible without generating any single fault. Advanced circuits contained so many components and connecting them were virtually impossible to build. This problem was known as **tranny of numbers**. In the summer of 1958 Jack Kilby at Texas Instruments found the solution of this problem. Kilby's idea was to make all the electronic components and chip out of the same block of semiconductor material. The resistor could be made from bulk semiconductors and capacitors could be made from p-n junctions. In September 1958, he had his first integrated circuit ready. Although the first integrated circuit was pretty crude and had some problems, but the idea was groundbreaking. By making all the parts out of the same block of material and adding material to connect them, now there was no need for individual discrete components any more.

Robert Noyce came up with his own idea for the IC, half a year later than Jack Kilby. Noyce's circuit solved several practical problems that Kilby's circuit had, mainly the problem of interconnecting all the components on the chip. This made the integrated circuit more suitable for the mass production.

The integrated circuit is the heart of all the electronic equipments today and has revolutionized the way we live: navigational system, computers, pocket calculators, industrial monitoring and control systems, digital watches, digital sound systems, word processors, communication network and others. An IC chip is smaller and thinner than a baby's finger nail, yet it is equivalent to the thousands of electronic components all operating simultaneously. The properties that encourage the wide use of ICs are its small size, negligible weight and reliable performance. Due to mass production and manufacturing experience, the price of ICs have been lowered in modern days. So that, now they are used in all kinds of consumer products, from toys to home computers. The IC represents the first great invention that deals with the storing, processing and interpretation of information rather than the manipulation of the physical environment. The success with which the IC performs these functions has given technology an entirely new dimension.

There are several ways of classifying the integrated circuits as to their use and method of fabrication. The most common classification are linear (analog) or digital according to applications, and monolithic or hybrid according to fabrication process. Biploar integrated circuit contains bipolar junction transistors as their principal elements. MOS integrated contains MOS transistors as their principal elements. Some of ICs contain both types of transistors. Integrated circuits are also categorized according to the number of transistors or other active circuit devices they contain. In this chapter, the fundamentals of MOS chip fabrication will be discussed and major steps of the process flow will be examined. It is not the aim of this chapter to present a detailed discussion of the silicon fabrication technology. Rather, the emphasis will be on the general outline of the process flow and on the interaction of various processing steps, which ultimately determine the device and the circuit performance characteristics. There are very strong links between the fabrication process, the circuit design process and performance of the resulting chip. Hence, the circuit designers must have a working knowledge of chip fabrication to create effective designs and in order to optimize the circuits with respect to various manufacturing parameters. Also, the circuit designer must have a clear understanding of the roles of various masks used in the fabrication process and how the masks are used to define various features of the devices on-chip.

11.2 TERMINOLOGY

11.2.1 Microelectronics

Microelectronics is the branch of electronic technology devoted to the design and development of extremely small electronic devices that consume very little electric power. Although, the term is sometimes used to describe discrete electronic components assembled in an extremely small and compact form, it is often taken as a synonym for integrated circuit technology. Discrete components that can be made using microelectronic techniques include resistors, capacitors and transistors.

Hybrid microcircuits contain a mixture of discrete and integrated circuits on a single substrate.

11.2.2 Semiconductor Device Fabrication

It is the process used to create chips, the integrated circuits that are present in everyday electrical and electronic devices. It is a multiple step sequence of photographic and chemical processing steps during which electronic circuits are gradually created on a wafer made of pure semiconducting material.

11.2.3 Wafers

A wafer is made out of extremely purified silicon that is grown into monocrystalline cylindrical ingots, using Czochralski process. These ingots are then sliced into 0.75 mm thick wafers and polished to obtain a very regular and flat surface, as shown in Fig. 11.1.

Once the wafers are prepared, many process steps are necessary to produce the desired semiconductor integrated circuit. In general, the steps can be grouped into four areas:

(a) Front end processing (b) Test
(c) Back end processing (d) Packaging

In the front end, one makes the components of the circuit whereas in the back end one has to add metal to connect the components and then test and package the chip.

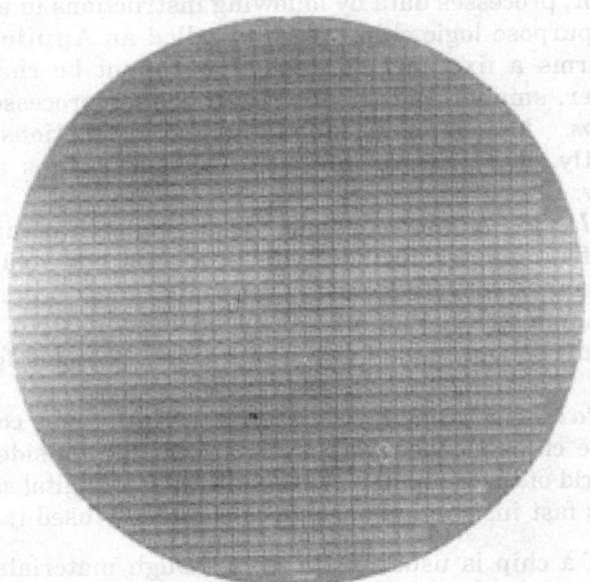

FIGURE 11.1 A view of wafer.

Two terms, wafer size and feature size, are important for chip fabrication. The wafer size is the diameter of the wafer used in the semiconductor manufacturing process. For a given fabrication facility, the wafer size is fixed, because it would cost so much money to retool the entire fab's process for a new wafer size. It is always required that the wafer size should be large because from bigger wafer one can produce chips in larger batches. However, wafers have a very specific property that keeps manufacturers from making them arbitrarily large. The way wafers are made, defects are much more likely to occur nearer the edge of the wafer than near the centre. So, the number of defects increases as we move outwards from the centre of the wafer. The feature size is the size of the smallest feature that the fab's equipment can etch onto the surface of a wafer. A feature could be wire or a transistor. The terms *feature size* and *transistor size* are used interchangeably. Like wafer size, the feature size for a given fab is also fixed. The entire fab is designed to produce chips with a certain feature size.

11.2.4 The Chip

Several hundred identical integrated circuits (ICs) are made at a time on a thin wafer several centimetres wide, and the wafer is subsequently sliced into individual ICs called **chips** as shown in Fig. 11.1. The terms *chip*, *silicon chip*, *microchip* and *integrated circuit* (IC) are synonymous.

Types of chips by function

(a) *Logic chips*.   A logic chip processes data. A general purpose logic chip, called microprocessor; processes data by following instructions in a software program.

A special-purpose logic chip, which is called an **Application Specific IC** (ASIC), performs a fixed set of steps that cannot be changed. An ASIC is typically faster, smaller and cheaper than a microprocessor.

(b) *Memory chips*.   Memory chips store data and instructions either temporarily or permanently. RAM (Read and Write Memory) chips are the computer's main memory.

(c) *Microcontrollers*.   A microcontroller is a single chip that contains all the components of a computer, including the processor, non-volatile memory (ROM: read only memory), volatile memory (RAM), I/O control unit and timing clock.

(d) *Bio chip*.   A microchip that uses tiny strands of DNA to latch onto and quickly recognize thousands of genes at a time, intended for use in biological environment.

(e) *Analog/digital and signal processing chips*.   A/D converters and D/A converters are chips that convert signals from the outside world (analog) to the digital world of the computer. A related chip is a Digital signal process (DSP) that performs fast instruction sequences commonly used in such applications.

The base material of a chip is usually silicon, although materials such as sapphire and gallium arsenide are also used. Silicon is found in quartz rocks and is purified in a molten state. It is then chemically combined (doped) with other materials to alter its electrical properties. The result is a silicon crystal ingot upto eight inches in diameter, as shown in Fig. 11.2, that is either positively (p-type) or negatively (n-type). Slices of the ingot approximately (1/30)th of an inch thick are cut from this

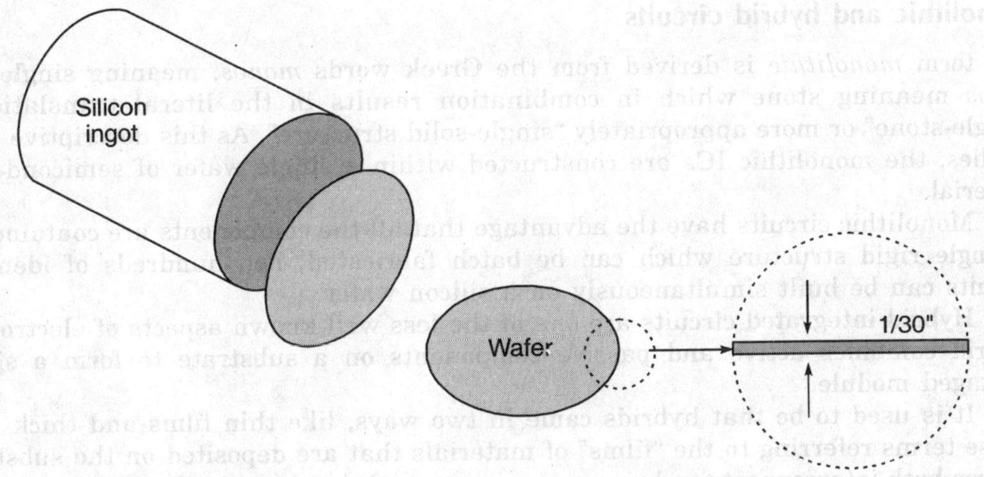

FIGURE 11.2 Representation of slicing ingot to get wafer.

"crystal salami". The slices are called **wafers**. The chips are sliced out of the wafer and good ones are placed into packages. The chip is connected to the package with tiny wires, then sealed and tested as a complete unit. Chip making is extremely precise. The operations are performed in a clean room, as the dust particles can mix with the microscopic mixtures and easily render a chip worthless.

11.2.5 The Clean Room

The sizes of the components on chip produced in a modern chip fabrication plant are extremely small. To understand how small they are, pick up a hair from your head and then cut it into two parts. Now look at the cross-section. This tiny area is hard to see with naked eyes, but one can fit thousands of modern transistors on this area. With this small sizes, the production of modern chip demands precision at an atomic level. Tiny particles like hair, a speck of dust, a bacteria or even the single particles in tobacco smoke become huge objects that are big enough to ruin a chip. Therefore, chip production takes place in a clean room. This is a specially designed room, where extremely effective air filters and air circulation systems change the air completely upto ten times a minute.

To prevent any further contamination, workers wear special suits called **bunny suits**. These protective outfits are made up of ultra clean material and sometimes have their own air filtering systems.

11.2.6 Types of Integrated Circuits

There are ways to categorize the integrated circuits depending on their use and method of fabrication. The most common categories are linear (analog) or digital according to their application and monolithic or hybrid according to fabrication.

Monolithic and hybrid circuits

The term *monolithic* is derived from the Greek words *monos*, meaning single and *lithos* meaning stone which in combination results in the literal translational, "single-stone" or more appropriately "single-solid structure". As this descriptive term implies, the monolithic ICs are constructed within a single wafer of semiconductor material.

Monolithic circuits have the advantage that all the components are contained in a single rigid structure which can be batch fabricated, i.e., hundreds of identical circuits can be built simultaneously on a silicon wafer.

Hybrid integrated circuits are one of the less well-known aspects of electronics. Hybrid combines active and passive components on a substrate to form a single packaged module.

It is used to be that hybrids came in two ways, like thin films and thick film. These terms referring to the "films" of materials that are deposited on the substrate to form both interconnects and passive components. Nowadays the distinction between two types is blurred.

Hybrids may be analog or digital and are often used in specialized or custom applications.

The processing steps for the two hybrid techniques are quite different. In thick film circuits, the resistors and interconnection patterns are "printed" on a ceramic substrate by silk-screen or similar process. One advantage of this method is that the resistors can be made below the rated values and then trimmed by abrasion.

Thin-film technology allows for greater precision and miniaturization and generally preferred when space is an important consideration. Thin-film interconnection patterns and resistors can be vacuum deposited on a glass substrate.

11.3 THE MAKING OF A CHIP

Computer circuits carry electrical signals (pulses) from one point to another. The pulses flow through transistors which act as switches (ON/OFF). The current flowing through one switch effects the opening or closing of another and so on. Transistors are wired together in patterns of Boolean logic, logic gates construct circuits. Circuits make up CPUs and other electronic systems.

All circuits were first originally designed by humans. Today, many logic functions reside in libraries, and designers pick up and choose modules from a menu. There is always a little bit of "glue logic" necessary to interconnect them which must be done logic gate by logic gate. If required function is not predesigned then this part will have to be created gate by gate. The computer converts the logic circuit design into transistors, diodes and resistors. Now, it is required to connect millions of components together. After inspection, the electronic images are transferred to machinery that creates glass, lithographic plates, called **photomasks**. The photomask is the actual size of the chip, replicated many times to fit on a round silicon wafer upto 12 inch in diameter.

Modern chip production is based on the photolithography. In photolithography a highly energetic ultraviolet light is shone through a mask, (the mask describes the parts of the chip) and the UV-light will only hit the areas that are not covered by the mask. When the film is developed, the areas hit by light are removed. Now, the chip has unprotected and protected areas forming a pattern that is the first step to the final components of the chip.

Next, the unprotected areas are processed so that their electrical properties can change. A new layer of material is added and then the entire process is repeated to build the circuit, layer by layer. When all the components have been made and circuit is complete then a layer of metal is added. Just as before, a layer of photosensitive film is applied and exposed through a mask. However, this time the mask used describes the layout of the wires connecting all the parts of the chip. Again UV is shone through this mask. The light hits the photoresist material that is not protected by the mask. Now, the chemicals are used to remove the photoresist hit by UV light called **etching**. Another step of etching is to remove the unprotected metal. This leaves a pattern of metal as described by mask. Now, the chip has layer of wires that connect the different components.

In modern technology, ICs need more than one layer of wires. When the final layer of connecting metal wires have been added, the chips on the silicon wafer are tested to see whether it is performing the requirement or not. The chips are separated from the wafer with a diamond saw to form individual ICs. Finally, each chip is packaged into the protective casing and subjected to another series of testing. After testing, the chip is ready to be shipped to manufacturer of digital (or analog) devices.

11.4 PASSIVE COMPONENTS

11.4.1 The Integrated Circuit Resistor

To fabricate a passive resistance on the chip one can deposit a resistive layer on silicon substrate, then pattern the layer by lithography and etching processes. A silicon dioxide is grown thermally on the silicon substrate and a window is defined in this oxide layer. Now, implant (or diffuse) impurities of the opposite conductivity types into the wafer through window. Figure 11.3 shows the top and cross-sectional views of resistors having a meander shape and bar shape. The resistance of the bar shaped resistor is given as:

$$R = \rho \frac{L}{A} = R_\square \frac{L}{W} \tag{11.1}$$

where $R_\square$ is called **sheet resistance**, having unit of ohms per square ($\Omega/\square$). The sheet resistance ($R_\square$) is determined by the implantation or diffusion process and the ratio (L/W) is determined by the pattern dimensions. Once the value of $R_\square$ is known, the resistance is given by the ratio (L/W) or the number of squares (each square has an area of $W \times W$) in the resistor pattern. The end contact areas will introduce additional resistance to the IC resistors. For the meander-shape resistor, the electric field lines

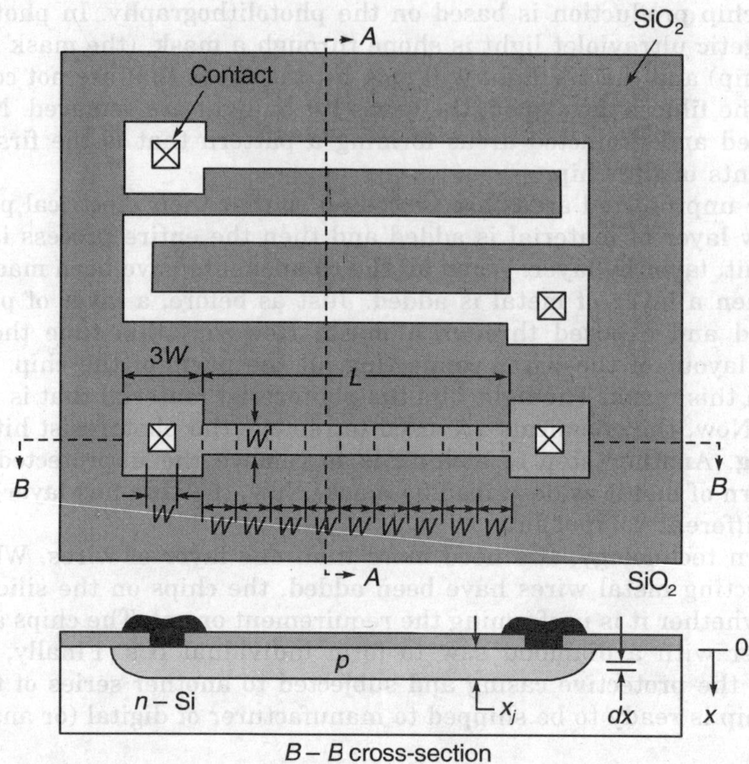

FIGURE 11.3 Integrated circuit-resistor.

at the bends are not spaced uniformly across the width of the resistor, but are crowded toward inside the corner.

11.4.2 The Integrated Circuit Capacitor

There are mainly two types of capacitors used in integrated circuits. One is MOS capacitors and other is *p-n* junction capacitors. The MOS capacitor can be fabricated by using a heavily doped region as one plate, the top metal electrode as the other plate and the intervening oxide layer as the dielectric. Figure 11.4(a) shows the top and cross-sectional views of a MOS capacitor. Initially, a thick layer is grown thermally on silicon substrate, then a window is defined by lithography and then etched in the oxide. Diffusion or ion implantation is used to form a p^+-region in the window area, whereas the surrounding thick oxide serves as a mask. Now a thin oxide layer is grown thermally in window area and then metallization is performed.

The capacitance per unit area is:

$$C = \frac{\varepsilon_{ox}}{d} \ \text{F/cm}^2 \tag{11.2}$$

where ε_{OX} is the dielectric permittivity of silicon dioxide and d is the thickness of the oxide. From Eq. (11.2), it is clear that the capacitance value can be increased by either increasing ε_{OX} or decreasing the oxide thickness d.

A p-n junction is sometimes also used as a capacitor in an integrated circuit. Figure 11.4(b) shows the top and cross-sectional view of a n^+-p junction capacitor. The p-n junction device should be operated in reverse-bias and in this case, capacitor is not constant but varies with applied reverse-bias V_R as $(V_R + V_{bi})^{-1/2}$ where V_{bi} is the built-in potential.

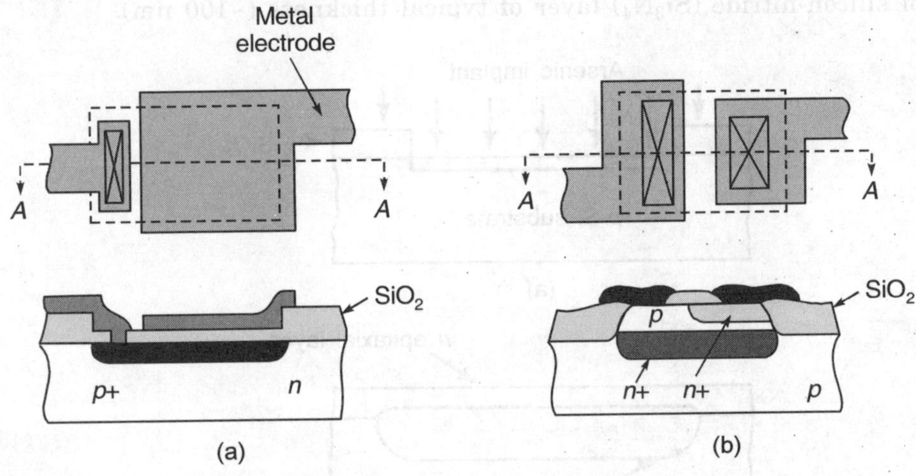

FIGURE 11.4 Integrated circuit-capacitor.

11.4.3 Diodes

Design of p-n junction in a monolithic circuit is very simple. It is sometimes usual practice to use transistors (BJT) to perform diode actions. Since, many transistors are included in the monolithic circuit and hence no special diffusion step is required to fabricate the diode. The most common method to realize the diode is to use the emitter junction as a diode by shorting collector and base.

11.4.4 The Basic Fabrication Process of Bipolar Junction Transistor (BJT)

The most commonly used bipolar junction transistor (BJT) in ICs are n-p-n transistor because of higher mobility of electrons. For fabrication of n-p-n transistor, following steps are required:

(i) Starting material is a p-type lightly doped ($\sim 10^{15}/cm^2$), <111> or <100> oriented polished silicon wafer.

(ii) A thick oxide of order of 0.5–1 µm is grown thermally on the wafer and a window is then opened in the oxide. A precisely controlled amount of low energy arsenic ions (~ 30 keV, $\sim 10^{15}/cm^2$) is implanted into the window region, as shown in Fig. 11.5(a).

(iii) Next is a high temperature (~1100°C) drive-in step forms the n^+-buried layer, having a typical sheet resistance of 20 Ω as shown in Fig. 11.5(b). Main purpose of this layer is to reduce the series resistance of the collector.

(iv) Another step is to deposit a n-type epitaxial layer. The oxide is removed and the wafer is placed in an epitaxial reactor for epitaxial growth. The thickness and doping concentration of the epitaxial layer is determined by the ultimate use of device. This step is shown in Fig. 11.5(b).

(v) Next step is then to form a lateral oxide isolation region. A thin oxide-layer (~50 nm) is thermally grown on the epitaxial layer, followed by a deposition of silicon-nitride (Si_3N_4) layer of typical thickness (~100 nm).

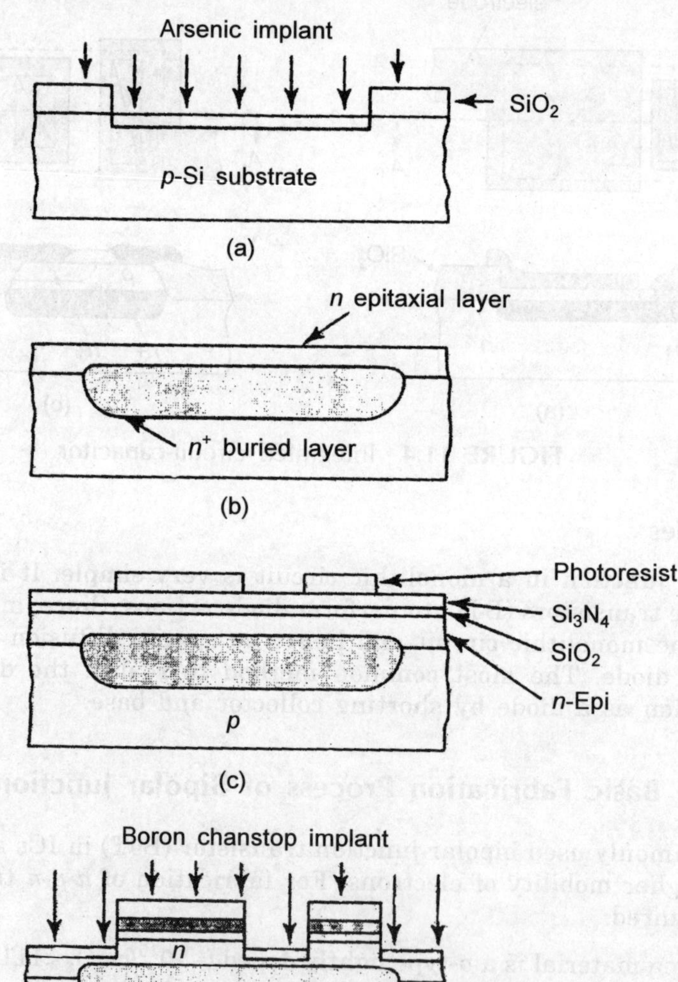

FIGURE 11.5 (Contd.)

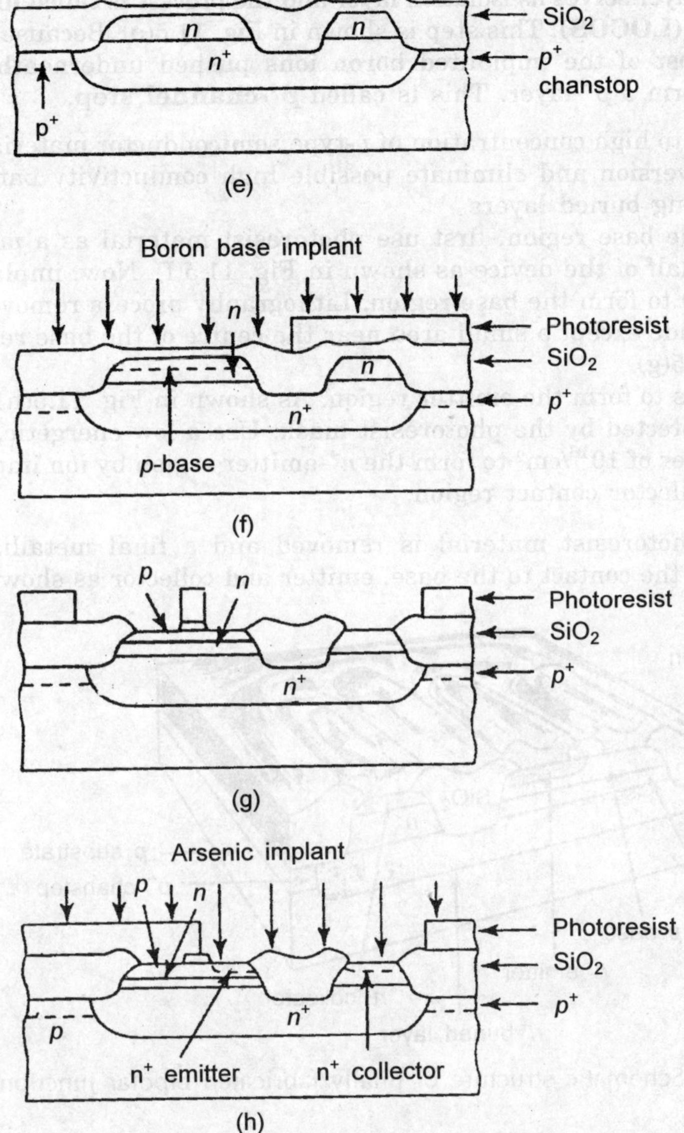

FIGURE 11.5 Various fabrication steps for bipolar junction transistor.

(vi) Now etched out the nitride-oxide layer and about half of the epitaxial layer by using a photoresist material as a mask as shown in Fig. 11.5(c).

(vii) Boron ions are now implanted into the exposed silicon area as shown in Fig. 11.5(d).

(viii) Remove the photoresist material and place the wafer in an oxidal furnace. Since, the nitride layer has a very low oxidation rate, therefore, a thick oxide will grow only in the areas which are not protected by the nitride layer. This oxidation layer serves as isolation layer and the process is called **local oxidation of silicon (LOCOS)**. This step is shown in Fig. 11.5(e). Because of segregation effects, most of the implanted boron ions pushed underneath the isolation oxide to form a p^+-layer. This is called **p^+-channel step**.

Note: Due to high concentration of p-type semiconductor material will prevent surface inversion and eliminate possible high conductivity paths among the neighbouring buried layers.

(ix) To form the base region, first use photoresist material as a mask to protect the right half of the device as shown in Fig. 11.5(f). Now, implant boron ions ($\sim 10^{12}$/cm²) to form the base region. Lithography process removes all the thin layer of oxide except a small area near the centre of the base region as shown in Fig. 11.5(g).

(x) Last step is to form the emitter region. As shown in Fig. 11.5(h) that the base area is protected by the photoresist mask. Use a low-energetic, high-arsenic-dose of order of 10^{16}/cm² to form the n^+-emitter region by ion implantation and then n^+-collector contact region.

Lastly the photoresist material is removed and a final metallization step is performed to form the contact to the base, emitter and collector as shown in Fig. 11.6.

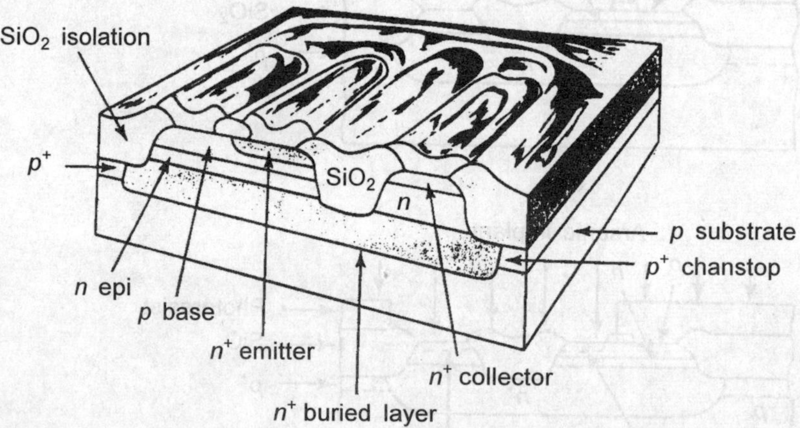

FIGURE 11.6 Schematic structure of finally fabricated bipolar Junction transistor.

Note: 1. By using epitaxial layer one can realize high frequency, high voltage and low saturation voltage transistors.

2. A high quality SiO_2 is recommended in order to reduce the radiation damage.

3. Nowadays the Al contact is often replaced with one in Al/polysilicon. This leads to faster devices and eliminate a recommendation. Current at Al/Si interface thus increasing the gain and the radiation tolerance.

4. Chip production is divided into two major parts: front end and back end. In the front end, one can make the components of the circuit. In the back end, one adds the metal to connect the components and then one can test them and pack them.

EXAMPLE 11.1

Find the length of the semiconductor bar to design a resistance of the value 12 kΩ whose width is 10 μm. Take the sheet resistance 1 k$\Omega/\square$.

Solution: Using Eq. (11.1),

$$R = R_\square \left(\frac{L}{W}\right)$$

or

$$\frac{12}{1} = \frac{L}{10\ \mu\text{m}}$$

or

$$120\ \mu\text{m} = L$$

EXAMPLE 11.2

What is the required oxide thickness for designing a capacitor of value $C = 10\ \mu f$?

Solution:

$$C = \frac{\varepsilon_{ox}}{d} \quad \text{or} \quad d = \frac{\varepsilon_{ox}}{c}$$

$$= \frac{3.9 \times 8.85 \times 10^{-14}}{10 \times 10^{-6}}$$

$$= 3.9 \times 8.85 \times 10^{-9} = 0.345\ \text{nm}$$

EXAMPLE 11.3

Why single-crystal silicon is used for integrated circuits than the polycrystalline silicon?

Solution: Because single crystal silicon substrate possess a high level of purity and perfection. It has no defects associated with grain boundaries. It is found in polysilicon. Such defects have been known to limit the lifetimes of minority carriers. Single silicon substrate may also have a high degree of chemical purity, a high degree of crystalline perfection and high structure uniformity.

EXAMPLE 11.4

How can one get a high grade starting single crystal silicon?

Solution: The acquisition of such high grade starting silicon material involvs two major steps:

(a) Refinement of raw material (such as quartzite) into electronic grade poly-crystalline silicon (EGS) using a complex multi-stage process.

(b) Growing a single-crystal silicon from this EGS either by Czochralski (CZ process) or float zone process.

11.4.5 Fabrication of *n*-MOS Transistor

The following steps are involved for fabrication of *n*-MOS:

1. To process an *n*-channel MOSFET, the starting material is a *p*-type lightly doped ($\sim 10^{15}/cm^3$), <100> -oriented silicon wafer. Now, first step is to grow SiO_2 layer on the surface by using LOCOS technology. The thickness of SiO_2 is typically of order of 1 μm thick, as shown in Fig. 11.7(a).

2. The surface is now covered with a photoresist material to define the active area of the device as shown in Fig. 11.7(b).

3. The photoresist layer is then exposed to ultraviolet light through a mask which defines those regions into which diffusion is to take place together with transistor channels. Depending on the type of photoresist material either positive resist or negative resist, the exposed portion of resist will be either hardened or softened. This is shown in Fig. 11.7(c).

4. These exposed areas are subsequently etched away together with the underlying silicon dioxide so that the wafer surface is exposed in the window defined by mask as shown in Fig. 11.7(d).

5. The remaining photoresist material is removed and a thin layer of SiO_2 (~ 0.1 μm typical) is grown over the entire chip surface and then polysilicon is deposited on the top of this to form the gate structure.

6. Further, photoresist coating and masking allows the polysilicon to be patterend as shown in Fig. 11.7(e) and then the thin oxide is removed to expose areas into which a *n*-type impurities are to be diffused to form the source and the drain regions as shown in Fig. 11.7(f).

7. Thick oxide (SiO_2) is grown over all again and then masked with photoresist and etched to expose selected areas of the polysilicon gate and drain, and source areas where connections are to be made. This is shown in Fig. 11.7(g).

8. Then metal is deposited over whole chip to the thickness of typically 1 μm. This metal layer is then masked and etched to form the required interconnection contact as seen from Fig. 11.7(h).

Note: 1. The <100> orientation is preferred over <111> because it has an interface-trap density that is about one-tenth that of <111>.

2. For an enhancement mode *n*-channel device, boron ions are implanted in the channel to increase the threshold voltage. For a depletion mode *n*-channel device, arsenic ions are implanted in the channel region to decrease the threshold voltage.

11.4.6 CMOS Fabrication

In the following sub-topic, we will focus our discussion on the well-established CMOS fabrication technology, which requires that both *n*-channel (*n*-MOS) and *p*-channel (*p*-MOS) transistors be built on the same substrate. To accommodate both *n*-MOS and *p*-MOS devices, special regions must be created in which the semiconductor type is opposite to the substrate type. These regions are called **wells** or **tubs**. A *p*-well is created in an *n*-type substrate or alternatively, an *n*-well is created in a *p*-type

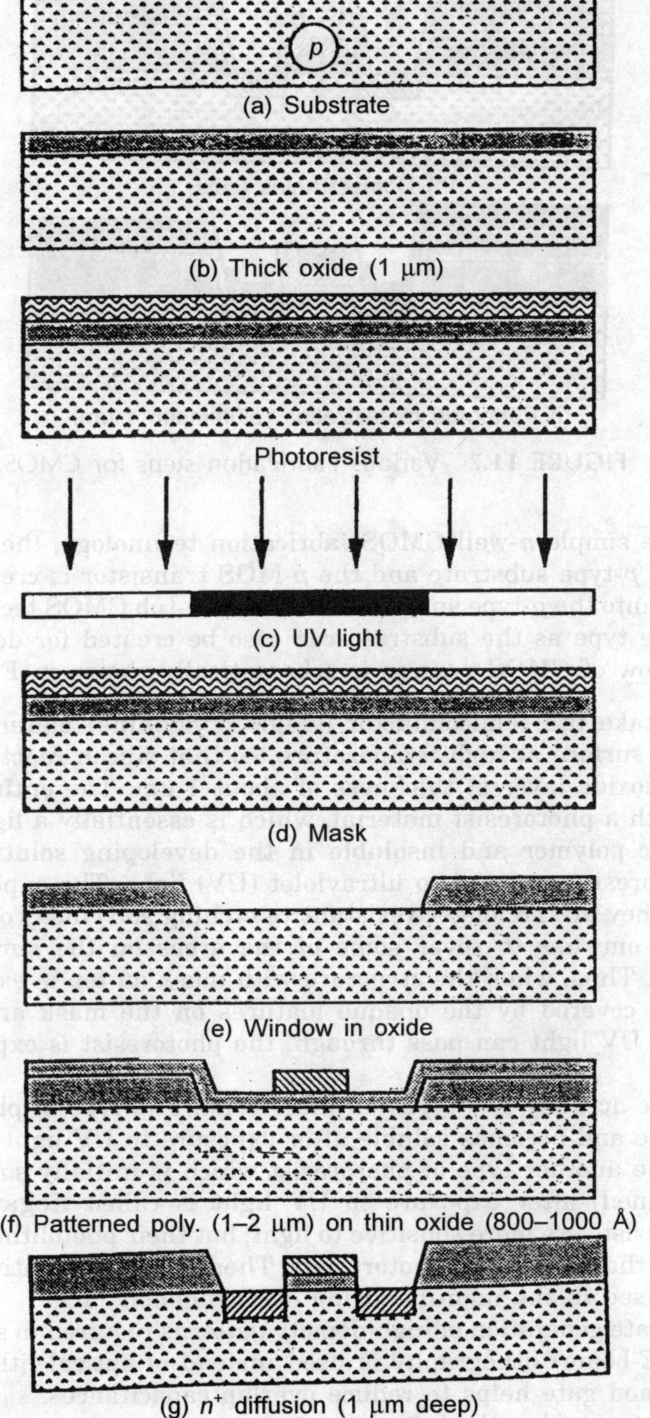

FIGURE 11.7 (Contd.)

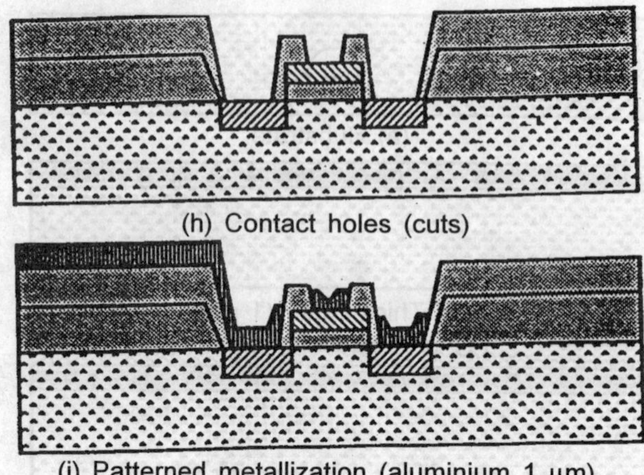

(h) Contact holes (cuts)

(i) Patterned metallization (aluminium 1 μm)

FIGURE 11.7   Various Fabrication steps for CMOS.

substrate. In the simple n-well CMOS fabrication technology, the n-MOS transistor is created in the p-type substrate and the p-MOS transistor is created in the n-well, which is built-in into the p-type substrate. In the twin-tub CMOS technology, additional tubs of the same type as the substrate can also be created for device optimization.

A simple flow of CMOS process is schematically shown in Fig. 11.8.

Note: 1. First take the semiconductor (single crystalline) material and then pass oxygen over the surface at high temperature, so that silicon reacts with oxygen and form a silicon dioxide layer of thickness of about 1 μm. The entire oxide surface is then covered with a photoresist material, which is essentially a light-sensitive, acid-resistant organic polymer and insoluble in the developing solution initially. Now, expose the photoresist material to ultraviolet (UV) light. The exposed areas become soluble so that they are no longer resistant to etching solvents. To selectively expose the photoresist, one has to cover some of the areas on the surface with a mask during exposure. Thus, when the structure with mask on top is exposed to UV light, areas which are covered by the opaque features on the mask are shielded. In the areas where the UV light can pass through, the photoresist is exposed and becomes soluble.

2. There are actually two types of photoresist. The type of photoresist which is initially insoluble and becomes soluble after exposure to UV light is called **positive photoresist**. The another type of photoresist which is initially soluble and becomes insoluble (hardened) after exposure to UV light is called **negative photoresist**. Negative photoresists are more sensitive to light, but their photolithographic resolution is not as high as that of positive photoresists. Therefore, the negative photoresists are less commonly used in the manufacturing of high-density ICs.

3. For accurate generation of high-density patterns required in sub-micron devices, electron beam (E-beam) lithography is used instead of optical lithography.

4. Self aligned gate helps to reduce overlap capacitances.

5. Important considerations for layout are:

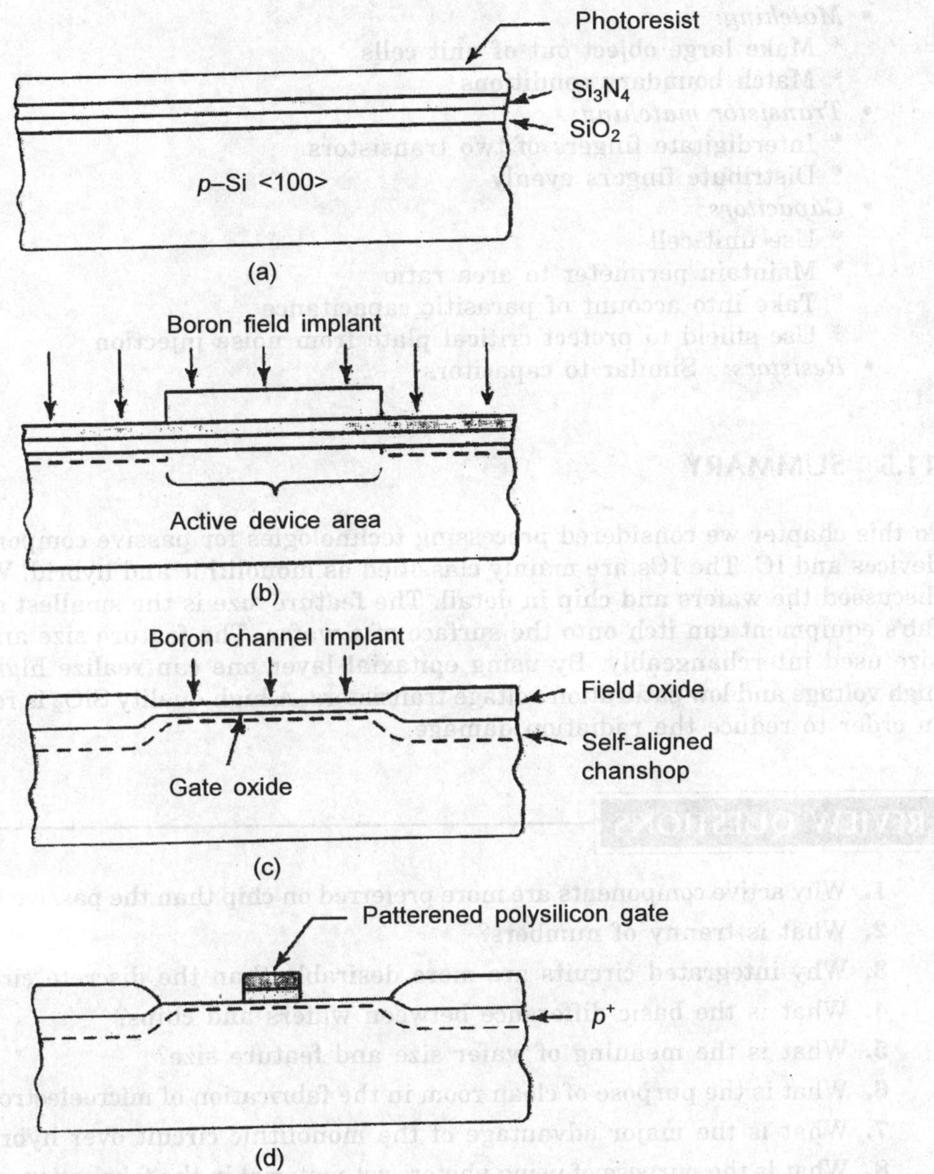

FIGURE 11.8   Schematic representation of simple flow of CMOS process.

- *Diffusion resistors:* When designing mask, it is needed to take into account lateral diffusion that lead to different widths and lengths.
- *Layer misalignment:* Some misalignment lead to catastrophic failure, e.g., short between source and gate.
- *Reducing junction capacitances:*
  * If it is possible, share junctions of multiple transistors
  * Reduce perimeter of junctions

- *Matching:*
  * Make large object out of unit cells
  * Match boundary conditions
- *Transistor matching:*
  * Interdigitate fingers of two transistors
  * Distribute fingers evenly
- *Capacitors:*
  * Use unit cell
  * Maintain perimeter to area ratio
  * Take into account of parasitic capacitance
  * Use shield to protect critical plate from noise injection
- *Resistors:* Similar to capacitors

11.5 SUMMARY

In this chapter we considered processing technologies for passive components, active devices and IC. The ICs are mainly classified as monolithic and hybrid. We have also discussed the wafers and chip in detail. The feature size is the smallest size that the fab's equipment can itch onto the surface of a wafer. The feature size and transistor size used interchangeably. By using epitaxial layer one can realize high frequency, high voltage and low saturation voltage transistors. A high quality SiO_2 is recommneded in order to reduce the radiation damage.

REVIEW QUESTIONS

1. Why active components are more preferred on chip than the passive components?
2. What is tranny of numbers?
3. Why integrated circuits are more desirable than the discrete circuits?
4. What is the basic difference between wafers and chips?
5. What is the meaning of wafer size and feature size?
6. What is the purpose of clean room in the fabrication of microelectronic circuits?
7. What is the major advantage of the monolithic circuit over hybrid circuit?
8. What is the purpose of using photoresist material in the fabrication of the device?
9. Discuss the importance of the mask in fabrication.
10. Why ultraviolet rays are allowed to fall on the photoresist material?
11. Why etching is performed during fabrication process?
12. Why is <100>-orientation preferred in n-MOS fabrication? What are the disadvantages if 100 thin a field oxide is used in n-MOS devices?
13. Why do we use a p^+-polysilicon gate for p-MOS?

NUMERICAL PROBLEMS

1. For an equivalent oxide thickness of 2 nm, what will be the physical thickness when high k-material nitride ($\varepsilon_i/\varepsilon_o = 7$), Ta_2O_5 (25) or TiO_2(80) are used?

2. For the sheet resistance of 1 kΩ/$\square$, find the maximum resistance that can be fabricated on a 2.5×2.5 μm chip using 2 μm lines with a 4 μm pitch, i.e., distance between centres of parallel lines.

3. Design a mask set for a 5 pf MOS capacitor. The oxide thickness is 30 nm. Assume that the minimum window size is 2×10 μm and the maximum registration errors are 2 μm.

A

Physical Constants

Avogadro's constant	N_{AVO}	6.022×10^{23}	Mol^{-1}
Boltzmann's constant	k_B	1.38×10^{-23}	J/K
		8.62×10^{-5}	eV/K
Electron charge	q	1.6×10^{-19}	C
Electron rest mass	m_0	0.511×10^6	eV/C^2
		9.11×10^{-31}	kg
Permeability – free space	μ_0	1.2566×10^{-8}	H/cm
Permittivity – free space	ε_0	8.85×10^{-14}	F/cm
Planck's constant	h	4.14×10^{-15}	eVs
		6.63×10^{-34}	Js
Reduced Planck's constant	$\hbar$	6.58×10^{-16}	eVs
		1.055×10^{-34}	Js
Speed of light	c	3.0×10^{10}	cm/s
Thermal voltage	$k_B T/q$	0.0259	V

Material Parameters for Important Semiconductors, Si and GaAs

Bulk Material Parameters for Silicon

Lattice constant (Å)	$a = 5.43$
Dielectric constant	11.9
Intrinsic carrier concentration (cm^{-3})	1.0×10^{10}
Energy band gap (eV)	1.12
Sound velocity (cm/s)	9.04×10^5
Density (g cm^{-3})	2.33
Effective mass along X (m^*/m_0) – transverse	0.19
Effective mass along X (m^*/m_0) – longitudinal	0.916
Effective mass along L (m^*/m_0) – transverse	0.12
Effective mass along L (m^*/m_0) – longitudinal	1.59
Heavy hole mass (m^*/m_0)	0.537
Electron mobility at 300 K (cm^2/(Vs))	1450
Hole mobility at 300 K (cm^2/(Vs))	500
Thermal conductivity at 300 K(W/(cm°C))	1.5
Effective density of states in conduction band (cm^{-3})	2.8×10^{19}
Effective density of states in valence band (cm^{-3})	1.04×10^{19}
Non-parabolicity along X (eV^{-1})	0.5
Intravalley acoustic deformation potential (eV)	9.5
Optical phonon energy at Γ (eV)	0.062
Intervalley separation energy, X–L (eV)	1.17

Bulk Material Parameters for GaAS

Lattice constant (Å)	$a = 5.65$
Low frequency dielectric constant	12.90
High frequency dielectric constant	10.92
Energy band gap at 300 K (eV)	1.425
Intrinsic carrier concentration (cm^{-3})	2.1×10^6
Electron mobility at 300 K (cm^2/(Vs))	8500
Hole mobility at 300 K (cm^2/(Vs))	400
Longitudinal sound velocity (cm/s) along (100) direction	4.73×10^5
Density (g/cm^3)	5.36
Effective mass at Γ (m^*/m_0)	0.067
Effective mass along L (m^*/m_0)	0.56
Effective mass along X (m^*/m_0)	0.85
Heavy hole mass (m^*/m_0)	0.62
Effective density of states conduction band (cm^{-3})	4.7×10^{17}
Effective density of states valence band (cm^{-3})	7.0×10^{18}
Thermal conductivity at 300 K (W/(cm °C))	0.46
Non-parabolicity at Γ (eV^{-1})	0.690
Intravalley acoustic deformation potential (eV)	8.0
Optical phonon energy at Γ (eV)	0.035
Intervalley separation energy, Γ–L (eV)	0.284
Intervalley separation energy, Γ–X (eV)	0.476

Note: Γ designates a point in k-spare, X and L designate directions in k-space. Γ refers to the $k = 0$ point at the centre of the Brillouin zone. X refers to the {100} directions and L to the {111} directions.

APPENDIX
C
SPICE Model

C.1 *p-n* JUNCTIONS

An introduction to *p*-SPICE for the *p-n* junction diode has been given by Banzhof. The element line for a *p-n* junction diode is:

$$\text{D XXX N}_+ \text{ N}_- \text{ Mode Name } <\text{Area}> <\text{OFF}> <\text{IC = VD}>$$

where D XXX is the name of the *p-n* junction diode, N+ is the anode node and N− is the cathode node.

MOD NAME is the model name which is used in an associated. .MODEL control line. These items are required in the *p-n* junction diode element line. The quantities in the < ... > are the optional parameters. An element line may be continued by entering +a sign at the start of the next line.

The *p*-SPICE *p-n* junction diode parameters are given in Table C.1.

Table C.1 *p*-SPICE parameters for *p-n* junction diode

No.	SPICE keyword	Parameter name	Default value	Units
1.	IS	Saturation current	1.0 e−14	A
2.	2N	Ideality factor	1.0	–
3.	RS	Ohmic resistance	0	ohm
4.	VJ	Built-in voltage	1	V
5.	EG	Energy-gap	1.11	eV
6.	XTI	Saturation current temperature exponent	3.0	–
7.	BV	Reverse breakdown voltage	∞	V

(Contd.)

Table C.1 *p*-SPICE parameters for *p-n* junction diode (*Contd.*)

No.	SPICE key word	Parameter Name	Default value	Units
8.	IBV	Current at breakdown voltage	1.0e–3	A
9.	TT	Transit time or minority carrier life time	0	S
10.	CJO	Zero-bias depletion capacitance	0	F
11.	M	Grading coefficient	0.5	–
12.	FC	Coefficient for forward-bias depletion capacitance expression	0.5	–
13.	KF	Flicker noise coefficient	0	–
14.	AF	Flicker noise exponent	1	–

C.2 MESFET MODEL

For the pSPICE MSFET model, equivalent circuit is shown in Fig. C.1.

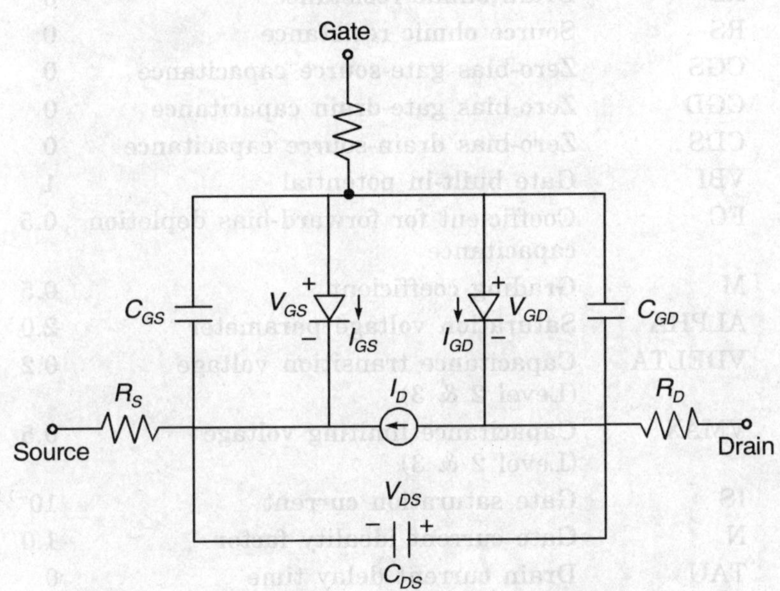

FIGURE C.1 *p*-SPICE equivalent model of MESFET.

The general form of the element line for MESFET is:

B XXXX   ND   NG   NS   MODNAME   <Area>

where B XXXX is the name of the MESFET with seven or less alphanumeric characters. The drain, gate and source nodes are represented by ND, NG and NS respectively.

The model name is given by MODNAME and is used again in the ·MODEL control line. The AREA factor gives how many of the MESFET are put in parallel to give one B XXXX. The MODNAME is the model name given to the MESFET in the element line. The *p*-SPICE parameters are given in Table C.2.

Note: Several *p*-SPICE MESFET models are used and are designated as level 1, 2 and 3. The difference in the models are largely the representation of the drain current and the parasite capacitances.

Table C.2 SPICE MESFET Model Parameters

No.	Text symbol	SPICE keyword	Parameter name	Default value	Units
1.	—	LEVEL	Model index (1, 2 or 3)	1	–
2.	V_p	VTO	Pinch-off voltage	– 2.5	V
3.	β	BETA	Transconductance parameter	0.1	A/V^2
4.	λ	LAMBDA	Channel-length modulation	0	V^{-1}
5.	R_G	RG	Gate ohmic resistance	0	Ω
6.	R_D	RD	Drain ohmic resistance	0	Ω
7.	R_S	RS	Source ohmic resistance	0	Ω
8.	C_{GS}	CGS	Zero-bias gate-source capacitance	0	F
9.	C_{GD}	CGD	Zero-bias gate-drain capacitance	0	F
10.	C_{DS}	CDS	Zero-bias drain-source capacitance	0	F
11.	V_{bi}	VBI	Gate built-in potential	1	
12.	FC	FC	Coefficient for forward-bias depletion capacitance	0.5	
13.	m	M	Grading coefficient	0.5	–
14.	α	ALPHA	Saturation voltage parameter	2.0	1/V
15.	–	VDELTA	Capacitance transition voltage (Level 2 & 3)	0.2	V
16.	–	VMAX	Capacitance limiting voltage (Level 2 & 3)	0.5	V
17.	I_s	IS	Gate saturation current	10^{-14}	A
18.	n	N	Gate current ideality factor	1.0	–
19.	τ	TAU	Drain current delay time	0	sec
20.	B	B	Doping tail parameter (Level 2)	0.3	V^{-1}
21.	γ	GAMMA	Static feedback parameter (Level 3)	0	–
22.	δ	DELTA	Output feedback parameter (Level 3)	0	AV^{-1}
23.	Q	Q	Power law parameter (Level 3)	2	–
24.	E_g	EG	Energy gap (barrier height)	1.11	eV
25.	–	XTI	IS temperature exponent	0	–

(Contd.)

Table C.2 SPICE MESFET Model Parameters (*Contd.*)

No.	Text symbol	SPICE keyword	Parameter name	Default value	Units
26.	–	VTOTC	VTO temperature coefficient	0	V/°C
27.	–	BETATCE	BETA exponential temperature coefficient	0	%/°C
28.	–	TRG1	RG temperature coefficient (linear)	0	°C^{-1}
29.	–	TRD1	RD temperature coefficient (linear)	0	°C^{-1}
30.	–	TRS1	RS temperature coefficient (linear)	0	°C^{-1}
31.	k_f	KF	Flicker-noise coefficient	0	
32.	α_f	AF	Flicker-noise exponent	1	

C.3 MOSFET MODEL

The general form of the elemental line for a MOSFET is:

M XXXX ND NG NS NB MODNAME <L = VAL> <W = VA + <AD = VAL> <AS = VAL> <PD = VAL> <PS = VAL> <NRD = VAL> <NRS = VAL> + <OFF> <IC = VDS, VGS, VBS>.

where M XXXX is the name of MOSFET.

ND, NG, NS and NB are the drain, gate, source and bulk nodes. MODNAME is the model name which is used in an associated. .MOD control line. These items are required in the MOSFET element line. The optional parameters are the quantities in < ... > and an element line may be continued by entering a+ sign at the start of the next line. The meanings of the optional parameters are:

L is the length of the channel. (in metres)
W is the width of the channel. (in metres)
AD is the area of the drain diffusion. (in square metres)
AS is the source diffusion area. (in square metres)

If anyone of these parameters are not specified, default value used. The remaining optional parameters are:

PD is the perimeter of the drain junction (in metres)
PS is the perimeter of the source junction (in metres)
NRD is the number of squares for the drain region
NRS is the number of square for the source region

OFF indicates an initial condition for dc analysis. IC sets the initial conditions for V_{GS}, V_{DS} and V_{BS} for us with a transient analysis. The model form for the MOSFET are:

- MODEL MODNAME NMOS <(PAR1 = PVAL1. PAR2 = PVAL....)>.

MODNAME is the model name given to a MOSFET in the element line and NMOS for *p*-MOS denotes whether the MOSFET is an *n*-channel or a *p*-channel device. PAR is the parameter name of one of the optional parameter as listed in Table C.3. PVAL is the value of the designated parameter.

Table C.3 SPICE2 and *p*SPICE MOSFET dc Model Parameters

No.	Text symbol	SPICE keyword	Level	Parameter name	Default value	Units
1.	−	LEVEL	1–3	SPICE model 1, 2 or 3	1	−
2.	V_T	VTO	1–3	Zero-bias threshold voltage	0.0	V
3.	γ	GAMMA	1–3	Bulk space-charge parameter	0.0	$V^{0.5}$
4.	ψ_s	PHI	1–3	Surface potential	0.6	V
5.	KP	KP	1–3	Transconductance parameter	2.0E–5	A/V^2
6.	λ	LAMBDA	1, 2	Channel-length modulation	0	V^{-1}
7.	t_{ox}	TOX	1–3	Gate-oxide thickness	1.0E–7	metre
8.	N_b	NSUB	1–3	Substrate doping	0.0	cm^{-3}
9.	N_f	NSS	2, 3	Fixed oxide charge	0.0	cm^{-2}
10.	N_{it}	NFS	2, 3	Interface-trapped charge	0.0	cm^{-2}
11.	−	TPG	2, 3	Type of gate material + 1 opp. to substrate − 1 same as substrate 0 AI gate	1	−
12.	μ	UO	1–3	Surface mobility	600	cm^2/Vs
13.	U_c	UCRIT	2	Critical electric field for mobility	1E4	V/cm
14.	U_e	UEXP	2	Exponential coefficient for mobility	0.0	−
15.	U_t	UTRA	2	Transverse field coefficient	0.0	−
16.	x_j	XJ	2, 3	Source or drain junction depth	0.0	metres
17.	x_{ji}	LD	1–3	Lateral diffusion	0.0	metres
18.	V_{max}	VMAX	2,3	Maximum carrier drift velocity	0.0	metres/s
19.	N_{eff}	NEFF	2	Total channel charge coefficient	1	−
20.	δ	DELTA	2, 3	Width effect on threshold voltage	0.0	−
21.	η	ETA	3	Static feedback on threshold voltage	0.0	−
22.	V_{bi}	PB	1–3	Source and drain junction built-in potential	0.80	V
23.	θ	THETA	3	Mobility modulation	0.0	−
24.	κ	KAPPA	3	Saturation field factor	0.2	−

EXAMPLE C.1

For an *n*-channel enhancement mode MOSFET, following parameters are given:

Substrate : $p = N_a^- = 10^{16}/\text{cm}^3$
Oxide thickness $t_{ox} = 50$ nm
Channel mobility $\mu_n = 1000$ cm^2/V-sec
At gate: Gate length $L = 3$ μm
Gate width $W = 12$ μm

Write the SPICE for level 1 to get the threshold voltage.

Solution: *MOSFET Threshold Voltage
V_{GS} 2 0
V_{BS} 3 0   DC 0
MOS1   2   2   0   3   MOSVT $L = 3u, W = 12u$
- MODEL MOSVT NMOS LEVEL = 1 NSUB = 4 E15
  +UO = 600 TOX = 50 E–9, TPG = 0
- DC V_{GS} 0 2 .01
- PROBE
- END

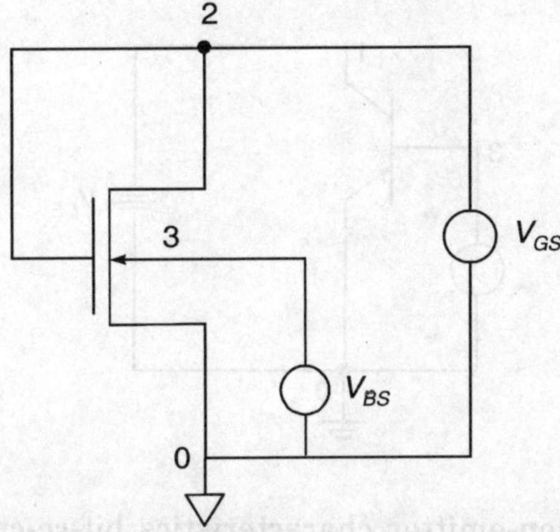

Note: * Indicates that first line or that particular line cannot be executed by SPICE.

DG V_{GS} 0 2 0.01 only indicates that gate source voltage varies from 0 to 2 volt dc in 0.01 steps. To get threshold voltage, plot $\sqrt{I_D}$ (drain current) with V_{GS} and see at what gate voltage current flows. This gate voltage corresponds to the threshold voltage.

C.4 MODEL FOR BJT

The general form of the element line for bipolar transistor is:

    Q XXX NC NB NE <NS> MODNAME <AREA> <OFF> <$I_C = V_{BE}, V_{CE}$>

where Q XXX is the name of the bipolar transistor.

NC, NB, NE are the collector, base and emitter NODES respectively. MOD NAME is the model name which is used in an associated ·MODEL control line. The optional parameters are the quantities in the < ... > and the element line may be continued by entering + a sign at the start of the next line.

The model form for bipolar transistor is:

$$\text{MODEL MODNAME NPN} \quad <(\text{PAR1} = \text{PVAL1} \quad \text{PAR2} = \text{PVAL2} ...)>$$

where MODNAME is the model name given to a bipolar transistor in the element line and NPN or PNP denotes the type of transistor used. The optional parameters for BJT are listed in Table C.4.

EXAMPLE C.2

Use *p*SPICE to obtain the $I_C - V_{CE}$ output characteristics for a *p-n-p* transistor in the common emitter configuration with $I_S = 1.85 \times 10^{-14}$ A, $\beta_F = 65$, $\beta_R = 0.1$ and an early voltage of 100 V. Let I_B vary from 0 to 50 μA in steps of 10 μA and use 0.05 voltage steps for V_{CE} between 0 to 5 V.

Solution: First draw a circuit diagram as shown, then assign the numbers to the emitter, base and collector nodes. Assign number 0 to ground.

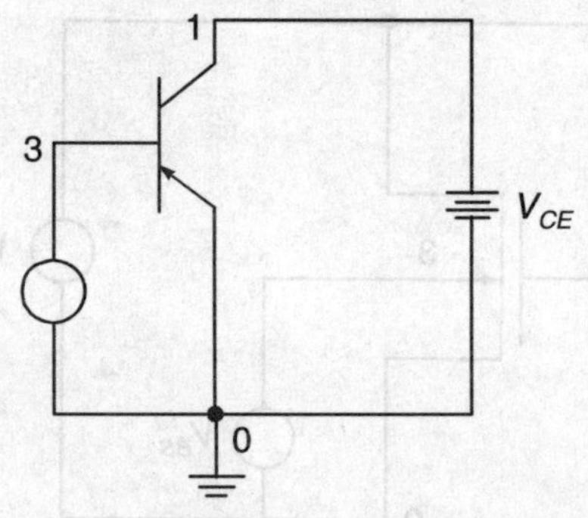

- BJT *p-n-p* common-emitter characteristics bjt-ce-ctr.
  VCE 0 1
  JBI 3 0
  Q1 1 3 0 Q216
- MODEL Q216 PNP BF = 65 BR = 0.1 IS = 1.85 E–14 VAF = 10
- DC VCE 0.5 0.05 JBI 0.0 50.0U, 10.0U
- PROBE
- END

Table C.4 Overall p-SPICE parameter

No.	Text symbol	SPICE keyword	Parameter name	Default value	Units
1.	I_S	IS	Saturation current	1.0E–14	A
2.	β_F	BF	Ideal maximum forward-current gain	100	–
3.	β_R	BR	Ideal maximum reverse-current gain	1	–
4.	n_F	NF	Forward-current ideality factor (ranges from 1.0 to 2.0)	1.0	–
5.	n_R	NR	Reverse-current ideality factor (ranges from 1.0 to 2.0)	1.0	–
6.	$I_{S_{src}}$	ISE	Emitter-base space-charge and/or surface-recombination saturation current	1.0E–13	A
7.	n_E	NE	Emitter-base ideality factor (ranges from 1.0 to 2.0)	1.0	–
8.	$I_{S_{src}}$	ISC	Collector-base space-charge and/or surface-recombination saturation current	1.0E–13	A
9.	n_c	NC	Collector-base ideality factor (ranges from 1.0 to 2.0)	1.0	–
10.	–	IKF	High-injection "knee" current for β_F roll-off	∞	A
11.	–	IKR	High-injection "knee" current for β_R roll-off	∞	A
12.	V_A	VAF	Forward early voltage	∞	V
13.	V_B	VAR	Reverse early voltage	∞	V
14.	E_g	EG	Energy gap	1.11	eV
15.	–	XTI	Saturation current temperature exponent	3.0	–
16.	–	XTB	Forward and reverse β temperature exponent	0	–
17.	r_B	RB	Zero-bias base resistance	0	Ω
18.	r_{BM}	RBM	Minimum base resistance at high currents	RB	Ω
19.	I_{rB}	IRB	Current where base resistance falls to half its minimum value	∞	A
20.	R_E	RE	Emitter resistance	0	Ω
21.	R_C	RC	Collector resistance	0	Ω
22.	$C_{jE}(0)$	CJE	Zero-bias emitter-base depletion capacitance	0	F
23.	$V_{bi_{EB}}$	VJE	Emitter-base built-in voltage	0.75	V
24.	m_E	MJE	Emitter-base grading coefficient	0.33	–
25.	–	FC	Coefficient for forward-bias depletion capacitance expression	0.5	–

(Contd.)

Table C.4 Overall p-SPICE parameter (*Contd.*)

No.	Text symbol	SPICE keyword	Parameter name	Default value	Units
26.	$C_{JC}(0)$	CJC	Zero-bias collector-base depletion capacitance	0	F
27.	V_{biCB}	VJC	Collector-base built-in voltage	0.75	V
28.	m_c	MJC	Collector-base grading coefficient	0.33	–
29.	$C_{js}(0)$	CJS	Zero-bias collector-substrate depletion capacitance	0	F
30.	V_{bi_s}	VJS	Collector-substrate built-in voltage	0.75	V
31.	m_s	MJS	Collector-substrate grading coefficient	0	–
32.	X_{CJC}	XCJC	Fraction of collector-base depletion capacitance connected to internal base node	1	–
33.	τ_F	TF	Ideal forward transit time	0	s
34.	τ_R	TR	Ideal reverse transit time	0	s
35.	X_{tF}	STF	Coefficient for the bias dependence of TF	0	–
36.	V_{tF}	VTF	Voltage describing V_{CB} dependence of TF	∞	V
37.	I_{tF}	ITF	High-current parameter for effect on TF	0	A
38.	P_{tF}	PTF	Excess phase at $f = 1/2\pi\tau_D$	0	°
39.	–	NK	High-current roll-off coefficient	0.5	–
40.	I_{ss}	ISS	Substrate p-n junction saturation current		A
41.	n_s	NS	Substrate p-n junction ideality factor	1.0	–
42.	R_{co}	RCO	Epitaxial region resistance	0	Ω
43.	V_0	VO	Carrier mobility "knee" voltage	10	V
44.	Q_{co}	QCO	Epitaxial region charge factor	0	C
45.	γ	GAMMA	Epitaxial doping factor	1E–11	–
46.	–	TRE1	RE resistance linear temperature coefficient	0	°C^{-1}
47.	–	TRE2	RE resistance quadratic temperature coefficient	0	°C^{-2}
48.	–	TRB1	RB resistance linear temperature coefficient	0	°C^{-1}
49.	–	TRB2	RB resistance quadratic temperature coefficient	0	°C^{-2}
50.	–	TRM1	RBM resistance linear temperature coefficient	0	°C^{-1}
51.	–	TRM2	RBM resistance quadratic temperature coefficient	0	°C^{-2}
52.	–	TRC1	RC resistance linear temperature coefficient	0	°C^{-1}
53.	–	TRC2	RC resistance quadratic temperature coefficient	0	°C^{-2}

Index